Pollutant Effects in Freshwater

Pollutant Effects in Freshwater

Applied limnology

E. B. Welch
Professor Emeritus of Environmental Science
Department of Civil and Environmental Engineering
University of Washington
Seattle, Washington, USA

J. M. Jacoby
Professor of Environmental Science and Engineering
Department of Civil and Environmental Engineering
Seattle University
Seattle, Washington, USA

With a contribution on Hydrographic Characteristics by

T. Lindell
Centre for Image Analysis
Uppsala University
Uppsala, Sweden

Taylor & Francis
Taylor & Francis Group
LONDON AND NEW YORK

First published 2004
by Taylor & Francis
2 Park Square, Milton Park, Abingdon, Oxon, OX14 4RN

Simultaneously published in the USA and Canada
by Taylor & Francis
270 Madison Ave, New York NY 10016

Taylor & Francis is an imprint of the Taylor & Francis Group

Transferred to Digital Printing 2007

First edition 1980
Second edition 1992
© 1980 Cambridge University Press; 1992 Eugene B. Welch;
2004 Eugene B. Welch and Jean M. Jacoby

Typeset in Times by
Integra Software Services Pvt. Ltd, Pondicherry, India

All rights reserved. No part of this book may be reprinted or reproduced or utilised in any form or by any electronic, mechanical, or other means, now known or hereafter invented, including photocopying and recording, or in any information storage or retrieval system, without permission in writing from the publishers.

British Library Cataloguing in Publication Data
A catalogue record for this book is available
from the British Library

Library of Congress Cataloging in Publication Data
Welch, E. B. (Eugene B.)
  Pollutant effects in freshwater : applied limnology / Eugene B. Welch and Jean M. Jacoby.
    p.   cm.
  Rev. ed. of: Ecological effects of wastewater. 2nd ed. 1992.
  Includes bibliographical references and index.
  1. Sewage disposal in rivers, lakes, etc.—Environmental aspects.
  2. Water—Pollution—Environmental aspects.   I. Jacoby, Jean M., 1956–   II. Welch, E. B. (Eugene B.) Ecological effects of wastewater. III. Title.
  QH545.S49W54 2004
  628.1′68—dc22
                                                                                    2003061721

ISBN10: 0–415–27991–7 (hbk)
ISBN10: 0–415–42990–0 (pbk)

ISBN13: 978–0–415–27991–8 (hbk)
ISBN13: 978–0–415–42990–0 (pbk)

Contents

Preface ix

PART I
General concepts of aquatic ecosystems 1

1 Ecosystem function and management 3
  1.1 *Ecosystem composition and energy sources* 4
  1.2 *Energy flow and nutrient cycling* 5
  1.3 *Efficiency of energy and nutrient use* 8
  1.4 *Management of ecosystems* 10

2 Standing crop, productivity and growth limitation 12
  2.1 *Pyramid of biomass* 12
  2.2 *Productivity* 13
  2.3 *Efficiency of energy transfer* 14
  2.4 *Population growth and limitation* 15

3 Hydrographic characteristics 21
  3.1 *Types of flow* 21
  3.2 *Boundary layers* 23
  3.3 *Physical factors controlling wastewater distribution* 25
  3.4 *Density effects* 25
  3.5 *Density currents* 27
  3.6 *Thermal properties of lakes* 28
  3.7 *Origin of lakes* 28
  3.8 *Thermal cycle in a typical mid-latitude lake* 29
  3.9 *Classification of lakes by temperature regime* 32
  3.10 *Water movements* 36
  3.11 *Seiches* 36
  3.12 *Surface waves* 37
  3.13 *Wind-driven currents* 38

3.14 Tidal effects 39
3.15 Water movements in large lakes 40
3.16 Thermal bar 42
3.17 Measurement of water movements 43
3.18 Dispersal processes 44
3.19 Sedimentation 55

4 Nutrient cycles   56
   4.1 Phosphorus 56
   4.2 Nitrogen 75
   4.3 Sulphur 79
   4.4 Carbon 81

5 Characteristics of pollutants   86
   5.1 Domestic sources 86
   5.2 Urban runoff 89
   5.3 Industrial wastes 90
   5.4 Agricultural wastes 92
   5.5 Biological pollution 93

PART II
Effects of pollutants in standing water   95

6 Phytoplankton   97
   6.1 Seasonal pattern 99
   6.2 Population growth kinetics and concept of limitation 100
   6.3 Effects of light and mixing 108
   6.4 Effects of temperature 128
   6.5 Nutrient limitation 136
   6.6 Cyanobacterial toxicity 166
   6.7 Cyanobacteria and water supplies 172

7 Eutrophication   176
   7.1 Definition 176
   7.2 Phosphorus mass balance models 177
   7.3 Trophic state criteria 186
   7.4 Eutrophication of coastal marine waters 210

8 Zooplankton   212
   8.1 Population characteristics 212
   8.2 Population dynamics 213
   8.3 Filtering and grazing 218

8.4 Zooplankton grazing and eutrophication 219
8.5 Trophic cascades 222
8.6 Temperature and oxygen 225
8.7 Invasive, non-native zooplankton 227

9 Macrophytes 228
9.1 Habitats 228
9.2 Significance of macrophytes 229
9.3 Effects of light 232
9.4 Effects of temperature 235
9.5 Effects of nutrients 235
9.6 Nutrient recycling 241
9.7 Alternate stable states 242
9.8 Invasive, non-native macrophytes 246

10 Lake and reservoir restoration 248
10.1 Pre-restoration data 249
10.2 External controls on P 251
10.3 In-lake controls on P 260
10.4 In-lake controls on biomass 265

PART III
Effects of pollutants in running water 271

11 Periphyton 273
11.1 Significance to productivity 274
11.2 Methods of measurement 276
11.3 Factors affecting growth of periphytic algae 277
11.4 Grazing 296
11.5 Organic nutrients 298
11.6 Periphyton community change as an index of waste type 303
11.7 Effects of toxicants 306
11.8 Nuisance 308

12 Benthic macroinvertebrates 311
12.1 Sampling for benthic macroinvertebrates 313
12.2 Natural factors effecting community change 314
12.3 Oxygen as a factor affecting community change 318
12.4 Temperature 320
12.5 Effect of food supply on macroinvertebrates 326
12.6 Effect of organic matter 327

	12.7	Effects of toxic wastes 334	
	12.8	Suspended sediment 336	
	12.9	Recovery 338	
	12.10	Assessment of water quality 339	
	12.11	Invasive, non-native invertebrates 350	

13 Fish 353

	13.1	Dissolved oxygen criteria 353
	13.2	Temperature criteria 362
	13.3	Temperature standards 370
	13.4	Comments on standards 371
	13.5	Global warming 374
	13.6	Toxicants and toxicity 376
	13.7	Invasive, non-native fish 397

Appendices 401

Appendix A	Description of benthic macroinvertebrates and comments on their biology	403
Appendix B	**Study questions and answers**	406
Appendix C	**Glossary**	427

References	440
Index	494

Preface

The purpose of this book is to convey a broad, but in many respects still detailed, account of the effects of the major pollutants from point and nonpoint sources on freshwater aquatic ecosystems. To cope with expanding knowledge, there is an increasing need for synthesis and integration of pertinent information in many areas. This is especially true with effects of pollutants in aquatic ecosystems which is a prime example of an area requiring broad, interdisciplinary study. However, many pollution problems may each require a separate treatise in order to cover the recent literature fully. Although much detail must be sacrificed in a book such as this, its value is in introducing the reader to the basic concepts of limnology with application to several pollution problems all under one cover. The emphasis is on conveying an understanding of cause and effect in assessment rather than an assessment of the national or world-wide state of the various problems.

Material is presented at a level that hopefully appeals to scientists and engineers with and without a biological background. In this regard, a course using the material in this book has been successful in drawing from upper division undergraduate and graduate students from several different disciplines at the University of Washington for thirty years and is offered conjointly between Civil and Environmental Engineering and Fisheries. A similar course using this book has been taught for ten years at Seattle University. Many of the concepts presented have been formulated over that time and this third edition is an update of some sections adding pertinent literature published over the past twelve years, as well as a rewriting and restructuring of other parts. Scientists and engineers with pollution enforcement and resource agencies, and in industries involved in pollution assessment/mitigation, should find this new edition worthwhile.

Although much progress has occurred in the control of water pollution and improvement in water quality in developed countries over the past three decades, the degradation of water resources by waste substances continues in developed and developing countries alike. Eutrophication, the problem to which much of this book is devoted, is much better understood and has been managed by controlling point sources to several, well-publicized larger lakes. However, degradation continues in countless streams, lakes and reservoirs in developed and developing countries through non-point sources of nutrients such as urban and agricultural runoff. Because of the emphasis on eutrophication and the coverage of basic limnological concepts, the title is now Applied Limnology. While thermal effects due to waste heat from power plants never reached a problem of large magnitude due to the insistence on cooling facilities, it threatens to reappear through global warming. Acidification of freshwater is a near global problem although reductions in emissions have recently occurred in developed countries. Toxicants in the environment have received the major attention in the past two decades, largely due to the problems of hazardous waste. Although groundwater is most impacted by toxicant seepage from landfills, some surface waters are also affected. Problem emphasis changes, but the response of ecosystems to various levels of substances does not change. For instance, trout will respond in a predictable way to increased temperature whether the cause is a power plant effluent or global warming, and algae to nutrients whether from sewage effluent or urban runoff. Thus, there is need for people in environmental protection in general and water quality control in particular to understand the basic concepts regarding the assessment of aquatic ecosystem degradation. It is to this end that we have revised this book.

We are indebted to many for discussions and suggestions including Gunnel and Ingemar Ahlgren (temperature and eutrophication), Barry Biggs (periphyton), Mike Brett (zooplankton, eutrophication, hydrographics), Jim Buckley (toxicity), Dennis Cooke (eutrophication and restoration), Rich Horner (periphyton), Brian Mar (modelling), Gertrude Nürnberg (eutrophication), John Quinn (periphyton), Geoff Schladow (hydrographics), and Dimitri Spyridakis (phosphorus and carbon).

Eugene B. Welch and Jean M. Jacoby
Seattle,
Washington, USA

Part I

General concepts of aquatic ecosystems

Chapter 1

Ecosystem function and management

An ecosystem can be described as some unit of the biosphere, or the entire biosphere itself, within which chemical substances are cycled and recycled while the energy transported as part of those substances continually passes through the system. Although every ecosystem must ultimately obey the laws of thermodynamics and degrade to complete randomness (wind down), consistent with the universe, energy may be accumulated momentarily in an ecosystem. However, without a continuously renewed input of energy, the accumulation would be exhausted and the system would ultimately wind down.

The single continuous input to the world's ecosystems is from solar energy, and the conversion of that electromagnetic energy into chemical energy and then into work is what allows an ecosystem and the organisms that make up that system to function.

Because the processes of energy flow and nutrient cycling are quite variable, the assignment of definite boundaries between what might otherwise seem like clearly separate systems is not easy. However, for the sake of practicality and manageability, boundaries are cast that allow ease of study and process measurement. Thus, a stream and its immediate watershed, as well as a lake and its watershed inputs, may be considered ecosystems. The system could be considered closed under some conditions if one is describing chemical nutrients, but never if one is referring to energy, because there is always an input to and a loss from the system. Because an ecosystem responds to inputs as an integrated system, the study of whole systems is useful for management.

Limnology is the study of lakes and streams that integrates the interactions among their physical, chemical and biological components, while *applied limnology* has its focus on the effects of human activities on these aquatic ecosystems. Water is an absolute requirement for humans and other organisms on earth. Although water is one of the most abundant compounds found in nature, covering approximately three-fourths of the earth's surface, over 97% of the total water supply is contained in the oceans and other saline water bodies (Todd, 1970). Of the remaining 3%, a little more than

2% is tied up in ice caps and glaciers, as well as in soil and atmospheric moisture. The remaining 0.6% is found in freshwater lakes, rivers and shallow groundwaters (Todd, 1970). It is upon these freshwater systems that humans most heavily rely, but, ironically, most seriously alter.

Effective management of watersheds is critical to controlling pollutant inputs to surface and groundwaters and maintaining an acceptable level of quality and availability of freshwater. The decreasing availability of high-quality freshwater has become the most serious environmental problem in many areas of the world. The World Health Organization (WHO) estimates that 1.4 billion people do not have access to safe drinking water and 2.9 billion people lack adequate sewage disposal means. More than 250 million new cases of water-borne disease are reported annually, about 10 million of them resulting in death (with about half of those deaths among children younger than 5 years). Despite improved water quality in some areas of North America and Europe, pollutant inputs to aquatic ecosystems from industrial, urban and agricultural sources have increased globally. With the human population growing at a rate greater than 200 000 people per day (annual increase of 1.4%) (United Nations, 2001), the degradation of the world's aquatic resources continues to increase. Understanding the complexities of ecosystem structure and function must be the foundation for developing water policies for the long-term sustainability of the world's aquatic resources.

1.1 Ecosystem composition and energy sources

Each ecosystem has a structure that determines how it functions in the transfer of energy and the cycling of nutrients. This structure can be thought of as the organization of the internal groupings of chemical nutrients and energy through which the functioning occurs. This matter is living and dead, the living being represented best by the trophic levels of organisms leading from algae and/or rooted plants to fourth-level carnivores, for instance, in some grazing food webs. Outside organic matter is processed and decomposed by insects, fungi and bacteria through to the carnivores in detritus-based food webs.

The watershed has come to be regarded as an integral and inseparable part of the ecosystem (Borman and Likens, 1967; Likens and Borman, 1974). The character of the watershed determines whether the stream's energy source is mostly autochthonous (produced within) or allochthonous (produced outside). In a comparative study of areas in the northeastern USA, Likens (personal communication) has shown that an autochthonous-producing forest results in the domination of the drainage stream by allochthonous inputs, whereas the watershed lake is autochthonous. Wissmar et al. (1977) found a high mountain lake in a coniferous forest to be dominated by allochthonous inputs (Table 1.1). Lakes and streams in poorly vegetated watersheds (Minshall, 1978) and relatively large lakes tend to be dominated

Table 1.1 Comparison of energy sources in two ecosystems

Energy source	Energy (g carbon m^{-2} year^{-1})			
	Deciduous forest			Coniferous forest
	Forest	Streams	Lake	Lake
Autochthonous	941	1	88	4.8
Allochthonous	3	615	18	9.4

Sources: Data on deciduous forest from G. E. Likens, personal communication; data on coniferous forest from Wissmar et al. (1977).

by autochthonous sources. According to the river continuum concept, low order streams in forested watersheds, as well as high order lower stretches of often turbid rivers, tend to be heterotrophic or detritus based, while intermediate order streams are autotrophic (Vannote et al., 1980). Chapter 11 includes further discussion of stream productivity.

The relative proportions of allochthonous to authochthonous energy sources in ecosystems is useful for management decisions affecting aquatic ecosystems. To protect and use aquatic systems effectively, one must know how systems use and respond to different amounts and varying compositions of natural energy.

1.2 Energy flow and nutrient cycling

If one could measure all the organisms and their main processes in an ecosystem and if their energy-consuming and energy-processing characteristics fell neatly into separate levels, one would be able to arrange a flow diagram as shown in Figure 1.1. Such a diagram shows several important points: (1) energy flows through the system and does not return, because by the second law of thermodynamics matter moves toward randomness, from states of high concentration to states of low concentration, and when that happens the energy contained becomes less (note entropic heat loss); (2) loss in energy as heat occurs at each step in the transfer process; and (3) allochthonous energy moves through the heterotrophic microorganisms or decomposers. A significant role in net production of usable energy for consumers is attributed to these microorganisms (note arrow from decomposers to consumers) in addition to decomposition (Pomeroy, 1974).

Even for allochthonous sources from surrounding forests or from cultural input of organic waste, the ultimate source must be either photosynthesis or to a lesser extent chemosynthesis.

The process of photosynthesis includes two phases: (1) a light reaction that traps solar energy, releases molecular O_2 and produces ATP (adenosine

6 General concepts of aquatic ecosystems

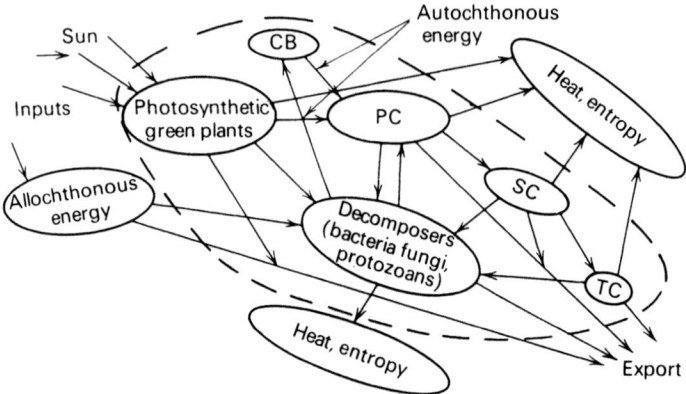

Figure 1.1 Energy flow and nutrient cycling in an aquatic ecosystem (Boundary indicated by dashed line). CB, chemosynthetic bacteria; PC, primary consumers; SC, secondary consumers; TC, tertiary consumers.

triphosphate) through photophosphorylation; and (2) a dark reaction (light not needed) that utilizes the trapped energy as ATP and fixes CO_2 or HCO_3^- into reduced organic cell material. The process yields energy and synthesizes new cells and can be summarized by:

$$6CO_2 + 12H_2O + \text{light} + (\text{chlorophyll } a \text{ and accessory pigments})$$
$$\rightarrow C_6H_{12}O_6 + 6O_2 + 6H_2O$$

The organic compound is glucose – the compound that holds the sun's energy trapped by photosynthetic organisms and allows the maintenance of the biosphere (that thin film of atmosphere, water and land around the planet Earth that supports life). The process is performed by green plants and a few pigmented bacteria. Only the green plants release O_2 as a byproduct.

Chemosynthesis is the other process through which organisms are totally self-sufficient in trapping energy and building cell material. This process yields energy through the oxidation of reduced inorganic compounds and thus requires no previous biological mediation by the various bacteria utilizing this process. The primitive Earth was rich in reduced inorganic compounds that are currently used by some bacteria for energy. An example is nitrification:

$$2NH_3 + 3O_2 \rightarrow 2HNO_2 + 2H_2O + \text{energy}$$

This process is less important to ecosystems as an energy yielder and more important as a material recycler. Chemosynthetic processes by bacteria are heavily involved in the cycling of such nutrients as nitrogen and sulphur.

Nutrients, such as N, C, P and S, follow the same pathways as energy in the system; they are consumed by autotrophs (green plants and chemosynthetic bacteria) from inorganic pools and fixed into organic compounds. These nutrients are then transferred through trophic levels just as energy is, because the reduced organic compounds, of which organisms are composed, are the carrier of the entrapped chemical energy. For example, CO_2 is the most oxidized state of C, but when fixed into glucose through photosynthesis, 1 mole $(C_6H_{12}O_6)$ contains 674 kilocalories (kcal) of energy releasable through respiration by plants, animals, or decomposer microorganisms by the same biochemical pathways. The energy content of whole organisms in the various trophic levels ranges from 4 to 6 kcal g^{-1} dry weight.

The principal difference between energy and nutrient transport is that once organisms have utilized the energy from complex compounds and oxidized them completely (e.g. glucose to CO_2 and H_2O), the compounds are recycled through the inorganic pool(s) and are almost totally reusable by the community. There is no permanent loss, but, practically, a certain fraction may be lost to the sediments and require tectonic uplift for recycling. The recycling process is shown in Figure 1.2 (note heavy arrow from decomposers to inorganic pool). There are also chemical–physical processes involved in nutrient cycling. Chapter 4 includes further discussion of nutrient cycling.

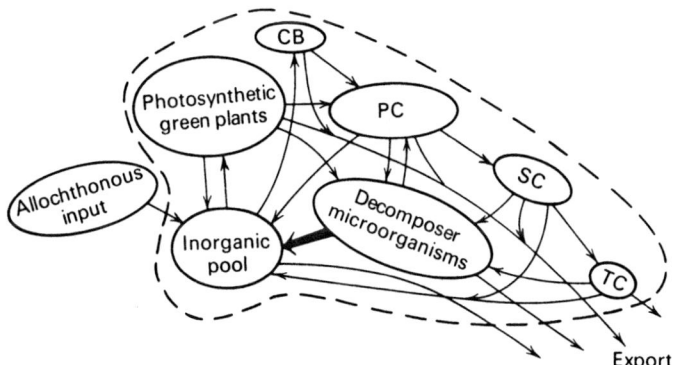

Figure 1.2 Transport and recycling of nutrients through an ecosystem (boundaries indicated by dashed line). CB, chemosynthetic bacteria; PC, primary consumers; SC, secondary consumers; TC, tertiary consumers.

1.3 Efficiency of energy and nutrient use

The efficiency of the transfer of energy and/or nutrients is usually measured by the ratio of net productivity of a trophic level to the net productivity available for its consumption (Russell-Hunter, 1970, pp. 25, 191). These values usually fall between 10 and 20%. The level of this efficiency depends greatly upon the structure of the food web through which the materials are moving.

Structure in ecosystems can be thought of as the organization of species populations into appropriate trophic levels. However, trophic level is rather artificial and few organisms conform to a single trophic level throughout their entire life cycle. Nevertheless, populations organize, in time, in such a way that energy usage is optimized. Ecosystems that are physically stable, such as tropical rain forests and coral reefs, maximize organization and complexity and remain rather constant in biomass, productivity, species diversity and, consequently, efficiency. Very simply, the greater the variety of energy users in an ecosystem (alternative pathways), the greater the chance that a quantity of energy packaged in a particular way will be intercepted and used before it leaves the system. Instability in ecosystem structure brought about by natural or culturally caused variability in the physical–chemical environment should result in decreased efficiency and instability in energy and nutrient usage, and nutrient recycling is, therefore, much 'looser'. Where systems approach steady state, that is, inputs equal outputs, nutrient recycling tends to be 'tighter' and 'more complete' (Borman and Likens, 1967; Bahr *et al.*, 1972).

Odum (1969) has illustrated the effect of maturity on the characteristics and processes of ecosystems (Table 1.2 and Figure 1.3). He proposed that diversity tends to be low in immature systems and high in mature systems. Stability is often suggested to increase with diversity, but a definition of

Table 1.2 Characteristic properties of ecosystems

Property	Type of ecosystem	
	Immature	Mature
Net production (yield)	High	Low
Food chains	Linear	Weblike
Nutrient exchange	Rapid	Slow
Nutrient conservation	Poor	Good
Species diversity	Low	High
Stability-resistance to perturbations	Low	High
Entropy	High	Low

Source: (Odum, 1969), with permission of the American Association for the Advancement of Science.

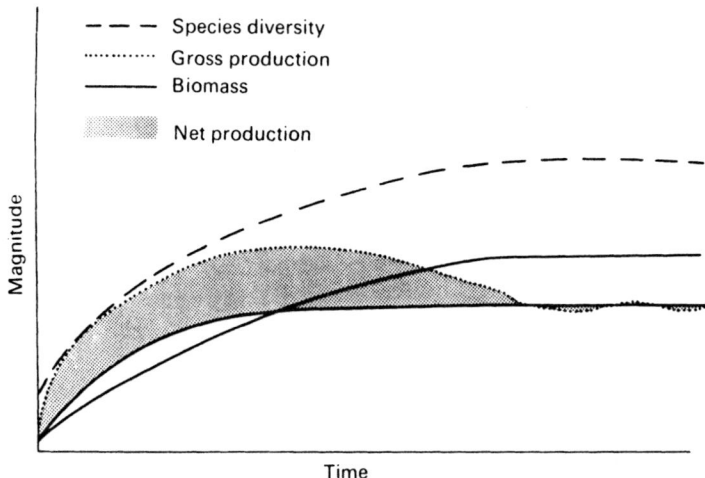

Figure 1.3 Conceptual plot demonstrating successional changes in species diversity, gross production (respiration + net production) and biomass. The shaded area representing net production approaches zero as the ecosystem matures (Bahr et al., 1972, modified from Odum, 1969).

stability and whether it is directly related to diversity is a matter of controversy. Washington (1984) reviewed the problem and cited Goodman (1975) who suggested that stability has generally included one or more of the following: a constancy of numbers, a resistance to perturbation, or ability of a system to return to a previous state after displacement. According to Figure 1.3, net production tends to be inversely related to diversity. Margalef (1969) suggested that at low diversity, fluctuations in production per unit biomass are quite large and diversity changes are small, indicating instability, whereas at high diversity, fluctuations in production per unit biomass tend to be small with considerable changes in diversity, indicating stability. Figure 1.3 suggests that with time ecosystems tend to develop into a rather steady-state condition, with little net productivity and maximized structure. If production is low and associated fluctuations are small in a mature ecosystem, it follows that nutrients are conserved and cycled more slowly and energy thus moves through the system at a slower rate and entropy is less – the system is more efficient in its functioning.

Examples of efficiency in ecosystems give some idea of their maturity. Fisher and Likens (1972) showed that Bear Brook, a northeastern forest stream, was only 34% efficient (total respiratory loss/total energy input). Thus, 66% of the energy left the system downstream unused, indicative of immaturity. Lindeman's (1942) study of Cedar Bog Lake showed it to be 54.5% efficient with 45.5% lost to sediments. Odum's (1956) results from

Silver Springs in Florida showed only 7.5% lost from the system downstream with 92.5% being utilized.

1.4 Management of ecosystems

What do humans want from aquatic ecosystems? Obviously they do not want to destroy them, but they are highly inclined to manipulate them to their purpose whether it be to increase or decrease their productivity, keep them natural, or have them accept (assimilate) wastes with a minimum adverse effect on other uses. The problem is that these uses are conflicting. For example, to produce more and bigger sport fish would usually require stimulation of primary production, which, in turn, may result in decreased structure (species diversity) with possibly less stability. The yield of particular game-fish populations may increase, but at the expense of the overall efficiency of energy-handling and nutrient-recycling capabilities of the system. That would be indicated by the occurrence of a large biomass of inedible algae along with a wider range in oxygen, pH and even toxic ammonia. Fish may be larger and more abundant, but they could also be subject to high mortality rates. In the same sense, ecosystems cannot assimilate wastes without some cost to their structure and stability, with consequent reduced efficiencies in nutrient cycling and energy utilization.

In reality, stable communities are probably not more able to resist change from waste input than unstable ones. Bahr *et al.* (1972) argue that if this were true, species diversity would decrease with waste input (abiotic change) at an increasing rate (curve B in Figure 1.4), rather than at a decreasing rate as is usually observed (Figure 1.4, curve A). Highly diverse communities may produce stable behaviour in ecosystems, but such a state is very sensitive in itself to change in the environment. Any change, or manipulation, often tends to lower that stability and decrease the efficiency of ecosystem functioning. Humans probably cannot have their cake and eat it too. That is, highly structured and diverse ecosystems that maximize recycling and energy-using efficiency are probably not highly resistant to disturbance. The use without abuse idea is discussed further, e.g. in Chapter 11.

The general characteristics and effects of various pollutants and other forms of disturbance on aquatic ecosystems and their organisms will be the focus of the following chapters. The pollutants of primary importance include nutrients (e.g. N, P), sediment, organic material, toxic substances (e.g. metals, synthetic organic chemicals) and pathogens. The leading causes of impairment in assessed freshwater of the USA include nutrients, metals (primarily mercury), bacteria and siltation (USEPA, 2002b). The primary sources of impairment were found to be runoff from agricultural lands, municipal point sources and hydrologic modifications (e.g. channelization, flow regulation, dredging). Primary causes of impairment in estuaries also included toxic substances (particularly mercury and pesticides) and

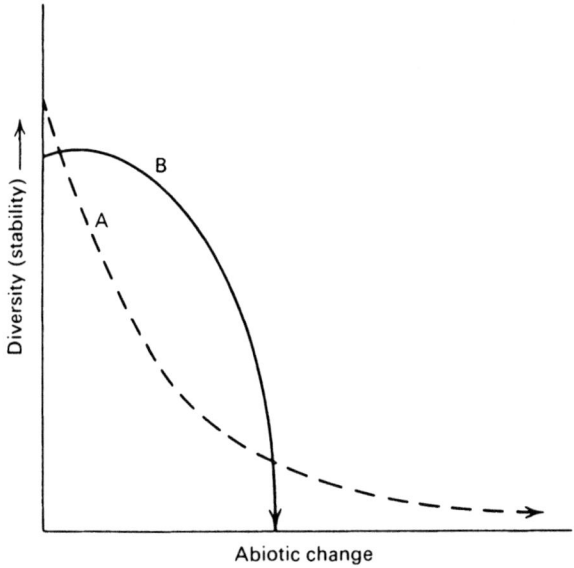

Figure 1.4 Ecosystem stability as a function of intensity in a stressful abiotic change. The two curves are discussed in the text (Bahr et al., 1972).

oxygen-depleting substances (i.e. organic material). USEPA (2002b) reported that approximately 40% of streams, 45% of lakes and 50% of estuaries that were assessed were not clean enough to support uses such as fishing and swimming. Introduced (i.e. non-native), invasive species also pose increasingly severe environmental problems in both aquatic and terrestrial ecosystems and can thus be considered as 'biological pollutants' or biopollution. The study of the effects of the above pollutants and disturbances in inland water bodies is the foundation of the discipline of applied limnology.

Chapter 2

Standing crop, productivity and growth limitation

Before population growth and control are discussed further, some important terms should be understood. Standing crop is often synonymous with biomass in that it represents the quantity per volume or area that is present and can be measured. The units of measure are wet weight, dry weight, numbers, or the quantity of some elemental constituent (N, P, C). Standing crop is the more general term; biomass is appropriate, of course, when the unit is mass.

Productivity is the rate of biomass formation and is a dynamic process. Productivity is often proportional to biomass, but in the case of fast-growing and efficiently grazed populations, biomass may be relatively small, whereas daily productivity is quite high. To understand that phenomenon, the concept of turnover rate is useful and can be defined as:

$$\text{Turnover rate} = \frac{\text{Productivity}(m\,l^{-3}t^{-1})}{\text{Biomass}(m\,l^{-3})}$$

where m = mass, l = length and t = time.

Turnover rate is thus the number of times that the biomass is theoretically replaced per unit time. Because they have the same units ($1/t$), turnover rate equals growth rate but only when biomass remains constant. When biomass is not constant from one time to another, the growth rate over that period is determined by the increase or decrease in biomass as well as the instantaneous measurement of productivity.

2.1 Pyramid of biomass

Productivity is directly proportional to biomass in a food chain only if the turnover rate at each trophic level is constant. A hypothetical example of this is shown in Table 2.1. Because energy is lost at each transfer in the food chain, productivity must always decrease from primary producer to secondary consumer. In this instance, productivity decreases by a factor of 10 at

Table 2.1 Hypothetical relation of biomass and productivity in a three-tiered food chain. Units are in mass (m), length (l) and time (t)

Trophic level	Biomass ($m\,l^{-2}$)	Productivity ($m\,l^{-2}t^{-1}$)	Turnover rate (t^{-1})
Secondary consumer	1	10	10
Primary consumer	10	100	10
Primary producer	100	1000	10

each level. There is no reason to believe turnover rate would remain constant. Biomass may be greater at consumer than producer levels because producers are smaller organisms with a higher ratio of cell surface to volume and therefore have more rapid turnover rates. In such a case the biomass pyramid may be inverted as cited by Odum (1959, p. 64). In Long Island Sound, zooplankton and bottom fauna biomass were double that of phytoplankton. In the English Channel the ratio was 3:1. Thus, from this theoretical standpoint, productivity may not, in fact, be very tightly related to biomass in nature, particularly in open water systems. If the biomass in the examples from Odum were multiplied by turnover rate, the pyramids of biomass would be pyramids of productivity (and energy flow) and become upright.

2.2 Productivity

Primary production is the rate of biomass formation of the autotrophs, which are primarily the chlorophyll-bearing micro- and macroorganisms. Net production rate is measured as the net amount of organic matter fixed (e.g. carbon assimilated) and transferrable to the next trophic level. Gross production equals total assimilation including respiration. These quantities can be represented as shown in Figure 2.1.

In an ecosystem with plants only, net production by primary producers equals net ecosystem production (NEP) and is actually the realized increase in biomass available for harvest or grazing by primary consumers (secondary producers). With consumers present, net ecosystem production (NEP) is the sum of the fraction of net production (NP) not consumed plus the realized growth of consumers (secondary production). The ratio of gross production to respiration (P:R) in these hypothetical examples is 2 with plants only (assuming $R = 0.5\,GP$) and between 1 and 2 with consumers present. Tropical rain forests have P:R ratios near 1 due to high rates of decomposition.

In heterotrophic, or allochthorous-fed ecosystems, P:R ratios are typically less than 1. The use of ratios of heterotrophic: autotrophic components is discussed further in Chapter 11.

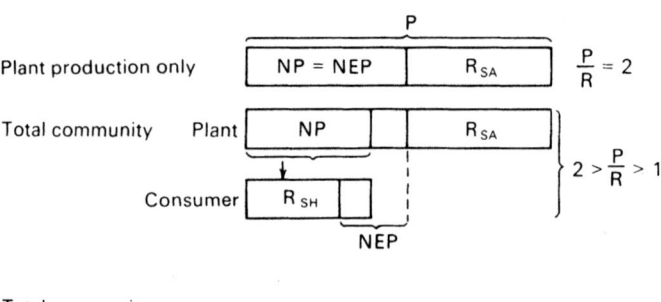

Figure 2.1 Hypothetical diagram of a simplified ecosystem energy budget. P, Gross production; NP, net production; R_{SA}, respiration autotrophs; R_{SH}, respiration heterotrophs; NEP, net ecosystem production (modified from Woodwell, 1970).

2.3 Efficiency of energy transfer

The energy efficiency of ecosystems has direct management implications, as pointed out earlier. At the first trophic transfer, photosynthesis by chlorophyll-bearing organisms is rather inefficient. For terrestrial systems, efficiency of light utilization is about 1%, but in water it is much lower: 0.1–0.4% (Kormondy, 1969). The lower efficiency in water results from the absorptive and scattering effects of light by water and its particles.

Utilization of primary productivity through secondary production is nominally about 10%, but may range from about 5–20%. The 90% that is lost at each trophic transfer is largely through respiration. The steps involved in handling energy at the grazer level are shown in Figure 2.2.

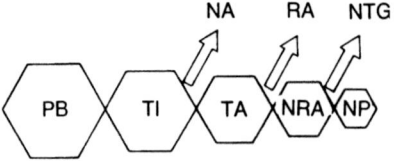

Figure 2.2 Energy budget of individual animal. PB, plant biomass; TI, total energy ingested; TA, total energy assimilated; NA, non-assimilation (excretion); RA, respired assimilation (maintenance and activity demand); NRA, non-respired assimilation; NTG, non-trophic growth; NP, net production (after Russell-Hunter, 1970).

Food energy assimilated is equal to that ingested minus excretion. Respiration and any non-trophic growth (e.g. clam shells) are the other losses that subsequently result with net production remaining for transfer to the next level.

In essence, the efficiency of energy utilization at each trophic level depends upon how individual consumers handle the energy ingested. Trophic transfer is relatively efficient or inefficient because of the individual populations that are present. If the structure of an ecosystem is disrupted by waste input such that many of its efficient energy utilizers are eliminated, then overall efficiency in the system would naturally decrease. Large populations of unconsumed producers could indicate the degrading effect on the dynamics of the system.

2.4 Population growth and limitation

2.4.1 Growth

Growth of any population of living organisms can be described as exponential in an environment where resources are unlimited. With exponential growth, individuals are added at an increasing rate as the population becomes larger. This unlimited growth can be described by the following first-order, differential equation:

$$dN/dt = rN \tag{2.1}$$

where r is the growth rate constant and N is population size.

The integrated form of this equation is:

$$N_t = N_0 e^{rt} \tag{2.2}$$

where N_t is population size at some time t after growth begins, N_0 is the initial size of the population and e is the base of natural logarithms.

Exponential growth of a population operates like compound interest. That is, the interest gained on an investment after some time interval is analogous to individuals added to a population and is dependent on the size of the investment (or population). By multiplying the interest rate by the new balance after each interval of time, the rate of increase in the total investment increases with time. This is shown graphically in Figure 2.3 and in Equation 2.1; as N increases, dN/dt increases.

The growth rate constant, r, has a maximum for each population that is consistent with its genetic capacity in an optimum environment. If the environment is less than optimum, r will not reach the genetically fixed maximum. Thus, an evaluation of r will indicate how optimum an environment is for a given population. In multicellular sexually reproducing organisms,

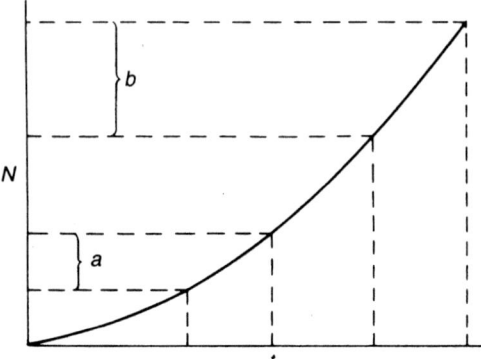

Figure 2.3 Unlimited population growth curve. Note that population increment *b* is much greater than *a* for an equal time increment, which is due to the larger population size. This is a result of the 'compound interest' phenomenon characteristic of populations.

r is affected by the age structure, or that fraction of the population that is reproductively active. If the age structure remains constant, r can be evaluated with respect to environmental variables. However, much of the change in r could be the result of altered age structure. This is not a problem in the analysis of unicellular asexually reproducing populations because theoretically all have an equal chance of reproduction. In both types of populations, r is the difference between birth and death rates, although with phytoplankton, as will be discussed later (Chapter 6), the growth rate is treated separately from loss (or death) rates. Typical growth rates range from 0.01 year^{-1} (1%y^{-1}) for a human population to 1.0 day^{-1}(100%d^{-1}) for algae and about 3.0 h^{-1} for bacteria.

In reality, population density cannot continue to increase exponentially because some resource will eventually limit further increase. The resource that exhausts first will slow and ultimately stop the process as its supply diminishes. An example of populations that grow rapidly, seemingly to 'explode', are spring phytoplankton blooms. They tend to exhaust the limiting resource and reach their maximum attainable biomass all at once, which is followed by a crash. This type of growth is often represented by J-shaped curves. On the other hand, if environmental resistance is gradual, the growth curve resembles an S-shape, in which specific stages in growth can be defined. Initially, population increase is slow as the organisms adjust to their environment (Figure 2.4). This is followed by exponential growth when resources are unlimited. When resources become limiting, population increase slows and ultimately stops if resource supply is exhausted. If additional resource is not forthcoming, the population will die and decomposition

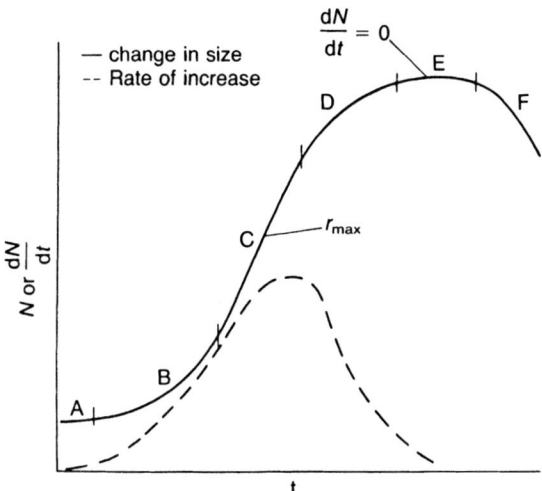

Figure 2.4 Population growth curve, environmentally limited, showing changes in size and rate of increase. Growth curve phases are: A, lag; B, accelerated growth; C, maximum rate; D, decelerated growth; E, stationary; F, death.

may recycle the limiting resource resulting in the start of a new growth curve.

The growth curve in Figure 2.4 is referred to as logistic and can be described, at least through the stable population phase, by the following equation;

$$dN/dt = rN[(K - N)/K] \qquad (2.3)$$

where K is the maximum attainable population size and r and N are as defined previously. Note that dN/dt, which is net productivity, reaches its peak consistent with the maximum slope (r) of the population curve (Figure 2.4).

The integrated form of Equation 2.3 is:

$$N_t = \frac{K}{1 + [(K - N_0)/N_0]e^{-rt}} \qquad (2.4)$$

2.4.2 Limitation

The concept of resource limitation will be briefly considered with respect to nutrients. Justus Liebig's law of the minimum is fundamental to understanding the control of population growth. It states that the yield of a crop will be limited by the essential nutrient that is most scarce in the environment relative to the needs of the organism. Relation to need is an important part of that definition. The element lowest in concentration may not be the limiting nutrient. For example, nutrient A may be twice as abundant in the environment as nutrients B and C, but if plant growth requires three times the amount of A compared to B or C, further growth will first be limited by a reduction in A. Growth can also be limited by more than one nutrient simultaneously, if all required nutrients are reduced proportionately, but remain in the required ratio for growth.

The yield limitation of a crop of microorganisms by the scarcity of a critical nutrient is illustrated in Figure 2.5. In this example, the growth rate (denoted by the maximum slope of the line; C in Figure 2.4) can vary with temperature, light intensity, nutrient concentration or species used in the test, but the maximum yield attained (in the Liebig sense) depends on the abundance of the most limiting nutrient. Note that in the case of a limiting nutrient, K in Equation 2.3 is analogous to the quantity of nutrient available for assimilation and converted to biomass. This concept is the basis for the algal growth potential (AGP) test (Skulberg, 1965; Miller *et al.*, 1978). AGP tests are also sensitive to inhibitors and have been used to evaluate toxicity of wastewater and hazardous waste sites (Peterson *et al.*, 1985).

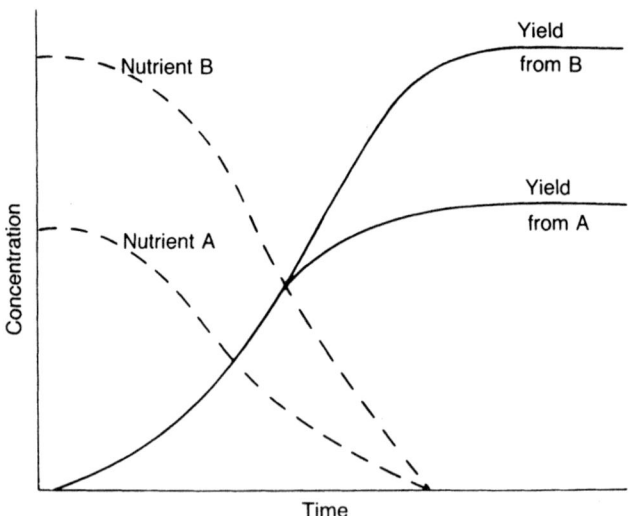

Figure 2.5 Growth curves for a microbial population showing the maximum yield as a function of initial nutrient (limiting) concentration.

2.4.3 Tolerance

Shelford's law of tolerance includes the concept of too much of a resource, as well as too little (Odum, 1959). The growth and survival of a population can be controlled by quantitative or quantitative deficiency or excess of any one of several environmental factors that may approach the limits of tolerance of that population. Aquatic environmental variables may include all or one of the following: toxic inhibitors, quality or quantity of food resources, micro- and macro-inorganic nutrients, temperature, dissolved solids (salinity), sediment, light, current velocity, dissolved oxygen, hydrogen ion (pH), etc.

2.4.4 Adaptation

Populations can be considered to have 'built-in', or genetic tolerances for environmental variables. Such heavy metals as Cu and Zn, which are usually thought of as only being toxic, are also required at low concentrations for metabolism. As such, they are referred to as micronutrients. However, at some higher concentration (depending on the population), they will become toxic.

This variation in genetically based tolerance is simply illustrated in Figure 2.6. An organism whose optimum for activity, growth and survival is denoted by the solid line would probably be lost from the system and replaced by a species whose optimum is denoted by the broken line if the factor increased in concentration range from A to B. The optimum range of environmental factors that describe the niche for a population include biotic as well as abiotic factors. However, physicochemical factors of water quality will be emphasized here.

Each species is adapted through genetic selection to a set of environmental conditions. Further genetic adaptation to changed environmental conditions proceeds slowly through several generations. However, within the limits of genetic tolerance, a species can adapt physiologically. For example, fish

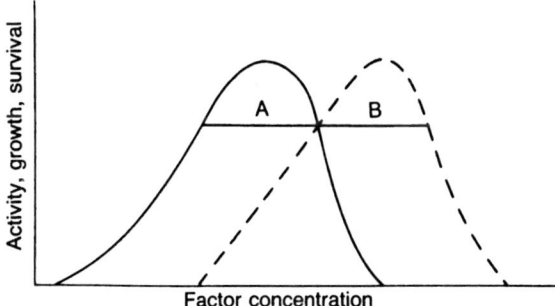

Figure 2.6. Hypotherical response of two populations to a required factor.

adapt to higher temperatures in summer, taking their cue from increased temperature and photoperiod. Algae adapt to low and high light by adjusting cell chlorophyll content. Overall, the species that has the greatest capacity to maintain the most growth to offset respiratory demands on energy, in the face of a changing environment beyond its optimum range, should be able to maintain the most stable population.

Chapter 3
Hydrographic characteristics
T. Lindell

The effects of wastewater in the aquatic environment are largely determined by the physical characteristics of the receiving water. However, the initial condition for the mixing process can be modified by adding the wastewater at variable depths or using different types of diffusers. The final concentration of a contaminant in the receiving water is thus determined by these initial manipulations plus the physical characteristics of the wastewater itself and the receiving water body. One type of environment, effective in dispersing the contaminant, may produce low concentrations; another environment, capable of very little dispersion may produce high concentrations. However, the best solution for water quality protection is not always to produce effective dispersion and low concentrations of a contaminant, but rather effective waste treatment. Sedimentation represents a loss of pollutants from the overlying water, and it is primarily a function of the pollutant concentration, nature and size distribution, as well as the physical characteristics of the water body. Thus, the movement of water and its entrained particles is fundamental to the understanding of the processes that control water quality.

3.1 Types of flow

Depending upon the movements of water, a flow may be classified as either laminar, turbulent or transitional in nature. Within a laminar flow the water movement is regular and 'plate-like', that is, the separate layers flow on top of each other without extensive mixing. The mixing that does occur is on the molecular level. Laminar flow, however, is a rare phenomenon in nature with the exception of those regions of the flow that are immediately adjacent to boundaries.

Water movement that is irregular, causing mixing eddies to develop on a macroscopic scale, is called turbulent. Within a turbulent water mass the water velocity varies considerably in both vertical and horizontal directions. The movement is characterized by erratic motions of the fluid and considerable lateral interchange of momentum. This type of movement is, therefore, most easily described statistically. Turbulent motion totally dominates in a natural environment.

When the flow is constant in space, it is termed uniform and is characterized by parallel stream lines. Even though a truly uniform flow is quite rare in nature, the assumption is often useful for purposes of modelling and analysis of pollutant transport and distribution. The opposite to a uniform flow is called a non-uniform or varying flow.

If the flow varies with time, it is termed unsteady; a flow that is at every point constant with respect to time is called steady. The steady-state condition is also a very useful generalization in modelling processes, as will be illustrated later with phytoplankton growth and lake phosphorus content.

Laminar flow occurs where the water velocity is low and/or close to an underlying solid, smooth surface. The main determinant of whether a flow is laminar or turbulent is the Reynolds number:

$$Re = \frac{vd}{\nu} \tag{3.1}$$

where v = flow velocity, d = a length scale of the system (for example, water depth) and ν = kinematic viscosity. As the kinematic viscosity changes little for water over the temperature and salinity ranges generally experienced, the flow velocity determines the Reynolds number for a given set of conditions. When one layer of fluid slides over another, the friction between the layers gives rise to a shear stress that, with increasing velocity, results in turbulent flow. This state is reached faster if the underlying bottom is rough. The shear stress for laminar flow is expressed as:

$$\tau = \mu \frac{du}{dz} \tag{3.2}$$

where τ = tangential force per unit area, μ = coefficient of viscosity, and du/dz = velocity gradient perpendicular to the direction of motion.

For turbulent motion, eddy viscosity is much larger than the molecular viscosity so the shear stress is often written as:

$$\tau = \varepsilon \frac{d\bar{u}}{dz} \tag{3.3}$$

where ε = eddy viscosity and it represents a time average of the velocity, where the averaging interval is significantly longer than the time scale of the turbulent fluctuations in velocity.

The magnitude of physical processes in water bodies greatly depends on the movement of the water. If the water is moving fast enough, as in some rivers, it is often impossible for a (gravity) wave to travel upstream. The flow

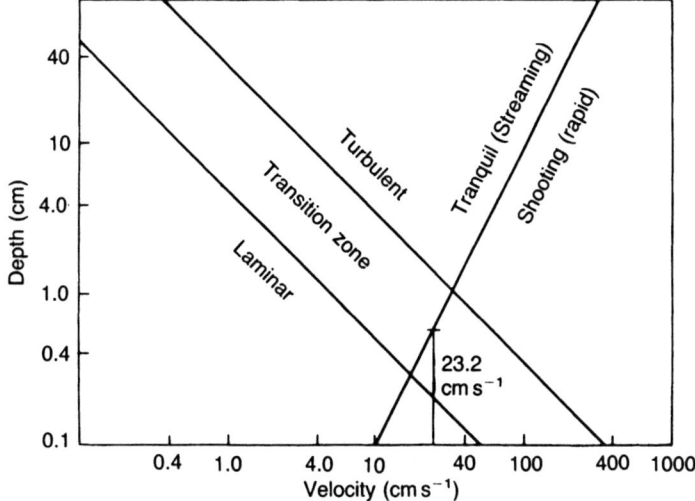

Figure 3.1 Types of water flow in a wide, open channel. See text for explanation (Sundborg, 1956).

in this case is termed shooting or supercritical. The velocity of propagation (c) of a gravity wave in this case is given by:

$$c = \sqrt{(gh)} \tag{3.4}$$

where g = gravity and h = depth of water.

If the flow is slower than the propagation of a gravity wave, the term tranquil (or streaming) is used. In this case a gravity wave could move upstream. The different regimes of flow can be separated according to water velocity and distance from a solid river bottom (Figure 3.1). There is a transition zone between a turbulent and a laminar flow within which the type of flow is essentially dependent on the water temperature.

3.2 Boundary layers

Figure 3.2 shows schematically the flow conditions in a stream above the first few centimetres from the bottom. However, closer to the bottom there is a thin layer where flow conditions undergo rapid changes. The water in immediate contact with the bottom tends to adhere to it, and there is, therefore, a sharp velocity gradient in this zone. This layer, in which the flow is appreciably retarded by the viscosity of the water and the friction against the bottom and within which there are shear stresses in the fluid, is called the boundary layer. Periphyton and benthic invertebrates inhabit this

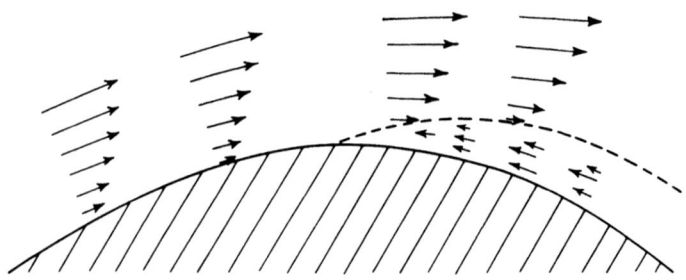

Figure 3.2 Boundary layer separation. See text for explanation.

reduced-velocity environment. The flow in the boundary layer is laminar immediately adjacent to the boundary, but at some distance away it is usually turbulent. Lake velocities can be low, but they are typically $>2\,cm\,s^{-1}$. As indicated in Figure 3.1, flow will be turbulent in lakes with depths $>10\,cm$.

The initiation of turbulence in the boundary layer is primarily a function of the water velocity and the stability of the flow (see the following discussion of density effects). The importance of the boundary layer and the effect of velocity and shear stress on nutrient availability to periphyton will be discussed later (Chapter 11).

In natural water courses, the boundary layer may be perturbed by the character of the bottom; for example, if the bottom is convex (e.g. surface of a boulder), the flow on the downstream side of the convex surface tends to create a reverse velocity component close to the bottom and causes a boundary layer separation (Figure 3.2). The boundary layer separation, as well as other effects of bottom irregularities and shear stresses between adjacent water layers that exist with normal velocity patterns, suggest that nearly all natural water movements are turbulent.

The interaction between the bottom and the overlying water is very important from the standpoint of water quality. The processes of erosion, deposition and resuspension of sediment are dependent upon the variations in the state of turbulence, the viscous shear and the form resistance of the bottom.

The velocity patterns and the state of flow with regard to the processes of deposition and resuspension are just as important as the dispersion and mixing of a pollutant entering a receiving water. For example, within a slow, near-laminar running stream a pollutant can accompany the stream for several kilometres without mixing on a macro scale; however, within a fast-moving turbulent stream the mixing may be practically complete in as little as 10 metres or so downstream, depending on stream size, manner of pollutant introduction, and so on.

3.3 Physical factors controlling wastewater distribution

There are several very important factors that control the responses of a lake to a waste input, e.g. size, shape, depth, residence time/flushing rate, temperature conditions (degree of stratification), hydrometeorological conditions and the hydrological regime. Any or all of these factors may influence the effects of a certain waste discharge in different ways and some of these effects will be discussed in detail below.

Water body size is important in defining the type of phenomena that will occur, e.g. a large enough lake will be dominated by the Coriolis force and thus circulate in a particular manner. A very deep lake, compared to a shallow lake, will show much less influence from the bottom, particularly with regard to short-term changes in water quality, i.e. there is none or little resuspension of particulate matter. A deeper lake, furthermore, generally has a long residence time or a low flushing rate, which is probably the most important single factor in governing physical processes in lakes. Variation in density with depth as a function of temperature, is also an important factor that can limit the water volumes involved in mixing and dispersion processes.

Meteorological, hydrological and hydro-meteorological factors strongly modify and govern the physical behaviour of water bodies and sometimes completely screen a positive or negative human impact. A period of sudden precipitation or snowmelt in spring (within a temperate climate) may cause a substantial nutrient load, which can completely hide and destroy the long-term effects of a wastewater treatment programme. Climatological effects (e.g. amount of solar energy) are also important for the overall functioning of lakes.

The importance of density effects in general will be discussed below before focusing on the effect of temperature on density. Finally, some physical properties of large lakes will be discussed, which are of importance in understanding the behaviour of pollutants in such lakes.

3.4 Density effects

The previous discussion assumes a water volume of uniform density. However, conditions could be markedly changed by density differences within the water mass. Vertical density differences are very common features of both lakes and estuaries and sometimes of rivers, and therefore the causes and effects of variations in density must be understood.

Density gradients in a water mass are caused by differences in water temperature and in dissolved or suspended material in the water. In freshwater lakes, temperature is the dominating cause of density stratification. Freshwater reaches its maximum density at about 4°C, while density decreases above and below that temperature (Figure 3.3). In estuaries, dissolved substances (salinity) are often the principal cause for stratification.

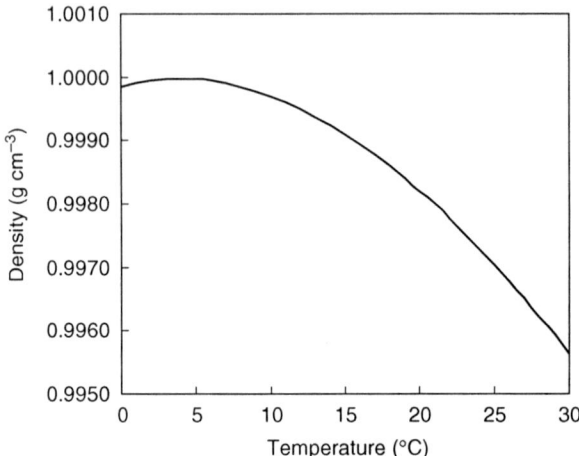

Figure 3.3 The density of distilled water as a function of temperature. The maximum density of water occurs at approximately 4°C.

In rivers and mainstream impoundments, suspended material may influence stratification, although this is quite rare.

Density usually increases towards the bottom of a water column, sometimes gradually over the entire vertical distance and sometimes rapidly within a certain depth interval. The relative effect of a given change in temperature over depth (ΔT/depth) on stratification, as temperature increases will be discussed in relation to phytoplankton, light and mixing in Chapter 6. When a strong vertical density gradient is present, which is common during summer in lakes, the turbulent velocity fluctuations are subdued or prevented and the turbulent exchange of water between contiguous layers becomes weaker or ceases. This occurs because the energy needed to move lighter layers downward and heavier layers upward increases with the density differences. Not only will the turbulent exchange be affected but also the vertical velocity distribution, the shear stress, and the transition from laminar to turbulent motion will be suppressed. Furthermore, if the stratification in a lake is sufficiently strong, it may even be useful to consider the lake as two bodies of water separated by the layer of greatest change in density with depth. This vertical distribution of water masses is quite similar to that of air masses above the earth.

The type of flow is often determined by the density gradient. A dimensionless parameter that describes the stability of the flow is the Richardson number, and is defined as:

$$R_i = -\frac{g}{\rho}\frac{d\rho}{dz} \bigg/ \left(\frac{d\bar{u}}{dz}\right)^2 \qquad (3.5)$$

where ρ = density of the water, $\bar{u}$ = mean water velocity, and g = acceleration resulting from gravity. When the liquid is homogeneous $R_i = 0$, the stratification may be called neutral. If $R_i > 0$, the stratification is considered stable; for $R_i < 0$, it is unstable. One can see that stratification increases as the density gradient increases.

There are some contradictory opinions on the minimum value of R_i for a transition from laminar to turbulent flow. However, far from boundary surfaces, a value of 0.25 (Prandtl, 1952) is usually accepted. Much lower values have been observed near a solid boundary.

3.5 Density currents

Density stratification can greatly influence the distribution of an incoming mass of water that could contain pollutants. The incoming volume of water may flow over, under, or through the main water mass as a result of gravity and the density characteristics of the respective water masses. Such a phenomenon is referred to as a *density current*. This phenomenon was described early in the classic work by Forel (1895). Although not often easily observed directly, density currents are nevertheless frequently present in lakes. For example, streams entering lakes invariably have different density characteristics than the lakes themselves. Density differences between a lake or reservoir and its inflowing water may be enhanced by man to create a density current.

The persistence of a density current in a lake is mainly dependent upon the state of flow which is indicated by the Richardson number. The density current will gradually decrease and disappear, but more rapid disappearance of the current will result with mixing across the interface if the Richardson value is low or a disturbance develops, such as a wavelike phenomenon similar to surface waves. If this wave travels with increasing amplitude, it finally breaks and extensive mixing occurs (Knapp, 1943). Small-scale mixing occurs when the density current leaves the water surface and continues as an undercurrent or when an obstacle on the bottom disturbs the current. If the high density of a current is mainly a function of suspended matter (usually associated with high turbulence), the phenomenon is called a turbidity current. Turbidity currents are most common in estuarine environments and in reservoirs on large rivers. They may also occur as artificially produced currents, for example, as a result of dredging.

In an estuary, a heavy suspended load causes a very rapid current, which is characteristic for most turbidity currents. These rapid currents are likely to be erosive, as evidenced by the large canyons formed in nearshore waters of the ocean that receive the world's largest river estuaries. When velocity of the inflow decreases, due to the topography of the lake or ocean bottom, the turbulence decreases and the suspended material settles out.

Rapidly moving turbidity currents are rare in lakes, but the slow-moving density displacements of volumes of water are common features.

3.6 Thermal properties of lakes

In addition to causing density stratification in freshwater lakes, temperature also affects the chemical and biological processes. Factors determining lake temperature are primarily latitude, altitude and continental location, and the clarity of the lake water itself. Within low-lying equatorial areas (between about 15°S and 15°N latitude), lakes are warm (27–30°C) all the year with essentially no yearly, but sometimes a daily, variation. Seasonal changes may occur in the subtropical regions (about 15–30°S and 15–30°N), especially in mountainous areas. Thornton (1987a) cites lakes throughout the tropics (30°S to 30°N) that undergo seasonal stratification patterns although they are less pronounced than in temperate regions. Near polar areas, lake temperature is low all year but usually with greater yearly than daily variations. Some lakes are frozen throughout the year.

Within the midlatitudes, lake temperature is strongly variable during most of the year. The sun is the principal dominating source of energy causing the seasonal variation. Incoming radiation, assuming a sufficiently large angle of incidence, is usually absorbed rather quickly within the upper water layers. The visible-range wavelengths are almost totally absorbed within the first 2 m of many lakes. However, in very clear, low productivity lakes, light can penetrate down many tens of meters and the absorption of heat changes accordingly. The attenuation of visible light in relation to phytoplankton photosynthesis is discussed in detail in Chapter 6.

Wind is the distributing force for the heated water within the basin. Variations in other factors – evaporation, inorganic material in the water, or loss of long-wave radiation from the water surface at night – also contribute to the distribution, but only to a small extent compared to the wind. The effect of wind and solar heating in determining the depth of mixing will be discussed later. Water's high specific heat is another important thermal property in lakes. This property of water allows for a large storage of energy and results in a much slower rate of heating and cooling of the lake than the surrounding land surface. When water cools, its density is increased. Most lakes are shallow and, to at least 100 m depth, the pressure effect on density caused by water depth can be ignored.

3.7 Origin of lakes

The origin of a lake is an important key to understanding its water quality and response to different pollutant inputs. The origin is also an important clue to the lake's form and its sediment characteristics. These factors will in turn determine the lake's detention time and affect its nutrient content. A detailed discussion of the classification of lakes according to their origin is found in classical textbooks (e.g. Hutchinson, 1957), but will only be discussed briefly here.

Lakes caused by tectonic movements (e.g. Lake Tahoe, Baikal, Tanganyika and Nyassa) and vulcanism (e.g. Lake Kivu and Crater Lake) are often deep with steep shores and frequently have a long detention time. Lakes formed by glacial activities, like most lakes in Canada, Scandinavia, the Great Lakes area, may appear in many different shapes due to different forces involved. However, they are mostly shallow and nutrient-poor, with rocky shallow shores and swash-zones. They often have medium to long detention times due to a complex morainic landscape.

Glacial lakes in flat environments which formerly were sea bottoms, or lakes caused by fluvial action (ox-bow, levee or coastal lagoon lakes), generally have high nutrient levels, high sediment loads and are shallow.

Lakes caused by chemical solution are often very complex and sediment-poor and the opposite is applied to deflation lakes caused by action of the wind.

Reservoirs are man-made, created by the damming of rivers. They are usually nutrient- and sediment-rich and have short detention times. Natural basins have usually developed slowly and continue to change very gradually in shape and physical characteristics, whereas reservoirs fill up rapidly and may continue to undergo marked changes in their characteristics.

3.8 Thermal cycle in a typical mid-latitude lake

Most lakes are located within the mid-latitudes (glacially sculptured or remoulded) and consequently undergo strong temperature variations during the year. Thus, discussion about temperature in lakes should begin with those in the mid-latitudes.

Consider a lake with ice cover in winter. When sunlight begins to have an effect in early spring, the ice cover melts. This process is very often accelerated in the final phases by wind, rain and darkening of the ice. The actual breaking up of the ice cover is often very fast, and usually occurs in a few hours, most often in the month of April. In the mid-latitudes, the spring flood often occurs at the same time and a period of strong vertical mixing and turbulence is initiated in lakes that have high flushing rates. When the spring flood ceases and the radiation from the sun has increased, the upper water layer begins to warm (generally in May). That process accelerates through mid-summer. Until mid-May, when the water is heated to about 6–8°C, little energy is usually required to mix the whole volume of water. The length of this circulation period is dependent upon several factors: the surface area, depth, topographic location, presence of winds, and so on. Those factors also determine the temperature the lake will reach before the vertical stratification begins. During calm and clear periods, stratification strengthens due to decreased wind mixing. The upper layer is mixed to a near-homogeneous temperature by turbulence, and below this surface layer, the vertical exchange of water is very limited, as previously discussed. The depth of the mixed layer (thermocline, see below) occurs where the force by the wind mixing the water downward is balanced by the resistance to that mixing. Resistance to mixing is due to buoyancy of the water, caused by heating and consequent density reduction.

The result of this process is the creation of three relatively well-defined separate layers in the water mass developed in the summer: an upper homogeneous layer (the epilimnion), a lower fairly homogeneous cold layer (the hypolimnion), and an intermediate layer characterized by a strong temperature gradient. The definition and description of the intermediate layer has always been a matter of controversy. The most commonly used terms are thermocline and metalimnion. However, this layer has little physical or chemical uniformity. The use of the term metalimnion, which implies a uniformity similar to the epi- and hypolimnions, may be misleading. Therefore, throughout this chapter the term thermocline is used.

A lake with a typical summer stratification has a vertical distribution of temperature as shown in Figure 3.4. The late-summer lowering of the thermocline is essentially an effect of wind and reduced solar heating. Although epilimnetic water is gradually heated until the latter part of July, the thermocline depth increases throughout the summer until it finally breaks up completely in the autumn (Figure 3.5). The initiation of autumn circulation (complete mixing) at that time is a direct function of the frequency and force of winds together with a decrease in temperature. The autumn circulation usually starts in September. The initiation, development, and termination of summer stratification is often very different from year to year (Figure 3.5).

Figure 3.5 also shows that the temperature of the hypolimnion gradually increases through the summer. The reason for this increase is still not satisfactorily explained. Solar radiation, turbulent transfer of heat and extensive mixing during upwelling events, as well as biological decomposition, certainly contribute heat. The concept of upwelling is best described as a situation in which the hypolimnetic water reaches the water surface because of strong wind effects that disturb the thermocline (discussed below). Hutchinson (1957) proposed that density currents are a main source of energy transferred to the hypolimnion (later adapted by Wetzel, 1975). This process has never clearly been established and extensive sampling in Swedish lakes with the aid of very sensitive instrumentation has so far not provided clear evidence for the importance of this process. There is, however, some indication that vertical entrainment occurs in summer and may be responsible for transporting phosphorus vertically from anoxic hypolimnia into the photic zone. This will be discussed in Chapter 4.

As solar radiation decreases in late summer, income of energy to the lake, and consequently its temperature, decreases. After the circulation in September, the water mass usually has a temperature of about 10°C. The temperature decreases further during total circulation reaching 4°C or

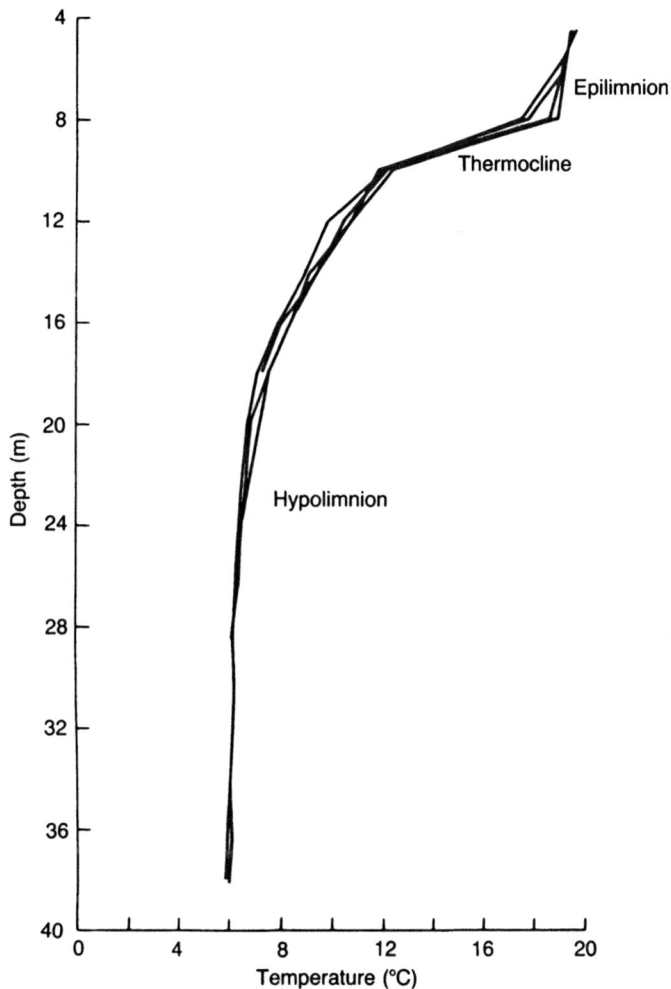

Figure 3.4 Five typical vertical temperature soundings along transects (Figure 3.17) from a typical dimictic lake, Lake Ekoln, Sweden, June 1969 (Kvarnäs and Lindell, 1970).

lower. As the density decreases at temperatures below 4°C and during a calm, cold night (usually in December) an ice cover will typically occur. The ice cover will continue to increase during winter, with a resulting vertical distribution of temperature in late March similar to that shown in Figure 3.6. The water temperature is close to 0°C in the surface layer and close to that of maximum density (4°C) in the bottom layers. The increase at depth in this case was caused by heating from the sediments plus inflowing water.

32 General concepts of aquatic ecosystems

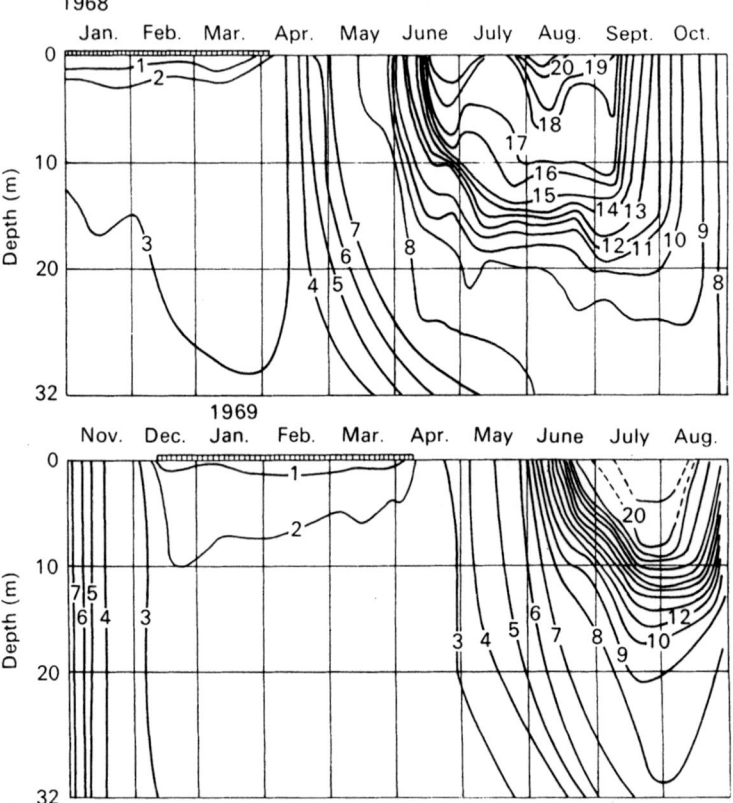

Figure 3.5 Yearly distribution of temperature as isotherms, Lake Ekoln, Sweden. Dashed line is approximated (Kvarnäs and Lindell, 1970).

The apparent slight instability indicated by the temperature difference between 8 and 22 m (i.e. denser water overlying lighter water) was actually compensated by a salinity gradient, which resulted in a stable water mass.

3.9 Classification of lakes by temperature regime

Much of the nomenclature of lakes is based on their thermal characteristics, which may be peculiar to a particular region and greatly affect the lake's physical, chemical and biological processes as shown in Chapters 4 and 6. Therefore, the different lake types are graphically described in Figures 3.7 and 3.8.

The mid-latitude lake discussed above is called dimictic, indicating the two circulation periods per year. Dimictic lakes also exist at high elevations within subtropical regions. The requirement of two circulation periods per

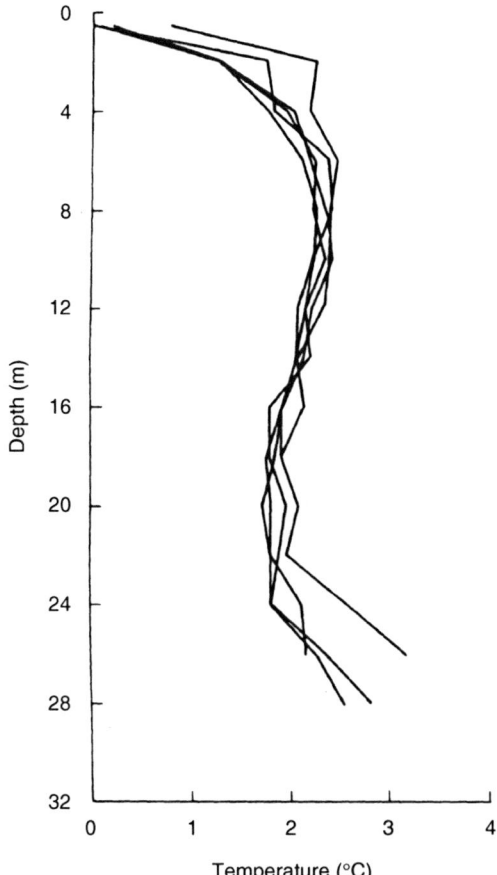

Figure 3.6 Five typical vertical temperature soundings from a dimictic lake, Lake Ekoln, Sweden, March 1970 (Kvarnäs and Lindell, 1970).

year is fulfilled in the mid-latitudes as long as the lake is continentally located. Within maritime areas in the mid-latitudes, the winter temperature does not reach a low enough level for the lake to freeze. Thus, only one circulation period occurs, resulting in a warm monomictic lake. The temperature during the winter is usually 4–7°C in a monomictic lake. Within cold regions, a corresponding lake type exists, with its one circulation period occurring during the summer when the temperature reaches 4°C. That lake is referred to as cold monomictic. If the climate is so extreme that the lake is always frozen, it is called amictic. Its opposite, a polymictic lake, is found within tropical areas where the yearly variation is insignificant, although short-term variations create weak stratification and periods of circulation in

34 General concepts of aquatic ecosystems

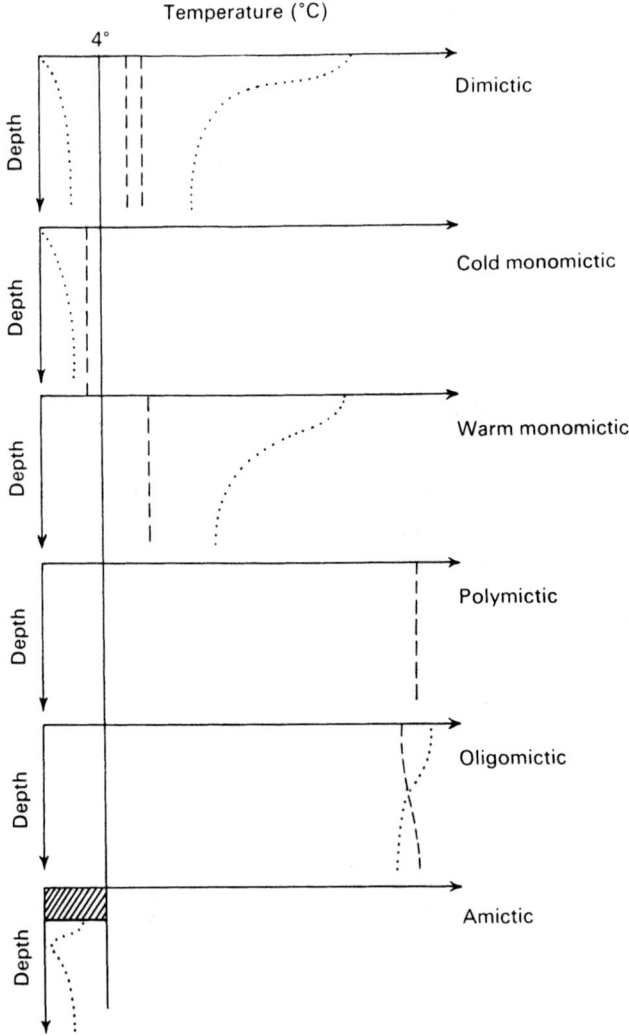

Figure 3.7 Lake types classified according to temperature criteria. Dotted lines indicate winter or summer stratification. Dashed lines indicate circulation temperatures. Shaded area indicates ice cover.

between. Polymixis can also occur in shallow-temperate lakes in summer and that feature will be shown later to greatly influence nutrient cycling (Chapter 4). Cold polymictic lakes also exist within high-altitude tropical areas with strong winds. Within the tropics, some lakes, called oligomictic, circulate only sporadically.

Hydrographic characteristics 35

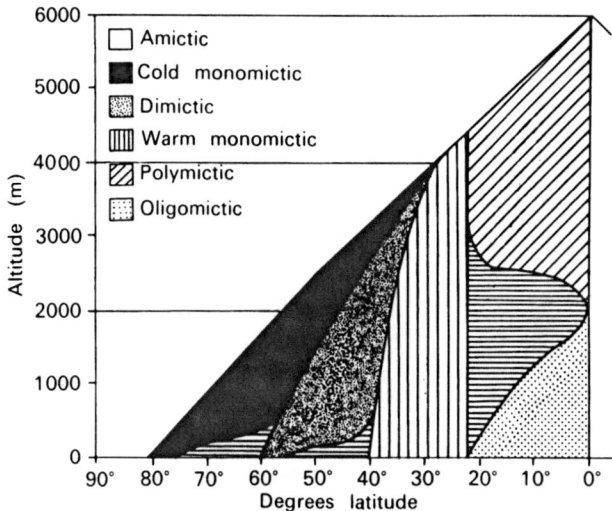

Figure 3.8 Classification of lake types, based on latitude and altitude. Horizontal lines in the figure are transitional zones (Modified from Wetzel, 1975).

In the above types of lakes, temperature variation is the main cause of stratification and, when circulation occurs, the whole vertical water column is included. A complete vertical circulation may in certain cases be suppressed by a permanent or a quasi-permanent chemical stratification (e.g. a salinity gradient) in a portion of the water mass. In those rare cases, lakes are termed meromictic and a very atypical temperature gradient occurs.

Dimictic and monomictic lakes permanently stratify during summer, while polymictic lakes stratify for short periods or not at all. Whether a lake stratifies thermally depends primarily on depth and wind. Considering that the area of a lake, and thus the wind fetch, is an index of wind energy, the depth necessary is a function of area. Gorham and Boyce (1989) evaluated a group of North American lakes, some that permanently stratified and some that did not, and found a line for the maximum depth that represented a threshold for stratification that increased with area. That is, as area increased, wind fetch also increased, so the depth at which wind energy equalled resistance to mixing due to density also increased. The equation for estimating on average the maximum depth for stratification is as follows:

$$Z_{max} = 0.34 A_s^{0.25} \tag{3.6}$$

where A_s = lake area in m^2. Thus, on average, a 20-ha lake needs to be >7.2 m and a 200-ha lake >12.8 m to permanently stratify. However, lakes too shallow to permanently stratify can nonetheless stratify during short periods of low wind conditions.

3.10 Water movements

The water movement of a river is caused by gravity. The water surface usually has a substantial inclination (gradient). Lakes normally have no gradient, so that factor is unimportant. The gradient is considered the cause of a lake current in only elongated, flow-through type lakes with short detention times.

The wind causes waves on the surface of the lake due to the friction against the water and the currents, which rapidly decrease with depth. An asymmetrical distribution is frequently created when water is transported through the lake. A continuous water transport in one direction at the surface is compensated by a transport in the opposite direction, which often occurs in deeper parts of the lake. This transport of water is especially interesting during the summer when the thermocline is inclined and the waters of the epilimnion and hypolimnion have essentially separate circulations because of the small amounts of friction between the two layers.

If the wind suddenly ceases, the thermocline will oscillate back and forth like a seesaw until its resting position in the horizontal plane is achieved. The rhythmic movement of the thermocline is called an internal seiche, and the corresponding movement on the water surface itself is referred to as a surface seiche (see following section). The oscillations of the surface seiches are considerably smaller.

3.11 Seiches

When the wind blows, it causes a piling up of the water on the windward side of the lake. The inclination of the water surface resulting from the wind is usually in the order of a few decimetres in large lakes and a few centimetres in small lakes. Morphometric conditions in lakes are, however, strongly modifying and, consequently, general rules are difficult to postulate. For a rectangular-shaped lake, Hellström (1941) suggested the inclination to be given by

$$s = \frac{3.2 \times 10^{-6}}{g\bar{z}} w^2 l \qquad (3.7)$$

where s is expressed as the total difference in height between the windward side and the leeside of the lake, g = acceleration resulting from gravity, $\bar{z}$ = mean depth, w = wind velocity and l = length of the basin. Thus, seiches should be greater in longer, shallower lakes that are exposed to the wind.

Oscillation of the water surface after the wind has ceased can be determined by the classical formula of Chrystal (1904):

$$t_n = \frac{1}{n}\frac{2l}{\sqrt{g\bar{z}}} \tag{3.8}$$

where the ratio of the periodicities for the uninodal, binodal and trinodal seiches is $t_1:t_2:t_3 = 100:50:33$, with respective nodes at $x = 0$ (uninodal), $x = \pm 0.25\,l$ (binodal), and $x = \pm 0.33\,l$ (trinodal); $l =$ length of the basin, $g =$ acceleration resulting from gravity, $\bar{z} =$ mean depth, and $n =$ number of nodes.

This general formula is derived from a rectangular basin with constant depth where the seiche can be considered a shallow water wave, but it has been shown to be applicable to most types of lakes. Surface seiches generally have a fairly small impact on the variability of water-quality characteristics. The internal seiche could, however, cause strong fluctuations in water quality both in time and space as the thermocline is separating two water masses that could be of very different quality. The internal seiche has an amplitude of a magnitude 100–1000 times larger than the surface seiche and is primarily determined by the density differences between the epilimnion and hypolimnion. The periodic motion of the internal seiche can be determined (Mortimer, 1952) by:

$$S_n = \frac{2l}{\left[g(\rho_h - \rho_e)/\left(\frac{\rho_h}{h_h} + \frac{\rho_e}{h_e}\right)\right]^{0.5}} \tag{3.9}$$

where $h_e =$ thickness of epilimnion, $h_h =$ thickness of hypolimnion, $\rho =$ density and $l =$ length of the basin.

The period of the internal seiche is often in the order of hours, whereas the surface seiche is calculated in minutes. The velocities associated with the internal seiche can be roughly estimated (Mortimer, 1952) as:

$$v_e = \frac{sl^2}{2Th_e} \tag{3.10}$$

$$v_h = \frac{sl^2}{2Th_h} \tag{3.11}$$

where $s =$ maximum slope of the thermocline, $l =$ length of the basin, $t =$ period, and $h_e, h_h =$ depth of epilimnion and hypolimnion respectively.

3.12 Surface waves

Surface waves are ecologically interesting primarily in near-shore areas. In the deep parts of lakes the waves are non-transporting and in general each water particle describes an orbital movement, the circle (wave amplitude) diameter of which is halved for each change in depth of λ/g, where λ = wavelength. The wave height is usually about 1/20 of the wavelength. The wave height is a function of the distance along the water surface in the direction of the wind (=fetch) and a function of wind speed and duration. In the nearshore zone where the wave breaks, the orbital movement is transferred to a horizontal water movement. The wave generates an erosive character that causes the bottom substrate to be unstable and highly turbid. Large waves in oceanic coastal zones or in large lakes influence the bottom down to great depths; however, the velocities resulting from the waves usually die out at a depth of approximately $\lambda/2$ below the water surface. In small lakes the wave action is limited, but even the leeward sides of large lakes are often shallow and heavily vegetated, whereas the windward sides are deeper and biologically bare, which is a result of the stronger wave action on the windward side.

3.13 Wind-driven currents

The direction and velocity of wind-driven currents are determined by the classical Ekman spiral. To be able to apply this law, however, one must assume a free water surface (no bottom and lateral friction). The Ekman current implies a surface current directed 45° to the right of the wind and a mean water transport 90° to the right of the wind because of the Coriolis force (Figure 3.9). In lakes, both bottom and lateral friction exist to a variable degree, and there is no generally accepted method of computing the currents from wind data in small lakes. Computations have been attempted based on the equations of continuity and motion, but they vary considerably in their final form.

For large lakes, the three-dimensional models for the Great Lakes by Simons (1973), based on classical oceanographic models but also incorporating the topography of the lake, have been very successful. Rather simple relations have also been proposed, as for Lake Ladoga (Witting, 1909), where a current velocity was calculated according to

$$v = 0.48/\sqrt{w} \tag{3.12}$$

where w = wind velocity. Witting observed a direction of the current slightly to the right of the wind. For some small Swedish lakes the following formula has been used:

$$v = chw \tag{3.13}$$

where $c = 2 \times 10^{-5}$ (for no stratification), h = mean depth, and $c = 1.5 \times 10^{-5}$ (for h = mean depth of the thermocline). As a rough estimate of surface currents in lakes, it is possible to consider them parallel to the wind

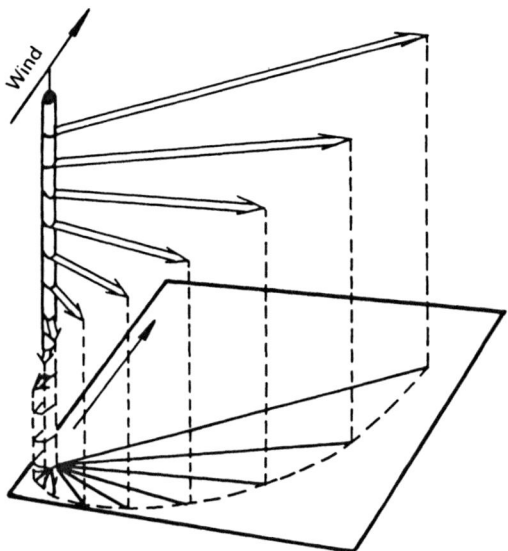

Figure 3.9 The distribution of water movements in a wind-driven ocean (the Ekman current) (Hellström, 1941).

and in the order of 2% of the wind velocity. In the deeper parts of lakes, water movements are slow, generally only a few centimetres per second. Previously this presented severe problems in recording water movements because the velocities were lower than the activating velocity of most impeller-type current meters. However, the widespread use of acoustic Doppler current meters in recent years has shed new light on the complexity of lake currents.

3.14 Tidal effects

The tide is caused mainly by the gravitational pull of the moon. The effect of the sun's gravitation can be observed in connection with extreme high and low tides in marine areas. Tides are believed to cause waves in the order of a few centimetres in large lakes. However, tides in lakes probably have little effect on the water quality.

In oceanic coastal zones the tides are highly variable because of the general distribution of the land masses and the local morphometry of the coastal zone. In some areas the tides could be close to zero, whereas in other places, especially in funnel-shaped estuaries, the tides may be of the order of several metres. The tides have up to two maxima and two minima per day for a lunar day that is 24 h 50 min. Such tides are referred to as semi-diurnal. An estuary with a large tidal range, therefore, has an exceptionally good transport mechanism as flushing occurs twice a day. See Chapter 6 for the effect of tidal prism thickness on phytoplankton.

3.15 Water movements in large lakes

The effect of the earth's rotation may be detectable in the inflow to small lakes, but almost never in the internal water movements. In large lakes and coastal waters the Coriolis force effect is visible with all water movements and may totally dominate the current patterns (Figures 3.10 and 3.11). The meaning of 'large' must be defined in relation to the latitude, as the Coriolis parameter, f, is defined as:

$$f = 2\Omega \sin \phi \tag{3.14}$$

where Ω = angular velocity of the earth and ϕ = latitude.

The Coriolis force is directed perpendicular to the right of the motion in the Northern Hemisphere and to the left in the Southern Hemisphere. The effect of the Coriolis force increases with latitude and is zero at the equator.

By using the definition of the Coriolis parameter, it is possible to calculate the radius of the inertia circle:

$$r = \frac{u}{f} \tag{3.15}$$

where u = water velocity. For a typical current velocity of $10\,\text{cm}\,\text{s}^{-1}$ at 45° latitude, for example, $r \approx 1$ km. If the width of the lake is five times the inertia circle radius, the effects of the earth's rotation can be visible, but if the width is 20 times the radius or more, the Coriolis force will dominate the currents.

A very typical feature of large lakes is rapid water movement in nearshore areas known as 'coastal jets'. When a wind initiates Ekman drift in the centre of a lake, the Coriolis force balances the wind stress. In a narrow coastal bank, however, the wind stress accelerates the water alongshore. This mechanism transports water to the windward side of the lake, and when the wind ceases, a return flow develops that is again confined to the nearshore bands. The coastal jets are sometimes connected to a type of deformed, so-called Kelvin wave (a seiche dominated by the Coriolis force), resulting in a very asymmetrical wave that progresses counterclockwise with a very rapid countercurrent (clockwise) on its rear side.

A persistent geostrophic counterclockwise circulation dominates during summer on the lakeward side of coastal jets (Figure 3.11). This circulation pattern is caused by a combination of wind, Coriolis force and temperature contrasts and may be so strong that the common and substantial vertical shifting of the thermocline resulting from wind never overcomes the effect of the circulation.

Figure 3.10 The Coriolis force strongly influences the distribution of Glomma River water in the Skagerrak Sea. Landsat-5 MSS, 14 June 1988.

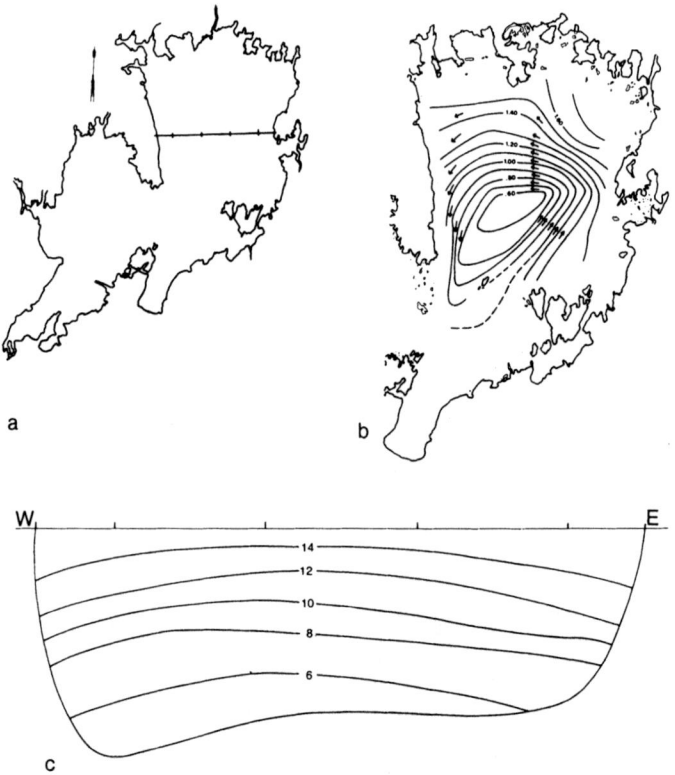

Figure 3.11 Geostrophic circulation of a large lake (Lake Vänern, Sweden). Along a cross section (a) of the lake the temperature pattern (b) is dome-shaped (c) The resulting dynamic pattern of height anomalies in centimetres (Lindell, 1975).

3.16 Thermal bar

A thermal bar is caused by the uneven heating of water and is common in large dimictic lakes during spring (and theoretically also in the autumn). If the temperature of the whole lake volume is well below 4°C and the heating during the spring is rapid, the rate of heating will be most pronounced in the nearshore shallow areas. The deep pelagic areas of the lake remain below or near 4°C. In the transitional zone between the high temperatures of the coastal areas and the cold open water there is a zone with a strong horizontal temperature gradient (Figure 3.12). This zone moves slowly from the nearshore areas toward the open waters. The duration of this process varies from days to months, depending on the differences in heating and size of the lake. The typical thermal bar phenomenon is transformed to the temperature

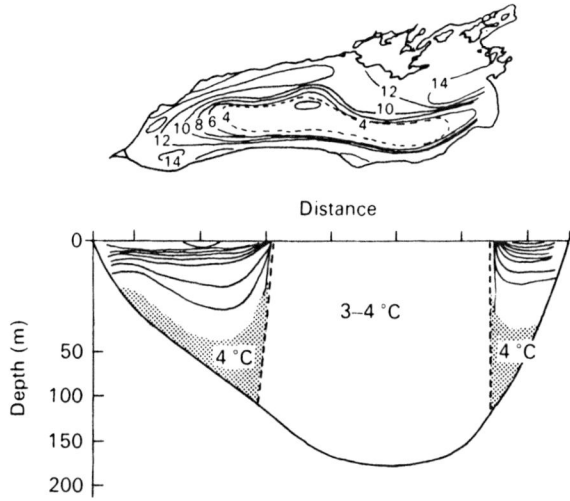

Figure 3.12 The temperature distribution in Lake Ontario (surface and longitudinal section) during 7–10 June 1965, showing the position of a thermal bar (Rodgers, 1966).

pattern typical for geostrophic water movements. The currents associated with these phenomena are characterized by a predominant alongshore counterclockwise circulation in the nearshore stratified zone. Modelling of the thermal bar has shown downward movements on the lakeside of the bar and upwelling in the warm water. Its development from early spring to the summer circulation is not yet satisfactorily explained.

The thermal bar has relevance for the dispersion of wastewater because the nearshore warm waters are trapped close to the coastline. The resulting high nutrient concentrations could in turn trigger algal blooms. Such a nearshore effect has been suggested in the Great Lakes (Beeton and Edmondson, 1972).

3.17 Measurement of water movements

Water velocities in natural bodies vary from several $m\,s^{-1}$ in rapidly running rivers through several $dm\,s^{-1}$ in coastal areas to commonly less than $30\,cm\,s^{-1}$ in lakes. To record horizontal moderately high velocities in water is fairly straightforward and could be performed by current meters, drifters or dye. However, water movements can be complex in nature and especially slow in lakes, even slower than the starting velocity of common current meters. In that case special vanes have been used, where the inclination of the vane compared to the vertical direction is recorded.

The most common technique for measuring currents is based on the Doppler shift principle. These accurate acoustic doppler instruments work on the principle of measuring the frequency shift produced when an acoustic beam reflects off a particle or bubble in the flow. Resolution as low as 1 mm s^{-1} and simultaneous measurement through the entire water column are possible.

Alternatives to current meters are drifters or drogues which could be placed at different depths and often have a surface float. The float can be followed by optical instruments from land, indicating direction and distance, or simply followed by radar.

Dyes have frequently been used for studies of diffusion in lakes and slowly moving rivers. Most practical from the measuring point of view is the use of fluorescent and radioactive tracers, but those may cause environmental hazards and permission to use them may be complicated. Remote sensing has been widely applied in large and medium scale systems, e.g. to study coastal circulation, or for detecting and defining natural inputs or waste discharges (suspended matter or temperature) in lakes or coastal waters. By using near infrared bands from satellites or infrared films from aircraft, the movements of algae-laden water have been estimated (Figure 3.13). An example of the use of satellite imagery in tracing pulp mill discharges into the Baltic Sea is shown in Figure 3.14.

3.18 Dispersal processes

In principle it is possible to separate the dispersion processes among small-scale and large-scale phenomena. The small-scale diffusion processes appear in the neighbourhood of the sources of pollutants or water courses, whereas the large-scale processes are closely associated with the pelagic areas of lakes. The parameters that influence the dispersion processes are water velocity and eddy diffusivity in the horizontal (K_y) and vertical (K_z) direction. K_y and K_z are dependent upon the character of the turbulence, which in turn is dependent upon water velocity and stability, as has been mentioned earlier. They are also dependent on the scale of the phenomenon.

3.18.1 Horizontal diffusion

The horizontal eddy diffusion along the horizontal plane (x) may be defined as

$$K_y = ku\frac{\mathrm{d}w^2}{\mathrm{d}x} \tag{3.16}$$

where u = water velocity, in the x direction; w = width of the plume, in the y direction; and k is a constant. Csanady (1970) has proposed a value of 0.03 for the constant in the equation.

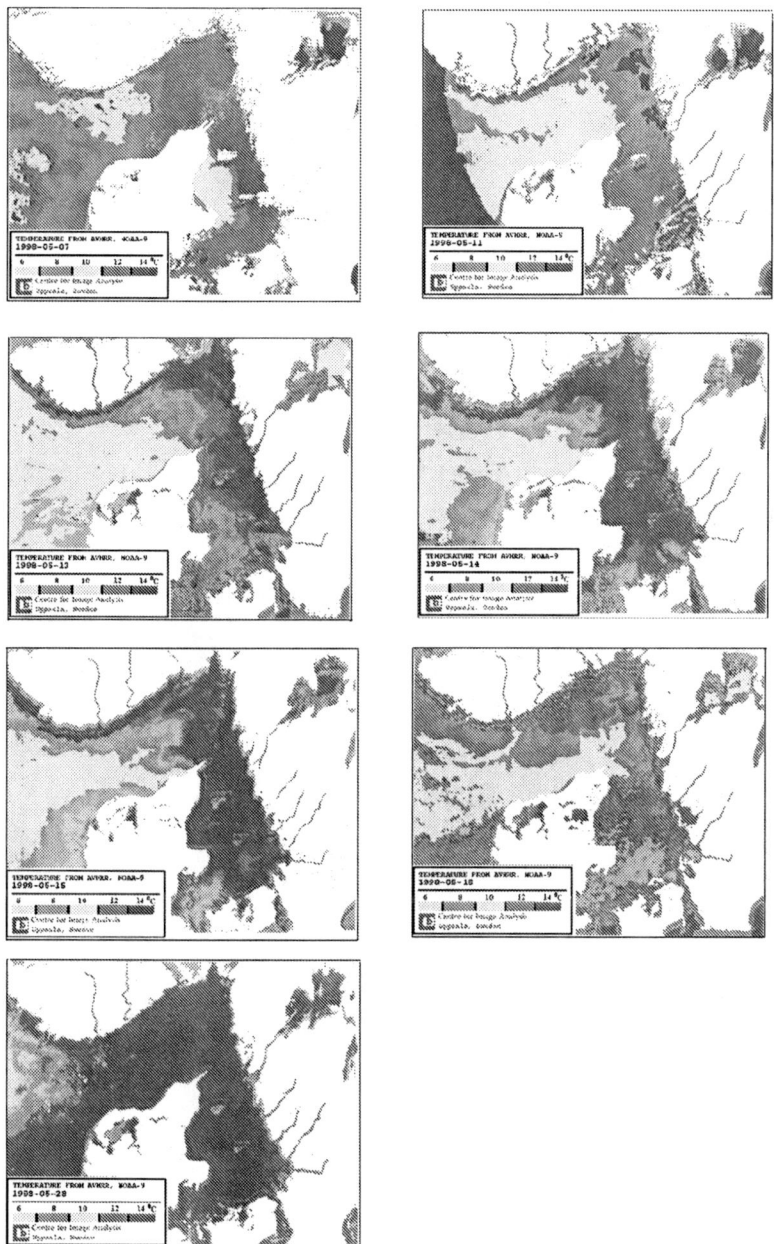

Figure 3.13 The movement of the toxic algea *Chrysochromulina polylepsis* which was entrained with the warm water, as indicated by the temperature distribution. Recorded by NOAA-9 A VHRR satellite sensor.

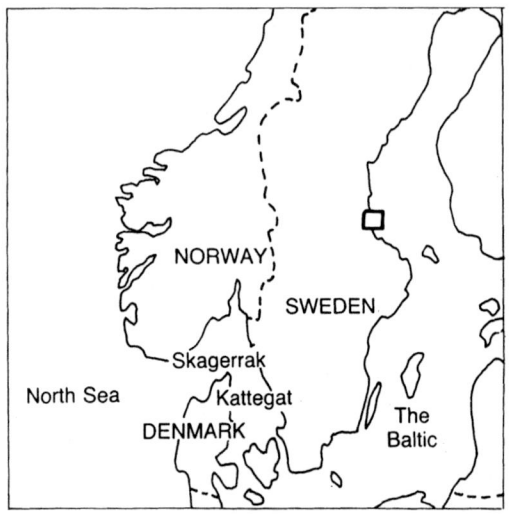

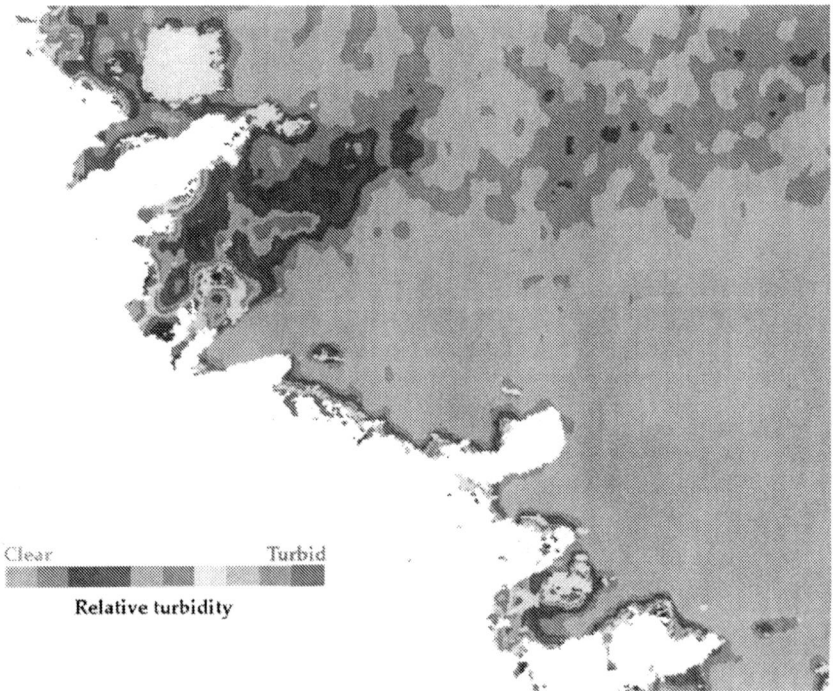

Figure 3.14 The relative turbidity of a wastewater discharge from a paper industry along the coastline of central Sweden, 6 October 1981, from Landsat-2 MSS.

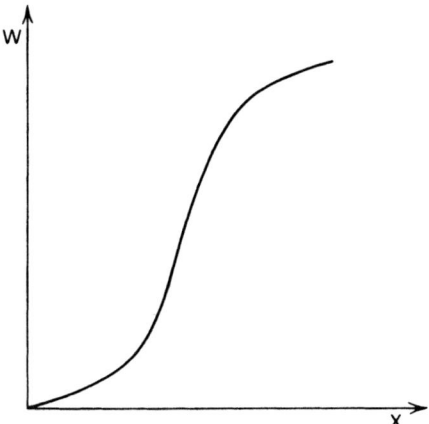

Figure 3.15 Horizontal diffusion. Changes in the width of a plume (W) in the direction of the length axis (X).

Horizontal diffusion is of interest in the dispersal of a wastewater plume into a lake or coastal water. Also, satellite imagery could assist in verifying results from diffusion modelling (Figure 3.13). The diffusivity increases rapidly with size of the plume. The increase in the width of the plume exceeds that of a simple linear increase. Although resulting from turbulence, the dispersion process at sufficient distances from the source becomes more similar to molecular diffusion. This development is the consequence of the growing plume, causing an increase in the size of turbulent eddies, which become increasingly important. The size of the turbulent eddies does have an upper limit, which is why this type of diffusion eventually becomes similar to that of molecular diffusion. Although the processes are non-linear with complicated interrelationships, a linear diffusion model is often sufficiently good for most purposes. A typical development of a plume from a dye injection will be related to distance as shown in Figure 3.15 Csanady (1970) defines plume diameter as $\approx 10\%$ of the width (w) of the cross section at the surface. The horizontal diffusion coefficient in lakes is usually in the order of $100-1000 \, cm^2 \, s^{-1}$.

3.18.2 Vertical diffusion

Vertical diffusion has certain similarities with horizontal diffusion, but the magnitude of vertical diffusion can be affected by many important variables, such as density, water velocity and current direction, which can change rapidly with depth. Vertical diffusion is especially different in the presence of a thermocline. As noted earlier in discussing density, it is sometimes useful

48 General concepts of aquatic ecosystems

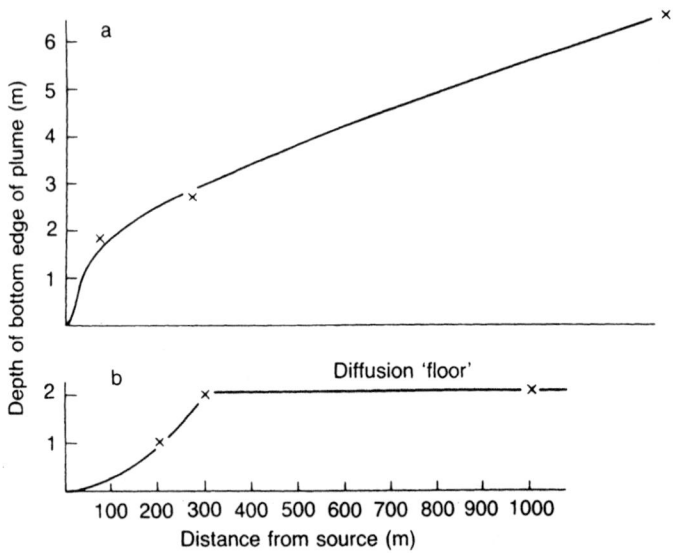

Figure 3.16 Vertical diffusion. Typical patterns of diffusion in Lake Huron in the absence (a) and presence (b) of a thermocline (from Csanady, 1970).

to consider a lake as two lakes with the epilimnion as the 'upper lake'. The lower limit of the epilimnion acts as a floor in the process of diffusion. The two main types of vertical diffusion may, therefore, be generalized according to Figure 3.16. Typical vertical diffusion values are $\approx 5 \, \text{cm}^2 \, \text{s}^{-1}$, but initially much higher values (up to $30 \, \text{cm}^2 \, \text{s}^{-1}$) are common (Csanady, 1970). Compared to horizontal diffusion, vertical diffusion is much smaller in magnitude because of the differences in the horizontal and vertical limitations to turbulence.

3.18.3 *Large-scale phenomena*

The most important large-scale diffusion process for most sizes of lakes is the meandering type that follows a large inflow. Meander diffusion is similar in appearance to the meandering of rivers; that is, the movement is sinusoidal in plan view with a helicoidal component. There are no general rules for predicting the occurrence and frequency of meandering in lakes.

In large lakes where thermal bars and coastal jets exist, the dispersion of pollutants often leads to concentrated levels in the coastal zones, thus restricting horizontal diffusion. It should be noted, however, that the jets and thermal bars vary substantially with time and the diffusivities, therefore, could be very different from one day to the next.

3.18.4 Typical dispersion patterns of plumes

The following types of dispersion may occur.

1 Complete mixing;
2 The pollutant usually disperses horizontally;
   (a) At the surface;
   (b) At an intermediate level;
   (c) Along the bottom;
3 Combinations of 1 and 2.

Complete mixing is almost entirely a function of the velocity of the receiving water because diffusion coefficients depend on velocity. A rapidly flowing stream always has a high degree of turbulence and, consequently, an input of wastewater from a point or non-point source can be homogeneously distributed within the whole river width usually after only a few hundred metres. However, the distance will depend on how the pollutant enters the stream and the stream size. It is easy to be misled by this fact when considering the water quality of a river system.

To determine the water quality in a downstream lake, in addition to observing pollutant effects in the stream near the input, the total transport per unit time of the pollutant should be calculated. To calculate the transport of a substance requires knowledge of the quantity of water as well as the average concentration. In a strongly turbulent stream there is little interest in whether the pollutant is dispersed as a concentrated jet or by the diffusion process. The turbulence of the stream will dominate the dispersion of material immediately and the resulting cross-sectional concentration may become homogeneous a few hundred metres downstream in a moderate-width stream. Consequently, adverse effects are noted in turbulent waters downstream from the mixing zone only at relatively higher pollutant-loading rates, whereas in a non-turbulent stream adverse effects (or 'hot spots') may occur.

If the densities of the receiving stream and added wastewater are different and the stream velocity is low, the pollutant will mix into the stream very slowly. High concentrations of the discharged pollutant often remain unmixed very far downstream, sometimes several kilometres, if the degree of turbulence is low, the density gradients are strong and the stream width is substantial. Considerable variability exists in the mixing processes because of density gradients and bottom topography. Another form of dispersion occurs, however, when the usually denser wastewater progresses downstream close to the bottom. This is the result of high dissolved solids content causing the incoming wastewater to be extremely dense. In that case the temperature has little effect on density and, hence, on dispersion.

When the density differences are considerably less, the situation is equally critical. The incoming water mass tends to hold to the entering side of the river for a long distance, still assuming that fairly low river water velocities exist. This process can frequently be observed, for instance, at the confluence of two tributaries carrying very different loads of suspended sediment.

In streams with this type of incomplete mixing of wastewater, the ecological effects may be severe on one side of the river and non-existent on the other. The monitoring of stream water quality seldom includes samples from the entire river width. An incomplete mixing could, therefore, easily lead to non-representative water-quality sampling and local samples would not provide spatially averaged concentrations, which would occur if the pollutant were uniformly distributed over the cross section.

A wide variety of different patterns of dispersion that are mainly a function of water velocity and, to some extent, density gradients, also occur. To accurately monitor stream water quality usually requires a careful study of the patterns of turbulence during different conditions of stream flow. This can be accomplished with the aid of tracer studies.

Although dispersion and mixing in estuaries and lakes are more restricted by large differences in density, the mixing processes, on the other hand, are accelerated by wind and currents in the open water. The diffusion process in a lake or an estuary is fairly simple to define from a number of theoretical linear models, but the complete modelling and forecasting of the dispersion of a wastewater plume is very complicated, not only because of the complexity of the model itself, but also because of the stochastic nature of the varying meteorological and hydrological conditions. However, a few descriptive cases of mixing will be discussed.

An estuary[1] in a lake, formed at the mouth of an entering stream, is similar to a point-source input to a stream. The main types of mixing are complete mixing and mixing that is density-dependent at the surface, bottom, or intermediate inflow level.

The basic condition governing complete mixing is a non-stratified water mass (spring and autumn for a dimictic lake). This is most often an adequate assumption because the density of the stream seldom deviates much from that of the lake. Only very dense incoming water, close to the density maximum of 4°C, or a flow of dense suspended or dissolved material, will dive as a density current toward the bottom of the lake and remain there for some time. The best opportunity to obtain a low concentration of a pollutant in a lake occurs during the normal spring–autumn mixing periods.

During summer stratification, essentially three types of dispersion and mixing occur. The inflow of water stays at the surface of the lake if the density of the incoming water is lower than that of the lake water. That process is most simply illustrated in Figure 3.17. Early summer is usually the time when this

---

1 Estuary in this freshwater case is defined as a drowned river mouth, rather than where seawater is measurably diluted with freshwater.

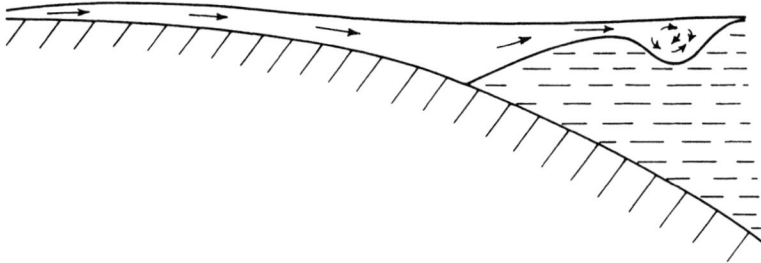

Figure 3.17 Surface inflow in a lake.

condition first occurs. During that part of the year, the incoming water is warmer than the lake water (unless the river water is meltwater) because the land mass warms up more rapidly than the water and the volume of incoming water is usually small relative to the lake volume. This phase often starts at the end of the spring flood in dimictic lakes and ends as the incoming water begins to dive later in the summer. In the beginning of this phase, the initial mixing is quite strong because of the high velocity of the incoming water. The stream water gradually becomes more concentrated as evaporation processes become significant. Stream velocities gradually decrease and the inflowing water dives to a depth of equal density, usually the upper level of the thermocline. This inflow at the top of the thermocline is generally very distinct with little initial mixing due to low velocities. Dispersion of the inflowing water is mostly in the horizontal and longitudinal directions. In some lakes it is possible to follow an inflowing stream for a long distance. A heavily polluted river is entering the lake close to section 01 (Figure 3.18). The high specific conductance of the river could be traced on top of the thermocline all through the lake to the outlet in the south-eastern part.

The remaining type of dispersion/mixing, in which the inflow dives to the near-bottom layers of a lake, usually exists in winter. This type is analogous to the inflow moving along the bottom in rivers as described earlier. Occasionally, when small differences in temperature exist between the lake and inflowing water, only a slight increase in dissolved substances is sufficient to cause the inflow to move to the bottom (Figure 3.19). Because river discharge is low in winter, there is little initial mixing in dimictic lakes. Ice cover also protects the lake from direct wind effects minimizing mixing. It is a common feature in dimictic lakes in winter for inflowing water to accumulate in the deeper parts of the lake and not mix with the rest of the lake water until the spring flood occurs.

In low-elevation estuaries there is often an interaction between the water masses of the sea or the lake and the incoming water. The initial mixing zone

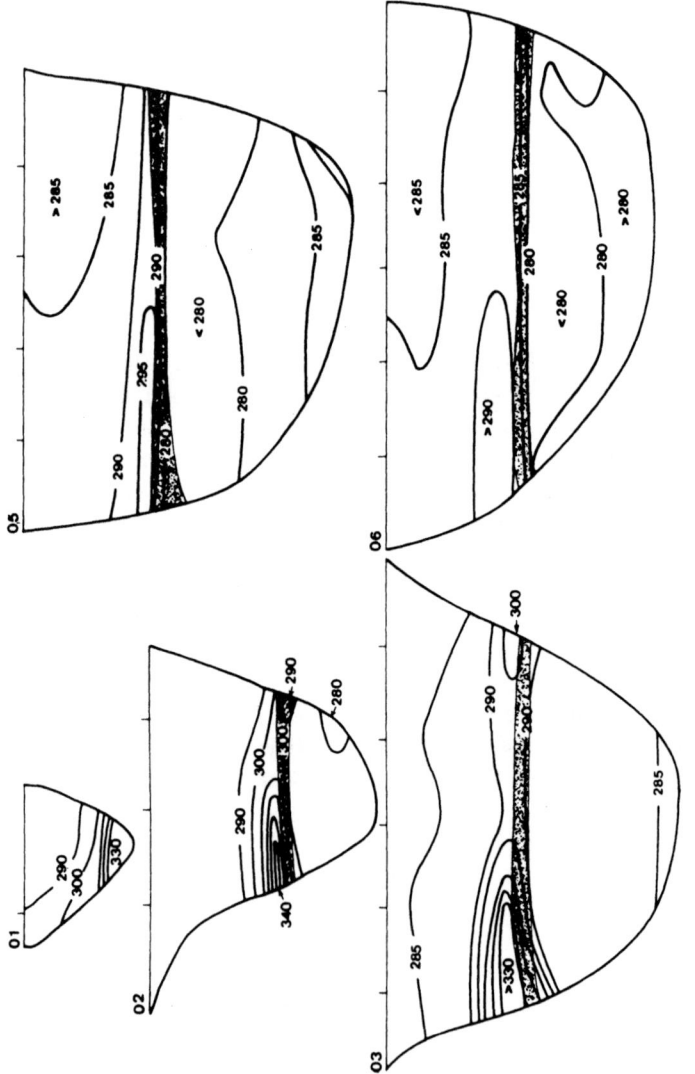

Figure 3.18 (Continued).

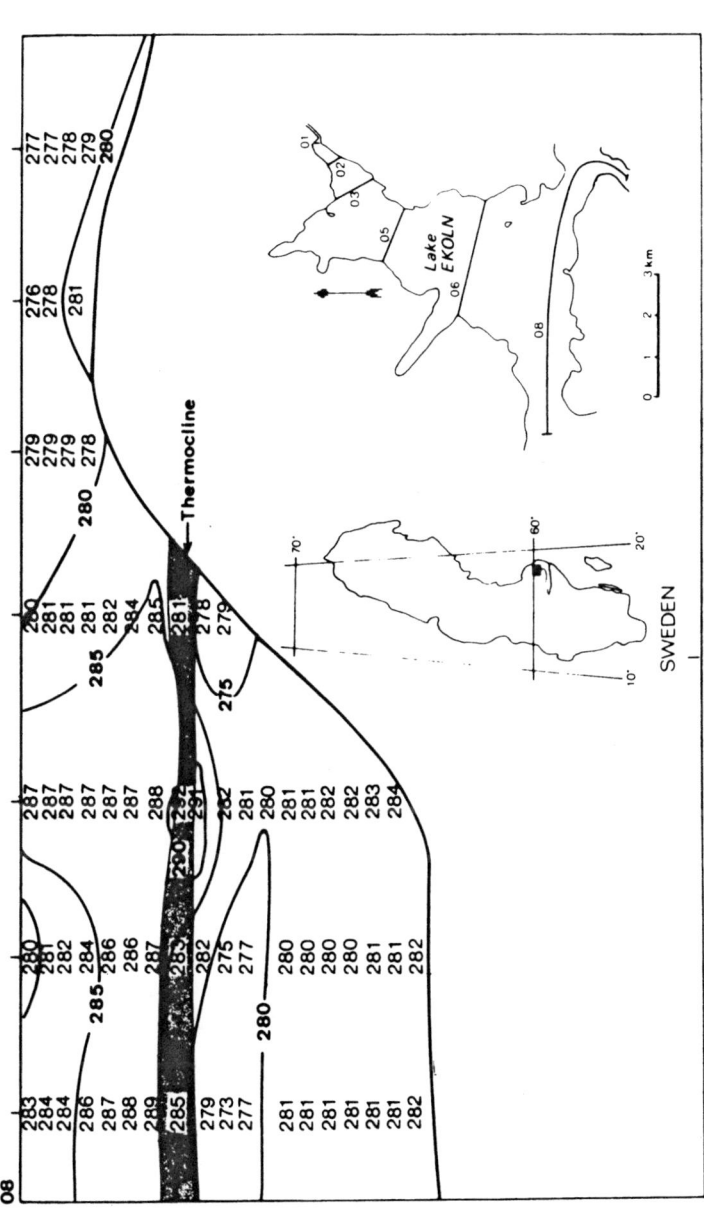

Figure 3.18 Cross-sectional recordings of specific conductance in the River Fyris within Lake Ekoln, Sweden, August 1970.

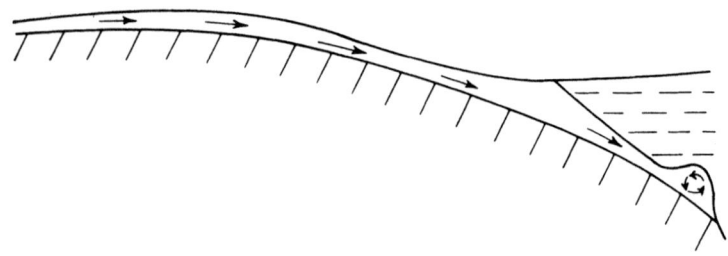

Figure 3.19 Bottom inflow in a lake.

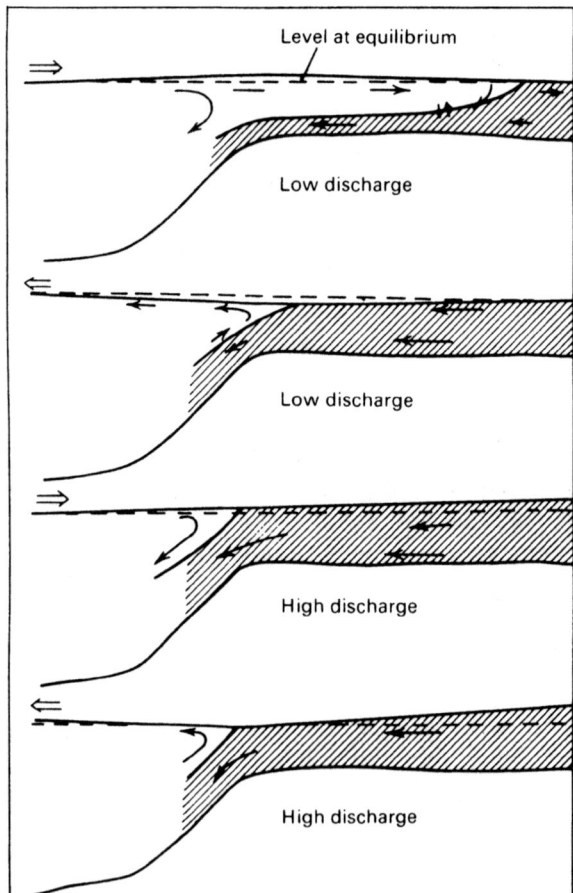

Figure 3.20 Interaction between river water and lake water at different flow conditions and different wind directions. Open arrows indicate wind, solid arrows indicate flow of water. The river water is shaded.

could vary greatly because of differences in tides, discharges and wind effects. Lake water, for example, may be forced far upstream from the estuary under a strong onshore wind and the initial mixing thus occurs not at the river mouth but several hundred metres upstream (Figure 3.20).

In an estuary facing the sea, low-density inflowing water flows on top of the salt water. The initial mixing zone, as well as the degree of mixing, is dependent primarily upon the ratio of tidal prism volume and volume of river inflow per tidal cycle. A salt-water estuary tends to be more highly stratified when the river inflow is large compared to the tidal prism and hence mixing associated with tidal energy is not so large. In very exposed shallow estuaries, however, the wind is as important a factor as the tide.

3.19 Sedimentation

In a system in equilibrium between erosion, transport and sedimentation, the plant and animal life has adapted to the stable conditions. When one or more of these processes is altered by natural or human causes, life in the system will be directly influenced. Sedimentation processes may be particularly affected by increased sheet erosion of nutrient-rich soils from watersheds or resuspension of bottom material with high concentrations of nutrients, pesticides, metals or other pollutants, having significant ecological effects. These effects will be dealt with later, but a few facts concerning the physical conditions governing the sedimentological processes in lakes will be mentioned here.

There are essentially two main areas where sedimentological processes are important in lakes: the littoral and pelagic zones. The innermost part of the littoral is affected by a large transport of sediment caused by wave erosion and is called the swash-zone. River deltas are also important in that sense.

Deltas and the swash-zone are characterized by coarse sediment while the deeper central parts of the lakes are depositional areas, usually containing very fine sediments. This non-uniform distribution of sediments, where fine material is continually removed from shallow areas and transported to deep areas is referred to as focusing. The usual differences in physical processes between small and large lakes will also govern the characteristics of the sediments. In large lakes, the previously discussed dominating forces affecting sedimentation are wind and wind-induced currents, but in small lakes, inflows of rivers may dominate the process. The common requirement for sedimentation of very fine material, however, is a period in which the watermass is completely calm (e.g. during the ice-covered winter for dimictic lakes) or if physiochemical conditions favour flocculation of material.

Chapter 4

Nutrient cycles

4.1 Phosphorus

A discussion of the cycles of nutrient elements will begin with phosphorus (P), because it is the most limiting nutrient, or key nutrient, in freshwater. The annual productivity of freshwater lakes is controlled more often by P than by any other nutrient or environmental factor. The forms of P and a general cycle will be described first and the more important processes will be indicated. Because of the importance of the P store in the sediments as a long-term source to the overlying water in lakes, even after external inputs have been curtailed, the reactions at the sediment–water interface will be discussed in detail. The principal processes involved in P cycling will be applied to laboratory and open-lake conditions by reviewing pertinent experimental results.

4.1.1 Forms of P

P is separated into several forms in the analytical process through primarily mechanical filtration. The only form that can be determined with the acid molybdate method for P analysis (APHA, 1985) is orthophosphate. Orthophosphate is the form available for uptake and is determined in the filtrate after a portion of the water sample is passed through a $0.45\,\mu m$ pore size filter. Because colloidal particles with sorbed P may pass through such filters, the analysis may actually measure some of the sorbed P and hence, more than orthophosphate. Thus, the results from the undigested filtrate fraction is referred to most commonly as SRP (soluble reactive P) and is analogous to DIP (dissolved inorganic P). Total P (TP) is determined by complete digestion of the unfiltered water with persulphate, or other strong oxidant, and autoclaved. This process releases all sorbed and complexed inorganic and organic P. If digestion is also performed on the filtrate, then the difference between SRP and TP in the filtrate (which is TDP) is dissolved organic P (DOP) and the difference between TP in the unfiltered sample and TDP is total particulate P (TPP). TP and SRP are usually the only two determinations made routinely and the difference between them can be taken as an estimate of TPP, because DOP is usually in relatively low concentration. TP ranges from ≈5 mg l^{-1} in sewage effluent (most of which is SRP) to as little as several µg l^{-1} in remote oligotrophic lakes. SRP varies greatly with abundance and demands of the algae and recycling activity, often occurring at levels that stretch the capability of the methods. A detection level of at least 2 µg l^{-1} is mandatory in lake research and requires a 10 cm spectrophotometric cell length for best results.

4.1.2 General cycle for P

Pools and transfers of P are depicted generally in Figure 4.1. The ultimate source of P to aquatic ecosystems is from phosphate rock, whether through erosion from developed or undeveloped land or through wastewater inputs. P is initially utilized through plant and microbial uptake of dissolved inorganic P (DIP) associated with the processes of photosynthesis, chemosynthesis and decomposition. All organisms require P for metabolism and structure. Photosynthesis by green plants, which are the phytoplankton in the open water of lakes and estuaries, is most responsible for the uptake of DIP. Macrophytes in the shallow areas and bacteria in the water column and in the surficial sediments can also be effective in removing DIP from the water. Subsequently, phytoplankton and bacteria are consumed by animal

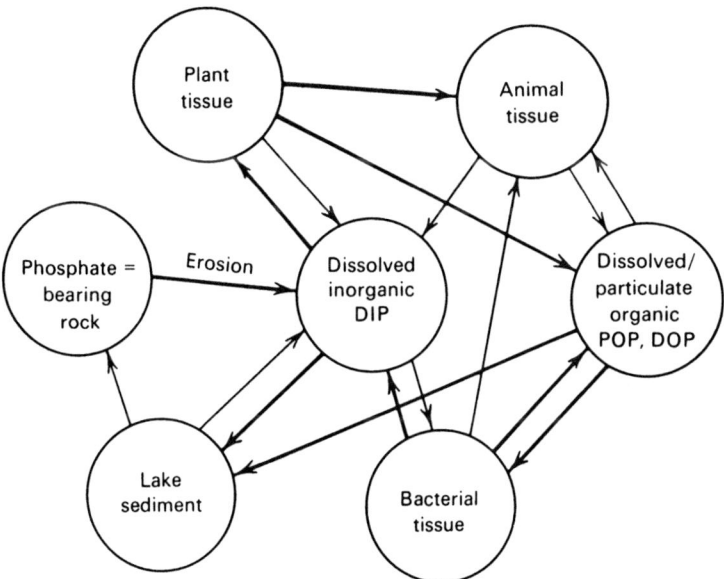

Figure 4.1 The aquatic phosphorus cycle with the transfers of greatest magnitude emphasized.

grazers, which are in turn consumed by predators. A fraction of the DIP taken up can enter the dissolved and particulate organic P pools (DOP, POP) through, respectively, excretion and death. DOP and POP can be recycled to DIP by bacteria through decomposition of organic matter. DIP can be released through autolysis (Golterman, 1972) without microorganisms and through excretion by animals.

The remaining processes include sedimentation of PIP and POP and sediment release of DIP. Sedimentation of detrital (dead) and living phytoplankton and zooplankton fecal pellets are important processes in the loss of P from the open water. Much of the sedimented P is refractory and is retained as part of the permanent sediments. However, depending on the physical–chemical conditions at the sediment–water interface and in the surficial sediments, a portion of sedimented P can be released back into the overlying water as DIP.

4.1.3 Sediment–water interface processes

An important characteristic of lake sediments is the extent to which they retain P. Lakes in equilibrium with their external supply of nutrients usually have a net retention in their sediment, at least on an annual basis. That is, the rate of sedimentation exceeds the rate of sediment release, which is commonly referred to as internal loading. Even lakes that have a net annual retention of P may experience a net release, or net internal loading, during part of the year, usually summer (Welch and Jacoby, 2001).

The exchange of P between sediment and water in a lake can depend on several factors acting separately or together. The more important ones are: oxidation–reduction potential (redox), which is dependent mostly on the oxygen concentration; pH; water exchange, as it affects diffusion and transport; temperature, as it affects microbial activity and decomposition; and the relative fractions of P in the sediment that are bound with iron, aluminium, calcium and organic matter, and, in some cases, loosely bound. The quantity in these fractions can vary greatly from lake to lake. The relative importance of these processes will vary with depth and degree of thermal stratification. The annual cycle in a typical temperate, stratified, anoxic (hypolimnion) lake will be considered first.

At the onset of thermal stratification, oxygen (DO) declines in the hypolimnion due to microbial decomposition of organic matter in the water column and in the surficial sediments. DO concentrations decline faster near the sediment surface due to the dominant effect of sediment DO demand over that in the water column (Livingstone and Imboden, 1996). The area of the anoxic hypolimnion is practically defined as sediment overlain by measured DO as $\leq 1\,\text{mg}\,\text{l}^{-1}$ (Nürnberg, 1995b).

As DO approaches zero over the sediment surface, reducing conditions prevail in the surficial sediments and iron is reduced from its ferric (Fe^{3+}) to its ferrous form (Fe^{2+}). P, that was bound to the hydroxy complexes of ferric iron (e.g. $FeOOH-H_3PO_4$), is now solubilized and released into the interstitial pore water of the sediments and is available for diffusion into the overlying, anoxic water. The rate of diffusion is a function of the concentration gradient in SRP between the interstitial pore water and the overlying water (Penn et al., 2000). Adequate exchange of low P water overlying the sediment can maintain a high concentration gradient and enhance the release rate. As a result of a reasonably constant rate of release, the hypolimnetic P content increases more or less linearly throughout the stratified period. Rates of release attributed to the iron-redox process are quite variable with values from $<1\,mg\,m^{-2}\,day^{-1}$ to as high as $52\,mg\,m^{-2}\,day^{-1}$ (Löfgren, 1987; Nürnberg, 1987a), Nürnberg (1984) reported a range from 1.5 to $34\,mg\,m^{-2}\,day^{-1}$ with a mean of $16\,mg\,m^{-2}\,day^{-1}$ for 21 lakes.

While P is increasing in the hypolimnion, it is decreasing in the epilimnion through uptake and sedimentation via phytoplankton cells and zooplankton fecal pellets. While some of the P increase in the hypolimnion may be due to microbial decomposition of settling phytoplankton cells and detritus, results by Gächter and Mares (1985) show that SRP is readily sorbed by particulate matter as it settles through the hypolimnion regardless of the lake's trophic state and that the source of increased SRP in the hypolimnion is the sediments. They conclude this because in the upper hypolimnion P is not released in proportion to O_2 loss, while in the deeper portion, P increased in excess of the expected stoichiometric gain of one mole P for each 138 mole loss in O_2.

When the lake destratifies in the autumn, the whole water column and surficial sediments are replenished with DO. As that occurs, ferrous iron is oxidized to the ferric state and P is once again sorbed to its hydroxy complexes and returned to the sediment. Birch (1976) showed that 90% of the P released during the stratified anaerobic period in monomictic Lake Sammamish, Washington was quickly resedimented in November following turnover. The net result of these processes can be seen in the seasonal changes in DO, TFe and TP concentrations in the hypolimnion of Lake Sammamish (Figure 4.2). Although hypolimnetic mean DO does not show an anoxic condition, a large area overlying the sediment was $<1\,mg\,l^{-1}$.

P may again increase during the winter in dimictic lakes if DO reaches low levels under the ice. However, if the water overlying the surficial sediments remains oxic, then P will not be released, even if the deeper sediments are highly reducing and the lake is eutrophic. Figure 4.3 shows the redox potential in sediments and immediately overlying water during winter in Esthwaite water and Ennerdale, two English lakes (Mortimer, 1941, 1942). Although of widely different trophy, as indicated by the redox level in the sediments, the overlying water was well oxidized in both lakes. At near zero DO ($<1\,mg\,l^{-1}$) redox potential will decrease to about 0.2 V and iron will be reduced (Søndergaard et al., 2002). Under these reducing conditions in the

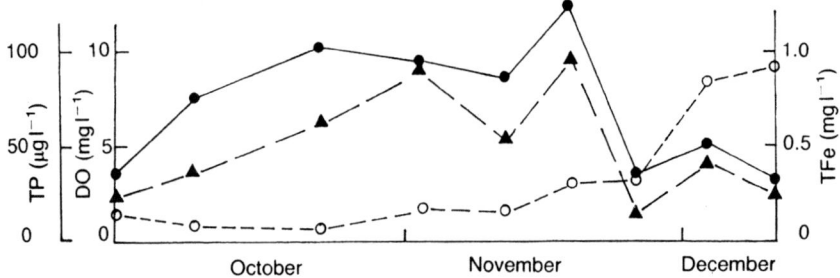

Figure 4.2 Hypolimnetic mean concentrations in Lake Sammamish before and after autumn turnover (Rock, 1974). Closed circles = TP; triangles = TFe; open circles = DO.

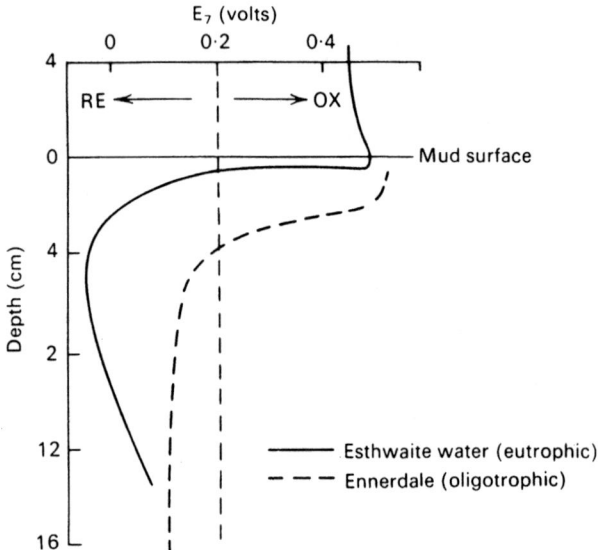

Figure 4.3 Vertical distribution of oxidized and reduced conditions in the sediments of two English lakes in winter (modified from Mortimer, 1942 with permission of Blackwell Scientific Publishers Ltd).

sediment, P can diffuse upward to replace P that was previously released from surficial sediments, but if an oxidized layer exists, as shown in Figure 4.2, P will be readily sorbed in the surficial layer (Søndergaard et al., 2002).

The iron redox cycle in lake sediments and the implications to P exchange with the overlying water were described initially by Einsele (1936) and Mortimer (1941, 1971), but have been an intriguing and important topic among limnologists ever since. Lake sediments are very complicated chemically and iron redox may not be controlling P exchange, as, for example, in sediments where high concentrations of the live cyanobacterium *Microcystis* in the surficial sediments and high carbon content in deeper sediment favour the reduction of carbon instead of iron (Davison et al., 1985). Gächter et al. (1988) found that one half the dry matter in surficial sediments of Lake Sempoch, Switzerland was contained in bacterial cells. Laboratory and field experiments indicated that uptake of P by bacteria during oxic conditions and release during anoxia is an important process in the exchange of P between sediments and water. They and others (Golterman, 2001) suggested that there tends to be an uncoupling of Fe and P when those elements are released following the onset of anoxia in several lakes and was noted, although to a lesser degree, in Lake Sammamish. (Figure 4.2). In highly eutrophic lakes, with high Si:Fe ratios in sediment, the formation of insoluble FeS via bacterial reduction of SO_4^{2-} (see S cycle) may limit the ability of Fe to sorb P, resulting in high rates of P release, even under oxic conditions (Søndergaard et al., 2002; Gächter and Müller, 2003).

Nevertheless, there still seems to be general agreement that iron redox is the principal mechanism controlling P exchange in the deep-water sediments of most stratified lakes. In support of that, analysis of sediment P fractions has been useful in defining those forms that represent mobile P, which is composed of loosely sorbed P and Fe-bound P, as opposed to P bound by Al, Ca and organic matter (Hieltjes and Lijklema, 1980; Boström et al., 1982; Psenner et al., 1988). Nürnberg (1988) has shown that the Fe–P fraction (as BD-P, indicating the extraction reagent) in anoxic lake sediment was highly correlated with the sediment P release rate. Moreover, decreases in the Fe–P fraction in sediments were proportional to increments in TP in overlying water of the cores. Also, decreases in the Al- and Fe-bound P fractions in sediments during summer were comparable to mass balance determinations of P internal loading (Jacoby et al., 1982). The loosely sorbed fraction is considered fully mobile and may be relatively high in lakes enriched by sewage effluent. Al–P is also sensitive to pH, as is Fe–P, but not to low DO, so that the fraction would remain bound in anoxic stratified lakes.

Numerous investigations have shown that internal P loading from sediments in shallow 'oxic' lakes can represent a large fraction of the total TP loading (Jacoby et al., 1982; Lennox, 1984; Riley and Prepas, 1984; Ryding, 1985; Søndergaard et al., 1999; Welch and Jacoby, 2001). Internal loading in shallow, eutrophic Danish lakes accounted for summer TP concentrations being 2–4 times higher than winter values (Søndergaard et al., 1999). But what are the principal mechanisms operating in shallow, unstratified, apparently oxic lakes and littoral zones of stratified lakes?

There are several mechanisms that explain internal P loading in shallow, unstratified lakes (Boström et al., 1982; Welch and Cooke, 1995):

- Dissolution of Fe- and Al-bound P by high pH, which is caused by high photosynthetic activity.
- Dissolution of Fe–P under reduced conditions during periods of calm weather when anoxia develops at the sediment–water interface, with later entrainment due to wind mixing.
- Mineralization of organic P into dissolved P by microbial metabolism driven by temperature.
- Release of dissolved P directly from bacterial cells or through metabolism of dissolved organic P excreted from algal cells in the sediments.
- Wind- and wave-caused resuspension of particulate P, with soluble P desorbing due to the high sediment–water concentration gradient or due to high pH caused by photosynthetic activity.

Internal loading in most shallow lakes likely results from the simultaneous occurrence of several of the above mechanisms. Determining the relative importance of these mechanisms is difficult and has not actually been done in a specific lake. However, the magnitude of individual mechanisms has been demonstrated in many lakes.

The solubility of iron is controlled by pH under oxic conditions (Stumm and Leckie, 1971). At a pH of ≈ 6, which is rather typical for the sediments of soft-water lakes, the solubility of ferric iron is minimal and therefore P can be effectively sorbed and removed from the water column (Ohle, 1953; Figure 4.4). This is the principal process by which P is retained in lakes as discussed for stratified lakes. With increasing pH the solubility of iron increases and P that was sorbed to ferric iron hydroxy complexes is released (Figure 4.5). The high photosynthetic rates in nutrient-rich lakes can result in mid-day pH values of 10, which could produce very high rates of P release from sediments in littoral regions in deep lakes and throughout in shallow lakes (Jacoby et al., 1982; Andersen, 1975; Boers, 1991). However, it is not clear that such high pH levels persist near the well-buffered sediments, which due to high decomposition rates normally have a pH ≈ 6. Photosynthetically caused high pH could, nevertheless, maintain P solubilized in the water column and available for algal uptake (Ryding, 1985; Löfgren, 1987).

High pH due to photosynthesis may also combine with wind-caused resuspended particles to result in high water column TP (Lijklema et al., 1986; Søndergaard et al., 1992). Suspended particles normally sorb P and remove it from the water column when they settle, but P can also be mobilized from resuspended particles at low particle concentrations and high pH (Koski-Vähälä and Hartikainen, 2001). Whether high-pH water reaches the sediment surface in shallow lakes or desorbs P from resuspended sediment and maintains high concentrations in the water column, internal P

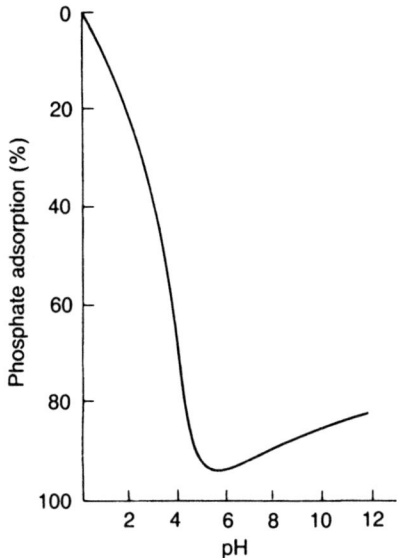

Figure 4.4 Relative removal of PO_4^{3-} from solution with $Fe(OH)_3$ versus pH (Ohle, 1953).

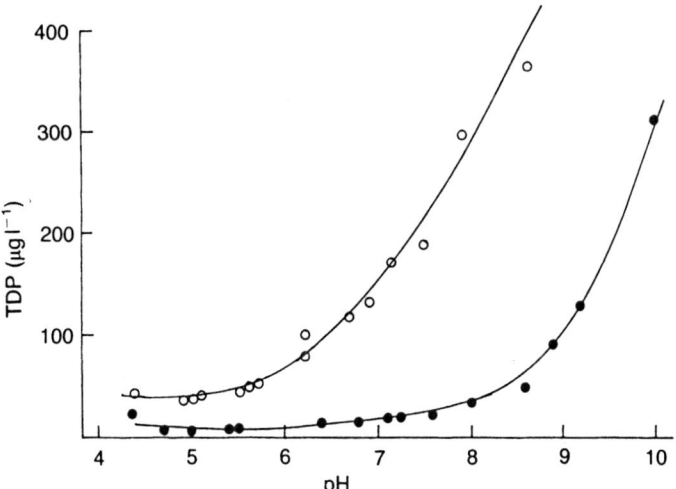

Figure 4.5 Relationship between equilibrium total dissolved P and pH in surficial sediments at two sites in Long Lake, WA (Jacoby et al., 1982). ○, Midlake (3.5 m); ●, Lilies (1 m).

loading has nonetheless been directly related to pH, as was the case for the shallow (2 m mean depth) Upper Klamath Lake, Oregon (Figure 4.6). That photosynthesis was the cause for high pH was shown by strong relations between lake pH (>10) and chl *a* (Kann and Smith, 1999).

Wind-caused resuspension of particles with high loosely sorbed P has been shown to produce high internal loading in Lake Arresø, Denmark (Søndergaard, 1988; Kristensen *et al.*, 1992) and Lake Balaton, Hungary (Luettich *et al.*, 1990; Koncsos and Somlyódy, 1994).

Temperature can be an important driving force in the release of P especially from organic sediments. Kamp-Nielsen (1974) has shown that P release from Lake Esrom sediments was directly related to an increase in temperature. Also, Ryding (1985) found the rise in water temperature to be a key factor in the release of P in 16 shallow, eutrophic Swedish lakes. The role of temperature is related to the stimulation of bacterial activity, which in turn creates anoxic conditions and the reduction of iron. Bacterial activity can increase dramatically in response to normally moderate increases in temperature as shown by Boström *et al.* (1985) for highly organic Lake Vallentunasjön sediments. They found a Q_{10} of 10.6 over a temperature range of 4–20°C. Water near the sediment surface would undergo such

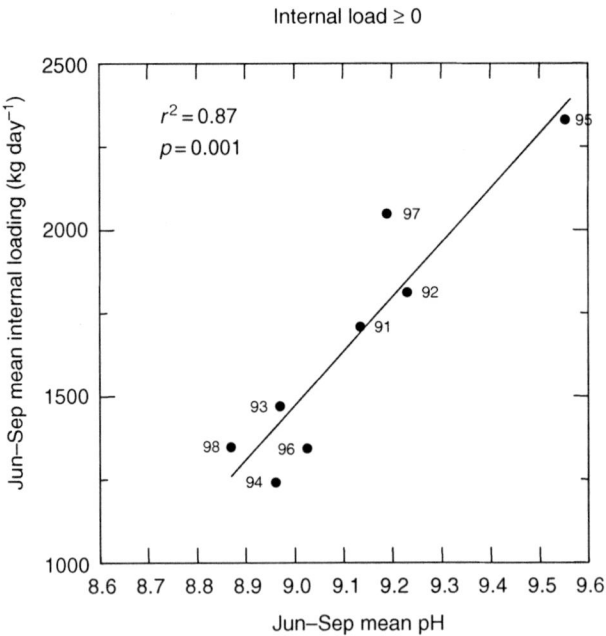

Figure 4.6 Summer net internal P loading (positive values) related to pH in Upper Klamath Lake, Oregon (J. Kann, May 2003, personal communication).

temperature increases only in shallow unstratified lakes and in littoral regions. But wind mixing apparently maintains nearly saturated DO conditions throughout the water column, which should not favour the iron redox mechanism as an explanation for P release in such environments. However, Fe redox reactions may be important in shallow lakes under certain conditions.

Although oxygen-temperature profiles in shallow environments normally show largely uniform temperature and saturated or supersaturated DO concentrations throughout the water column, these conditions may not persist at the sediment–water interface. Slight thermal stratification can occur temporarily under quiescent conditions permitting DO to deplete in the water layer immediately overlying the sediment (Stefan and Hanson, 1981; Riley and Prepas, 1984; Löfgren, 1987). Even if no thermal stratification is present, DO can deplete in the immediate overlying water under quiescent conditions (Welch et al., 1988a). Moreover, even in stratified lakes iron has been observed to increase before DO in the overlying water is completely depleted (Mortimer, 1941, 1942; Davison et al., 1985). Thus, large release rates of P from sediments during summer in unstratified lakes may often be attributed to the iron redox mechanism. In addition to bacteria causing the reduction of iron indirectly through DO depletion via their metabolic activity, evidence exists that bacteria may reduce iron directly (Jones et al., 1983). This is supported by strong positive correlations between iron and TP during the summer in shallow, eutrophic Lake Finjasjön in Sweden (Löfgren, 1987). TP increase in shallow Danish lakes during summer was related to increased TFe if the TFe:TP ratios in sediments were >10 by weight, indicating that P was controlled by Fe, although the sediment P release could have been due to high pH as well as Fe redox (Jensen et al., 1992).

Increased wind mixing, following the quiescent period, would then entrain the layer overlying the sediment and transport the P throughout the water column. Although the entrained iron would oxidize upon mixing, the entrained P could remain in the water column under conditions of photosynthetically caused high pH. Ahlgren (1980) has shown a positive relationship between the wind-fetch distance per unit mean depth and the summer amplitude in TP in some shallow, eutrophic Swedish lakes. In a similar fashion, Osgood (1988a) found that internal loading in lakes around the Twin Cities, Minnesota, increased when the mean depth/area$^{0.5}$ (m/km) was less than about seven. The relative thermal resistance to mixing, and to a lesser extent flushing rate, was inversely correlated ($r^2 = 0.92$) with internal loading over a 9-year period in Moses Lake (mean depth 5.6 m), Washington (Jones and Welch, 1990).

Also, it is clear that P released from sediments in enriched, shallow, polymictic lakes would immediately be available in the lighted zone for uptake by phytoplankton. Rather than sedimented P and sediment-released P being trapped in the hypolimnion until turnover, as is often the case in deep stratified lakes, P can be continually recycled throughout the water column during summer in unstratified lakes.

The dredging of Lake Trummen can be used to illustrate the importance of sediment–water exchange of P in shallow, eutrophic, polymictic lakes. Lake Trummen, in Växjö, Sweden, has gained international fame as the first, large sediment dredging project to restore a lake (Björk, 1972). The lake's maximum depth was only 2 m prior to dredging and core studies showed that although the sedimentation rate had decreased over its 10 000 year history, it had recently increased from 0.2 to 8 mm year^{-1}, in response to the discharge of sewage effluent from the city. The sediment P content correspondingly increased during 1920–1958, from 50 to 800 mg l^{-1} ($\approx 0.5-8$ mg g^{-1} dry matter, assuming 10% water content). Zn and Cu increased as well. Sediment release experiments showed that the most recently deposited sediments were obviously significant sources for nutrients, compared to the deeper, historical sediments (Table 4.1). Note that P was released under aerobic conditions, either signifying low Fe:P ratio or high loosely sorbed P. In an effort to recover the lake, the sewage effluent was diverted, but no improvement was noted over the subsequent 10-year period. Removal of the top 1 m of rich sediment quickly resulted in the recovery of the lake. Thus, the sediment–water exchange of P was sufficient to maintain the high productivity and poor water quality in spite of the removal of the sewage input. Following removal of the rich layer of sediment, the release rate of P declined (Table 4.1). More details about this project will be covered in Chapter 10.

Although sediment-released P in unstratified lakes is readily available to phytoplankton in the lighted zone, this is not usually the case in stratified lakes. Sediment-released P in the hypolimnion can become available to phytoplankton in the lighted epilimnion through diffusion from high to low SRP concentrations and erosion of the thermocline and entrainment of hypolimnetic water into the epilimnion as a result of summer storms. This

Table 4.1 Release of N and P in containers with water and sediment under aerobic and anaerobic conditions (mg m^{-2} day^{-1})

	Aerobic		Anaerobic	
	PO_4-P	NH_4-N	PO_4-P	NH_4-N
Black gytta deposited in 1960s	1.7	0.0	14.0	73.0
Brown gytta deposited 1000 yr BP, depth >40 cm	0.0	0.0	1.5	0.0
After dredging[a]			1.24	

Source: Björk et al. (1972).

Note
a Bengtsson et al. (1975).

has been shown for several lakes. For example, an August storm resulted in disruption of the thermocline in Lake Ballinger ($Z_{max} = 12$ m) near Seattle, apparently causing an increase in SRP, which stimulated a phytoplankton increase (Figure 4.7). This phenomenon was first demonstrated by Stauffer and Lee (1973), who showed that a thermocline sinking of 1 m in Lake Mendota, Wisconsin resulted in a doubling of the eplimnetic P content and contributed significantly to summer algal blooms. Analysis of continuous temperature data from Lake Ekoln, Sweden showed seven events of thermocline erosion during the stratified period in 1970. An example is shown in Figure 4.8, when the TP content in the epilimnion was calculated to have increased by 15%. Good evidence for thermocline migration and entrainment of hypolimnetic P was shown by Larsen et al. (1981) for four wind-mixing events during summer in Shagawa Lake, Minnesota. Eplimnetic P increased from 74 to 130 µg l^{-1} and that coincidentally there was nearly as great a decline in the hypolimnetic P content. In addition to entrainment, they also estimated that diffusion could account for a vertical transport of 4.5–9 mg m^{-2} day^{-1} of P into the epilimnion. These processes can be incorporated into a two-layer dynamic TP model that will demonstrate the effectiveness (i.e. availability) of internal loading in stratified lakes (see P modelling in Eutrophication).

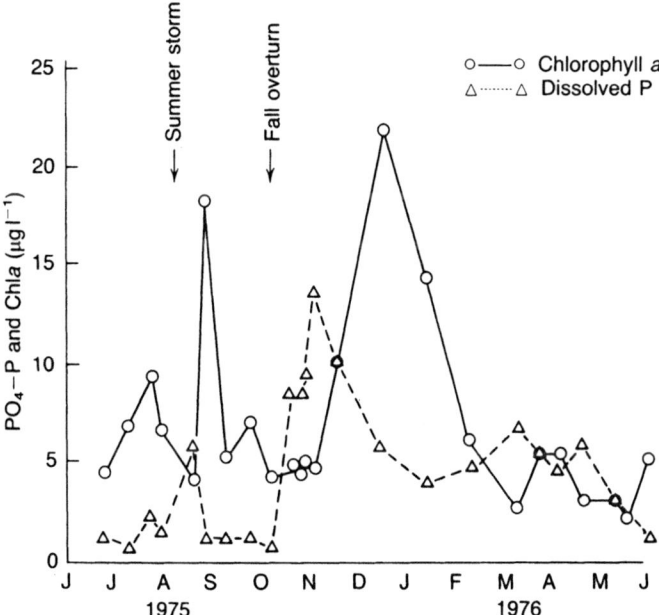

Figure 4.7 Dissolved P increase in the epilimnion of Lake Ballinger, Washington, as a result of an August storm with a subsequent response in plankton chlorophyll.

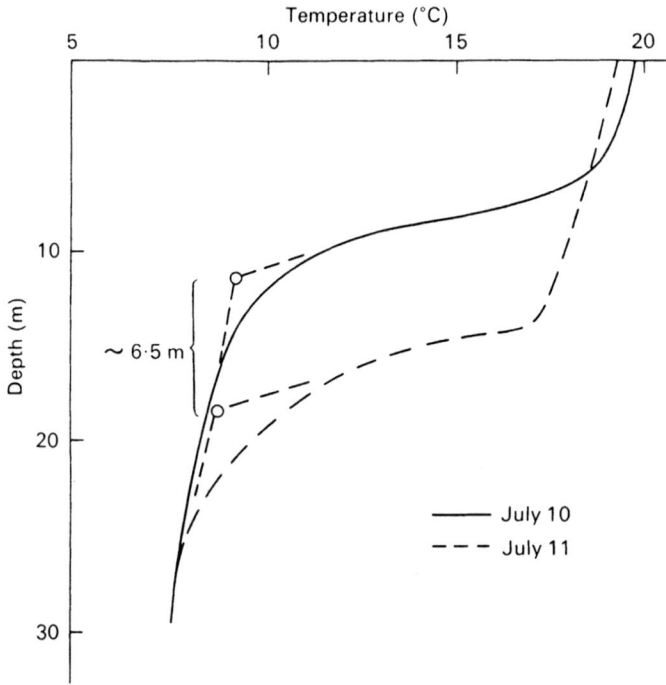

Figure 4.8 Temperature profiles from Lake Ekoln. Sweden preceding and following one of seven thermocline disturbances during the summer of 1970. Thermocline position estimated by extrapolation to approximate the overall thermocline displacement (6.5 m). Temperature was continuously monitored at 1 m intervals (data from National Swedish Environmental Protection Board).

There are other processes that can account for the transport of sediment P to lake water. Gas ebullition from sediments can serve as a transport process for interstitial SRP if the surficial sediments are reduced (Enell and Löfgren, 1988). Bioturbation of the surficial sediments by benthic invertebrates may also be an important transport process (Boström et al., 1982). Excretion by benthivorous fish feeding on detritus and sedimented algae and disrupting sediment can be substantial (Shapiro et al., 1975; Havens, 1991, 1993; Breukelaar et al., 1994). Death and decomposition of rooted macrophytes, which depend largely on sediments for their P supply, is a process by which sediment P can be transported to lake water (Schultz and Malueg, 1971; Carpenter and Adams, 1977; Landers, 1982; Rorslett et al., 1986). The relative significance of that source will be discussed further in Chapter 9. Vertical migration of buoyant and mobile phytoplankton is yet another transport process that has been demonstrated in several lakes. Trimbee and Harris (1984) found that the blue-greens (cyanobacteria) *Microcystis aeruginosa*, *Gomphosphaeria lacustris* and *Lyngbya birgei* may have originated largely from sediments in Guelph Reservoir, Ontario because downward facing traps at 10 m captured 2–4% of the bloom population that could have been produced by only five or six divisions of the 2–4% inoculum. TP was also shown to migrate with the algae. Transport by algae and blue-greens within the water column has been shown for vacuolate *Oscillatoria* (Klemer et al., 1985) and for a motile alga, *Cryptomonas marssonii* by Salonen et al. (1984). Extensive work on Green Lake, Washington, showed that translocation of P from sediments to water via *Gloeotrichia echinulata* varied from <1 to >100 mg P m^{-2} during five summers (Barbiero and Welch, 1992; Perakis et al., 1996; Sonnichsen et al., 1997). The high release contributed 65% of the summer internal P load. Translocation by other cyanobacteria in Green Lake was relatively small, but rates were more substantial for *Aphanizomenon flos-aquae* in Upper Klamath Lake (Barbiero and Kann, 1994). See 'Succession' for further discussion of this process (Section 6.5.6).

4.1.4 Recycling and the fate of added P

Lakes tend to reestablish equilibrium levels of P in the water in a relatively short time following manipulation of the inputs. Permanently increasing the P concentration in lakes is a very difficult process; many frequent and small additions are necessary in order to raise and maintain a higher concentration, as has been shown in experimental enrichment of lakes (Schindler, 1974). The rapid cycling of P has been demonstrated in laboratory microcosms using radiophosphorus (^{32}P). The principal results of this early work by Hays and Phillips (1958) were as follows (Figure 4.9).

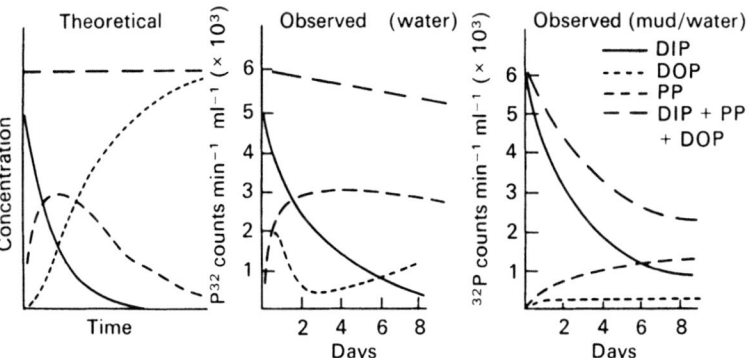

Figure 4.9 Distribution of ^{32}P among three forms after addition to microcosms. DIP, dissolved inorganic P; DOP, dissolved organic P; PP, particulate P (Hays and Phillips, 1958).

1 DIP was rapidly removed from the water phase at an exponential rate, but the reaction series did not reach completion as theoretically predicted, i.e. all DIP was not converted to DOP.
2 Bacteria took up and held much of the DIP as PP, which was recycled back to the water phase as DIP.
3 The loss of total P was much greater with mud, which competed with bacteria for DIP and was much slower to release DIP.
4 An equilibrium was established between water and solid phases in which total ^{32}P was only ≈ 0.1 of the initial level.

The experiment was repeated in a small Nova Scotia lake to which 11 mCi of ^{32}P was added as $KH_2{}^{32}PO_4$. The level of ^{32}P decreased to 80% of the initial after 2 weeks. The concentration peaked in plankton after 4–16 h and in fish after 3–4 days (Hays and Phillips, 1958).

Similar experiments on the fate of P were performed in Crecy Lake (20 ha, mean depth 2.4 m) in New Brunswick, but with inorganic fertilizer. Nevertheless, results were comparable with the ^{32}P work. The fertilization theoretically raised P to 390 μg l^{-1}, N to 210 μg l^{-1} and K to 270 μg l^{-1}. The results from fertilizations in 1951 and 1959 and successive years are shown in Figure 4.10 (Smith, 1969).

From such experiments, the turnover rates can be estimated by assuming first-order kinetics and calculating the rate of disappearance of P in the various phases following fertilization. The loss rate for DIP was between 5–6% day^{-1} for the three fertilizations (Table 4.2). The turn over times (time required for loss from a phase) could then be calculated as the ratio of the amount in a particular phase to the loss rate for that phase. P in the solid phase underwent a 10–20-fold slower turnover than did P in the dissolved phase, which illustrates how P was conserved. Nevertheless, by the next year there was no evidence of the previous year's fertilization, thus supporting the earlier statement that permanently increasing the P content in lakes is a difficult process.

Phosphorus pool size and transfer among pools within the water column can be modelled for a system if enough data are available. The definitions of pools and transfer rates are often grossly oversimplified for the sake of practicality and lack of knowledge. However, transfer rates are variable and can be highly dependent on such factors as light, temperature and species composition of the trophic level. As a result, dynamic models of lake ecosystems are more useful for investigating how the system works than for prediction for management purposes.

To illustrate how a model could be constructed, Stumm and Leckie (1971) developed a hypothetical, steady-state model for the water column of a stratified lake (Figure 4.11). Any substance can be modelled, but usually either C, N or P is used. Conversions from one element to another are made using an appropriate elemental ratio, which is another simplification.

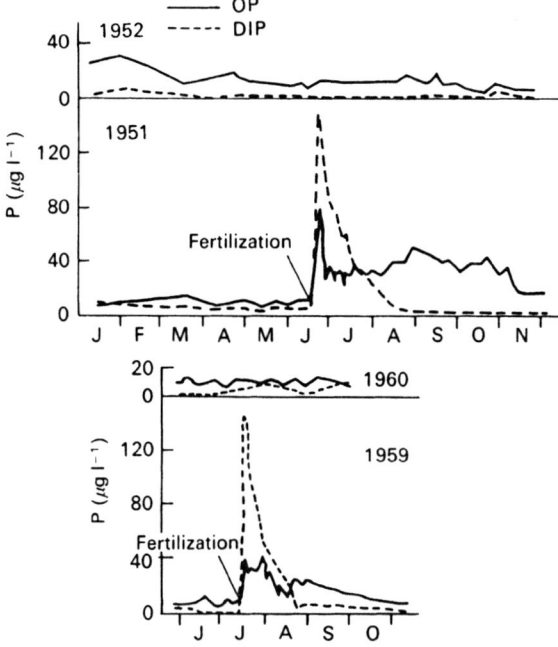

Figure 4.10 Effect of fertilizing Crecy Lake with P, N, and K. Concentrations of dissolved inorganic and organic phosphorus (DIP, OP) during years of fertilizing and succeeding years are shown (after Smith, 1969).

Table 4.2 Turnover time of P in water and solids of Crecy Lake estimated from observed loss rates

Fertilization	Loss rate (% day^{-1})	Turnover time (days)	
		Water	Solids
First	5.9	17	176
Second	5.6	18	248
Third	5.3	19	394

Source: Smith (1969).

Note
Turnover time (days) = (amount in phase)/[loss rate (per day)].

P is modelled in Figure 4.11 and the transfer rates are expressed in $\mu g\,l^{-1}\,day^{-1}$ and pool sizes in $\mu g\,l^{-1}$. The model suggests that $\approx$30% of the uptake demand (1.1/3.5) is supplied by bacterial decomposition in the epilimnion, whereas 20% is diffused from the hypolimnion and the remainder

72 General concepts of aquatic ecosystems

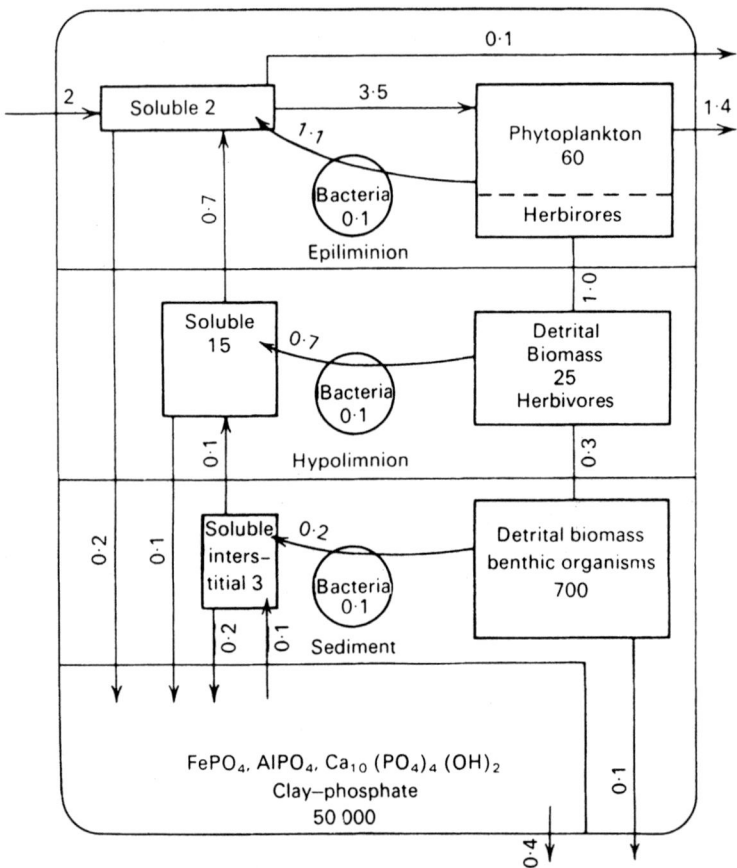

Figure 4.11 Hypothetical steady-state model of phosphorus cycling in a lake (after Stumm and Leckie, 1971).

comes from the inflow. The hypolimnetic source is reasonable considering the earlier discussion of thermocline erosion due to storms. About one third of the P loss from the plankton rains into the hypolimnion via sinking fecal pellets and dead phytoplankton cells. There are also losses from the soluble phase to the sediments due to inorganic precipitation. Chemical regenerative processes from the sediment account for only ≈5% of the biological contribution to the soluble P pool (0.1 versus 2.0 $\mu g\, l^{-1}\, day^{-1}$). As indicated earlier, and to be discussed further in Chapter 7, contributions of P from chemical exchange across the sediment–water interface in shallow stratified and unstratified lakes alike can be much more significant.

Some omissions in Figure 4.11 are separate pathways for regeneration from dead cells via autolysis, which is not a function of bacterial activity, and excretion of DIP direct from phytoplankton and grazing zooplankton, the latter being one of the more important regenerative processes. These processes were indicated separately in Figure 4.1.

Another important process often omitted in lake models of the water column and sediments is the interchange with the littoral zone. Although contributions from the littoral zone have been emphasized recently, there are pertinent results from an early study conducted by Hutchinson (1957, p. 748; Hutchinson and Bowen, 1950) on Linsley Pond, Connecticut. ^{32}P was added on 21 June 1946 and results were observed during August 1–15 by noting the change in the amount of tracer in the various pools. This whole-lake experiment illustrates the interaction in the transfer of P among the epilimnion, littoral region and the hypolimnion. The transfer rates of P in kg week^{-1} were as follows for the epilimnion: epilimnetic increase, 0.26; loss to hypolimnion, 1.55; transport from littoral to epilimnion (sum), 1.81; and for the hypolimnium: hypolimnion increase, 3.75; gain from epilimnion, 1.55; gain from sediment (difference), 2.20. The hypolimnion was assumed to act as a sink in this balance analysis. Some of the epilimnetic gain may have come from the hypolimnion due to processes mentioned earlier. However, the littoral was probably the major contributor as indicated.

Rich and Wetzel (1978) have stressed the importance of the littoral region in supplying energy and materials to the economy of lakes. Recent studies have confirmed the importance of the littoral in supplying P and other substances to the pelagic region. Cooke *et al.* (1978, 1982) showed by mass balance calculations that internal loading of P was reduced but not eliminated in West Twin Lake, Ohio after an alum treatment of the anoxic hypolimnetic sediments. Internal loading in East Twin, the control lake, remained about double that of West Twin after treatment. The conclusion was that the continued internal supply in treated West Twin must have come from the littoral region, because the alum floc had blocked P release from the sediments.

The littoral processes could include macrophyte decomposition, bioturbation or aerobic release from sediments. Decomposition of *Myriophyllum spicatum* in the littoral zone was found to be a significant recycling process in Lake Wingra, Wisconsin, contributing an amount equal to $\approx 60\%$ of the external P load and 30% of the total load (Carpenter and Adams, 1977; Prentki *et al.*, 1979; Carpenter, 1980a). Loss of P from healthy shoots was considered nil, but the net (minus uptake) release from the littoral through senescence was estimated at $2 \text{ g m}^{-2} \text{ year}^{-1}$, which is $17 \text{ mg m}^{-2} \text{ day}^{-1}$ (Smith and Adams, 1986, see Chapter 9). For comparison, Nürnberg (1984) reported that the release rate from anoxic hypolimnia of 21 lakes ranged from 1.5 to $34 \text{ mg m}^{-2} \text{ day}^{-1}$ with a mean of 16. Thus, lakes with extensive beds of rooted macrophytes (at least. *M. spicatum*) may have a significant portion of internal loading coming from the littoral zone.

4.1.5 Sediments as a source or sink for P

The question as to when and to what extent sediments are a source or sink for P is important for understanding lake productivity as well as for the management of lake quality. Most of the processes involved in the sediment-water exchange of P have been covered in the previous discussion. While Fe redox is usually the controlling process in stratified anoxic lakes, there are several processes that may be operating coincidentally in shallow lakes and estimates of the relative contribution of each of those processes in a given lake are lacking and may be impossible to obtain. Some general summary comments may help clarify the issue.

1 Lakes are nearly always a sink for P in the long term, whether they are shallow, deep, oxic or anoxic. Unless a fraction of the input has been recently diverted, the input mass of P will usually be greater than the output on a year-to-year equilibrium basis. The fraction retained is usually greater than 0.5, but can vary widely (Vollenweider, 1975). Another way to verify this is to compare the retention times of P and water; they are, respectively, 0.77 and 2.0 years in Lake Sammamish, calculated by the ratio of lake quantity to inflow rate. Retention is also higher in wetlands on an annual basis, hence their significance in watershed management (Kadlec and Knight, 1996).

2 Although sediments act as an ultimate sink, they can act as significant sources during a portion of the year. That is true for wetlands as well as lakes. Probably the most important process determining when and to what extent sediments will act as a source is iron redox, whether the lake is stratified, unstratified, oxic or anoxic. Stratified, oxic lakes generally yield little P from sediments because iron in surficial sediments is always oxidized and effectively sorbs P assuming that the Fe:P ratio is sufficiently high. However, in stratified lakes that have anoxic hypolimnia and in unstratified lakes that experience alternating quiescent and windy periods during the summer, the release of sediment P, due to iron reduction, results in lake output of P exceeding the input. Photosynthetically caused high pH and recycling processes in the water column tend to retard the resettlement of P. Wind resuspension can also be important, but the extent to which P remains in the water column is still dependent on pH and the Fe:P ratio. The other processes that supply P from sediments, such as macrophyte decomposition, bioturbation, excretion by benthivorous fish, gas ebullition, algae migration, etc., have been demonstrated to be significant in some instances, but generally have been less important than iron redox.

3 Whether the sediment-released P in stratified lakes actually reaches the photic zone and is available for algal uptake is important to lake productivity, but not with respect to the source/sink issue. So long as P reaches the

hypolimnion, even if unavailable to algae, the sediments are technically a source for that P. However, there is ample evidence that some of that hypolimnetic P can be transported into the epilimnion via diffusion and entrainment, although in some instances the metalimnion can be an effective barrier.

4.2 Nitrogen

The general cycle of N in an aquatic ecosystem is shown in Figure 4.12. The cycle is very complex, and many of the transfer processes and associated pool sizes are important not only to aquatic productivity but also to environmental quality in general and human health. Two large biological source and sink processes occur with N that do not occur with P, which greatly influences the significance of N in the control of productivity and waste treatment. These processes involve the transfer of N from atmospheric N_2 through microbial fixation and its return to the atmosphere via N_2O and N_2 through denitrification. The processes of nitrification (oxidation of ammonia) and denitrification are reactions that would proceed without biological mediation but at a slow rate. Microorganisms greatly speed the rate of reaction and at the same time capture the energy available in the reduced compounds through an ordered series of cellular enzyme catalysed reactions. Because the energy sources are inorganic, the organisms are referred to as chemolithotrophs (Brock, 1970).

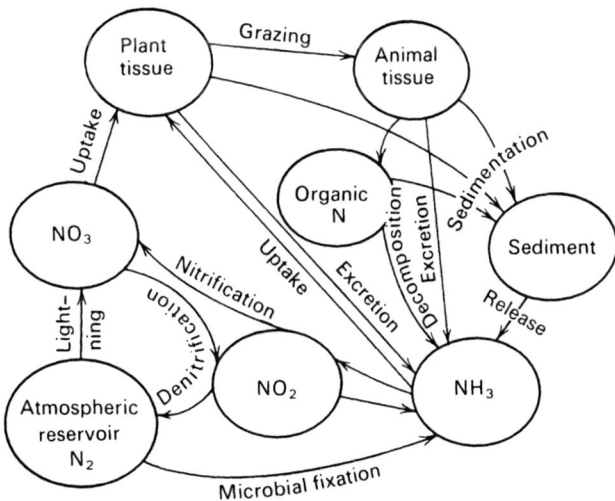

Figure 4.12 The aquatic nitrogen cycle.

4.2.1 Pool concentrations

Nitrogen is most abundant as N_2, comprising nearly 80% of the atmospheric constituents. However, before this form is available to plants it must be fixed either biologically or through a light-energized reaction. Nitrate (NO_3^-) is the most common form that can be used by aquatic plants, and its content ranges from a trace when productivity is high, and all available N has been absorbed into plant tissue, to concentrations usually between 500 and 1000 $\mu g\, l^{-1}$ during periods of no utilization. Concentrations well over 1 $mg\, l^{-1}$ can occur but are usually associated with inputs. Ammonia (NH_3) or ammonium (NH_4^+), which is the principal form in water, becomes abundant in the absence of oxygen or in very enriched waters, but it is usually less abundant than NO_3^-. Because NH_4^+ is a more reduced form, it is often preferred over NO_3^- by plants. NO_2^- is not used and in fact can be rather toxic. Because it readily oxidizes to NO_3^-, it is usually not present in appreciable amounts.

4.2.2 Nitrification

Nitrification is the process by which NH_3 is transformed into NO_2^- and finally into NO_3^-. The process occurs only under aerobic conditions. Organisms that commonly perform the transformations are *Nitrosomonas* and *Nitrobacter*. Although the processes are energy yielding, as shown below, the yield is rather low compared to other transformations in the cycle (Delwiche, 1970). The reactions in nitrification are:

$$2NH_4^+ + 3O_2 \rightarrow 2NO_2^- + 2H_2O + 4H^+ + \text{energy}$$
$$2NO_2^- + O_2 \rightarrow 2NO_3^- + \text{energy}$$

4.2.3 Denitrification

This process occurs only in the absence, or near absence, of oxygen. A common denitrifying organism is *Thiobacillus denitrificans* which is a chemolithotroph and the associated reaction is:

$$5S^{2-} + 6NO_3^- + 2H_2O \rightarrow 5SO_4^{2-} + 3N_2 + 4H^+ + \text{energy}$$

Heterotrophic bacteria, such as *Micrococcus, Serratia, Pseudomonas* and *Achromobacter*, are also denitrifiers when oxygen concentrations are low. As they are facultative anaerobes, they can also carry on aerobic respiration and are part of the normal flora in sewage (Christensen and Harremoes, 1972).

In denitrification, the reverse process of nitrification, the bacteria reduce NO_3^- to NO_2^- and in turn form molecular gaseous N_2 or N_2O (nitrous oxide). This represents the mechanism of N loss from ecosystems and can be used as a treatment method for the removal of N from wastewater. However, the environment must be aerobic for nitrification and anaerobic for denitification. The necessary alternation of aerobiosis/anaerobiosis in order for denitrification to represent a significant N loss has implications in the N balance and nutrient limitation in lakes as well as in wastewater treatment. The more productive a lake, the greater is the chance for oxygen to reach low levels, which would allow denitrification. For wastewater treatment, an aerobic period must precede the anaerobic period in order for NH_4^+ to be converted to NO_3^-, which can then be denitrified without oxygen. N reduction to NH_4^+ from NO_3^- also occurs.

4.2.4 N fixation

Nitrogen fixation is an energy-consuming aerobic process carried on in aquatic environments by such bacteria as *Azotobacter* and *Clostridium* and by the cyanobacteria *Nostoc*, *Anabaenopsis*, *Anabaena*, *Gleotrichia* and *Aphanizomenon*. N-fixing cyanobacteria usually dominate in enriched lakes in summer. Nitrogen fixation can represent a significant input of N to an ecosystem. Measured rates for *Anabaena* range from 0.04 to $72 \mu g\,l^{-1}\,day^{-1}$. Such rates could represent a several fold turnover in dissolved inorganic N in enriched systems, because NO_3^- is usually depleted to very low levels or is undetectable when blue-green fixers appear. Horne and Goldman (1972) have shown that blue-green fixation contributed 43% of the annual N input to Clear Lake, California.

N fixation is performed in the heterocysts of blue-greens. The abundance of these extra large cells in the filaments or chains of the above N-fixing blue-greens usually increases with NO_3^- depletion (Horne and Goldman, 1972). Because fixation is an energy-demanding process, it becomes advantageous only when NO_3^- and NH_4^+ are no longer available.

This source of N for the N fixers also becomes available for other phytoplankton as the cells of the N fixers decompose. In many eutrophic lakes *Aphanizomenon* and *Microcystis* usually do not occur together in large masses but often seem to alternate in abundance. N availability may be the cause for that alternation. Also, the availability of N can be a cause for the succession to blue-greens, particularly the heterocystous blue-greens (see Chapter 6 on succession).

Just as N loss through denitrification can increase in significance as productivity increases (because of oxygen depletion), fixation also increases with productivity due to depletion of NO_3^-. Fixation rates of the planktonic blue-greens in enriched Lake Erie have been determined to range from 4.2 to 230 nmol acetylene $mg^{-1}\,N\,h^{-1}$ and from 0.69 to 25 nmol $mg^{-1}\,h^{-1}$ in less rich Lake Michigan (Howard *et al.*, 1970). Horne and Goldman (1972) have measured a maximum rate of several hundred $nmol\,l^{-1}\,h^{-1}$ in highly enriched Clear Lake.

Table 4.3 N balance from Lake Mendota, Wisconsin

Income		Loss	
Source	(%)	Sink	(%)
Waste water	8.1	Outflow	16.4
Surface water	14.7	Denitrification	11.1
Precipitation	17.5	Fish catch	4.5
Groundwater	45.0	Weeds	1.3
Fixation	14.4	Sedimentation	66.7
Total	100.0	Total	100.0

Source: Modified from Brezonik and Lee (1968).

Complete budgets for N in lakes are constructed much less frequently than for P. Brezonik and Lee (1968) developed a complete budget of N income and loss in Lake Mendota (Table 4.3). In that case the gain from fixation was very similar to that lost by denitrification, although neither was the most important source or sink. Sedimentation was estimated by differences between loss and income. Another source and sink, respectively, were marsh drainage and groundwater recharge for which they had no values. Messer and Brezonik (1983) cite several N budgets that show much larger fractions (>50%) of incoming N lost through denitrification.

4.2.5 Implications to nutrient limitation

The availability of N and P to the primary producers in freshwater ecosystems varies markedly because of unique processes in the N cycle, together with the differing chemical behaviour of N and P. Reiteration of some of the contrasts between the N and P cycles should give better insight as to which nutrient is apt to be most limiting.

1 In a freshwater lake the residence time for an incoming quantity of N is longer in the water than that for P, because the strong sorptive capacity of inorganic metal complexes and organic particulate matter for PO_4^{3-} tends to remove P to the sediments. Although some sedimented P can be recycled via several processes (e.g. iron redox, macrophyte decomposition, etc.) in most cases the overall efficiency of such recyling does not seem to be very high, i.e. even anoxic lakes are still net sinks for P. N, on the other hand, in the form of NO_3^- and NH_4^+ is much more soluble and is not readily sorbed by inorganic complexes as is PO_4^{3-}. Besides NH_3 can also be released from anaerobic sediments.

2 Although significant fractions of N income can be lost from aquatic systems via denitrification, this occurs only in anaerobic and thus, highly enriched waters. As a result, the process contributes to N limitation in highly

enriched lakes, but should not greatly contribute to an N shortage in waters of low or moderate enrichment.

3 The atmosphere provides a ready source of N to N-depleted systems through the biological fixation of N_2. This occurs in aerobic environments and requires only the presence of heterocystous cyanobacteria. When available N is depleted, these organisms can dominate and thus provide a very large part of the N supply to enriched systems. In low or moderately productive systems, however, inorganic N has not declined enough to offer an advantage to N fixation. Because it is an energy-consuming process, fixation is advantageous only if NO_3^- and NH_4^+ are scarce. Another factor that restricts the supply of N through fixation is a rather low maximum rate of N replacement in cells, i.e. growth rate, of 0.05 day^{-1} (Horne and Goldman, 1972).

4 Yet another source of N that does not occur in the P cycle is via precipitation. Although rainwater contains P and has been shown to be important in lakes in phosphorus-poor watersheds (Schindler, 1974), NO_3^- in rainfall is a rather constant phenomenon, having been transformed from the N_2 reservoir in the atmosphere by lightning. In a temperate area the N in rain is about 6 kg ha^{-1} year^{-1} for a 75 cm level of precipitation (Hutchinson, 1957).

Overall, there are obviously fewer sources for P than for N and sedimentation is probably a more efficient remover of P than N in most aquatic ecosystems. Therefore, P should usually be less available than N and considering the potential for P recycling and loss of N through denitrification in anoxic environments, N should probably limit in highly enriched systems. Nutrient limitation will be discussed further in Chapter 6.

4.3 Sulphur

The sulphur cycle is interesting from the standpoint that several significant water-quality changes result from waste inputs and involve processes in the S cycle. However, S itself is almost never a limiting nutrient in aquatic ecosystems. The normal levels as SO_4^{2-} are more than adequate to meet plant needs. The cycle is shown in Figure 4.13 with some of the bacteria and associated processes that are responsible for the indicated (numbered) transformations.

Odorous conditions are easily created when waters are overloaded with organic waste to the point that O_2 is removed. Then SO_4^{2-} is the electron acceptor often used for the breakdown of organic matter. Step 1 in the cycle shows the production of H_2S, which has a rotten-egg smell. If NO_3^- is available, N-reducing bacteria will dominate and odours will be minimal. The production of H_2S, which is toxic, may create a problem for fish. However, it does not persist long in the presence of O_2.

Thiobacillus ferroxidans contributes to acid mine water by oxidizing FeS (ferrous sulphide), resulting in H_2SO_4 and a pH that may be as low as 1.0.

80 General concepts of aquatic ecosystems

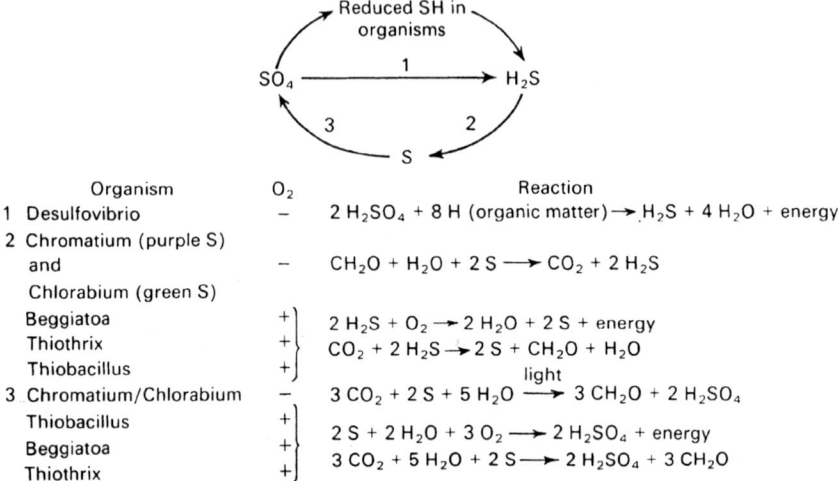

Figure 4.13 The sulphur cycle (modified from Brock, 1970) and associated bacteria-mediated reactions (Klein, 1962).

Bacterial activity may cause 80% of such acidity. The same process occurs with *T. thiooxidans* and is responsible for pipe corrosion (Brock, 1970).

Of considerable interest is the occurrence of 'plates' of sulphur bacteria in lakes (Brock, 1970), composed of either photosynthetic, facultative anaerobic purple or green sulphur bacteria or aerobic, colourless sulphur bacteria. Regardless of type, they are rather restricted to intermediate layers, usually the metalimnion. Photosynthetic bacteria need light as well as H_2S. Although H_2S may be abundant under anoxic conditions at depth, light is not abundant. The abundant light in the surface layer is of no help to these organisms because H_2S is unstable in the presence of O_2. So these organisms occur at a rather restricted depth interval (plate) where both light and H_2S occur. Therefore they are unlikely to occur in aerobic lakes.

The aerobic, non-photosynthetic bacteria have much the same problem of restriction. Because they require both O_2 and H_2S, their appearance is restricted to the interface of declining O_2 and H_2S levels, often at the bottom of the metalimnion.

SO_4^{2-} enters aquatic ecosystems through atmospheric depositions of sea salt and as a combustion product of fossil fuels, as well as through natural weathering processes in the watershed. The acidity of precipitation downwind of industrialized areas is caused primarily by H_2SO_4. While lakes are acidified by deposition of such anthropogenic SO_4^{2-}, a significant fraction of SO_4^{2-} may be consumed through step 1, resulting in a production of alkalinity, even in unproductive oxic lakes (Schindler, 1986). Apparently there is

4.4 Carbon

Carbon is the element used most frequently to study productivity and trophic transfer of energy in aquatic ecosystems. This is logical because C comprises nearly 50% of the dry organic matter in living organisms. Also, the C content tends to be more stable, per unit dry matter, than N or P, because C usually is not limiting to growth in the long term. The pertinent pools, sources and transfers for C are shown in Figure 4.14.

The sediment pool, although omitted for clarity, is a sink as in the P and N cycles. In contrast to P, the atmosphere is a source as in the case of N. However, the rate of input of CO_2 from the atmosphere into the cycle is dependent solely on a physical process: diffusion across the air-water interface as defined by Henry's Law. CO_2 is the principal form of C that is utilized by plants, although bicarbonate (HCO_3^-) is used by some species (Goldman et al., 1971). Another difference in the C cycle, compared to the N and P cycles, is that the concentration of free CO_2 is controlled predictably by pH, temperature and the total C content ($C_T = CO_2 + HCO_3^- + CO_3^{2-}$) and, because it is a gas, the diffusion rate from the atmosphere where it comprises 0.03%.

The reactions and equilibria of the CO_2 system will be described briefly in order to clarify the availability of C for plant uptake and the effect of plant photosynthesis on water quality. CO_2 diffuses into water from the atmosphere when the water is undersaturated relative to the air and from water to air when supersaturated. The amount of CO_2 in water at equilibrium with air is given by Henry's Law:

$$CO_2 = P_{CO_2} \times K_H \quad (4.1)$$

where CO_2 is expressed as $mol\, l^{-1}$, P_{CO_2} is partial pressure in atmospheres (atm) and K_H is the Henry's Law constant in $mol\, l^{-1}\, atm^{-1}$ ($10^{-1.37}$ at 25°C). CO_2 can be highly supersaturated in lakes following ice cover, especially in oligotrophic lakes where photosynthetic activity is low, reaching values of several hundred per cent saturation or more (Wright, 1983). In productive eutrophic lakes, CO_2 can by depleted to very low levels causing the pH to rise to 10 or more (Andersen, 1975).

CO_2 reacts with water to form carbonic acid, H_2CO_3, as follows:

$$CO_2 + H_2O \rightarrow H_2CO_3^* \quad (4.2)$$

where $H_2CO_3^*$ indicates the sum of aqueous CO_2 and H_2CO_3, where the latter is only $\approx 1\%$ of the total. Free aqueous CO_2 is actually calculated as $H_2CO_3^*$.

82 General concepts of aquatic ecosystems

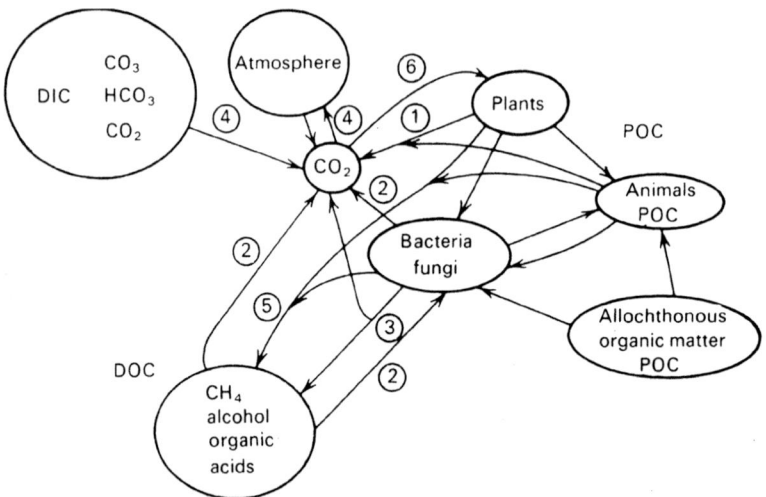

Figure 4.14 The carbon cycle with some important biomediated process.
1. This process is normal aerobic respiration: $CH_2O + O_2 \rightarrow CO_2 + H_2O$.
2. The same respiratory process is performed by microorganisms through decomposition.
3. Bacterial decomposition is also performed by anaerobic respiration. Since anaerobic metabolism is incomplete, only a portion of the carbon goes to CO_2. The remainder is contributed to the dissolved organic pool (DOC), which is slowly metabolized to CO_2. Some anaerobic processes in which a carbon by-product other than CO_2 is produced are performed by methane bacteria, which can carry on the following overall reaction (Brock, 1970):

$$C_6H_{12}O_6 + X\,H_2O \rightarrow Y\,CO_2 + 2CH_4$$

4. The principal source of free CO_2 is dissociated from the dissolved inorganic carbon (DIC) pool. This supply depends on the total inorganic carbon content, temperature, and pH.
5. Organism excretion contributes to the DOC pool.
6. CO_2 is assimilated through photosynthesis into particulate organic carbon (POC) to initiate the cycle.

H_2CO_3 dissociates in water according to the following reactions:

$$H_2CO_3^* \rightarrow H^+ + HCO_3^- \tag{4.3}$$

$$HCO_3^- \rightarrow H^+ + CO_3^{2-} \tag{4.4}$$

These reactions attain an equilibrium that is dependent on pH and temperature as follows:

$$K_1 = [H^+][HCO_3^-]/[H_2CO_3^*] \tag{4.5}$$

$$K_2 = [H^+][CO_3^{2-}]/[HCO_3^-] \tag{4.6}$$

where K_1 and K_2 are, respectively, the first and second dissociation constants for the system and $= 10^{-6.3}$ mol l^{-1} and $10^{-10.3}$ mol l^{-1} at 25°C. These equilibria are shown graphically in Figure 4.15 as a function of pH. From the equilibrium expressions above it is clear that at a pH of 6.3, half of the C_T is $H_2CO_3^*$ and half is HCO_3^- and at a pH of 10.3 the same is true for HCO_3^- and CO_3^{2-}. This is also evident in Figure 4.15 where C_T is constant and the system is closed to the atmosphere. If that were not so, $H_2CO_3^*$ would remain constant at any pH if in equilibrium with the atmosphere.

Photosynthesis and respiration are two major factors that cause a significant departure from equilibrium of the system with the atmosphere. As algae photosynthesize, depleted CO_2 can be replaced in water by the following two reactions:

$$H_2CO_3 \rightarrow CO_2 + H_2O \tag{4.7}$$

$$HCO_3^- \rightarrow CO_2 + OH^- \tag{4.8}$$

If CO_2 is replenished from the atmosphere as fast as it is removed by algae, the pH will not change, because H_2CO_3 and H^+ in the first equilibrium expression remain constant. However, pH will decrease if CO_2 production through respiration is in excess of CO_2 loss through diffusion to the

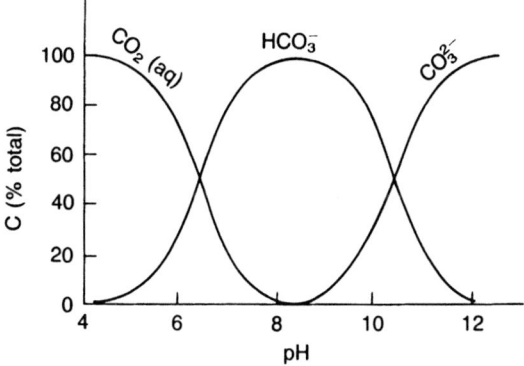

Figure 4.15 The distribution of the three forms of total C, when C_T is constant, as a function of pH (modified from Golterman, 1972).

atmosphere. That is, H_2CO_3 would increase above saturation and according to the first equilibrium expression, so would H^+.

From the first equilibrium expression, it is also clear that as algae consume CO_2 faster than it can be replaced by diffusion from the atmosphere, H^+ will decrease and, thus, pH will rise. That is because HCO_3^- can be considered to represent alkalinity at pH < 8.3, and would tend to remain constant. HCO_3^- can represent alkalinity at pH < 8.3 because as Figure 4.15 shows, CO_3^{2-} does not exist to any appreciable extent below that pH.

In reality, however, HCO_3^- is not constant as CO_2 is utilized but is dissociated to H_2CO_3, CO_2 and H_2O, as shown earlier. This, as well as the second dissociation reaction, are several times faster than necessary to match any conceivable algal demand (Goldman et al., 1971). Considering the change in HCO_3^- as pH raises, alkalinity at pH < 8.3 can be defined as:

$$\text{Alkalinity} = HCO_3^- - H^+ + OH^- \qquad (4.9)$$

Alkalinity, however, will remain constant as CO_2 (and HCO_3^-) is consumed, because as HCO_3^- and H^+ decrease, OH^- will increase in order to maintain electrical neutrality.

The CO_2 system provides a buffer against the addition of H^+ and OH^- through the first and second dissociation reactions. Thus, pH remains relatively constant so long as C_T remains constant and in equilibrium with the atmosphere. However, the effect of photosynthesis is to remove CO_2 and, thus, C_T. If CO_2 were replaced from the atmosphere as fast as algae removed it, the pH would not rise. But the mere fact that C_T is observed to decrease and pH rise is evidence that algal uptake does exceed atmospheric resupply (Schindler, 1971b; King, 1972). In this way, photosynthesis tends to be a self-limiting process; by effectively reducing the CO_2 concentration, HCO_3^- and C_T are depleted and the pH rises. As the pH rises, the free CO_2 concentration decreases and both effects can contribute to inhibition of photosynthesis.

A hypothetical diurnal pattern of the effect of photosynthesis and respiration on DO and pH, and the self-limiting effect of pH, is shown in Figure 4.16. In this example, a very high but realistic photosynthetic rate of C uptake of $3.7 \, g \, m^{-3} \, day^{-1}$ is assumed. A rate of about this magnitude was measured in Lake Trummen, Sweden before restoration (Cronberg et al., 1975). The ratio of photosynthesis to respiration was assumed to be 2/1, also reasonable. The uptake of CO_2 (photosynthetic rate) progressively decreases as the residual CO_2 concentration declines, completely ceasing as light and CO_2 extinguish at 6 p.m. At that point DO and pH reach their maximums at $\approx 175\%$ saturation and 9.0, respectively. The first dissociation constant at 25°C was used to calculate pH.

DO and pH frequently exceed such values even if photosynthetic rates are lower. This is because reaeration and respiration are not sufficient to

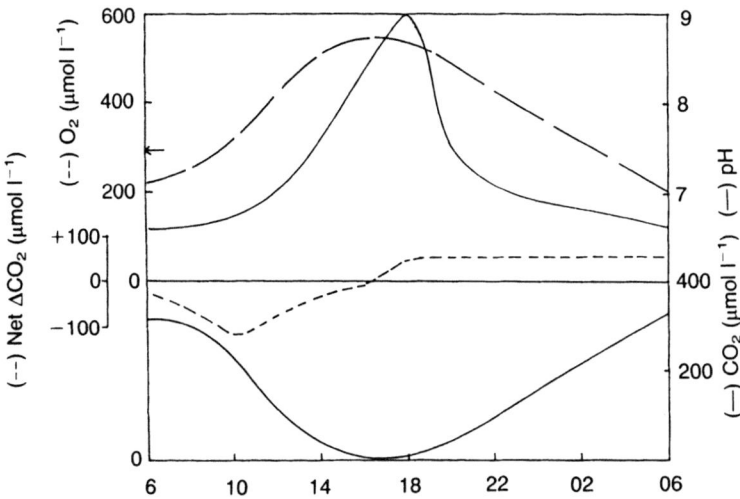

Figure 4.16 Diurnal changes in net CO_2 uptake/release, through photosynthesis/respiration, and resulting concentrations of CO_2 (aq) and O_2 and pH in a system with 30 mg l^{-1} $CaCO_3$ alkalinity (e.g. Lake Washington). No exchange with the atmosphere is assumed, as well as a 2:1 ratio of daily photosynthesis (3780 mg C m^{-3} day^{-1}) to respiration, which is not unrealistic. Photosynthetic rate is reduced when CO_2 becomes depleted. pH was calculated using $K_1 = 10^{-6.3}$, 25°C and alkalinity remains constant. Arrow shows 100% saturation.

completely replenish CO_2 and deplete DO during the night. Thus, the initial concentrations of CO_2 and DO are lower and higher, respectively, with each successive day of calm, sunny weather. As a result high pH and DO values persist for increasingly longer periods.

Chapter 5

Characteristics of pollutants

Much of the waste of civilization enters water bodies through the discharge of water-borne wastes, termed wastewater. That is, the water used by the urban population for drinking/washing or by industry for cooling/washing/processing is discharged carrying the unwanted and unrecovered substances. Wastewater can be treated or untreated and has conventionally referred primarily to that which enters water bodies at points of concentrated flow (through pipes), or point sources. Point sources include wastewater effluent from municipal and industrial sources. Although flow and pollutant loads can vary, this variability is not directly related to meteorological conditions (Novotny and Olem, 1994). Pollutants from diffuse or non-point sources enter water bodies in a diffuse manner at intermittent intervals that are generally related to meteorological events (e.g. precipitation). Diffuse sources are more difficult to monitor and to treat. Examples include urban runoff, agricultural and silvicultural runoff, flow from abandoned mines, and wet and dry atmospheric deposition over a water surface (Novotny and Olem, 1994). Point sources in the USA are regulated under the Clean Water Act of 1972 and the Water Quality Act of 1987 and have been, for the most part, effectively controlled through waste treatment processes. However, non-point sources remain a largely uncontrolled source of pollutants to surface waters.

Pollutants or pollution load are terms often used to describe wastes from civilization. Those terms imply some adverse effect and would be inappropriate in cases where wastes have no adverse, and may even have a beneficial effect. Non-native, invasive species that cause adverse environmental effects also may be considered as 'pollutants' and are commonly referred to as biopollution.

5.1 Domestic sources

Untreated domestic waste, or raw sewage, is usually greyish-brown, odiferous and relatively dilute (99% water). Those characteristics are often apparent in grossly polluted streams. There are four important constituents of domestic wastewater that are targeted for removal through treatment: total suspended solids (TSS), biochemical oxygen demand (BOD), nutrients nitrogen (N) and phosphorus (P), and pathogenic bacteria. The first four of these constituents have average concentrations (mg l^{-1}) in raw sewage of 200, 200, 40 and 10, respectively (Lager and Smith, 1975). Pathogenic bacteria are determined indirectly using the most probable number of bacteria from the coliform group as an index (5×10^7 most probable number (MPN)/100 ml in raw sewage).

BOD is an indirect determination of the readily oxidizable organic matter present. The reaction, by which organic matter is oxidized, can be described by first-order kinetics:

$$dL/dt = -kL \tag{5.1}$$

where L is BOD in mg l^{-1}, k is the reaction rate constant, and t is time. The integrated expression is:

$$L_t = L\, e^{-kt} \tag{5.2}$$

and if $y = L - L_t$, then:

$$y = L(1 - e^{-kt}) \tag{5.3}$$

so that $y = $ BOD at time t and L is the ultimate BOD (Figure 5.1). The 5-day BOD (BOD$_5$) is a common measurement of effluent strength.

Domestic wastewater usually contains high concentrations of ammonium N (15 mg N l^{-1}), because insufficient time and oxygen availability have prevented nitrification. Although carbonaceous BOD may be satisfied through

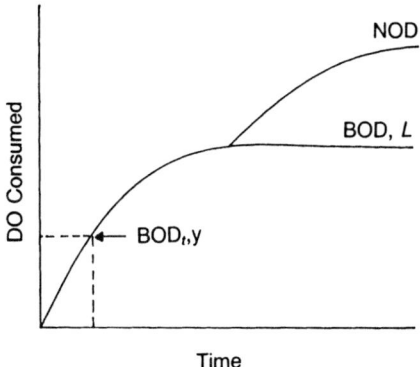

Figure 5.1 Relationship of BOD (carbonaceous DO demand) and NOD (nitrogenous DO demand) with time.

biological treatment, oxidation of the ammonium represents a nitrogenous oxygen demand (NOD) to be satisfied in the receiving water through nitrification.

Domestic wastewater can be treated by settling the heavy solids and skimming off the floatable materials. This is referred to as primary treatment and normally removes $\approx 35\%$ of the BOD, 30–50% of the TSS, but only ≈ 10–20% of the N and P (Viessman and Welty, 1985). Secondary treatment, which was mandated by the Clean Water Act of 1972, should remove at least 85% of TSS and BOD to concentrations of $30\,\text{mg}\,l^{-1}$ for each (USEPA, 1982). This is usually accomplished by the activated sludge process, which is a system providing large amounts of oxygen to actively growing bacteria ('activated sludge'), trickling filters, biological discs or aerated or unaerated lagoons (Viessman and Welty, 1985). What is accomplished with secondary treatment is the removal of dissolved and non-settleable BOD by bacterial decomposition and the finer particulate matter through chemical coagulation and clarification.

Advanced wastewater treatment is usually employed for the purpose of P removal, and sometimes N, because only ≈ 50 and 25% of these two nutrients, respectively, are removed through the primary and secondary processes (USEPA, 1982). Also, a large fraction of TN and TP is soluble. P removal can be efficiently accomplished by chemical precipitation with alum (aluminium sulphate), iron salts (e.g. $FeCl_3$), or lime. These processes are often $>90\%$ efficient, producing residual P concentrations of $\approx 0.1\,\text{mg}\,l^{-1}$ ($100\,\mu\text{g}\,l^{-1}$), which, unless sufficiently diluted, is still a rather high inflow concentration to a lake (see Chapter 7). N can be removed by biological denitrification, ammonia stripping, breakpoint chlorination or ion exchange (Viessman and Welty, 1985).

Coliform organisms are usually removed by chlorination, which can be employed after either primary or secondary treatment. This treatment is referred to as disinfection, which could be accomplished with other oxidants, such as ozone. The addition of free chlorine dissociates into molecular hypochlorite (HOCl) and the hypochlorite ion (OCl^-), which represent available free chlorine. Chlorine is consumed by organic matter, so dose is adjusted to maintain a free chlorine residual of $0.2\,\text{mg}\,l^{-1}$ and should remove coliforms to less than 200 MPN/100 ml (Viessman and Welty, 1985). Free chlorine reacts with ammonia to produce chloramine, which is also toxic, and together are referred to as total residual chlorine ($TRCl_2$). The problems caused by $TRCl_2$, if not removed by dechlorinating domestic wastewater effluents, will be discussed under toxicants and fish.

Domestic wastewater is also treated by septic-tank drainfields, or 'on-site' systems, in suburban and rural areas where sewer collection systems are not available. If the drainfield is not overloaded with wastewater, so that it clogs with organic matter, and if composed of soils that allow sufficient percolation, treatment (including P removal) can be highly effective (see Section 7.3.7).

5.2 Urban runoff

Runoff from urban watersheds is usually transported in what are called 'separated systems' (separated from sewage) or 'combined systems' (combined with sewage). During and following storms, stormwater flows and combined sewer overflows (CSOs) can contribute the same constituents as in sewage and in relatively high concentrations as well. Table 5.1 shows a comparison of the concentration and yield of the important constituents in raw sewage effluent and urban runoff from a typical urban area.

The interesting points about these average values are: the concentration and yield for TSS from stormwater runoff and sewage are similar. Streets and parking lots, etc. collect large amounts of solid matter, which washes off to streams and lakes; and BOD in urban runoff is much higher than in runoff from pristine areas, but much less than in raw sewage. Higher average concentrations for stormwater have also been given (TSS = 630 and BOD = 30 mg l^{-1}; USEPA, 1982). The important point is that streams can be composed entirely of stormwater following rains with BOD concentrations that would be the same as if a raw sewage effluent entered the stream with only a 6- to 12-fold dilution (200:30, 200:17). A stream BOD of 20 mg l^{-1} was observed during intermediate flow conditions in Lytle Creek, Ohio, in which gross biological effects were described (Chapter 12).

While TN and TP concentrations are much higher in runoff water than in runoff from pristine watersheds, they are less than in raw sewage. Metal (Zn, Cu and Pb) concentrations and areal yields (1 m runoff assumed), on the other hand, are often higher in runoff than in sewage effluent. That could vary, however, depending on how much industrial waste enters the municipal system. Much of the nutrients (N and P) come from dirt washed off vehicles or moved by wind, animal faeces and fertilizers applied to lawns. Sources of nutrients will be discussed further in Chapter 10.

Table 5.1 Comparison of the contributions of common constituents in sewage effluent between a natural pristine area and urban area of 1 km^2 with 2500 people (from Weibel, 1969; metal data from METRO, 1987)

Constituent	Concentration (mg l^{-1})			Yield (Metric tons km^{-2} yr^{-1})	
	Natural	Sewage	Runoff	Sewage	Runoff
TSS	0.8	200	227	68	64
BOD	1.0	200	17	68	5
TN	0.5	40	3.1	14	1
TP	0.02	10	0.4	3.5	0.1
Zn	0.002	0.16	0.3	0.06	0.3
Cu	0.002	0.11	0.2	0.04	0.2
Pb	0.002	0.04	0.7	0.01	0.7

The metals in urban runoff are sorbed to particles and come from corrosion of vehicles and piping systems, as well as from atmospheric emissions. Lead in runoff comes from gasoline and has declined since the advent of lead-free gasoline in the early 1970s. A large fraction of these metals are complexed (tied up) with particulate matter and are not in a form normally considered toxic.

Stormwater runoff can have rather high concentrations of coliforms (4×10^5 MPN/100 ml), although not as high as CSOs (5×10^6 MPN/100 ml; USEPA, 1982).

Urbanization results in an increase in the total impervious area (TIA) of a watershed due to the replacement of natural vegetation with hard surfaces such as parking lots, houses and roads. These impervious areas reduce infiltration of precipitation and increase the amount of surface water runoff that flows overland to water bodies. The effects of urbanization include modification of the natural hydrologic regime, increased pollutant loading and habitat degradation, all of which adversely affect stream quality and aquatic life. The altered hydrologic regime typically has higher and more frequent peak flows due to increased runoff and lower base flows due to loss of watershed wetlands. Higher peak flows also increase streambed scour, bank erosion and tributary incision.

In a study of the effects of urbanization on 22 lowland streams (120 reaches) in the Puget Sound basin, May *et al.* (1997) found that there was a measurable, significant decline in stream quality at a relatively low threshold of development (5–10% TIA). Coho salmon rearing habitat was significantly reduced as indicated by decreased large wood debris, lower intragravel DO, increased sediment in spawning gravel and declining water quality in the highly urbanized basins (%TIA > 45%). Biological integrity, as measured by the benthic invertebrate community, also declined as urbanization increased (Kleindl, 1995; see Section 12.10.2).

5.3 Industrial wastes

Wastewater from industrial processes are varied and complex often containing compounds not found in nature. Industrial wastewaters are often highly discoloured, turbid, alkaline or acid and unique to the industry in question. Wastewaters from some of the more common industries will be briefly described.

5.3.1 Pulping wastes

There are three pulping processes; Kraft, sulphite and ground wood. The magnitude of the waste problem from pulping is obvious from the fact that cellulose is the desirable commodity and cellulose comprises only about half of the tree. The structural portion of the tree, or the lignin, is separated from the cellulose by digestion with chemical additives. Alkaline solutions of sodium sulphate are used in the Kraft process, while acid solutions of sodium bisulphite are used in the sulphite process. After cooking, the pulp is washed, bleached and pressed. Each of those steps has a waste stream.

Sulphite plants produce very high BODs; effluents with 10–14% solids can have a BOD_5 of $30\,000\,mg\,l^{-1}$. Kraft effluents have lower BODs, but yield more odiferous gases to the atmosphere, such as sulphides and mercaptans, and higher suspended solids to wastewater. Toxicity of pulping wastes has been associated with the resin acid component, but more recently dioxin has been found in bleach plant wastewater.

5.3.2 Petroleum refinery wastes

There are several operations at refineries, such as oil storage, desalting and fractionation of crude oil, thermal and catalytic cracking, solvent refining, dewaxing, deasphalting, and wax and hydrogen manufacture. Each of these can produce wastes that have BOD, toxicity and/or produce taste/odour problems, due to the dissolved organics. Phenols are an important constituent in refinery wastewater. In addition, spills of fuel and crude oil from tankers at sea represent a substantial waste source from the petroleum industry.

5.3.3 Food processing wastes

This industry is also varied, but generally the wastewater consists primarily of TSS and BOD. Such industries as fruit and vegetable canneries, meal packing plants, fish canneries, dairies and breweries produce wastewaters with very high BODs, often ten times that of domestic sewage.

5.3.4 Mining wastes

Acid mine drainage and leaching from abandoned dredge spoils often produce 'wastewater' with high concentrations of toxic metal ions, such as Cu, Zn and Fe. The problem of acid mine drainage develops because the mined-out portions of coal veins fill with water, which becomes anoxic producing reduced iron and sulphur compounds (FeS). When the water drains from the veins it is aerated, promoting the biological oxidation of iron and sulphur, yielding sulphuric acid. The metal ions are more soluble in the acid water and hence become more toxic. There are many abandoned, uncontrolled dredge-spoil deposits that continually leach acidity and toxic metals to streams. One of these cases is discussed in Chapter 12.

5.3.5 Acid precipitation

In a sense, acid precipitation is analogous to acid mine drainage and urban stormwater runoff, except that this acid runoff creates the most serious effects in high elevation, remote areas, rather than in urban settings. That is because those areas have less capacity to neutralize the acidity. The acid precipitation is caused by strong acids (nitric and sulphuric) that originate as sulphur dioxide from primarily fossil-fuel burning power plants and nitrogen oxides from largely auto emissions that are later oxidized in atmospheric clouds. Although NO_3^- may be nearly as important as SO_4^{2-} in causing a low pH in rain, SO_4^{2-} is considered more significant at acidifying surface waters because NO_3^- is usually a limiting nutrient in forested areas and is thus neutralized through biological uptake and growth. Although the equilibrium pH of pure water in contact with the atmosphere is about 5.6, precipitation is not considered acidified unless the pH is <5, because of the effect of natural background SO_4^{2-} (Charlson and Rodhe, 1982).

5.3.6 Toxic wastes

There are highly toxic, refractory organic compounds that have entered the environment via leachate from solid-waste landfills, direct discharge to surface water, atmospheric emissions that are deposited over broad areas, direct applications to control pests, and accidental spills. Some representatives of this group include polychlorinated biphenyls (PCBs), polybrominated biphenyls (PBBs), chlorinated hydrocarbon insecticides, and dioxins, which are chlorinated organic compounds. Many of these compounds were used because they continued to be effective over a long period of time, thereby making them economical. For the same reason, they are especially harmful if released into the environment. Use of PCBs and some chlorinated organic insecticides is banned in the USA, but because of their long life there are high concentrations still remaining in landfills, sediments of lakes and rivers and in the flesh of animals.

PCBs were used extensively as an insulating material in transformers and capacitors, primarily in the electrical industry. They have also been used in paint solvents and plastics. Concentrated residues still remain in landfills and sediments (see Section 13.5). PBBs were used in a fire retardant. Chlorinated hydrocarbon insecticides, of which DDT is the well-known example (including endrin, aldrin, heptachlor and dieldrin), were used extensively in the USA until the 1970s, when sale was banned. However, some use of residual stocks continued thereafter. They are still used in developing countries and often returned to the USA in imported food (Revelle and Revelle, 1988). Many insecticide substitutes of shorter longevity have been developed, such as the pyrethrins. Dioxins are a group of chlorinated hydrocarbons that can be formed through combustion and were a byproduct in the manufacture of 2,4,5-T, a brush control herbicide used in the Vietnam War.

5.4 Agricultural wastes

The wastewater from agriculture is more varied and extensive than any of the above types. Cultivation results in increased erosion and high TSS loads in receiving waters. Along with the TSS come nutrients and pesticides that were applied for fertilization and insect control. The increased waste loads can also occur in forested areas through timber harvesting, which is in a sense agriculture. BOD/TSS wastes from livestock enclosures can be highly concentrated. Return flows from rill (or flood) irrigation can carry high concentrations of TSS, fertilizers and pesticides.

There are many attempts underway to control the wastewater load from agriculture through best management practices (BMPs). These involve minimizing erosion through improved cultivation and harvesting practices, treatment of wastewater from livestock enclosures, minimizing fertilizer applications through soil-testing procedures and conversion to spray irrigation, which yields much less TSS in the runoff water because return flow ditches are eliminated.

Agricultural non-point source control is implemented in the USA mostly through voluntary programs that primarily use BMPs. The effectiveness of BMPs is extremely variable and may not have a direct benefit to the land or water adjacent to where they are applied (Novotny and Olem, 1994). Pollution in agricultural runoff (sediment, N, P, pesticides, bacteria, pathogens) and the physical changes that occur in or adjacent to riparian areas are responsible for much of the non-attainment of water quality standards and associated beneficial uses reported by the states (USEPA, 2002b).

5.5 Biological pollution

Humans have introduced species to ecosystems where they do not naturally occur for thousands of years. Some of these introductions were intentional and others were accidental. Introductions of non-native (also referred to as 'exotic') species have been beneficial to humans, in the case of both crops and livestock, but other introduced species have had negative effects. Adverse consequences of the introduction of species include destruction of habitat, extirpation of native species, loss of biodiversity, spread of diseases, alteration of ecosystem processes and economic costs due to degradation of water resources, crops, rangelands and forests. An introduced species that has negative environmental consequences can be considered a biological pollutant.

Characteristics of invasive species include high reproductive rates, high genetic and phenotypic variability, and habitat generalism with broad dietary requirements (Elton, 1958; Williamson, 1997; Kolar and Lodge, 2001). They are often associated with humans, taking advantage of habitats disturbed by people or early successional habitats with low species diversity or no predators (Moyle and Light, 1996). Not all species that are introduced to a non-native habitat will become established or invasive. The 'Tens Rule' proposed by Williamson (1997) is that approximately 10% of introduced species become established and that 10% of these established non-natives become pests. However, once established, eradicating an invasive species is almost impossible.

Biotic homogenization refers to the process by which regional differences among floras and faunas are reduced by the establishment of exotic species and loss of native species. Biological invasions, and subsequent homogenization of earth's biota, are accelerating primarily due to an increasingly mobile human society and increased international trade, and are an underappreciated aspect of global environmental change (Vitousek et al., 1996).

Islands are generally more vulnerable to invading species than are continents (Elton, 1958). Because lakes are like island habitats, they are also highly susceptible to invasion by non-native species (Magnuson, 1976). However, invasions are becoming increasingly widespread on continents. Although islands tend to have a greater proportion of non-native species (often as high as 50% of the total number), non-native species may comprise from a few per cent to more than 20% of the total species on continents (Vitousek et al., 1996, 1997). More than 1500 species of invasive non-native plants are established in the USA, Canada and Australia. European countries support several hundred invasive non-native plants (Vitousek et al., 1996). Worldwide, most regions are estimated to contain 10–30% non-native species. The Great Lakes of the USA alone have at least 139 non-native species. Hundreds of species have been introduced in aquatic systems in North America (Benson, 2000). The sources of most of these introductions include ballast water, aquaculture, aquarium trade, sportfishing and nurseries. Ballast water is likely to remain a primary source for future unintentional introductions (Benson, 2000).

Efforts to prevent and control the spread of invasive, non-native species include regulations, public education and surveillance to detect and eradicate new invaders. Since 1997, all ships that enter USA ports are required to empty their ballast tanks at sea and refill them with sea water in the hope of stemming the tide of aquatic invasions. The legal framework in the USA incudes statutes such as the Lacey Act of 1900, which authorizes the Secretary of the Interior to regulate the introduction of birds and animals in areas where they had not existed and prevents the importation of injurious plants or animals, and the Federal Noxious Weed Act of 1974, which seeks to prevent the introduction and spread of noxious weeds. More recently, the Non-indigenous Aquatic Nuisance Prevention and Control Act of 1990 mandates the development and implementation of a comprehensive national program to prevent and respond to problems caused by unintentional introductions of non-indigenous aquatic species into the waters of the USA. State and local noxious weed boards also have enacted laws and regulations that address the effects of invasive species and have implemented public education programs to increase public awareness of the problem. The successful management of established invasive species depends on understanding the processes by which the species is controlled in the ecosystem (Mooney and Drake, 1986; Mack et al., 2000). While eradication may be impossible, control of the spread of an invasive species may reduce its ecological and economic costs.

Part II

Effects of pollutants in standing water

Standing bodies of freshwater, namely lakes, ponds and reservoirs, are known as lentic environments. Although standing or lentic in nature, there is movement of water within their basins that is determined by morphometry (depth, shape, etc.), solar heat input and wind. Plankton algae, or phytoplankton, are at the mercy of water movement and are usually the most important producers in lentic water bodies. The extent to which water moves (is mixed) vertically and horizontally determines the distribution of plankton and the amount of light available to mixed plankton cells, so water movement indirectly determines the productivity of lentic water bodies.

The gradation among lakes, reservoirs, rivers and streams is large and the distinction is not always clear-cut physically. The principal factor separating these environments is water residence time. Algal abundance per unit phosphorus in the water was found to increase along a residence time gradient from rivers to impoundments to natural lakes (Soballe and Kimmel, 1987). However, large, slow-moving rivers and the heads of reservoirs and estuaries may have significant plankton populations even though their residence times are relatively short. The effect of residence time on the dominant producer organisms present is illustrated in Chapter 11.

While the emphasis here is on the ecological principles and effects of nutrients in lakes, those principles and effects apply to reservoirs as well. However, reservoirs generally have higher flushing rates (shorter residence times) and consequently higher nutrient loading than lakes (Walker, 1981). Main stem reservoirs usually show a zonal transition from a riverine to a lacustrine type environment (Kimmel and Groeger, 1984). Estuaries show similar transitions from freshwater to marine, but the boundaries can change diurnally as a result of tidal effects.

Lakes/reservoirs are affected by sediment, toxicant and BOD wastes, but dilution capacity is usually sufficient to avoid violations of standards for DO, pH and temperature, except in enclosed embayments near wastewater sources. Chlorinated organic and/or metal contamination of fish and sediments are serious problems in such water bodies as the Great Lakes (see Chapter 13), Puget Sound (Washington state) and the Baltic Sea (see Figure

3.13). However, effects of and controls for these types of waste have typically occurred in rivers. The degradation of whole lakes/reservoirs is more likely to occur from eutrophication, the causes, effects and controls for which are emphasized in Chapters 6–10, or from acidification, which is treated with toxicity in Chapter 13. Global warming may affect whole watersheds, through both thermal restriction of population distribution and hydrologic modification.

Chapter 6
Phytoplankton

The phytoplankton of lakes, estuaries and oceans is composed of single-celled algae. Cells of the plankton algae are largely microscopic in size ranging from a few microns to a few hundred microns in the longest dimension. Species may occur as single cells, or as colonies or filaments composed of many cells. The colonies/filaments of some species are often visible to the naked eye. For example, spherical colonies of the cyanobacterium *Gloeotrichia* are ≈ 1 mm in diameter and clearly visible. The filamentous cyanobacterium *Aphanizomenon* forms bundles of thichomes (filaments) that appear as grass clippings in the water. The phytoplankton are sometimes grouped by size into ultra-, nano- and net plankton (those caught with a net), with respective separations at about 10 and 50 µm (Wetzel, 1983). Phytoplankton are usually quantified by microscopic analysis of preserved water samples, rather than net samples, because of the significance of nanoplankton.

The taxonomic orders representing the major portion of the phytoplankton are the Chlorophyta (green algae), Chrysophyta (diatoms and yellow-green algae), Cyanobacteria (blue-green bacteria), Pyrrhophyta (dinoflagellates), Euglenophyta (euglenoids) and Cryptophyta (cryptomonads), although groupings may vary among authors. For general and taxonomic characteristics see Prescott (1954), Hutchinson (1967) or Wetzel (1983).

Cyanobacteria are of particular interest due to their effects on water quality and aquatic habitat. Cyanobacteria are microscopic unicellular, colonial or filamentous bacteria that primarily occur in freshwater systems. Because they share some characteristics of algae (e.g. cell wall structure, photosynthetic pigments) and conduct oxygenic photosynthesis, they are commonly (but inappropriately) referred to as blue-green algae. The term 'cyanobacteria' or, more simply, 'blue-greens' is used in this text. Many cyanobacteria are planktonic and are thus considered part of the phytoplankton while others are benthic, living attached to sediment or other substrata as part of the periphyton.

As discussed in later sections of this chapter, mass surface accumulations of cyanobacteria (often referred to as scums or blooms) in aquatic systems are primarily attributed to nutrient, particularly P and N, enrichment. Other environmental factors that promote cyanobacteria include high water temperatures, a stable water column, low light availability, high pH, low dissolved carbon dioxide (CO_2), low grazing pressure and low total nitrogen to total phosphorus (TN:TP) ratios (e.g. Reynolds, 1987; Paerl, 1988; Shapiro, 1990; Hyenstrand et al., 1998). High densities of cyanobacteria can cause unsightly surface scums, decreased water column transparency, unpalatable drinking water (i.e. taste and odour), toxic compounds and noxious odours. The decomposition of cyanobacteria can deplete dissolved oxygen and cause fish kills. Such problems can severely limit aquatic habitat, recreational activities, fisheries and use of a water body as a drinking water supply.

Most phytoplankton are maintained in the water column by turbulence caused by the wind, hence the term plankton. Many have irregular shapes hat increase their surface to volume ratio and accordingly decrease density, so as to resist sinking. However, except for cyanobacteria, which contain gas vesicles that render them buoyant, and some flagellated notile species, most will readily sink to the sediments under quiescent conditions. For example, *in situ* experiments conducted in plastic water columns have resulted in increased sinking loss rates because they are isolated from the normal water column turbulence. As the cells senesce they become denser, which increases their loss through sinking.

The three determinations usually made in characterizing the phytoplankton in standing waters are productivity, biomass and species composition. How much is being produced, how much is there and what is it? Productivity and biomass can be determined by chemical or physical procedures (Vollenweider, 1969b) and related to environmental factors without necessarily knowing the major organisms responsible. Although productivity and/or biomass may be sufficient to characterize the phytoplankton in certain situations, as will be seen later in this chapter, the taxa (species, genera, etc.) comprising most of the biomass may be equally important to water quality in many cases. In the same sense, the trophic state of a lake can be defined quantitatively, based on measurements of productivity and biomass (see Chapter 7). However, quantitative criteria for species composition have not been defined for that purpose.

Although detailed descriptions of seasonal changes in phytoplankton taxa, biomass and productivity in a particular lake are interesting and useful, it is usually more appropriate for water-quality management to know how the average state of lakes in general changes with an increasing or decreasing load of nutrient, acidity or sediment. That is, effort expended on developing relationships between substance input and average response, based on studies of large populations of lakes (Vollenweider, 1969a, 1976; Dillon and Rigler, 1974a; Jones and Bachmann, 1976; Chapra and Reckhow, 1979; Smith, 1982, 1986, 1990a,b; Prairie et al., 1989), has proven more valuable to lake management than if that effort had been expended on understanding the more detailed interactions within a given lake. Also, the value of whole-lake manipulations to limnology and lake management is apparent. Although small-container experiments (bottles, bags, etc.) give useful indications of lake response, they are largely inadequate because of the many factors operating in a whole-lake ecosystem that interact to buffer or magnify the response. Therefore, deliberate or accidental manipulations of whole-lake systems (Edmondson, 1972; Björk, 1974; Schindler, 1974; Edmondson and Litt, 1982) have presented more reliable information in that regard. Nevertheless, small-scale experiments are usually necessary to understand the underlying mechanisms that combine to produce the overall response.

6.1 Seasonal pattern

The general seasonal pattern shown for a northern temperate sea (Figure 6.1) is quite similar to that for a large temperate monomictic or dimictic lake. The nutrient content is normally high following autumn overturn and may remain high during winter in a monomictic lake; it may increase during spring overturn in a dimictic lake having declined some during winter ice cover. Both situations provide a large available nutrient supply for phytoplankton at a time when light is increasing. The large nutrient supply exists principally because there is little or no growth to remove nutrients during winter when light and temperature are low. A spring diatom 'outburst', or 'bloom', usually occurs when light intensity reaches a level so that gross photosynthesis exceeds respiration. Species with low temperature requirements are usually responsible for the spring outburst.

The spring bloom is followed by a mid-summer minimum in algal biomass and productivity, largely because nutrients have reached a low

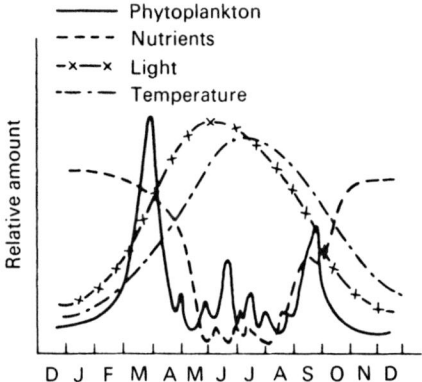

Figure 6.1 General seasonal cycle in north temperate oceanic phytoplankton and ecological factors (modified from Raymont, 1963).

and production-limiting level. The controlling effect of nutrients is particularly evident because light and temperature are at their maximums, thus the production potential is high. Grazing by herbivorous zooplankton is also occurring and may represent a significant loss of cells in addition to sedimentation.

In autumn, the significance of nutrients is again readily apparent because increased mixing, which results from lowered surface temperatures, entrains regenerated nutrients into the trophogenic (productive) zone. More nutrients from depth may become available during late summer in lakes because of the continual deepening of the mixed layer. A fall outburst of diatoms often results in the temperate ocean, as is also the case for poorly enriched freshwater lakes. In highly enriched temperate lakes, however, blooms of cyanobacteria tend to develop in late summer and autumn or may occur throughout the summer, especially in shallow unstratified lakes. Seasonal cycles of stratification and algae also occur in tropical lakes, although less pronounced than in their temperate counterparts (Thornton, 1987a).

6.2 Population growth kinetics and concept of limitation

Growth of a population of microorganisms can be described as exponential if resources are unlimited as shown in Chapter 2. This unlimited growth can be represented by a first-order expression where the rate of increase in the population, dX/dt, which is also net productivity, is dependent on the size of the population, X, and the growth rate constant, μ, is symbolized by:

$$dX/dt = \mu X \tag{6.1}$$

The same relationship of population size with time shown in Figure 2.3 holds here; only the symbols for biomass and growth rate have changed to conform with those commonly used to describe growth of microorganisms.

The growth rate, μ, is controlled by the concentration of most limiting nutrient at any time, as well as other environmental factors. The control exerted on growth by a limiting nutrient can be described by the Michaelis–Menten relationship developed from enzyme kinetics; bacterial growth being dependent of the concentration of an organic substrate and algal growth on that of a limiting inorganic nutrient (Herbert et al., 1956; Droop, 1973). The equation for growth rate of a microorganism limited by a single nutrient is:

$$\mu = \mu_{max} N/(K_N + N) \tag{6.2}$$

where μ_{max} is the maximum rate attainable for that population under optimal environmental conditions and is determined by its genetic character, N is the concentration of limiting nutrient, or in the case of bacterial growth it

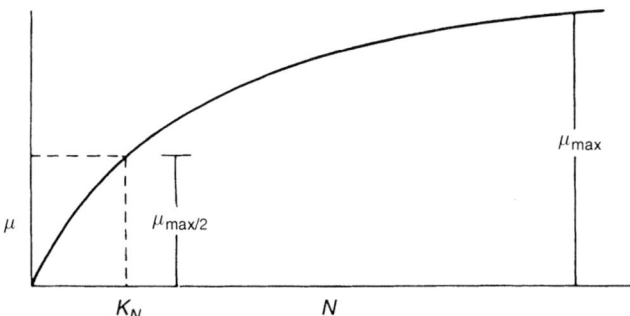

Figure 6.2 A Michaelis–Menton type relationship between microorganism growth rate and the concentration of limiting nutrient.

is indicated by S for substrate, and K_N (or K_S) is the half-saturation constant or the nutrient (or substrate) concentration at one half the maximum rate of growth. This relationship, and a definition of the parameters, is shown in Figure 6.2. The expression shows that at very high concentrations of nutrient and a relatively low K_N, growth rate approaches the maximum. As N decreases, especially when it drops below K_N, μ becomes increasingly dependent on N.

Substituting Equation 6.2 into the first-order population growth Equation 6.1 gives the following expression for the change in population size controlled by a limiting nutrient:

$$dX/dt = \mu_{max} X N/(K_N + N) \qquad (6.3)$$

6.2.1 Mixed reactor

To understand better how population growth is controlled and the relationship of biomass to nutrient concentration, growth in a continuously stirred tank reactor (CSTR) or chemostat will be considered. If algae are grown in such a mixed reactor, there will be an inflow of water and dissolved nutrient (N_i) and an outflow of nutrient (N), water and algal cells (X). The population growth in the reactor is given by:

$$dX/dt = \mu_{max} X N/(K_N + N) - DX \qquad (6.4)$$

where D is the dilution rate determined by the inflow rate of water/reactor volume. That parameter is also known as the water exchange rate or flushing rate when referring to lakes. When nutrient-rich water is added to a algal-seeded chemostat, the population growth and loss begin, the loss being the fraction of reactor volume replaced per unit time (D) multiplied by the concentration of algal biomass in the reactor (X). The population will continue to increase faster than cells are lost through the outlet until the concentration of cells (X) becomes large enough that the loss equals the growth. At that point, $dX/dt = 0$ and a steady-state condition is reached. The population will continue to grow and maintain a steady-state biomass so long as the inflow rate of water and inflow nutrient concentration (N_i) remain constant. The dynamics of phytoplankton growth in lakes can begin to be understood by considering further the steady-state growth in a chemostat.

The change in nutrient concentration (N) in the reactor with time is given by:

$$dN/dt = DN_i - DN - [\mu_{max} X/YN/(K_N + N)] \qquad (6.5)$$

where Y is the yield coefficient or the mass of cells per unit mass of nutrient. The equation simply says that the rate of change of N in the reactor is equal to the difference between the inflow rate of nutrient and the uptake rate by algae. To illustrate the steady-state phenomenon, the calculated response of algal biomass and nutrient concentration in a reactor, starting from initial conditions of $N_i = 5.0\,\text{mg}\,l^{-1}$, $X = 5.0\,\text{mg}\,l^{-1}$, $D = 0.5\,\text{day}^{-1}$, $\mu_{max} = 1.0\,\text{day}^{-1}$, $Y = 64$ and $K_N = 0.3\,\text{mg}\,l^{-1}$, is shown in Figure 6.3. N is observed to increase initially as the inflow of nutrient exceeds the removal rate by the algae. But as the algal biomass increases the removal rate of N increases and the concentration falls. The system reaches a steady state for biomass and nutrient after about 12 days.

The steady-state result of Equation 6.5 for N in a reactor is:

$$N = K_N D/(\mu_{max} - D) \qquad (6.6)$$

This equation shows that at a very low dilution rate there is ample time for algal uptake to reduce N to low levels. As D increases, however, there is less

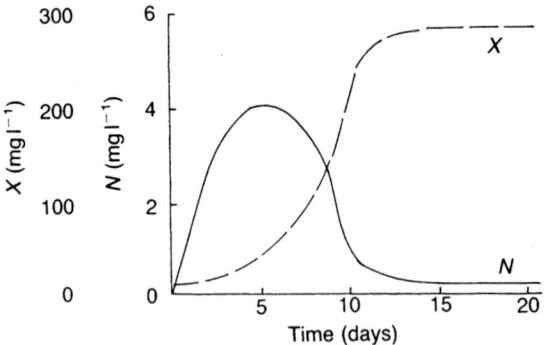

Figure 6.3 Calculated response of microorganism biomass (X) to nutrient content (N) in a mixed reactor over time.

and less time for utilization of the incoming nutrient and the residual N in the reactor increases. At the point where $D = 0.5\,\mu_{max}$, $N = K_N$. When $D > \mu_{max}$, the loss exceeds the growth, washout occurs and $N = N_i$.

The steady-state biomass is determined by the difference between the inflow (N_i) and reactor (N) concentrations of nutrient:

$$X = Y[N_i - K_N D/(\mu_{max} - D)] \qquad (6.7)$$

or, substituting from Equation 6.6:

$$X = Y(N_i - N) \qquad (6.8)$$

Multiplying both sides by the dilution rate one has steady-state productivity as follows:

$$DX = DY(N_i - N) \qquad (6.9)$$

The expected response of steady-state biomass, residual nutrient and productivity to changing dilution rate is shown in Figure 6.4. The results in Figure 6.4 are easily explained with the aid of Equations 6.6, 6.8 and 6.9. The residual nutrient concentration (N) is independent of the inflow concentration, but is controlled by the dilution rate. This is because algal cells rapidly take up any increase in nutrient that results from continually adding the nutrient-rich inflow and turn it into biomass. N begins to increase only when D increases to a point where there is insufficient time for the algae to absorb the difference between N_i and N. According to Equation 6.6, this increasingly happens when D exceeds $0.5\,\mu_{max}$. As washout is approached, N approaches N_i. This illustrates why the soluble form of the limiting nutrient is usually in very low concentration in lakes.

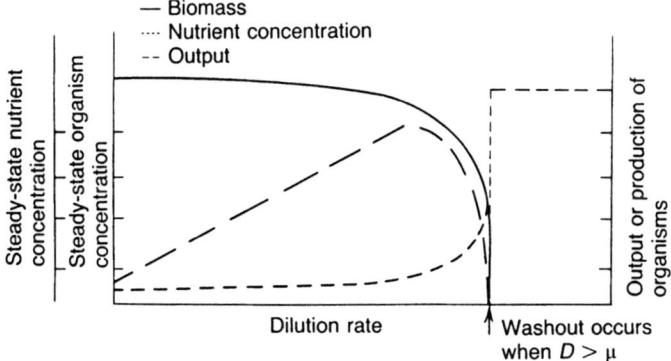

Figure 6.4 Dilution rate versus biomass, production and nutrient concentration in a chemostat culture (Herbert et al., 1956).

Biomass in Figure 6.4 remains rather constant over the lower range of dilution rate; again, dilution rate is slow enough to permit near complete removal of nutrient and biomass is simply proportional to the difference between inflow and in-reactor concentrations of nutrient (Equation 6.8). As D increases and μ_{max} is approached, the utilization of N is less complete and biomass declines with washout occurring when $D > \mu_{max}$. If the dilution rate is held constant and the inflow nutrient concentration is increased, the time for uptake is the same so the added nutrient is simply converted to biomass and the latter increases proportionately. This process has great significance for managing the biomass of algae in lakes; all other loss rates being constant, biomass should be a function of the inflow concentration of available (soluble) nutrient.

As Equation 6.9 shows, productivity or output of algal cells is simply the product of dilution rate and biomass in the reactor. Hence, productivity increases in proportion to dilution rate until the loss rate through dilution begins to approach washout. Productivity, or efficiency of the system to convert nutrients or substrate, is more efficient at high dilution rates than the simple product of X and D may indicate.

Example problem

Calculate the steady-state biomass in a continuous flow culture of algae for each of three dilution rates ($D = 0.3, 0.6$ and $0.9 \, day^{-1}$) at each of three inflow concentrations of limiting nutrient ($N_i = 2.0$, 4.0 and $6.0 \, mg\,l^{-1}$). Assume $\mu_{max} = 1.0 \, day^{-1}$, $K_N = 0.3 \, day^{-1}$ and a yield ratio $= 64$. When is biomass most dependent on N_i and when on D?

Using Equation 6.7, because N varies with D and has some impact on X:

$$X = Y[N_i - K_N D/(\mu_{max} - D)]$$

gives the following X values:

	D		
N_i	0.3	0.6	0.9
2.0	120	99	−45
4.0	248	227	83
6.0	376	355	211

At D values substantially below μ_{max}, when washout is not occurring, X is linearly dependent on N_i (at low D, N is reduced to very low levels). As D approaches μ_{max} (near washout), this linear relationship ceases to exist and D has a larger control on X (and residual N increases, since time is insufficient for complete uptake).

Changes in other variables as dilution rate increases are also of interest. The growth rate, μ, also increases as D increases. This is because residual N increases allowing a higher growth rate (Equation 6.2). Growth rate (μ) must increase as D increases because at steady state $(\mathrm{d}X/\mathrm{d}t = 0), D = \mu$. This is obvious from the following expression for constant biomass (X):

$$\mathrm{d}X/\mathrm{d}t = \mu X - DX \tag{6.10}$$

The dependence of μ on D is apparent from the Michaelis–Menten relationship. As D increases, N increases, especially after $D > 0.5\,\mu_{\max}$ (see Equation 6.6), and the $\Delta\mu/\Delta N$ decreases. At low D ($D < 0.5\,\mu_{\max}$), $\Delta\mu/\Delta N$ is large, representing the steeper slope in the Michaelis–Menten relation.

In summary, an increase in limiting nutrient concentration in the inflow increases biomass and production, whereas an increase in the dilution rate only increases production. Environments where dilution or flushing rate can have a large influence on biomass and soluble nutrient control are small lakes or bays with large watersheds or large volume tributary streams, constricted estuaries and sewage lagoons. These are ecosystems with detention times of the order of 10 days or less, or expressed as dilution rate, 0.1 day^{-1} or more. For example, Soballe and Threlkeld (1985) observed that plankton algal biomass varied directly with detention time in a small reservoir with a mean detention time of 2.4 days (0.4 day$^{-1} = \rho$) over a range up to 12 days detention time (0.08 day^{-1}). Advection controlled algal biomass, because nutrient content was high, supporting a relatively high growth rate.

The concept of continuous-culture kinetics has been employed to predict the algal biomass leaving reservoirs with detention times less than two weeks (Pridmore and McBride, 1984; McBride and Pridmore, 1988). A continuous culture, logistic model, in which growth rate was restricted by a chlorophyll a (chl a) maximum based on total P concentration, gave very reliable predictions of steady state, growing season chl a concentrations at the outlet of reservoirs with detention times of 1.3 and 5.6 days. In the short detention time reservoir, outflow chl a was only $\approx 14\,\mu\mathrm{g\,l}^{-1}$, while the TP content of $48\,\mu\mathrm{g\,l}^{-1}$ would have predicted a chl a of $39\,\mu\mathrm{g\,l}^{-1}$, clearly a situation where detention time was most limiting. Their model predicted a chl a of $\approx 15\,\mu\mathrm{g\,l}^{-1}$.

6.2.2 Batch reactor

Population growth in ecosystems with longer detention times is more typical of growth in batch cultures or closed containers. For this system, the dynamic equations for biomass (Equations 6.1–6.3) still apply, but because there is not a constant loss of biomass, a steady state cannot be reached. The batch culture is more typical of a natural system in which biomass and nutrient concentration continually change. Nutrient concentration declines

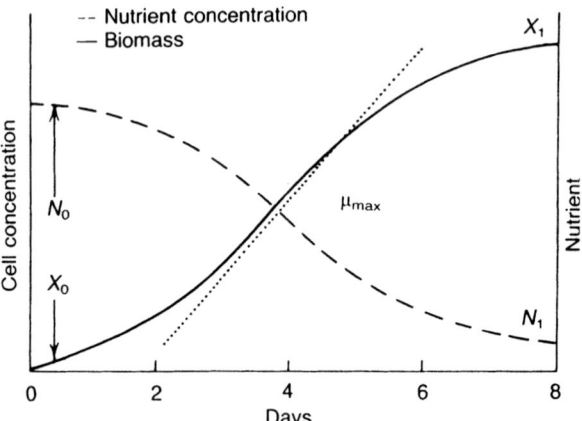

Figure 6.5 Nutrient concentration and biomass in an algal batch culture showing initial and final levels (after McGauhey et al., 1968, p. 8).

as biomass increases as shown in Figure 6.5. As N is depleted, the resulting, growth rate declines causing a slowing in the biomass increase, which ultimately ceases as N reaches the minimum concentration. There is some cell loss through death and decay, which increases as growth slows. If conducted at optimum light and temperature for the species involved, nutrient utilization and growth will be rapid so that maximum biomass or yield will be consistently proportional to the initial nutrient concentration. Figure 6.5 is a typical result of an algal growth potential (AGP) test with *Selenastrum* or *Scenedesmus*.

6.2.3 Other losses

There are other losses of phytoplankton in lakes, such as grazing by zooplankton and sedimentation through sinking, which are usually more important than losses through dilution. The following equation shows a simplified version of how these losses are frequently considered:

$$dX/dt = \mu X - DX - GX - SX \qquad (6.11)$$

where G and S are, respectively, the grazing and sinking rates. These rates can be represented by functional submodels, such as described for μ (Equation 6.2). To illustrate the significance of loss rates in controlling phytoplankton biomass, biomass change in response to three different rates of combined loss have been calculated and are shown in Figure 6.6. For the example, μ was held constant at 1.0 day^{-1} as were G and S at 0.5 day^{-1} and

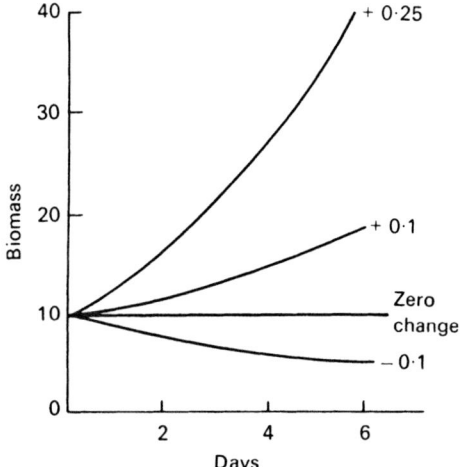

Figure 6.6 Biomass rates of change in a hypothetical algal population with $\mu = 1.0$ day^{-1} and combined loss rates of 0.75, 0.9 and 1.1 day^{-1}.

0.2 day^{-1}, respectively. The effect of D was tested at 0.5, 0.2 and 0.4 day^{-1}. The resulting biomass change rates were +0.25, +0.1 and −0.1 day^{-1} and the biomass change in Figure 6.6 was calculated with the integrated equation below:

$$X_t = X_0 e^{(\mu - D - G - S)t} \qquad (6.12)$$

Although grossly oversimplified, this example, with reasonable values for the rates involved, shows how sensitive phytoplankton biomass is to loss rates. Changes in loss rates can actually have as much or more influence on phytoplankton biomass than changes in limiting nutrient concentration. For example, consider perturbing the simplified, steady-state system shown in Figure 6.7. An increase in nutrient supply rate (N/t) to an actively growing population should result in increased productivity

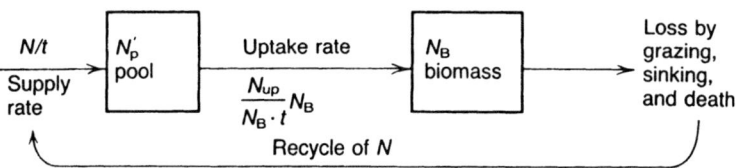

Figure 6.7 Idealized steady-state system for plankton limited by nitrogen (after Dugdale, 1967).

(nutrient uptake rate ($N_B N_{up}/N_b t$)) and possibly biomass, if loss rates remain low, without a noticeable increase in the nutrient pool. Supply rate could increase by either increasing the recycle rate or the inflow rate (e.g. dilution rate in the chemostat). The nutrient pool size could remain relatively low because of increased uptake rate. Biomass may not increase appreciably if loss rates, especially grazing, would also increase. However, some increase in nutrient pool size and biomass are necessary if uptake rate and grazing loss rate are to increase, because such rates are concentration dependent.

6.3 Effects of light and mixing

6.3.1 Light quality

Lakes may appear blue, green, red-brown or yellow. This is due to the selective transmission of different wavelengths of light. Dissolved and particulate matter scatters, absorbs and reflects light at some wavelengths more than at others. Pure water transmits blue light better than green and ultraviolet (UV) better than red and infrared (IR). In water with dissolved material, green light transmits farthest, then blue, red, UV and IR (Goldman and Horne, 1983). This is why pristine lakes, with very little particulate matter, appear blue or blue-green. Other wavelengths are absorbed by water first allowing blue light to penetrate farthest (Figure 6.8). Also, short wavelengths are reflected most and thus blue light is reflected back to the surface, further

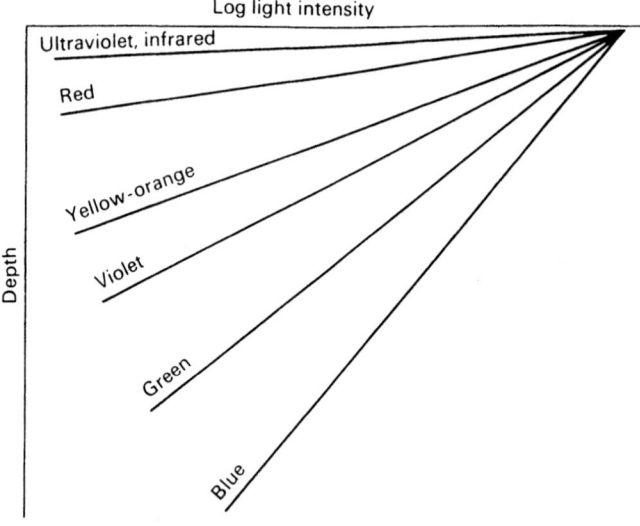

Figure 6.8 Relative peneltation of different wavelengths of light through lake water.

enhancing the blueness of pristine lakes. However, with increasing dissolved matter, lake water will appear more green than blue. Lake water that is brown or reddish-brown has high concentrations of particulate matter that absorb the shorter blue and green wavelengths first.

Chlorophyll a is the active pigment in photosynthesis, absorbing energy in the short and long wavelengths within the visible spectrum, which represents roughly one half the incident light energy over the range of 400–700 nm (Figure 6.8). Although chl a absorbs most in the red and violet areas, those wavelengths penetrate least (Figures 6.8 and 6.9). The energy from blue and green wavelengths, however, is available for photosynthesis through its absorbance by accessory pigments and transport to chl a (Schiff, 1964):

Cartonoids → Phycocyanins → Phycoerythrins → Chlorophyll

This is consistent with the observation that water column photosynthesis is predictable from the available quantity of the most penetrating wavelength, which is most often green light (Rodhe, 1966). In a comparison of photosynthetic rates in twelve European lakes with a range of light transmission characteristics, Rodhe found that relationships between photosynthetic rate and the intensity of the most penetrating wavelength were very similar for all the lakes for depths below saturation where light is the limiting factor (Figure 6.10). Thus, the total incident visible light is used reliably as photosynthetically available radiation (PAR) and is generally related to photosynthetic rate as shown in Figure 6.11. Three regions can be depicted in the relationship as a light-limited or direct-response portion, a light-saturated

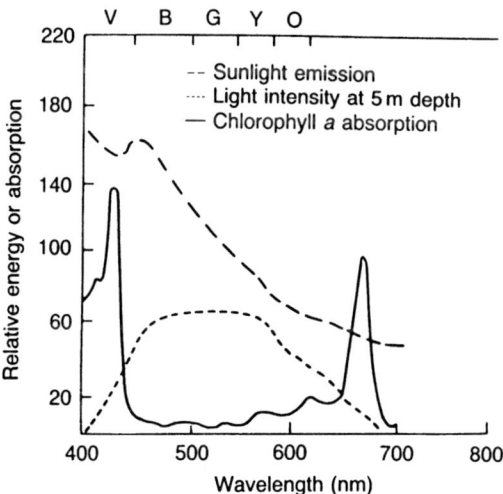

Figure 6.9 Absorption spectrum for chlorophyll compared with the penetration in water and emission from sunlight. Letters indicate colours (Schiff, 1964).

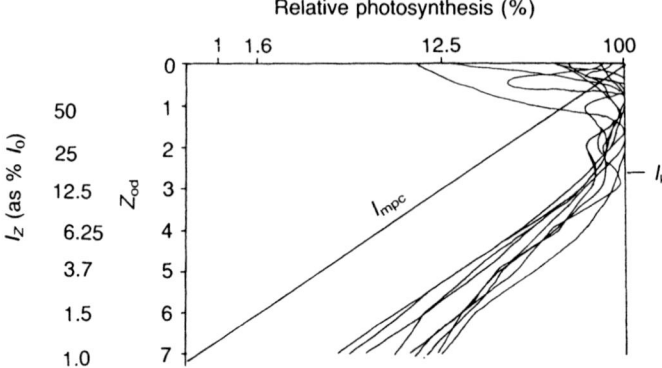

Figure 6.10 Relative photosynthesis versus optical depth (Z_{od}) and light intensity $I_z(=100\% \ Z^{-Z_{od}})$ in a group of 12 European lakes compared to the most penetrating component (I_{mpc}, straight line) (modified from Rodhe, 1966) I_k = saturation.

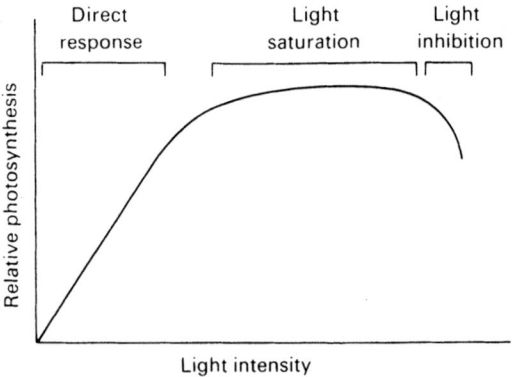

Figure 6.11 Hypothetical relationship between photosynthesis and light intensity.

portion and a light-inhibited portion where photooxidative destruction of enzymes occurs (Wetzel, 1983). The relationship could be expected to vary with species, temperature and adaptation.

6.3.2 Light quantity and limitation to photosynthesis

Light of any wavelength (colour) is attenuated with depth, by water and dissolved and particulate matter, according to the following equation:

$$I_Z = I_0 \, e^{-K_t Z} \tag{6.13}$$

where I_0 is the incident light intensity, I_Z is the remaining light intensity at depth, K_t is the attenuation coefficient for water and dissolved and particulate matter combined, and Z is depth. According to this equation, light penetrating the lake surface is reduced very rapidly at first and then more slowly as depth increases. Also, the greater the dissolved and particulate matter, and hence the attenuation coefficient, the more quickly light will decrease with depth.

An important source of particulate matter in lakes is plankton algae, whose abundance can be estimated by measuring the chl a concentration. Light transmission can be determined with a Secchi disc, the depth of which is a function of particulate matter (chl a) and attenuation due to water itself (K_w) and dissolved matter (K_d). By including chl a as the measure of particulate matter and adding attenuation by water and dissolved matter, i.e. K_{wd}, the general attenuation equation can be written as follows:

$$Z_{SD} = (\ln I_0/I_Z)/(K_{wd} + K_p \text{ chl } a) \tag{6.14}$$

If the Secchi disc is assumed to disappear at about 10% of surface intensity (100%), then the numerator on the right side becomes 2.3. Chlorophyll a is in mg m^{-3} and K_p is expressed as m^2 mg^{-1} chl a, which can be taken as 0.025. There are several empirical relationships in the literature that were derived from lake data sets and that approximate the above equation. Reckhow and Chapra (1983) approximated the effect of K_{wd} in their equation by estimating maximum Secchi depth (SD) without chl a (8.7 m), while Carlson assumed non-algal attenuation was negligible:

$$SD = 8.7/(1 + 0.47 \text{ chl } a) \text{ Reckhow and Chapra (1983)} \tag{6.15}$$

$$SD = 7.7/\text{chl } a^{0.68} \text{ Carlson (1977)} \tag{6.16}$$

where chl a is expressed as µg l^{-1} and SD is expressed as metres.

These relationships, and one derived to fit data from Green Lake, are shown in Figure 6.12. Equations 6.15 and 6.16 produce very similar relationships, but substantially underestimate transparency in Green Lake. Developing such specific relationships of SD-chl a is frequently possible to do for a given lake, even with one or two years' data, while it is not possible for other trophic state variables with so little data (see Section 7.3).

Light quantity can be expressed in several units. Flux of visible (PAR) light is expressed as g cal cm^{-2} min^{-1} or Ly min^{-1} or µE m^{-2} s^{-1} (Ly = Langley, E = Einstein). Light saturation has been suggested to occur as low as $\approx$1.7 Ly h^{-1}, or 100 µE m^{-2} s^{-1} or 4800 lux as a measure of intensity (Talling, 1957a, 1965). Inhibition should begin to occur at $\approx$8.6–12.9 Ly h^{-1} or 510 – 770 µE m^{-2} s^{-1} and 25000–37000 lux. The bounds for light saturation are thus 7% and 35–50% of the full visible range of average daily June sunlight (25 Ly h^{-1}, 1500 µE m^{-2} s^{-1} and 71000 lux).

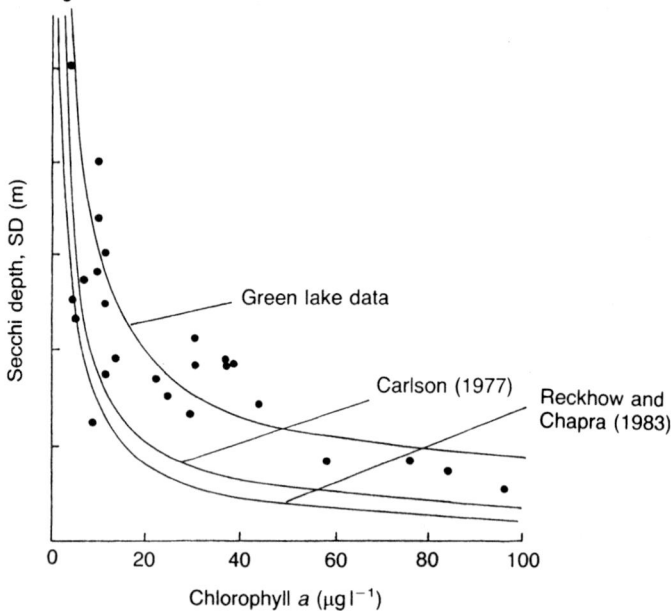

Figure 6.12 The relationship between surface chl a and Secchi depth (SD). Data from Green Lake during June–September 1959, 1965 and 1981 (SD = 10.3/chl $a^{0.56}$; $r = -0.87$). The other lines are based on data from several lakes (Carlson, 1977; Reckhow and Chapra, 1983).

The photosynthesis profile shown in Figure 6.13 is a summation of the curves of light attenuation with depth and the photosynthesis–light relationship.

Light intensity may be so great at the surface that photosynthetic rate is inhibited. At some greater depth the optimum occurs or the photosynthetic rate is saturated at that temperature. Photosynthesis decreases exponentially from that point in proportion to light and usually has decreased to insignificant levels at $\approx 1\%$ of surface intensity, which defines the photic-zone depth. In many clear lakes significant photosynthesis extends below 1% intensity. The compensation depth occurs where respiration equals photosynthesis and no growth results below that depth. This model applies only to a well-mixed photic zone, which unfortunately is not a typical situation in lakes, although it is more so in the ocean. Complete mixing is an acceptable assumption for explaining the process, but it must be remembered that the stratification of nutrients and algal biomass, which occurs in most lakes, will affect the photosynthetic profile. This process, however, does occur to the extent that light is the controlling factor and, consequently, these kinds of profiles will usually be observed. This pattern was evident in Chester Morse Lake, an oligotrophic lake in western Washington (Figure 6.14).

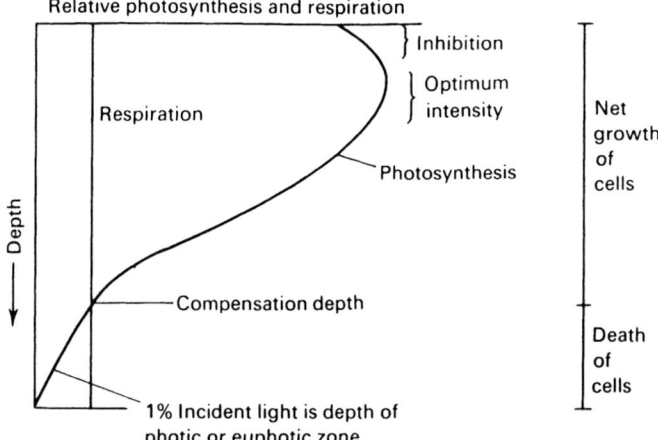

Figure 6.13 Relative photosynthesis and respiration versus lake depth, showing various characteristics of the water column.

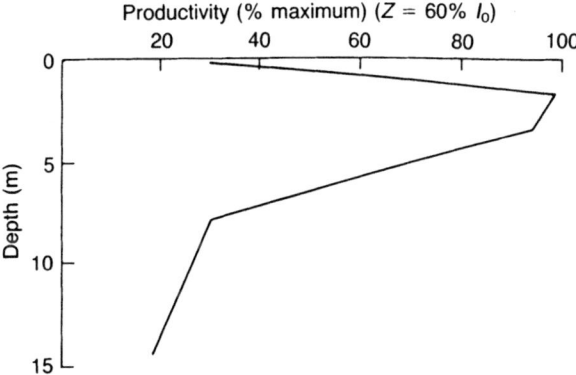

Figure 6.14 Distribution of productivity (relative) with depth in Chester Morse Lake, Washington. Values are based on means over a 1.5-year period (after Hendrey, 1973).

6.3.3 Efficiency of photosynthesis

Only a small percentage of available light is actually used because there are too few algae and too much light. About 1% of total incident light is fixed in photosynthesis as a biosphere mean, or ≈2% of the visible light because the visible fraction represents about one-half of the total light energy. In water this efficiency drops below 1%; Riley reported 0.18% for an oceanic mean (Odum, 1959).

114 Effects of pollutants in standing water

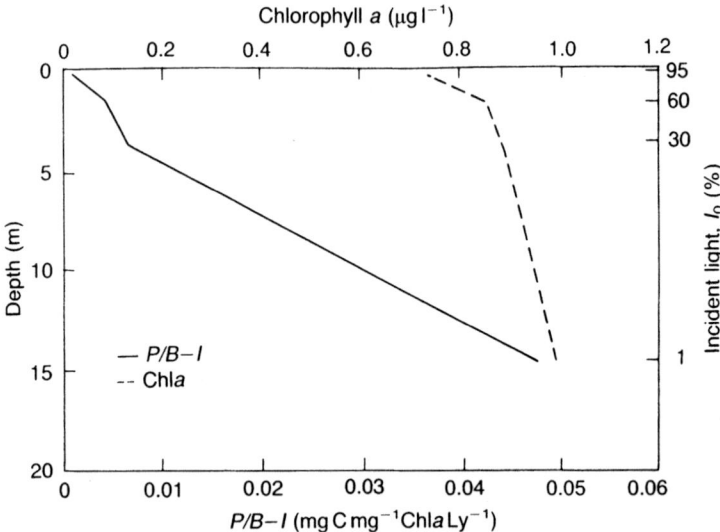

Figure 6.15 Distribution of chl *a* and normalized productivity in Chester Morse Lake. Values are means over a 1.5-year period (after Hendrey, 1973).

Efficiency can be increased greatly by decreasing the light intensity and maximizing the cells' exposure to it. Cultures of *Scenedesmus* and *Chlorella* have utilized as much as 50% of the incident light under these conditions (Brock, 1970). The cell itself can increase efficiency by adjusting its chlorophyll content – shade-adapted cells tend to contain more chlorophyll than light-adapted cells. For example, in Chester Morse Lake a photosynthetic efficiency index ($P/B - I$) was shown to increase greatly with depth, whereas chl *a* increased only slightly (Figure 6.15).

6.3.4 Predicting light-limited photosynthesis

The photosynthesis–depth profile can be predicted reasonably well by knowing P_{max}, the maximum photosynthetic rate, and the depth at one-half the light intensity at which P_{max} would occur without saturation (I_k). I_k and 0.5 I_k are defined in the light–photosynthesis relationship shown in Figure 6.16 (after Talling, 1957b). In this case, intensity is indicated by the percentage of incident light. As photosynthesis increases with increased light intensity, saturation begins to occur. That is, light becomes more abundant than the photosynthetic mechanism is able to utilize and the relationship departs from linearity, i.e. one unit of photosynthesis per unit of light. I_k is defined as the intensity at which P_{max} would occur if this saturation phenomenon did

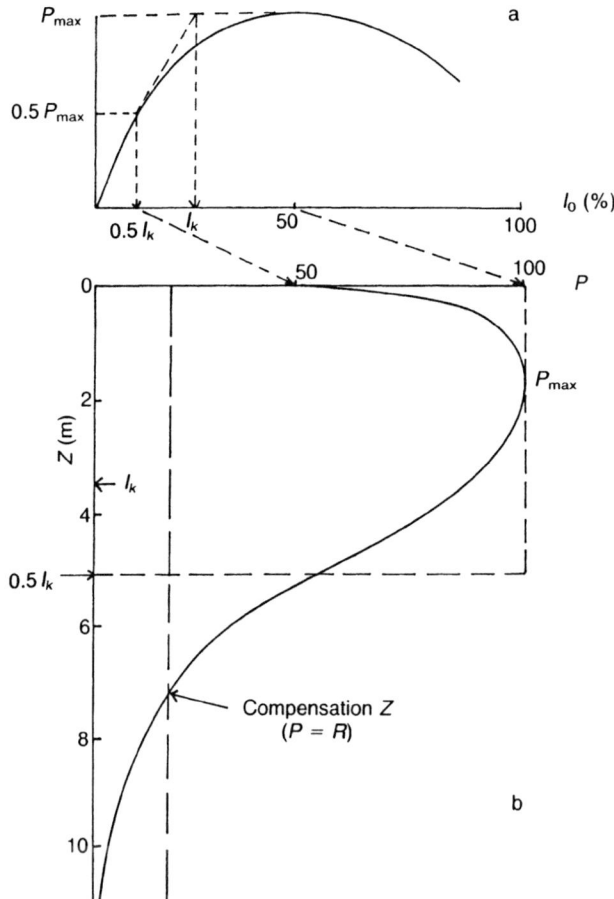

Figure 6.16 A hypothetical profile of gross productivity (P) with depth (Z) and light intensity (I) showing points of I_k, 0.5 I_k and compensation (after Talling, 1957b).

not exist. Rather than saturation occurring abruptly with increased light, the approach to saturation is gradual, as shown in Figure 6.16a, ultimately reaching a maximum at 50% I_0. Beyond 50%, P declines due to inhibition, reaching 0.5 P_{max} at 100% I_0. I_k, according to this model, occurs at 16% I_0 and, therefore, 0.5 I_k at 8% I_0.

The profile of photosynthesis with depth (Figure 6.16b) is simply the light response relationship that is tipped on end and reversed. The depth profile has a more constricted shape compared to the curve in Figure 6.16a because light intensity decreases exponentially with depth. The depths of I_k and 0.5 I_k can be located by knowing the extinction coefficients for water and chlorophyll ($K_w + K_c$) assuming K_w includes all non-chlorophyll attenuating material. Areal gross photosynthetic rate can be estimated by:

$$\text{Areal } P_{\text{gross}} = Z_{0.5I_k} \times P_{\text{max}} \tag{6.17}$$

The product of $Z_{0.5I_k}$ and P_{max} produces that area of a rectangle in the profile that approximates the area under the photosynthetic curve including the tail below $Z_{0.5I_k}$ (Figure 6.16b). For example, if $P_{\text{max}} = 50 \text{ mg m}^{-3} \text{ day}^{-1}$ and $K_w + K_c = 0.5 \text{ m}^{-1}$, $Z_{0.5I_k}$ would occur at 5.1 m [$\ln(I_0/I_{0.5I_k})/0.5 \text{ m}^{-1}$]. Thus, gross productivity would be, $5.1 \text{ m} \times 50 \text{ mg m}^{-3} \text{ day}^{-1} = 255 \text{ mg m}^{-2} \text{ day}^{-1}$.

This procedure resulted in predicted rates of productivity within 5% of observed for a group of 12 European lakes (Rodhe, 1966). Relative photosynthetic rate versus percentage surface light intensity, compared on log scales, for these lakes shows that at light intensities less than an average I_k, photosynthesis tends to be linearly related to the most penetrating wavelength (Figure 6.10). The success of this method is illustrated by the rather consistent shape of the photosynthetic curves, in spite of widely varying areal rates of productivity and P_{max}. It also says that 50% and 16% are reasonably good estimates for P_{max} and I_k, respectively.

To predict photosynthetic rate at depths throughout the water column, a model that represents the relationships in Figure 6.16 is needed. Steele's (1962) model, which relates photosynthetic rate to relative light intensity, with P_{max} occurring at 50% I_0 is often used for that purpose:

$$P = P_{\text{max}} 2I/I_0 e^{1-2I/I_0} \tag{6.18}$$

Where $I = 50\% I_0$, $P = P_{\text{max}}$, which is consistent with the previous model. Where $P = 0.5 P_{\text{max}}$, $I = 12\% I_0$, which is analogous to $0.5I_k = 8\% I_0$ in the formulation by Talling (1957b). However, this model reflects inhibition such that where $I = I_0$, $P = 74\% P_{\text{max}}$. An absolute rate for P_{max} is also required for this model. Also because P_{max} occurs at light-saturated conditions, its magnitude is dependent upon temperature and nutrient content.

The depth of mixing may limit the light available to a mixing population of plankton cells. The average amount of light available to mixing cells is given by:

$$I = I_0(1 - e^{-kz})/(kz) \tag{6.19}$$

where z is the mixed layer depth. Thus, by incorporating both mixed layer depth and light extinction, this equation provides a much better indication of light available in the water column over 24 h and under various mixing scenarios than would exist from incident light measurements alone. For example, I would be 20 and 43% I_0 for K_t values of 0.5 and 0.2 m^{-1}, respectively.

Most plankton algae have a cell density greater than that of water and will sink from the water column unless continually mixed upward. If they are not returned to the photic zone frequently enough, growth will cease, which may also occur if mixing is too strong and/or at too great a depth, relative to light availability.

The settling velocity (W_s) of an algal cell can be estimated by assuming conformity to Stokes' Law:

$$W_s = 2gr^2(\rho' - \rho)/(9\eta\phi) \tag{6.20}$$

where g is gravitational acceleration, ρ is the density of water, ρ' is density of the plankton cell, r is the radius of a sphere of equal volume as the cell, η is viscosity and ϕ is the 'coefficient of form resistance' (Reynolds, 1984). Thus, one can see that species-to-species differences in cell size (r), shape (ϕ) and density difference ($\rho' - \rho$) can account for the relative success and dominance of one species over another depending on strength and depth of mixing, which can determine the amount of light available. Some algae can resist settling through motility (flagellated green algae and dinoflagellates) and gas vacuoles (some blue-greens), while diatoms, with cell densities on the order of 1.003 g cm^{-3}, have little resistance and will sink on the order of a metre per day.

Under non-limiting nutrient conditions, i.e. during the well-mixed, pregrowth period in the spring in both deep temperate lakes and the northern ocean, the onset of the spring diatom bloom can be predicted relatively well. The approach involves the prediction of the critical depth, which is that depth below which the water column total of algal cells will not produce net growth over the duration of a day. To describe the critical-depth concept, it is appropriate to begin with the model first developed by Sverdrup (1953) as follows (see also Murphy, 1962):

$$P = m/k \; I_0(1 - e^{-kz})t - ntz \tag{6.21}$$

where P is a factor proportional to daily net production, m is a factor related to O_2 production, n is a factor related to O_2 consumption, k is the vertical extinction coefficient, z is the mixing depth in metres, I_0 is the effective radiation passing the sea surface on a daily basis and t is time. At the compensation light intensity (I_c), $n = mI_c$, then:

$$P = I_0(1 - e^{-kz})/k - I_c Z \tag{6.22}$$

At the critical depth ($D_{cr} = z$), net photosynthesis is zero so $P = 0$, thus:

$$D_{cr} = I_0/I_c(1 - e^{-kD_{cr}})/k \tag{6.23}$$

And if $I_c = 4.3\,\text{Ly day}^{-1}$, then:

$$D_{cr} = I_0/4.3(1 - e^{-kD_{cr}})/k \qquad (6.24)$$

Thus, the necessary information to know in order to predict D_{cr} for a lake, estuary or ocean area is k, I_0 and I_c. From the last equation it is clear that D_{cr} is primarily a function of I_0. As incident light increases, the compensation depth, and hence D_{cr}, increases as shown in Figure 6.17. At the time when D_{cr} exceeds the mixing depth, net productivity is possible and a plankton bloom can occur.

The critical depth may be confused with compensation depth. Compensation depth represents a light intensity where photosynthesis equals respiration, i.e. if a plankton cell remained stationary at that depth it would receive just enough light to exist, but no growth could occur. However, that is an unrealistic condition, because cells sink and are mixed vertically in the water column, such that some of the time they are below the compensation depth, and essentially dying, while some of the time they are above the compensation depth where the population is growing. The question is, what is happening in the water column overall on a daily basis? How does the amount of light available above the compensation depth compare with the depth of the mixed layer, most of which may be below the compensation depth, through which the respiration demand must be met? The first term on the right-hand side in Equation 6.21 represents gross productivity, or total available light converted into O_2, while the most right-hand term is respiration throughout the depth of the mixed layer or O_2 consumed. Thus, considered from another perspective, critical depth is clearly the depth that would be necessary to generate enough respiration to exactly use up the photosynthate produced from the amount of light available above the mixing depth.

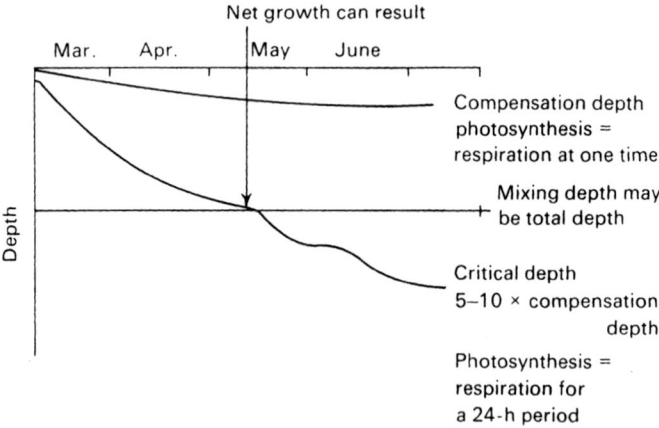

Figure 6.17 Illustration of the critical depth concept (modified from Marshall, 1958).

The net effect is that if the average light intensity received by a mixing plankton population is greater than the compensation depth light intensity, net production will result. In many clear-water bodies, such as the ocean, that sustainable average light intensity can be received if cells are mixed to depths not greater than 5–10 times the compensation depth. In very clear lakes, the compensation depth may be from 50 to 100 m. Few lakes are that deep over much of their area, so in spring the mixing depth may be, in effect, the lake bottom and as light intensity increases, the critical depth may exceed the maximum lake depth, allowing net growth and a bloom, as early as February. In other cases, such as when turbidity is high, the decreasing mixing depth as thermal stratification increases can determine the bloom timing. Examples of the effect of mixing depth on productivity will follow.

Although the critical depth concept developed for oceanic phytoplankton may not be entirely appropriate for some lakes, because they are usually shallower than Z_{cr} (D_{cr}), the ratio of $Z_{eu}:Z_m$ (euphotic:mixing) is nonetheless an important concept for lake productivity (Talling, 1971). This can be demonstrated with the following hypothetical example. Given two morphologically different lakes with the same mixing and critical depths, the lake with the greatest ratio of $Z_{eu}:Z_m$ will have the greatest productivity, all other things being equal (Figure 6.18). To illustrate better the morphometric differences, use of mean depths of the euphotic and mixing depths would be more appropriate. Thus, for mixing depth 1 the potential productivity ratio would be about 1.0:

$$\frac{\text{Productivity lake A}}{\text{Productivity lake B}} = \frac{\overline{Z}_{eu}/\overline{Z}_{m1}}{\overline{Z}_{eu}/\overline{Z}_{m1}} = \frac{0.5/1.0}{0.4/0.8} \approx 1.0$$

However, with a doubling of the mixing depth, productivity in lake B should exceed that in lake A because of the increased shallow area in lake B.

$$\frac{\text{Productivity lake A}}{\text{Productivity lake B}} = \frac{\overline{Z}_{eu}/\overline{Z}_{m2}}{\overline{Z}_{eu}/\overline{Z}_{m2}} = \frac{0.5/2.0}{0.4/1.3} \approx 0.8$$

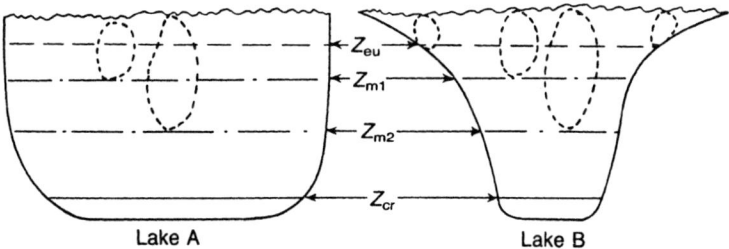

Figure 6.18 Two hypothetical lakes with different morphometry but similar euphotic, mixing and critical depths. The dashed ovals represent mixing patterns for plankton under the two mixing depths.

and therefore as mixing depth increases in the two lakes, the time spent by plankton in the photic zone will be relatively longer in lake B than in lake A. This concept was clearly demonstrated with respect to decreasing surface water elevation for Lake Chapala, Mexico (Lind and Dávalos-Lind, 2002). As lake elevation was lowered below 152 m, the region where the ratio $Z_{eu}:Z_m$ was >0.5 proportionately increased and promoted algal blooms. Thus, although the concept of Z_{cr} may not strictly apply it can help to explain why some lakes are more productive than others, notwithstanding similar enrichment levels.

The effect of mixing depth may also be examined using the model of Oskam (1978):

$$P_{net} = P_{max}C[F_i\lambda/(K_w + K_cC) - 24rZ_m] \tag{6.25}$$

where P_{net} is expressed as mg C m^{-2} day^{-1}, P_{max} as mg C mg^{-1} chl a h^{-1}, C as mg m^{-3} chl a, K_w is the extinction coefficient for water and non-algal matter (actually K_{wd}) in m^{-1} and K_c for algae in m^2 mg^{-1} chl a (nominally 0.025), F_i is a dimensionless function of light intensity to expand P_{max} to areal productivity (nominally 2.7), λ is photosynthetic hours, r is the respiratory fraction of photosynthesis, Z_m is the depth of mixing and 24 is the number of hours for respiration. As an example, a Z_m of 10, 20 and 30 m would give potential P_{net} productivities of about 139, 19 and -101 mg m^{-2} day^{-1}, assuming a K_w of 0.5 m^{-1}, a P_{max} of 1 mg C mg^{-1} chl a h^{-1}, a C (chl a) of 5 mg m^{-3} and 12 h day^{-1} of photosynthesis. In addition to illustrating the balance between respiration as a function of mixing depth and photosynthesis in the water column, this model also shows the effect of self absorption by plankton algae themselves.

By setting $P_{net} = 0$ and rearranging, Equation 6.25 can also predict the critical depth ($Z_m = Z_{cr}$):

$$Z_{cr} = F_i\lambda/(K_w + K_cC)24r \tag{6.26}$$

In contrast to Equation 6.23 for critical depth, this equation does not contain light intensity. Therefore, it is not useful for predicting the timing of the spring bloom, but rather to estimate the potential of mixing depth to control productivity given the clarity of the water. Thus, critical depth is a function of the extinction coefficients. For the above example, the Z_{cr} is 43 m. Thus, if the lake approaches that depth over much of its area, then mixing the lake artificially may control productivity and biomass as well.

Example problem

Given a lake with a mean depth of 10 m and a non-algal attenuation coefficient of 0.5 m^{-1}, would it be possible to control algal biomass to a level less than 10 mg m^{-3} chl a by complete mixing? Why? Assume an absorption

coefficient for chl a of 0.025, a 10 h photosynthetic day and a respiration: P_{max} quotient of 0.1.
From Equation 6.25:

$$Z_{cr} = \frac{2.7 \times \lambda}{24r(K_w + K_cC)} = \frac{2.7 \times 10}{24 \times 0.1(0.5 + 0.25)}$$
$$= 15\,\text{m}$$

Biomass would probably not be controlled below 10 mg m^{-3} chl a, because most of the lake is shallower than the critical depth that would allow a maximum of 10 mg m^{-3}. Therefore, the potential maximum biomass possible with available light could be >10 mg m^{-3}, assuming nutrients were not limiting.

The model can be rearranged to predict maximum biomass possible under non-nutrient-limited conditions before being limited by self absorption:

$$C_{max} = (1/K_c)[F_i\lambda/(24rZ_m) - K_w] \quad (6.27)$$

This model tends to give values consistent with maximum possible chl a of 250 mg m^{-2} (also in mg m^{-3} in a 1 m water column) (Wetzel, 1975; p. 337). Oskam (1978) showed that for a K_w of 1.0 m^{-1}, the biomass of some highly enriched German reservoirs should be controlled if the mixing depth were maintained at 15 m. In reality there was less biomass realized than predicted by the model presumably due to zooplankton grazing. The potential maximum biomass as a function of mixing depth is shown in Figure 6.19.

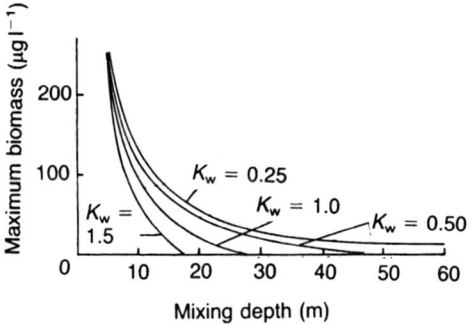

Figure 6.19 Relation of maximum phytoplankton biomass (chl a) to mixing depth for different levels of non-algal attenuation of light (K_w) where $F_i = 2.7, \lambda = 12$ h, and $r = 0.05$ (see text, after Oskam, 1978).

6.3.5 Examples of light-mixing effects

Most relatively small temperate lakes greater than 6–8 m in depth will stratify thermally during summer. However, permanent stratification depends on surface area, which is a surrogate for wind fetch, as well as depth. Analysis of 108 north temperate lakes showed that permanent stratification resulted at a depth of 7 m for a 20-ha lake, but 19 m depth was necessary on average for a 1000-ha lake. The average depth for stratification could be estimated by $Z_{max} = 0.34 A_s^{0.25}$ (Gorham and Boyce, 1989). Thus, for relatively small lakes of about 100 ha (1 km^2), lack of permanent stratification would occur if the lake had maximum and average depths less than about 10 and 5 m, respectively. There are many shallow lakes with average depths around 3 m; these lakes would be polymictic even if relatively small in area. Lakes that do not stratify permanently during summer may stratify temporarily.

The depth of the thermocline can be considered as the depth of mixing, and that depth results from a balance between the force of the wind mixing the water downward and the resistance to that mixing due to the buoyancy of the water, caused by heating, and density reduction. The position of the thermocline varies from lake to lake, depending upon wind velocity and fetch, light intensity and water transparency. A stable thermocline will initially favour productivity because it results in increased available light to mixing algal populations and may stimulate a bloom. However, if stratification persists, nutrients become depleted, because the thermocline usually represents a barrier to circulation.

A common way to indicate the degree of stratification is by the change in density per unit depth or between surface and bottom. Density change can be indicated by temperature in freshwater, because dissolved materials usually are insufficient to influence density. However, in seawater, σ_t is used and is defined as the density of water resulting from salinity, temperature and pressure minus 1 multiplied by 10^3.

Another important aspect of stratification in freshwater is that the change in density of water becomes greater as temperature increases. Thus, a few degrees difference between surface and bottom at high temperature will provide more stability (resistance to mixing) than the same degree change at low temperature. This is illustrated by the term relative thermal resistance to mixing (RTRM), which compares the density gradient against the density gradient between 4 and 5°C:

$$\text{RTRM} = (D_b - D_s)/(D_4 - D_5) \tag{6.28}$$

where D_s is surface density, D_b is bottom density, D_4 is density at 4°C and D_5 density at 5°C. RTRM would be greater for a high compared to a low temperature range if the surface-to-bottom gradient were the same numerically. This has implications for shallow polymictic lakes, which could develop significant temporary stability with differences of only a couple of degrees between surface and bottom. Increased RTRM in Upper Klamath Lake, Oregon resulted in reduced atmospheric exchange and decreased off-bottom DO concentrations that was considered the principal cause for large fish kills (Perkins *et al.*, 2000).

In the Indian Arm, a fjord area in British Columbia, productivity was related to the degree of stability and to the compensation depth, which increased with time as incident light increased. When these two factors, which determine the amount of light received by a mixed plankton cell, were greatest, productivity was greatest (Figure 6.20). The periods of low stability were favourable to nutrient replenishment.

It is useful to compare the degree of stability in Indian Arm with that in Puget Sound. The unit ($10^5 \, m^{-1}$) is actually $(\rho - 1) \times 10^5 \, m^{-1}$. The values in Indian Arm ranged from 40 to 100 over the first 10 m, whereas in Puget Sound productivity was strongly affected by stability factors ranging from only 2–6 over 50 m depth (Winter *et al.*, 1975). Peaks in productivity and biomass usually occurred at minimum tidal-prism thickness (amplitude), which in turn allowed maximum stability and light availability during the spring months (Figure 6.21).

Water column productivity was related to available light in the mixed depth (Equation 6.19) within and outside the Frazier River plume in the Strait of Georgia, British Columbia (Parsons *et al.*, 1969). The strong stratification within the plume resulted in relatively shallow mixed depths and usually much

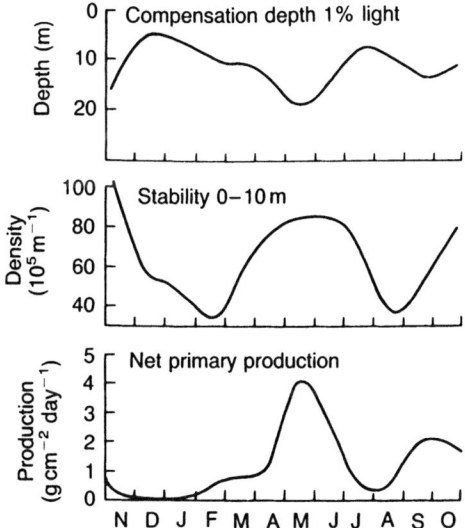

Figure 6.20 Seasonal relation among compensation depth, water column stability, and net primary production in a British Columbia fjord (modified from Gilmartin, 1964).

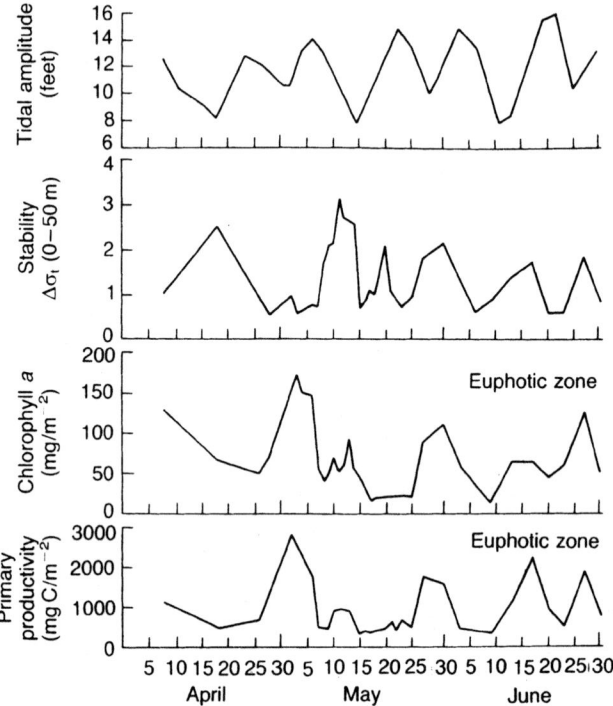

Figure 6.21 Productivity, biomass, chl *a*, stability and tidal prism thickness in Puget Sound, 1966 (Winter *et al.*, 1975).

greater available light to circulating plankton cells than was the case outside the plume where light was seldom sufficient for net productivity to occur.

In the Duwamish estuary at Seattle, there was no real phytoplankton activity until August when river flow was low and tidal flushing action was least. Such tidal conditions are characterized by low-high and high-low tides. The tidal conditions determine the turbulence and stability and are indicated, in addition to river flow, by tidal-prism thickness, which can be calculated as follows (units of metres):

$$\text{Unfavourable turbulence: TPT} = (\text{HH} + \text{LH}) - (\text{HL} + \text{LL})$$
$$= (3.6 + 3.0) - [(1.5 + (-0.9))] \quad (6.29)$$
$$= 6.0$$

Favourable stability: TPT = $(3.0 + 2.7) - (1.8 + 0.9) = 3.0$

where TPT = tidal prism thickness, HH = high-high, LH = low-high, HL = high-low and LL = low-low tides.

Figure 6.22 shows the timing of phytoplankton blooms at one point in the Duwamish related to freshwater discharge and tidal-prism thickness in 1965 and 1966. The water column reached maximum stability and minimum turbulence and flushing when tidal-prism thickness (difference between sum of two daily high tides and sum of two daily low tides) and freshwater discharge were least. At that time tidal excursions were also minimal and the net result was that surface water moved slowly up and down the estuary and was allowed to retain heat and become even more stable because of the increased temperature gradient. Primary productivity correlated with degree of stratification (temperature difference from surface to bottom) and river

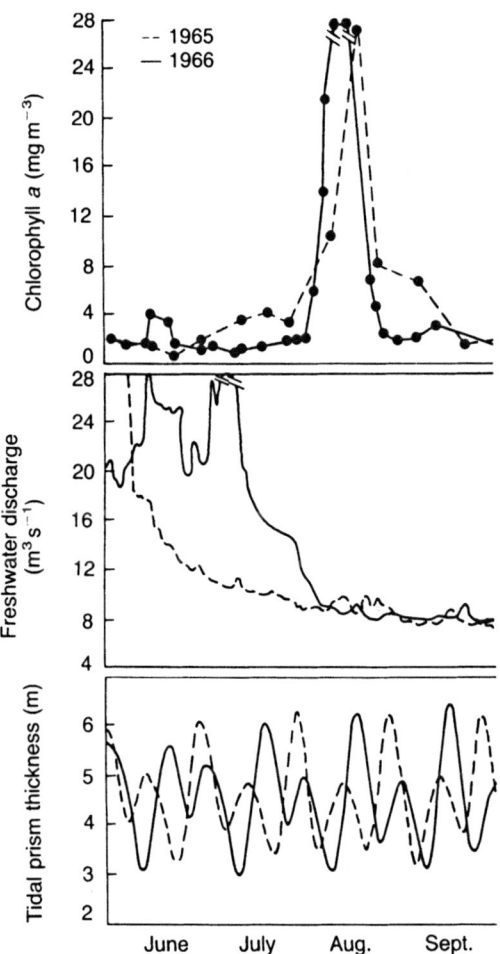

Figure 6.22 Tidal prism thickness, river discharge and chl *a* content in the Duwamish River estuary, Washington, 1965 and 1966 (Welch, 1969a).

discharge, and these two variables explained 65% of the variation in that parameter (Welch, 1969a).

Another way to examine the relationship between the physical limitation and productivity in the Duwamish is with a simple growth model. That is, the change in biomass X is a result of growth and loss (Equation 6.10):

$$\frac{dX}{dt} = \mu X - DX$$

where μX is growth and DX is loss through washout. Because nutrient content is very high and non-limiting (Welch, 1969a), the observed change in mixing depth can be hypothesized to affect the daily maximum growth rate based on ^{14}C productivity measurements made during the bloom. Thus, during neap-tide conditions (TPT = 3 m), the mixing depth was only 1 m and the average μ for a plankton cell mixed to that depth and average light exposure was about 0.6 day^{-1}. On the other hand, the spring-tide situation usually resulted in a mixing depth of 4 m and about one-half the μ (0.3 day^{-1}) because of the much reduced light per cell. The extinction coefficient was not considered in this case, but higher values usually occur with greater river discharge, which would cause further limitation during the spring runoff period.

The loss rate through flushing can be approximated from a dye study conducted in the estuary; which showed an average flushing (dilution) rate of about 0.2 day^{-1} over neap and spring tides alike. Knowing that spring tides allow more plankton loss than neap tides, the flushing rate was adjusted linearly so that a loss rate of 0.13 day^{-1} was assigned to the neap tides and 0.27 day^{-1} to the spring tides. Even so, it is clear from these values that mixing depth produced a much larger effect on population growth than flushing rate did on loss during the change from spring to neap tides. That is a greater growth of 0.3 day^{-1}, but a smaller loss of only 0.13 day^{-1}.

The net rates of population increase for the neap and spring tide situations, respectively, are:

$$\frac{dX}{dt} = 0.6X - 0.13X = 0.47X$$

and

$$\frac{dX}{dt} = 0.3X - 0.27X = 0.03X$$

Clearly a bloom would not be expected in the Duwamish when spring tidal conditions persist, even at low river flow, and this is precisely what has been observed. Moreover, if the μ of 0.47 day^{-1} and an initial biomass of 3 μg l^{-1} chl a are used to simulate the bloom initiation of 1966 (Figure 6.22), the bloom seems to be explicable. The time required for the biomass to reach the observed maximum of 70 µg l^{-1} chl a is 6.7 days from the first-order growth equation:

$$70\,\mu g\ l^{-1} = 3\,\mu g\ l^{-1} e^{(0.47)(6.7)}$$

Although the dynamics of the phytoplankton in that estuary are much more complex, this analysis nevertheless indicates how strong the effect is from physical factors alone, particularly as they affect the amount of light received.

Stratification can limit production, however, if the condition is prolonged, because of nutrient depletion in the epilimnion. Partial destratification during midsummer in a Thames River reservoir illustrates the part played in production and species succession by stratification and nutrient depletion (Taylor, 1966).

The typical plankton succession occurs during spring and summer, starting with diatoms dominating in the spring, greens and yellow-greens in the summer, blue-greens in late summer, and a small diatom bloom again in late autumn (Figure 6.23).

In mid-July a partial destratification eliminated the typical midsummer growth of green algae and stimulated *Asterionella formosa*, a diatom that usually blooms from February to March. The apparent cause for that summer bloom was the transfer to water at 9–12.5 m with 2.8 mg l^{-1} SiO$_2$ to the photic zone that contained only 0.5 mg l^{-1} SiO$_2$. Concentrations below that have been shown to be limiting to *Asterionella* (see Section 6.5).

The *Asterionella* bloom subsequently declined after a second destratification and transfer of fertile water that further increased nutrient levels. Blue-greens continued to increase during destratification probably due to their buoyancy. Population increase was more rapid for *A. formosa* in summer than in spring, but its persistence was much less, possibly a result of temperature that was higher than required for best population maintenance even though growth rate was higher (see Section 6.4). Nevertheless, *A. formosa* actually showed greater growth in summer if nutrients were available, possibly because of higher light-saturated photosynthetic rates.

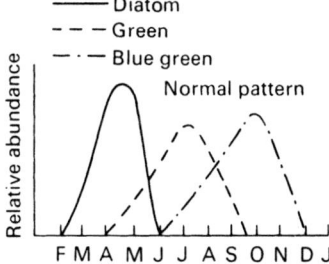
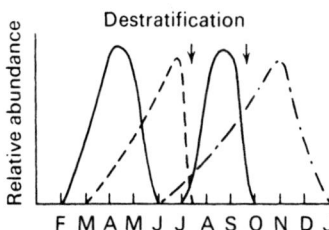

Figure 6.23 Succession among diatoms, green algae and blue-greens in a Thames River reservoir in years with and without destratification. Arrows indicate time of destratification (Taylor, 1966). Curves are idealized.

Thus the thermoclines in stratified lakes are usually a barrier to the recycling of hypolimnetic nutrients, and they can have a considerable range of effects on plankton algal production, depending on the permanency of the thermocline (see P cycling in Section 4.1 and TP two-layer models in Section 7.2).

Light-mixing depth effects are also important in shallow lakes that do not permanently stratify. The effect of lake level lowering was noted previously for Lake Chapala. The effect of depth on light availability and algal biomass was shown with a simple model of TP and light availability, and supported by data from 142 Dutch lakes (Scheffer, 1998). The model showed that chl $a > 100\,\mu g\,l^{-1}$ did not occur in lakes >3 m mean depth regardless of TP concentration, i.e. light limited maximum biomass, but biomass was strongly limited by TP at depths <3 m. A lake level increase of 1.5 m in eutrophic Lake Võrtsjärv, Estonia, between 1964 and 1993 decreased water column light availability, and algal production and biomass (Nõges and Järvet, 1995; Nõges et al., 1997). The 270-km^2 lake's mean depth increased from 2.1 to 3.2 m. As a result of the decreased algal productivity, pH declined by about one-half unit.

6.4 Effects of temperature

6.4.1 Growth rate

The question is how do phytoplankton respond to temperature and what is an appropriate model to describe that response? The most common description of the growth response of phytoplankton, and other microorganisms as well, is the Q_{10} rule:

$$Q_{10} = (\mu_2/\mu_1)^{10/t_2-t_1} \tag{6.30}$$

where μ_2 and μ_1 are growth rates in response to a temperature increase from t_1 to t_2. This is a simplified form of the Van't Hoff (1884)-Arrhenius (1889) equations and is also known as the RGT rule (Reaktidnsgeschwindigkeit-Temperatur-Regel), as reviewed by Ahlgren (1987). The equation says that the Q_{10} will be 2 if the growth rate doubles for a 10°C rise in temperature.

As shown in Figure 6.24, the light-saturated rate of photosynthesis (flat portions of the curves) is limited by temperature as it affects the rate of enzymatic processes. At low light intensities, photosynthesis increases in direct proportion to increased light, but reaches a maximum that depends upon temperature. This has been interpreted as a temperature-dependent limit of the organism's enzyme systems, in contrast to the photochemical pigment response (Wetzel, 1975). Photosynthesis is seen to roughly double in the mid-part of the range, from about 25–35°C, but the Q_{10} tends to decrease as temperature increases and, therefore, is not a true constant (Ahlgren, 1987).

Eppley (1972) summarized a large amount of data from experiments with a variety of phytoplankton species in batch culture and showed that the

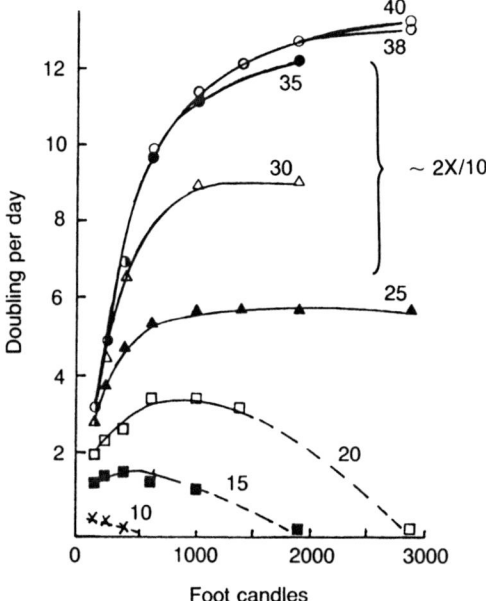

Figure 6.24 Relationship between growth rate and light intensity over a range in temperature for Chlorella pyrenoldosa (modified from Sorokin and Krauss, 1962).

maximum growth rates observed increased exponentially with temperature (Figure 6.25). From inspection, it can be seen that μ_{max} roughly doubles with each 10°C increase in temperature. The calculated Q_{10} for this relationship is 1.88 and agrees well with a similar relationship developed by Goldman and Carpenter (1974) from continuous culture experiments with a variety of species. These relationships may be more useful in describing the growth rate response of a mixture of species which have optima that vary over a seasonal range in temperature. However, the Q_{10} rule may not be appropriate for a single species over a seasonal temperature range.

Ahlgren (1987) compared several equations describing temperature-growth rate relationships using data from unialgal continuous culture experiments. She found that the more linear model of Belehrádek (1926) fit the experimental data more closely, and the same was generally true for data from other investigators if the linear model were used. That equation using velocity (v) to indicate growth rate is:

$$v = a(t - \alpha)^b \tag{6.31}$$

where a and b are empirical constants, t is temperature and α is biological zero.

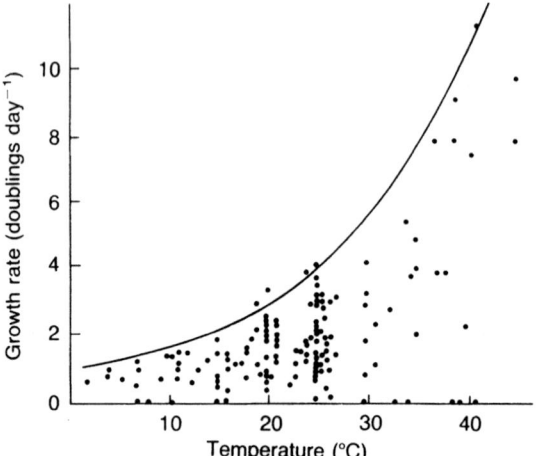

Figure 6.25 Relation of growth rate (μ_{max}) to temperature for a variety of culture experiments with different species (points) (Eppley, 1972).

Belehrádak (1957) argued that the Q_{10} rule is not a true constant over each species' thermal tolerance range and that it is based on chemical reaction rate processes, which take an exponential form as shown in Figure 6.25. He observed artificial 'breaks' in the temperature–growth rate response over a range in temperature if the data were expressed in the form of the exponential equations, but such problems were overcome if expressed in his more linear form.

According to Ahlgren, Belehrádek developed the more linear equation (Equation 6.31) because biological processes are highly complex and are probably controlled by physical processes at the cell surface, such as diffusion and viscosity. Ahlgren obtained very good correlations using Equation 6.31, with exponent b ranging between about 0.3 and 2.5. She found also that the half-saturation constant, K_s, and the yield coefficient, Y, were generally independent of temperature with *Scenedesmus quadricauda*.

The equation for temperature correction used in the Streeter–Phelps DO deficit model is given by:

$$K_1 = K_2 \theta^{(t_1 - t_2)} \tag{6.32}$$

where K_1 and K_2 are reaction rates at temperatures of t_1 and t_2, respectively, and θ is an empirical coefficient. This equation can be derived from Arrhenius' equation and, therefore, simulates chemical reaction rates. Although of an exponential form, its success may be because prediction is based on a mixture of species over the seasonal temperature range.

6.4.2 Seasonal succession

Lake phytoplankton contain many species with widely different thermal optima for growth rate. Examples for four species are shown inserted under the maximum growth rate curve from Eppley (1972) (Figure 6.26). With such a distribution of thermal optima within a phytoplankton community, changes in temperature alone from season to season should force succession among species according to those optima. If optima for light are also considered, a further dimension is added to account for succession. For example, the following pattern might be expected (Hutchinson, 1967). Winter: low light and low temperature (no growth); spring: high light and low temperature (diatoms); summer: high light and high temperature (greens); autumn: low light and high temperature (blue-greens).

The observed patterns of plankton algae succession are to some extent explicable by light and temperature changes. For example, Cairns (1956) has shown that algae in stream water from Darby Creek in Pennsylvania demonstrated the following shifts in response to raised and lowered temperature from ambient after sufficient time was allowed for population equilibration at each temperature:

$$20°C \leftrightarrow 30°C \leftrightarrow 35-40°C$$
$$\text{diatoms} \quad \text{green algae} \quad \text{blue-greens}$$

Adequate evidence exists that plankton algae do have readily definable thermal optima as shown by Rodhe (1948). He also showed that if such

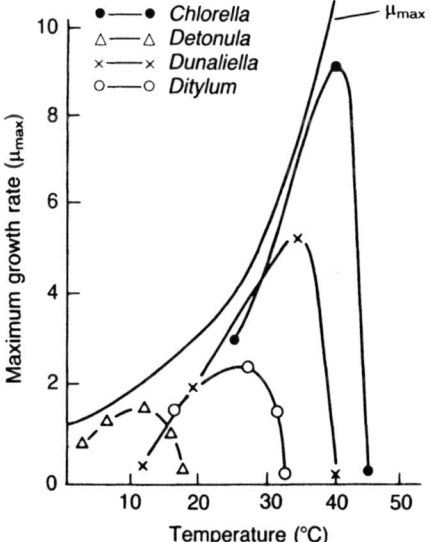

Figure 6.26 Maximum growth rate (μ_{max}) versus temperature with four species of phytoplankton (modified from Eppley, 1972).

thermal optima are designated by that temperature at which population growth is most prolonged, the temperature is lower than if the criterion of rapid growth were chosen. For *Melosira islandica*, Rodhe found that the most prolonged growth occurred at 5°C, growth was less prolonged at 10°C, and at 20°C growth was fast but death was rapid and the population did not persist. Typical optima are as follows: *Melosira islandica* (diatom) ≈5°C; *Synura uvella* (yellow-green) ≈5°C; *Asterionella formosa* (diatom) 10–20°C (≈15°C); *Fragilaria crotonensis* ≈15°C; *Ankistrodesmus falcatus* ≈25°C; *Scenedesmus, Chlorella, Pediastrum, Coelenastrum* 20–25°C.

A typical pattern of plankton algal succession is shown for Lake Sammamish (Figure 6.27). Note the spring diatom outburst and the summer and autumn blue-green dominance. The diatom increase was greatest at temperatures between 5 and 10°C, whereas blue-greens increased when temperature was between about 15 and 20°C. *Fragilaria* and *Asterionella* were important representatives in the diatom outburst.

An explanation for thermal optima being lower than the temperature for maximum growth rate is that as μ increases exponentially (or linearly) with increasing temperature, the death rate should also increase, but probably lags behind μ. Because biomass at any time is the difference between growth and loss (death) (see Equation 6.11), its maximum should occur at a lower temperature than for μ.

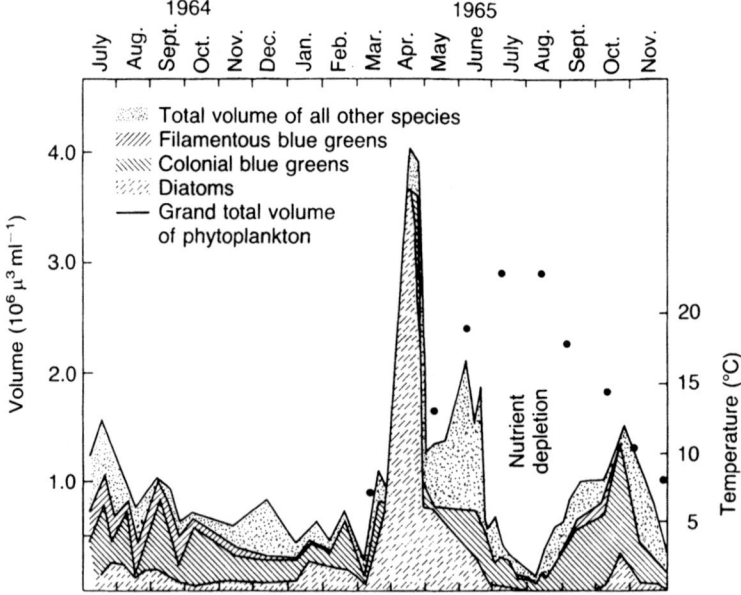

Figure 6.27 Surface phytoplankton composition at a centrally located station in Lake Sammamish. Temperature indicated by solid circles (after Isaac et al., 1966).

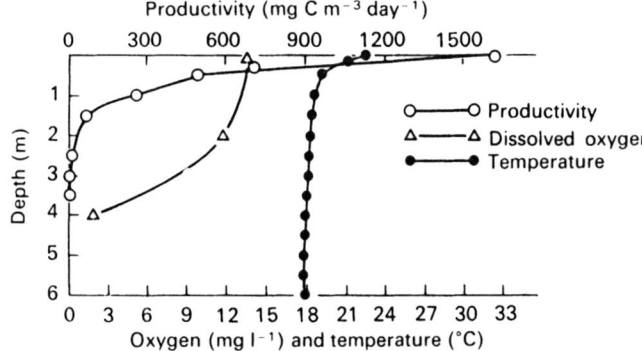

Figure 6.28 Observations of productivity and temperature, Clear Lake, California, on 15 October 1959 (a calm day). *Aphanizomenon* is the dominant species (modified from Goldman and Wetzel, 1963).

There are many exceptions to light and temperature as an explanation for such successional patterns in phytoplankton. *Oscillatoria rubescens* has occurred in Lake Washington at temperatures $>20°$ C and was the principal bloom species that first indicated the degradation in that lake, yet it prefers low light and low temperature, $\approx 10°C$ (Zevenboom and Mur, 1980). *Fragilaria crotonensis* should occur most prominently in spring and autumn because its optimum is from 12 to 15°C, yet often reaches a maximum at temperatures from 15–29°C. Moses Lake produces massive blooms of blue-greens of the genera *Anabaena*, *Aphanizomenon* and *Microcystis* (often referred to as 'Anny', 'Fanny', and 'Mike') at an average summer temperature of 22°C (Bush et al., 1972). Clear Lake, California shows the same situation of blue-green blooms occurring during moderate temperatures. Results are shown in Figure 6.28 for a calm autumn day during which *Aphanizomenon* was the dominant species (Goldman and Wetzel, 1963).

The occurrence of blue-greens tends to be initially confined to late summer or autumn in moderately enriched lakes, but as enrichment proceeds they begin to dominate in spring, even if temperature is low (e.g. *Aphizomenon* in Pine Lake, Washington, in early spring at a temperature of 8°C).

Why are light and temperature and their respective optima not adequate in predicting successional patterns within and among lakes?

6.4.3 Species adaptation

Algae are capable of adaptation to other temperature regimes even though somewhat removed from their optima. For example, the respiration rates of *Chlorella* show that an adaptation of strains can occur (Table 6.1). Although one strain obviously grows best at 30–35°C, as was indicated in Figure 6.26

Table 6.1 Effect of acclimation temperature on respiration in two strains of Chlorella pyrenidosa

Test temperature (°C)	Respiration for two acclimation temperatures (mm^3 O_2 mm^{-3} cells h^{-1})	
	25°C	38°C
25	4.5	8
39	1.6	18

Source: Sorokin (1959).

(same species), another strain nevertheless can be most active at lower temperatures.

Blooms of *Oscillatoria* can be observed below the ice in lakes during winter. Of course, their growth rate is low at that time, just as is the rate of *Chlorella* at less than optimum temperature (Table 6.1). However, if loss rates are also low, a bloom can result.

Thus, algae should be able to compete at temperatures slightly removed from their range and possibly dominate the community if some other factor(s) is(are) provided nearer their optimum and to some extent compensates the organism for the out-of-optimum temperature.

6.4.4 Interaction with other factors

If provided closer to a species optimum, nutrient supply may be one such factor that may allow that species to grow outside its temperature optimum. If the supply is great enough, even if the growth rate is relatively low, a bloom can result. This may, for example, explain blue-green blooms (*Aphanizomenon*) at 8°C in spring during periods of relatively high light intensities. Even if temperature is sub-optimum, blue-greens may be able to outcompete diatoms and greens if the nutrient supply is more to their liking. This may hold for *O. rubescens*, which adapted to the warm season in Lake Washington because the nutrient supply was increased. Such selection for higher P levels and low light is indicated in Table 6.2 for filamentous blue-greens.

The buoyancy mechanism allows blue-greens to compete with non-buoyant diatoms for light in warm quiescent conditions (Figure 6.28). Diatoms depend on turbulence to remain suspended in the lighted zone. Thus, with the lower sinking rate, the blue-greens are able to accumulate large masses when turbulence is minimal, assuming the nutrient supply is large (see Section 6.5.6).

Table 6.2 Variation in the relative abundance of greens, blue-greens and diatoms resulting from five levels of phosphorus enrichment at three light intensities in Lake Washington water after 2 weeks

Added P ($\mu g\, l^{-1}$)	Light intensity (lux)			Row means
	4000	2000	1000	
Greens (Number of cells)				
0	43	58	32	44
10	132	196	57	128
20	211	143	65	140
30	243	285	86	204
40	334	232	92	219
Column means	193	183	60	
Blue-greens (μm filament length)				
0	1220	1749	2680	1883
10	1860	4540	4160	3520
20	3120	5010	5620	4583
30	1840	3980	6000	3940
40	2840	6960	5440	5080
Column means	2176	4448	4780	
Diatoms (Number of cells)				
0	5	2	9	5
10	153	69	30	84
20	91	35	24	50
30	168	86	15	90
40	135	30	13	59
Column means	110	44	18	

Source: Hendrey (1973).

Although temperature and light contribute to species dominance and succession, they do not tell the whole story because the available nutrient supply and other factors are also elements in the adaptation of species to sub- or supra-optimum temperature and light. The blue-green species that usually prefer warm summer temperatures can apparently outcompete other algae for the available nutrients at that time. However, they can also adapt to other seasons when nutrient availability is high. In spite of sub-optimum temperature, blue-green species can dominate for most of the year in highly fertilized lakes. Species succession and blue-green dominance will be discussed further at the end of this chapter.

Although a given temperature is not entirely indicative of species dominance among different waters, succession from diatoms to greens to blue-greens is often reasonably predictable from a temperature increase imposed upon a given water (and nutrient supply) and temperature regime.

The interacting physiological effects of temperature and light on production could produce two results from a temperature increase.

1. A short-term effect of a temperature increase may result in some adaptation of existing species, with the possible occurrence of even a decrease in production if the original temperature optimum and the new temperature are too far apart.
2. A long-term species shift may occur with time, and species with optima at the new temperature will dominate and production will increase because the light-saturated rate of photosynthesis goes up with a temperature increase across the tolerable range for all species.

6.4.5 Heated waters from power plants

Not many examples exist to examine the effect of the addition of heated water, but results from some Polish lakes indicate that fairly extensive changes in productivity and species composition could develop from relatively small changes in temperature (Hawkes, 1969). The test lake was Lichen Lake which received heated waters from a power plant: annual temperature range 7.4–27.5°C; taxonomic components (species, varieties and forms) 285; dominant species *Melosira amgibua*, *Microcystis aeruginosa*; primary production $7.3\,\mathrm{g\,m^{-2}\,day^{-1}}$. The control lake was Slesin Lake which had no heated water added: annual temperature range 0.8–20.7°C; taxonomic components 198; dominant species *Stephanodiscus astraea* (no blue-greens); primary production $3.75\,\mathrm{g\,m^{-2}\,day^{-1}}$.

Global warming is likely to have effects on phytoplankton community composition, with blue-greens being favoured by increasing water temperatures. There was a switch in dominance from diatoms and cryptophytes to cyanobacteria in a eutrophic lake (Heiligensee) in Berlin coincident with an increase in ambient water temperature of 2.6°C between 1975 and 1994 (Adrian and Deneke, 1996).

6.5 Nutrient limitation

The concept of nutrient limitation has been defined previously in the section on growth kinetics. The role of specific nutrients, both micro and macro will be discussed here in terms of whether they limit growth rate or biomass, how limitation by a particular nutrient and its prediction can be determined, and how nutrient limitation affects species succession, especially to blue-greens.

6.5.1 Silica

Silica is considered a macronutrient (required in large amount by cells) for diatoms, which have cell frustules that are composed of silicious cell walls.

It is available for uptake as an undissociated silicic acid or silicate ion (Hutchinson, 1957), but is also adsorbed on particulate matter. The rapid removal of silica from the photic zone during spring diatom blooms indicates that uptake and sedimentation occur much faster than regeneration in the water column.

Cessation of the spring bloom of the diatom *Asterionella* has been closely linked to ambient silica concentration and thus represents a rather clear case for nutrient control. Early laboratory experiments (batch cultures) showed that $0.5\,\mathrm{mg\,l^{-1}}$ of SiO_2 represented the minimum concentration that would sustain maximum growth of *Asterionella* (Lund, 1950). A consistent year-to-year pattern of bloom cessation was observed to occur once SiO_2 fell below $0.5\,\mathrm{mg\,l^{-1}}$ (Figure 6.29). Increased loss through sedimentation could also be an explanation for biomass decline, as stratification increases in the spring. N and/or P are also macronutrients that can limit growth and terminate blooms, but a clear link between ambient concentration and cessation is not as easily observed with those nutrients.

Later experiments with *Asterionella* in continuous culture have supported the earlier work of Lund. Kilham (1975) demonstrated that the uptake rate of SiO_2 could be described by Michaelis–Menten kinetics and that concentrations accounting for 90% of μ_{max} were 0.82 and $0.39\,\mathrm{mg\,l^{-1}}$, very close to Lund's $0.5\,\mathrm{mg\,l^{-1}}$.

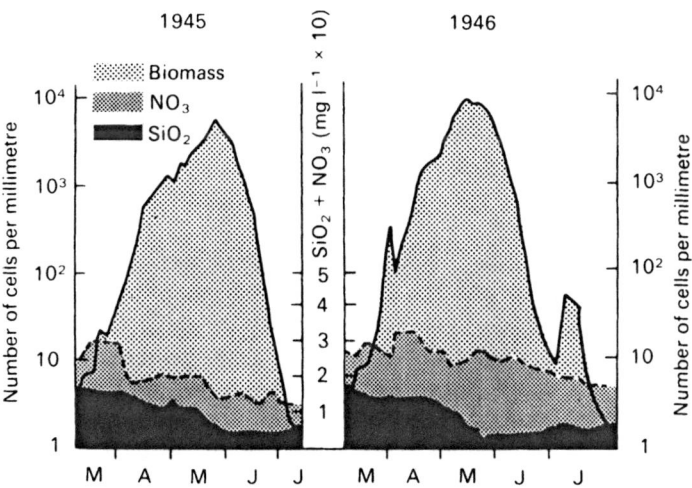

Figure 6.29 *Asterionella* biomass and nitrate and silica content in Lake Windermere in 1945 and 1946 (Lund, 1950 with permission of Blackwell Scientific Publications, Ltd).

6.5.2 Nitrogen and phosphorus

N and P are the macronutrients (required in large amount by cells) that are usually most scarce relative to the needs of organisms and therefore most likely to limit growth and biomass. Except in very low alkalinity waters, C is usually in sufficient supply so as not to be limiting at all or to be limiting only in the short term if N and P are plentiful. The special conditions for C limitation will be discussed later. Trace elements have been shown to limit growth in only a few instances and usually in oligotrophic lakes (Goldman, 1960a, 1960b, 1962).

The deficiency in cells for a macronutrient, and therefore its growth and biomass limitation by that nutrient, can be determined by examining the ratios in cell material. The Redfield ratio, or the average ratio found in marine plankton, is often used to judge the requirements of phytoplankton (Redfield et al., 1963; Sverdrup et al., 1942):

$$O:C:N:P = 212:106:16:1 \text{(by atoms)}$$
$$= 109:41:7.2:1 \text{(by weight)}$$

Applying this ratio to results of nutrient content in phytoplankton biomass from marine waters shows that N is usually in the shortest supply, or will be the nutrient to deplete first if all macronutrients are removed from water in the above ratios. Menzel and Ryther (1964) found the ratio of N:P in biomass to range from 5.4 to 17 by atoms and the C:P ratio by atoms to range from 67 to 91. Based on the Redfield ratio, these cells would range from deficient in N to just meeting the requirements for all nutrients.

In freshwater, on the other hand, P is usually the most limiting nutrient. Particulate matter, which normally in summer is predominately phytoplankton, in Lake Canadarago, New York showed a range in N:P (by atoms) from 17 to 24 during summer and a C:P (by atoms) of ≈ 200 (Fuhs et al., 1972). These results are in contrast to those from the marine environment above, and show that cells ranged from containing all nutrients in the appropriate ratio to being deficient in P.

The above comparisons only consider the cell content of nutrients and not the content in the ambient water. Vallentyne (1972) compiled data to evaluate the biospheric average of the demand ratio, which is the ratio of the average concentration of a given nutrient in plant tissue to the average concentration in the water. Thus, the higher the ratio for a given nutrient, the higher is the scarcity of that nutrient in the ambient environment, relative to the cellular needs. Vallentyne's data are summarized in Table 6.3 and show that P was the most limiting nutrient in both winter and summer.

The ratio of C:N:P in the ambient water is also often used to indicate which nutrient may be depleted or reach growth-rate-limiting concentrations first and thus could be identified as the potentially most limiting nutrient.

Table 6.3 Ratio of plant-to-water concentrations for important plant nutrients in a variety of freshwater habitats around the world

Nutrients	Winter	Summer
P	100 000	800 000
N	20–25 000	100–125 000
C	5–6 000	6–7 000

Source: After Vallentyne (1972).

Emphasis is on potential, because at the time of measurement concentrations of both N and P may be so high that neither is currently limiting growth. That process is evident in results of algal growth potential (AGP) experiments shown in Figure 6.30 (Skulberg, 1965). The initial N:P ratio in both experiments was 33:1 by atoms (15:1 by weight) and concentrations were so high as to not limit growth. Assuming that N and P were taken up by the algal cells in the required ratio of 16:1 (7.2:1 by weight), P should have been removed first and this is what happened; P reached undetectable amounts and ultimately stopped growth while N was still present at relatively high concentrations.

Under some conditions of low P and high N:P ratios in the ambient environment, cells may not actually be deficient in P, which is therefore

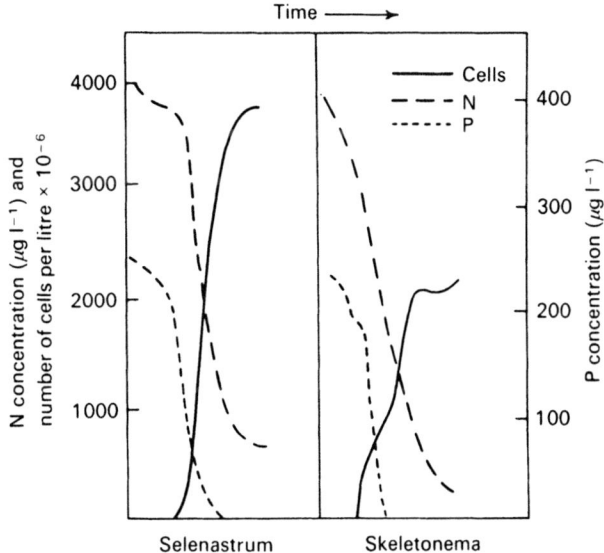

Figure 6.30 Growth of two species of algae related to N and P concentrations (soluble) in culture (Skulberg, 1965).

technically not limiting. If the recycling rate of P is sufficiently high, its supply rate may be adequate to meet the growth rate demands of the cell in spite of apparent limitation based on the ambient N:P ratio. Also, the cells may have taken P up in luxury amounts (above growth needs) or are releasing alkaline phosphatase enzymes to break down organic forms of P (the latter is used as a test for P limitation). However, the ambient ratio would still indicate the potentially most limiting nutrient even though growth is not currently limited. In Figure 6.30, *Skeletonema* appears to have grown after P was depleted, which may indicate that the cells had stored P in luxury amounts and subsequent growth utilized those luxury stores. In addition to using ambient and cellular N:P ratios, one may want to employ a test for alkaline phosphatase activity, which, if present in relatively high concentrations, would indicate current limitation (Fitzgerald and Nelson, 1966; Goldman and Horne, 1983).

Assuming that changes in cellular nutrient content are proportional to those in the ambient environment, response will be related to changing N or P in a manner similar to that shown in Figure 6.31 (Sakamoto, 1971). ^{14}C uptake in this case is proportional to biomass if experiment duration is several days. Limitation with P increase occurs when the ratio falls below the optimum level, nominally indicated by the Redfield ratio as 7:1 by weight. Limitation occurs with increasing N when the ratio exceeds the optimum.

Other ratios in cells may also indicate nutrient limitation, such as the amount of chl *a*. Healy (1978) has suggested bounds in the chl *a*:biomass ratio in chemostat cultures that indicate moderate and severe nutrient limitation. Using a C:dry weight ratio of 0.2, the boundary values for chl *a*:C range from 1 to 2% for severe and moderate limitation, respectively. Figure 6.32 shows the response of severely N-limited (ambient NO_3-N:SRP = 0.1) phytoplankton in sewage-polluted Pelican Horn of Moses Lake to the pumping of dilution water high in NO_3-N, but low in SRP. Prior to pumping in July, the chl *a*:C ratio was <1%, but after pumping it increased to >1% and even rose above 2% on occasion.

The limitation of growth by either N or P may vary seasonally as changes in environmental factors, such as available light, temperature, stratification,

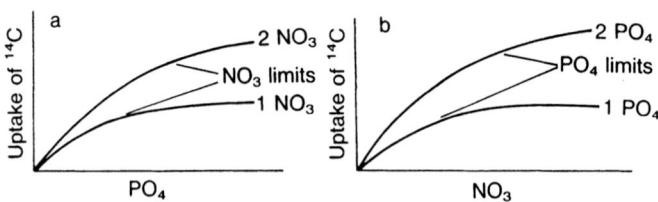

Figure 6.31 Response of phytoplankton ^{14}C uptake to P (a) and N (b) additions showing the significance of the ratio concept in determining which nutrient is limiting (generalized from Sakamoto, 1971).

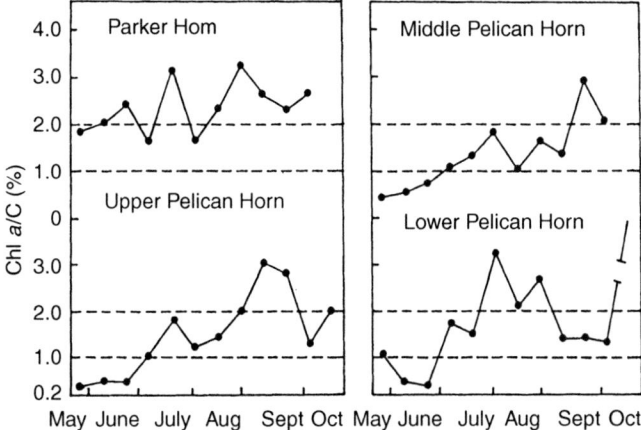

Figure 6.32 Ratios of chl a:cell carbon for phytoplankton in Moses Lake, Washington during May to October 1982. Horizontal lines indicate Healy's (1978) zone of 'moderate' nutrient deficiency (from Carlson, 1983).

internal loading, recycle rate, etc., affect the demand and availability of the two nutrients. There is a tendency for NO_3 to deplete more rapidly than SRP in the epilimnion during the growing season. This occurs in oligotrophic as well as eutrophic lakes and a large decrease in NO_3 is often accompanied by little or no change in SRP (Hendrey and Welch, 1974). From a consideration of ambient N:P ratios, limitation can change from P to N during the stratified period. In systems that have short detention times and that receive variable inputs (e.g. stormwater) or are subject to other events (thermocline erosion), the limiting nutrient can change even more frequently. For example, Smayda (1974) has shown with bioassays using a marine diatom (*Thalassiosira pseudonana*) that frequent changes occur in the limiting nutrient throughout the growing season in Narragansett Bay and that more than one nutrient could limit at once. N was usually the limiting nutrient and to a lesser extent Si, but limitation also resulted from the exclusion of EDTA (ethylenediamine tetraacetic acid) as a chelator of algal excretory products and only occasionally by exclusion of P and trace metals.

The nutrient that limits growth also tends to change as eutrophication increases and the lake's trophic state changes. Oligotrophic lakes usually have higher N:P ratios than eutrophic lakes (Table 6.4). As these example ratios suggest, the process of eutrophication tends to change the nutrient that limits growth from P in oligotrophic and mesotrophic states to N in the mid and later stages of eutrophy. The winter N:P ratio in Lake Washington changed from ≈10:1 during its most eutrophic condition, resulting from sewage effluent input, to values usually >20:1 during the years following

Table 6.4 Annual means of NO_3-N/PO_4-P (by weight) showing decreasing trends as eutrophication proceeds; unless specified, values are annual means

Kinds of lakes	Location	N/P ratios
Hypereutrophic	Hjälmaren (Sweden)	4/1
	Moses (WA)	1/1 (summer)
Eutrophic	Washington (WA)	10/1 (winter, 1963)
Mesotrophic	Washington	20/1 (winter, 1970)
	Mälaren (Sweden)	25/1
	Sammamish (WA)	27/1
Oligotrophic	Vättern (Sweden)	70/1
	Chester Morse (WA)	63/1

Source: Swedish lakes from National Environmental Protection Board.

diversion and recovery (Edmondson, 1978). The N:P ratio tended to be lower during the growing season, as indicated previously.

There are anomolies to the pattern of N or P limitation and trophic state. Bioassays using natural assemblages of phytoplankton species in oligotrophic lakes may often show no response, or even inhibition, to increased P in short-term exposures (Hamilton and Preslan, 1970; Hendrey, 1973), but with time for physiological adaptation and species selection, response to P in such waters is usually positive (Hendrey, 1973). There are also exceptions to the generalization that P will ultimately limit in oligotrophic lakes. N has limited growth in Lake Tahoe (Goldman and Carter, 1965; Goldman, 1981), where the spring–summer NO_3:SRP ratio is <3, and in Crater Lake where NO_3-N is $<1\,\mu g\,l^{-1}$ and SRP is $14\,\mu g\,l^{-1}$ during spring–summer (Larson et al., 1987). These lakes are ultra-oligotrophic by the conventional criteria.

N limitation in eutrophic and hypereutrophic lakes is a common occurrence. A typical example is represented by bioassays using water from two eutrophic Wisconsin lakes and the principal nuisance blue-green species, *Microcystis aeruginosa* (Gerloff and Skoog, 1954). Addition of the three elements together (N, P, Fe) resulted in growth about equal to that produced in the synthetic media, which contained a surplus of everything this cyanobacterium is known to require (Table 6.5). NO_3 addition produced the major portion of the biomass, as shown by results from those treatments in which each of the three elements was used alone. Although NO_3 was clearly the most limiting nutrient, there was added biomass produced from the three nutrients used in combination, with P being the second most important. *M. aeruginosa* is a non-N-fixing, blue-green species and, therefore, results could differ if an N fixer had been used.

Biomass increase of N-fixing blue greens should ultimately be controlled by P if sufficient time is allowed for N fixation. The depletion of NO_3, as a result of increasing SRP, would stimulate N fixation, which would allow

Table 6.5 Growth of *Microcystis aeruginosa* during 2 weeks in sterilized surface water from two lakes in Wisconsin containing various chemical additives

Additives	Lake Mendota water (cells mm^{-3})	Relative growth (%)	Lake Waubesa water (cells mm^{-3})	Relative growth (%)
Without additives	500	6	1 523	14
NO$_3$	5 780	64	7 737	73
PO$_4$	500	6	1 101	10
Fe	495	6	1 139	11
PO$_4$ + Fe	400	4	874	8
NO$_3$ + Fe	4 310	48	9 675	92
NO$_3$ + PO$_4$ + Fe	9 030	100	10 575	100
Synthetic nutrient solution	10 235	113	10 235	97

Source: Gerloff and Skoog (1954).

biomass to ultimately increase to meet the P supply (Hutchinson, 1970a). Smith (1990a) has shown for temperate and tropical lakes that N-fixation rate increases with TP concentration, except at very high TP. Schindler and Fee (1974) observed that P addition alone to eutrophic low-NO$_3$ lake water resulted in a biomass increase through N fixation requiring two to three weeks in plastic bag experiments *in situ*. Horne and Goldman (1972) found that N fixation contributed about one-half the N supply to Clear Lake, California. Although the growth rate of N-fixing blue-greens was relatively slow (growth rate 0.05 day^{-1}), fixation could supply the N to ultimately meet the available P supply.

N fixation does not always supply N to meet the ultimate supply of P, however. NO$_3$ addition to hypereutrophic Moses Lake water (plastic bottle bioassay *in situ*) in mid-summer, when NO$_3$ was low or undetectable, stimulated an increase in blue-green biomass and chl *a*. *Microcystis* was dominant and showed the greatest increase in biomass, as might be expected since it is a non-N fixer, although *Aphanizomenon* also increased (Welch *et al.*, 1972). However, the later (September) P-addition portion of the experiment had no effect on blue-greens; diatoms had become dominant and blue-greens were much reduced, possibly due to much lower lake temperature. Additional bioassays *in situ*, using combinations of N and P additions simultaneously, resulted in a substantial but delayed increase in biomass with P only (Figure 6.33). Presumably time was required for the necessary N to be supplied by fixation. Subsequent dilution of the lake with large amounts of low nutrient water showed N to be the controlling nutrient; summer chl *a* was directly related to the inflow NO$_3$ concentration (Welch *et al.*, 1984). Although N-fixing *Aphanizomenon* was the dominant blue-green, its N-fixation rate was apparently not fast enough to utilize substantially the available P at the

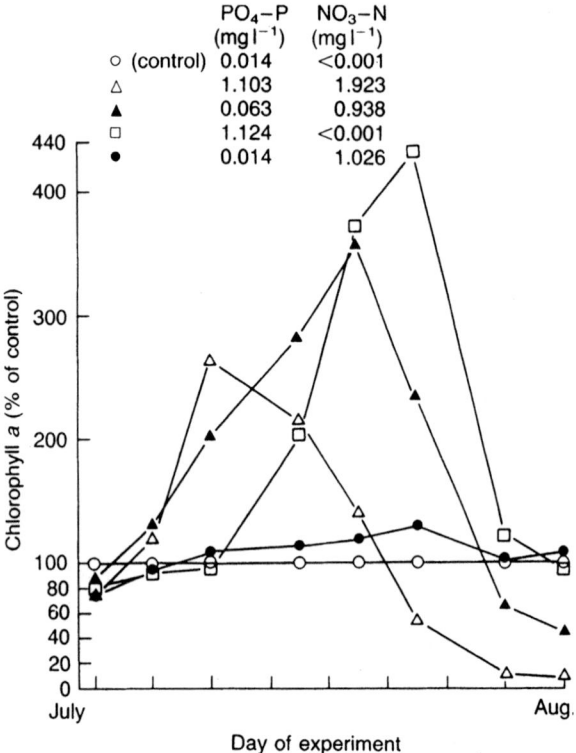

Figure 6.33 Effect of treated sewage on biomass of natural phytoplankton populations in Moses Lake (Welch et al., 1973).

relatively high, dilution-imposed, flushing rate (average 0.1 day^{-1}). Horne and Goldman (1972) have suggested that the maximum growth rate due to fixation is 0.05 day^{-1}.

The decreasing tendency for P to limit as eutrophication proceeds (trophic state increases) was shown by Miller *et al.* (1974) from bioassays using the green alga *Selenastrum* (AGPs) in water from 49 lakes in the USA. As the yield of *Selenastrum* increased, as an indication of the increasing trophic state, the percentage of P-limited lakes decreased (Figure 6.34). A similar pattern has been shown by a multiple regression analysis of TP, TN and chl *a* data collected *in situ* from 127 North American lakes (Smith, 1982). Although the regression and N:P ratios showed that P limited in most lakes, a better fit of the data was obtained when TN was included. That is obvious from the plot of TP on chl *a* where ≈20 of the high chl *a* observations are skewed substantially to the right of the prediction line and these represent lakes with TN:TP ratios <10, suggesting N limitation. However, a large

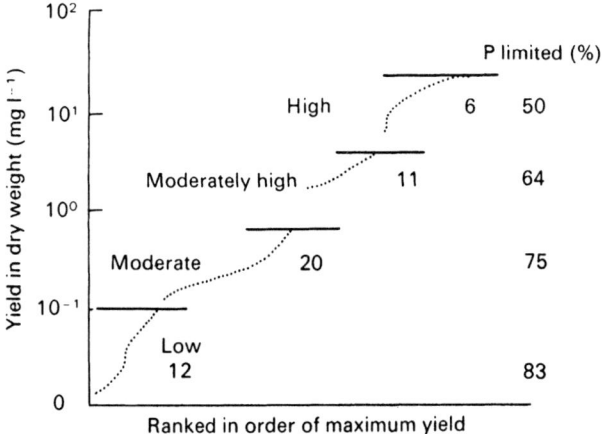

Figure 6.34 Yield of algal biomass in algal growth potential (AGP) experiments from 49 lakes in the USA showing a decrease in the percentage that were P-limited as the yield (and productivity) increased (modified from Miller et al., 1974).

data set from worldwide lakes does not show the separation of N- and P-limited lakes (Figure 6.35; Nürnberg, 1996, personal communication). Chlorophyll *a* seems to depend on P over a broad range up to $900\,\mu g\,l^{-1}$ TP and $110\,\mu g\,l^{-1}$ chl *a*, regardless of TN:TP ratio, although many of the low TN:TP ratio lakes are skewed to the right and the chl *a*:TP ratio (slope of the line) is rather low. Others have shown that chl *a* is dependent on TP at concentrations less than $200\,\mu g\,l^{-1}$ (Seip, 1994; Scheffer, 1998). Clearly then, P can be regarded as the controlling nutrient well into hypereutrophy, in spite of a tendency for N to become more likely to limit with increased trophic state. This is especially relevant in deciding which nutrient to control to effect a decrease in trophic state.

Explanations of why P limitation in lakes gives way to N limitation as eutrophication proceeds involves a consideration of the N and P cycling processes and the character of the inputs. According to Thomas (1969), NO_3 is always observable in oligotrophic lakes, but as PO_4 is added culturally, the 'NO_3 reserve' is depleted. One reason for this is probably that the ratios of N:P in wastewater sources are low, e.g. N:P in sewage effluent is about 2:1 to 4:1. Addition of effluent with this ratio would lower the lake ratio. Also, as eutrophication increases, light availability becomes less, due to attenuation from the increased organic matter. As light, rather than nutrients, becomes more limiting there are higher concentrations of unused nutrients remaining. Associated with that process is an increased potential for N loss from the lake through denitrification. As eutrophication increases, oxygen depletion,

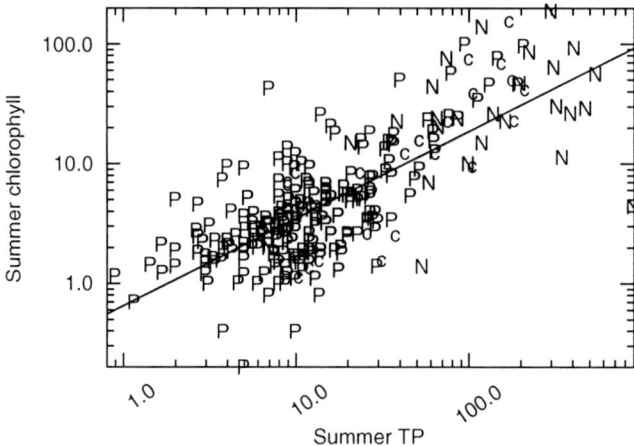

Figure 6.35 Mean summer chl *a* and total phosphorus concentrations from worldwide lakes. Symbols indicate nutrient limitation according to the N:P ratio by weight: P, P limitation above a ratio of 17; N, N limitation at ratio less than 7; C, co-limitation between these ratios. Based on data from Nürnberg (1996), it is redrawn with symbols showing the likely nutrient limitation indicated by N:P ratio (Nürnberg, personal communication).

especially at the sediment–water interface becomes more pronounced (see P cycle section). This should result in higher rates of denitrification, because that is an anaerobic process. As more nutrients remain unused due to increasing light limitation, and N is preferentially lost due to denitrification, the N:P ratio, especially the soluble fractions, would decline. In addition, the more extensive anoxic conditions at the sediment–water interface result in higher internal loading of P, which would lead to further lowering of the N:P ratios since N release does not increase with anoxic conditions. N fixation tends to restore the N:P ratio to a level of equal limitation, i.e. the ratio in unstarved cells. However, fixation is a rather slow, energy-demanding process compared to those processes that contribute high P relative to N. Thus, at some TP concentration ($\sim 200\,\mu g\,l^{-1}$), N fixation apparently becomes unable to completely restore the N resulting in consistently low TN:TP ratios in hypereutrophic lakes.

P is most apt to be limiting in oligotrophic through mesotrophic lakes because sediments are usually still oxic and represent an effective, relatively permanent sink for P because of the insoluble complexes formed with metals and its tendency to sorb to particulate matter. Soluble inorganic N, on the other hand, is not susceptible to such complexation and sorption and thus tends to be more conservative with respect to physical processes.

Furthermore, precipitation represents a relatively greater and more consistent source of N than P. Although rainwater contains some P (largely associated with dust) and has been shown to be a relatively important contributor in phosphorus-poor watersheds (Schindler, 1974), the NO_3 in rain is a rather constant phenomenon, having been transformed from the N_2 reservoir in the atmosphere by lightning.

Thus, there are fewer sources for P than for N and sedimentation is a more efficient remover of P than N from the water column. With few exceptions, N usually becomes limiting only in highly eutrophic or hypereutrophic lakes and even then fixation by cyanobacteria usually becomes an important source for N. With these processes operating, limitation of phytoplankton biomass to relatively low levels by reducing nutrient loading should be more easily attained in most lakes by restricting the availability of P rather than N, even if the TN:TP ratio indicates that N is currently limiting in highly eutrophic systems. Moreover, the fraction of cyanobacteria has been shown to decrease with decreasing TP concentration (Sas, 1989; Jeppesen et al., 1991; Seip et al., 1992; Seip, 1994), whether the decrease is due to decreasing TN:TP or other mechanisms (Shapiro, 1990). However, in lakes with high natural inputs of P, results may be more promising from N control. This is apparently the case in many lakes in New Zealand (White, 1983).

This reasoning is not supported in marine systems where N is usually the limiting nutrient, whether the system is oligotrophic or eutrophic (Ryther and Dunstan, 1971; Smayda, 1974). Why N is more consistently limiting in marine waters is not entirely clear. A partial explanation may relate to water column recycling for P, which is apparently faster than for N. Particulate matter, therefore, becomes richer in N than P as depth increases in the sea (Ryther and Dunstan, 1971), while the reverse occurs in lakes (Schindler, 1974; Birch, 1976; Gachter and Mares, 1985). Thus, as particulate matter sinks through a much deeper mixed layer in the sea, consistently lower soluble N:P ratios would result than in shallower lakes, where the high-P particulate matter would tend to be trapped by bottom sediments before recycling could occur. The N:P ratio would be lowered even more if N fixation were relatively unimportant locally or in the short term in the sea as suggested by Ryther and Dunstan (1971). Smith (1984) contends that there really is not that much difference between marine and freshwater environments with respect to which nutrient is limiting. Rather than basing the designation of limitation on bottle experiments, which is usually the case, whole ecosystem nutrient budget information should be used. He suggested that where N limits in the sea, the explanation may be due to high rates of hydrographic exchange that dominate nutrient availability permitting insufficient time for N fixation to restore N losses. N loading from wastewater and agriculture to marine near shore areas is considered the cause for large DO-depleted regions of the North Sea (Baden et al., 1990) and the Mississippi River delta (Rabalais et al., 2002).

Thus far the discussion of N and P limitation has considered conditions under which yield (biomass) is limited. That is, if the N:P ratio in the ambient environment indicates that N is ultimately limiting, then growth will stop when N is exhausted and the final yield will be proportional to the initial N concentration. This concept was demonstrated for dynamic systems in the section on growth kinetics, such that the steady-state biomass concentration is directly proportional to the inflow concentration of the limiting nutrient. Uptake rate, on the other hand, is dependent on the ambient concentration of nutrient, according to Michaelis–Menten kinetics. Growth rate is dependent on the cellular concentration, which may be higher than ambient if there is luxury storage and, therefore, the growth rate would be higher than the uptake rate. If cells have below quota concentrations, uptake could be higher than growth rate. If one assumes that cell concentrations are at quota amounts, then growth rate can be predicted, according to Michaelis–Menten kinetics, based on the ambient concentration.

There are several approaches to predicting growth rate of phytoplankton if there are changes in limitation between N and P, assuming a dependence on ambient concentrations. The first formulation is the 'multiplicative' model (Chen, 1970):

$$\mu = \mu_{max}[P/(K_P + P)][N/(K_N + N)] \tag{6.33}$$

This model produces a low estimate of μ when the concentrations of both nutrients are near the limiting level. For example, if K_P and K_N each $= 0.05\,\mathrm{mg\,l^{-1}}$, $P = 0.05\,\mathrm{mg\,l^{-1}}$ and $N = 0.25\,\mathrm{mg\,l^{-1}}$, $\mu = 0.42\,\mu_{max}$.

The second is an additive model (Patten *et al.*, 1975):

$$\mu = \mu_{max}/2[P/(K_P + P) + N/(K_N + N)] \tag{6.34}$$

Using the same values for N and P concentrations and the half saturation constants, $\mu = 0.67\,\mu_{max}$. By averaging the reduction ratios to account for the possible changing limitation, instead of multiplying them together, the estimate of μ increases.

Another form of averaging treats the reciprocal of the reduction terms (Bloomfield *et al.*, 1973):

$$\mu = \mu_{max}2/[1/(P/(K_P + P)) + 1/(N/(K_N + N))] \tag{6.35}$$

In this case, μ is estimated at $0.62\,\mu_{max}$. The final and possibly more supportable formulation is the 'threshold' model (Bierman and Dolan, personal communication, 1976):

$$\mu = \mu_{max}P/(K_P + P)\ldots \text{ if N:P} > 13.5:1 \text{ (by weight)} \tag{6.36}$$
$$\mu = \mu_{max}N/(K_N + N)\ldots \text{ if N:P} < 13.5:1 \text{ (by weight)} \tag{6.37}$$

With the same K_s and concentrations, this approach predicts a μ of 0.83 μ_{max}, because N:P is 5:1 and N should limit. Bierman and Dolan (personal communication) found that the threshold model predicted a μ closer to observed than the multiplicative model, when compared with chemostat culture data using *Scenedesmus*. They observed that the predicted ratios of μ/μ_{max} were 0.29 ± 0.3 for the multiplicative model and 0.43 ± 0.07 for the threshold model. The actual ratio, determined in cells from chemostat experiments, was 0.44 ± 0.02, thus supporting the threshold model. The data and threshold N:P ratio (30:1 by atoms; 13.5:1 by weight) were taken from Rhee (1978).

The light model of Steele and the temperature model of either Eppley or Bĕlehrádek (see Ahlgren, 1987) are often linked together with a multiple nutrient model to predict growth rate with depth and time. However, light, temperature and other nutrients may interact with N and P to produce variable growth responses, thus altering the half-saturation coefficients for N and P. For example, Young and King (1980) showed that growth rate of *Anacystis nidulans* tended to saturate at lower CO_2 concentrations as the combination of light and P content increased. Also, Hughes and Lund (1962) showed that the critical level for SiO_2 of 0.5 mg l^{-1}, with respect to *Asterionella* growth, could be effectively lowered with increased P addition.

6.5.3 Trace elements and organic growth factors

Many trace elements are required by algae for growth as co-factors for enzymes and there is a sound rationale as to how growth rate is affected by trace-element deficiency (Dugdale, 1967). Trace-element limitation in natural systems apparently does not occur frequently, however, in view of the fact that full response is nearly always attained with the addition of N and P. Experiments in four lakes, two of which were oligotrophic, in the Lake Washington drainage showed that trace-element additions alone never increased growth, although in some instances an additional enhancement was achieved by a complete medium addition over that of N and P (Hendrey, 1973). In view of these results, it would seem that trace elements are usually in adequate natural supply, particularly in eutrophic waters. A notable exception is the productivity response to Mo addition in oligotrophic Castle Lake, California (Goldman, 1960a) and some Alaskan lakes (Goldman, 1960b). Although many organisms have vitamin demands and may respond to dissolved organic compounds (e.g. glycolic acid, Fogg, 1965), these factors have seldom been demonstrated as important limiters to growth under natural conditions.

6.5.4 Chemical complexers

Organic complexers can either remove trace elements from solution and reduce inhibition or make them more soluble and available for growth.

The effect of complexers, such as EDTA, has been demonstrated by several workers (Wetzel, 1975).

Another factor linked to productivity control is the ratio of monovalent to divalent ions. As that ratio increased, productivity was observed to increase (Wetzel, 1972).

6.5.5 Carbon

Carbon is the macronutrient required by plankton in the greatest quantity, although it was not considered seriously as a principal cause for increasing eutrophication until 1969–1970. Essentially, the argument centred on the possibility that the organic carbon in waste effluents could cause eutrophication in many carbon-limited situations when the organic C was released by bacterial decomposition as CO_2 (Kuentzel, 1969; Kerr *et al.*, 1970). This idea was confronted by those who believed strongly that organic C from sewage or other waste effluents was not a principal cause for eutrophication, notwithstanding the potential limitation by C in some situations. Much was learned from this controversy and related experiments. The points and counterpoints made by the two schools of thought in 1970 are listed in Table 6.6.

Table 6.6 Two schools of thought clashed on many points as to whether C or P is the principal key to eutrophication control

School: carbon is key	School: phosphorus is key
Carbon controls algal growth	Phosphorus controls algal growth
Phosphorus is recycled again and again during and after each bloom	Recycling is inefficient: some of the phosphorus is lost to bottom sediment
Phosphorus in sediment is a vast reservoir always available to stimulate growth	Sediments are sinks for phosphorus, not sources
Massive blooms can occur even when dissolved phosphorus concentration is low	Phosphorus concentrations are low during massive blooms because phosphorus is in algal cells, not water
When large supplies of CO_2 and bicarbonate are present, very small amounts of phosphorus cause growth	No matter how much CO_2 is present, a certain minimum amount of phosphorus is needed for growth
CO_2 supplied by the bacterial decomposition of organic matter is the key source of carbon for algal growth	CO_2 produced by bacteria may be used in algal growth, but the main supply is from dissociation of bicarbonates
By and large, severe reduction in phosphorus discharges does not result in reduced algal growth	Reduction in phosphorus discharges materially curtails algal growth

Source: Great Phosphorus Controversy (1970).

To best clarify the problem, it would be most instructive to retrace the steps of this controversy, which originated in 1969 in the article by Kuentzel. He made the following points.

1 Inorganic C in the alkalinity system is insufficient to form a large algal bloom, which frequently occurs in enriched lakes. At pH of 7.5–9, the atmospheric input of CO_2 would amount, at most, to $1\,mg\,l^{-1}$ after several days of transport.
2 The $56\,mg\,l^{-1}$ (dry weight) of algae observed in a bloom in Lake Sebasticook, Maine would require $110\,mg\,l^{-1}$ CO_2, which would only come from a supply of organic matter decomposed by bacteria as a ready source of CO_2. About $30\,mg\,l^{-1}$ of organic carbon would be sufficient for that quantity.
3 The organic C in sewage effluents furnishes the needed nutrient (C) to form large algal blooms, because such low concentrations of dissolved PO_4-P ($<10\,\mu g\,l^{-1}$) are associated with large blooms. Therefore, it is obvious that sufficient PO_4 exists even at these low concentrations!

Shapiro (1970) responded to this article with the following points.

1 PO_4^{3-} is frequently, but not always, the cause of eutrophication.
2 Algae have an absolute requirement of P; $10\,\mu g\,l^{-1}$ dissolved at the time of a bloom does not indicate the supply, but only the difference between uptake and supply.
3 Sediments are not usually a source for P, but rather a sink.

A report by Kerr *et al.* (1970) was the next input to the controversy, demonstrating under experimental conditions that C could limit algal production in naturally soft water and that bacteria-originated CO_2 from organic matter serves as an effective source of C in such environments. More specifically, Kerr's findings were briefly as follows.

1 *Anacystis nidulans*, an obligate photoautotroph, was used as the test organism, which attained concentrations of 60×10^6 cells ml^{-1}.
2 Algae that received CO_2 from a dialysis bag with bacteria and organic matter grew to double the mass of that in the flask without such a CO_2 source; 6.5 versus 3.7×10^7 cells ml^{-1}. The same results were produced using CO_2-free air and 5% CO_2/air bubbled into flasks.
3 To determine if such bacteria-originated CO_2 was important in natural systems and, consequently, if the C from organic matter added to natural systems was important in bloom production, various enrichments with 'infertile' pond water ($<5\,\mu g\,l^{-1}$ P and N) were tested and population increases were recorded after 30 h of 'incubation' (Table 6.7).

Table 6.7 Relative response of *Anacystis nidulans* grown under various conditions

Conditions	Natural populations		Effect of E. coli	
	Bacteria	Algae	$CO_2 + HCO_3$ increase (mg l^{-1} 72 hr)	CO_2 produced (mg trapped)
Pond water	Decrease	No change	3.6	0
+N and P	Large increase	3 × increase	10.0	0.003
+Glucose	2 × increase	2.5 × increase	9.2	0.005
+Glucose +N and P	Large increase	6 × increase	12.0	0.030

Source: Kerr et al. (1970).

Kerr showed that CO_2 probably limited growth rate in the water tested, which was soft (alkalinity only 14 mg l^{-1} $CaCO_3$; dissolved inorganic carbon (DIC) = 5 mg l^{-1}), because the addition of organic C to this natural infertile water had about as significant an effect as N and P addition. Heterotrophs and autotrophs were stimulated by both N + P and glucose, but the greatest stimulation occurred with the addition of all three nutrients. Limitation of growth rate in these waters was clearly C > P + N. However, a question remained. Because glucose addition produced the smallest increase of any combination, would glucose addition produce the greatest long-term effect over one or more years in such infertile water? That is, which nutrient is most scarce (controlling) in the long term?

Schindler (1971b, 1974) answered the latter and most important question from the standpoint of lake management by a fertilization experiment in a low alkalinity (15 mg l^{-1} $CaCO_3$) lake (Lake 227) in Ontario, Canada. To that system, N and P were added weekly for three summers in an amount to reach 10 µg l^{-1} P. There was no change in total inorganic C in the water and the greatest change in lake algal content did not occur until the third year (Figure 6.36).

Dissolved P concentration showed very little change in response to fertilization, indicating the supply rate as the significant factor. Even in this low-carbon lake, in which short-term (hours) *in situ* photosynthesis experiments showed C to limit (Schindler, 1971b), a large mass of cyanobacteria (*Lyngbya* and *Oscillatoria*) can be produced if the N and P input is maintained long enough.

By following the CO_2 diffusion pattern with radium, Schindler (1971b, 1974) showed that the C supply originated in the atmosphere and that CO_2 was made to be limiting in the short term by adding N and P (as Kuentzel hypothesized and Kerr demonstrated). The pattern of CO_2 concentration in the fertilized and unfertilized lake over the summer can be illustrated as in Figure 6.37.

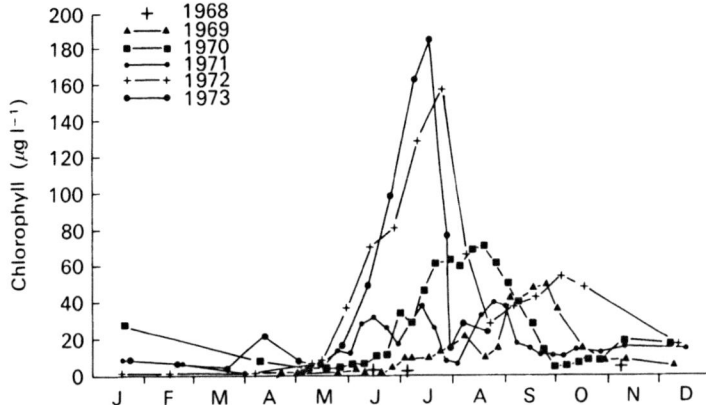

Figure 6.36 The response of phytoplankton standing crop (as chl *a*) in the epilimnion of lake 227, fertilized with 0.48 g P and 6.29 g N m^{-2} annually beginning in 1969. Chlorophyll concentration from six other unfertilized lakes never exceeded 10 μg l^{-1}. All lakes are less than 13 m deep. Note that lake 227 had a low standing crop (+) similar to that of the other lakes prior to fertilization. Large inputs of P and N will cause eutrophication problems regardless of how low carbon concentrations are, because the necessary carbon is drawn from the atmosphere (modified from Schindler and Fee, 1974).

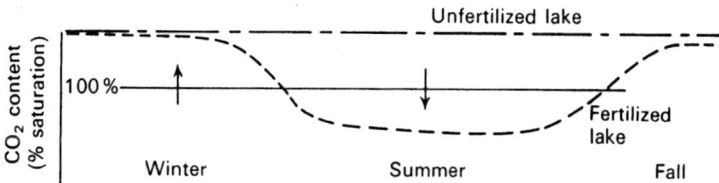

Figure 6.37 CO_2 level in fertilized and unfertilized lakes compared with the air saturation level of 100% (idealized from Schindler, 1971b).

The arrows, which indicate the determined direction of CO_2 transfer in the fertilized lake, show that additional amounts of N and P increased the supply of CO_2 from the atmosphere and eventually a large nuisance level of algae resulted. In a lake with higher alkalinity, the problem of a large algal mass may have occurred sooner because C would not have limited growth rate to such an extent in the short term.

In another lake (Lake 304), the same result occurred; that is, a gradual increase to a large algal biomass from enrichment with P, N and C. In this instance, however, biomass rapidly declined when P input was stopped in spite of continually added N and C (Figure 6.38), indicating that P was the key nutrient.

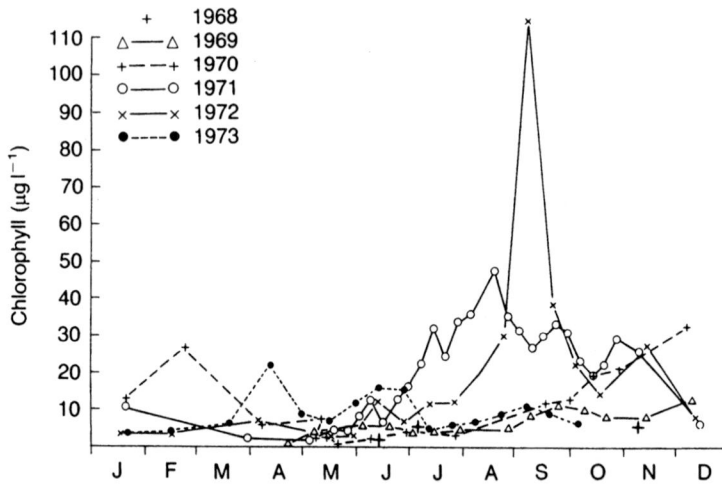

Figure 6.38 Phytoplankton standing crop (as chl *a*) in the epilimnion of lake 304 before fertilization (1968, 1969 and 1970); during fertilization with P, N and C (1971 and 1972); and with the elimination of P and continued fertilization with N and C (1973). Results show the effectiveness of phosphorus control in eutrophication management (Schindler and Fee, 1974).

The conclusion with regard to the roles of C, P and N in eutrophication is that P is usually more significant than N or C in the long term because the P supply relative to the needs for growth is usually less than for N or C. This is because the sources of P are more limited (with no atmospheric source), and PO_4 is more readily bound chemically, resulting in sediments usually being more of a sink for P than for N or C. The long-term effect would be more evident from biomass (Figure 6.38).

However, any nutrient (usually C, N or P) may limit growth rate in the short term. In this consideration the most limiting (controlling) nutrient can usually be indicated from the ratio of one concentration to another as indicated previously.

6.5.6 Species succession and dominance (why blue-greens?)

One of the most interesting and challenging subjects in limnology is understanding the succession of species in the phytoplankton. Although this is a problem in its own right, it is also a problem in water-quality management. Succession occurs seasonally in lakes, as has been discussed in general and previously in relation to temperature and light. Usually diatoms dominate in the spring, greens in summer, blue-greens (cyanobacteria) in later summer and possibly diatoms again in the autumn. However, there is much variability in this pattern. The variability that is of particular interest here is the tendency for cyanobacteria to dominate an increasing part of the growing season as eutrophication increases phosphorus content and algal biomass (Smith, 1990a,b). The objections to the blue-greens are many, but particularly they include: an objectionable appearance because they float and clump together causing scums, objectionable odours around beaches and taste/odours in drinking water supplies, a reduction in efficiency in predator–prey food chains because the filamentous and colonial nature of blue-greens, preclude effective grazing by filter-feeding zooplankton and a direct toxicity to mammals and fish when the blooms are very dense. (See Section 6.6 and Chapter 7 for more on toxicity.)

The challenge to water-quality management is to understand the cause(s) for the domination by blue-greens when eutrophication increases. Relationships between TP concentration or loading and per cent composition or the probability of bloom level have been developed and are useful for management (Walker, 1985; Sas, 1989; Lathrop et al., 1998). However, these relationships do not explain the mechanism(s) by which blue-greens gain and lose dominance in the phytoplankton. If something can be added biologically or chemically to more efficiently utilize the nutrients added to the system, the effects of eutrophication (obnoxious blooms of blue-greens) can be alleviated. This would be particularly valuable in situations where for practical or economic reasons the external nutrient supply cannot be reduced (Shapiro et al., 1975).

The role of light and temperature in the succession of phytoplankton, discussed previously, concluded that those factors, while important, could not begin to explain the phenomenon of blue-green dominance either seasonally or with an increase in eutrophication. This was also concluded by Shapiro (1990).

Varying requirements for certain nutrients was thought to contribute to succession. The succession from *Asterionella formosa* to *Dinobryon divergens* in the English lakes has been related to the different P requirements of the two organisms being met successively as the P content declines in the spring (Lund, 1950). The following ranges of dissolved P in $\mu g\,l^{-1}$ for optimum cell growth were suggested by Rodhe (1948): Low P, *Dinobryon divergens* and *Urogena americana*, lower limit < 20, upper limit < 20; Medium P, *Asterionella formosa*, lower limit < 20, upper limit > 20; High P, *Scenedesmus quadricauda*, lower limit > 20, upper limit > 20.

The Chlorococcales (e.g. *Chlorella*, *Scenedesmus*) apparently require high, extracellular concentrations of phosphorus to grow at maximum rates (see above) and probably also nitrogen and CO_2. The cyanobacteria (at least the nuisance-forming species), on the other hand, appear to grow well at low soluble concentrations, but require high supply rates (i.e. total concentrations) in order to produce a high nuisance biomass. These points will be apparent in the following discussion of the hypotheses that have been suggested to explain partially or wholly the dominance of blue-greens.

Hypothesis 1 is based on blue-greens being able to dominate at low CO_2 and high pH, which are conditions that occur more frequently as lakes become eutrophic. If CO_2 is increased and pH lowered, by adding either CO_2 or HCl, then the phytoplankton community should shift from blue-greens to green algae. Furthermore, by again raising pH (and lowering CO_2) blue-greens should once more dominate. This hypothesis was first proposed by King (1970, 1972) and further substantiated by Shapiro (1973, 1984, 1990).

Through laboratory and field work with sewage lagoons (King, 1970, 1972) originated the hypothesis that blue-green/green algal dominance is controlled by CO_2. The hypothesis is easily explained by Figure 6.39. Given a particular water with a constant alkalinity, the algal community would progress from one dominated by *Chlamydamonas* to a mixed assemblage of greens and finally to blue-green dominance as high photosynthetic rates and CO_2 consumption (stimulated by high N and P) raised the pH and lowered the CO_2 concentration. Furthermore, King contended that community shifts were associated with particular levels of CO_2, the critical level for blue-green dominance being $7.5 \mu mol\, l^{-1} (0.33\, mg\, l^{-1})$. Figure 6.39 also indicates that for a water of lower alkalinity the critical CO_2 concentration would be reached at a lower pH.

The concept is clarified further in Figures 6.40 and 6.41. Figure 6.40 shows the relationship between the pH, which provides the critical CO_2 level, and

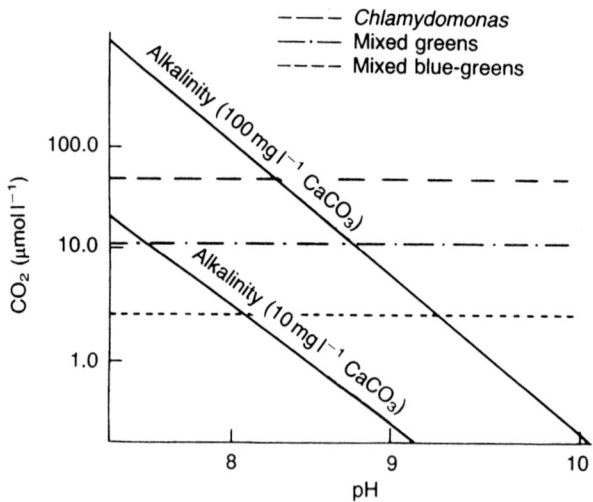

Figure 6.39 Boundaries for indicated algal communities in sewage lagoons in relation to alkalinity, pH and free CO_2 (modified from King, 1970).

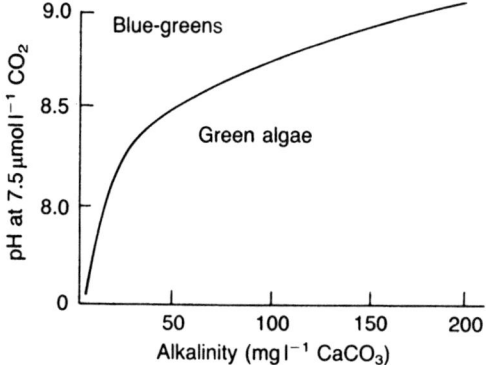

Figure 6.40 The pH at which the critical CO_2 concentration is reached, separating blue-green from green algae as a function of alkalinity (King, 1972).

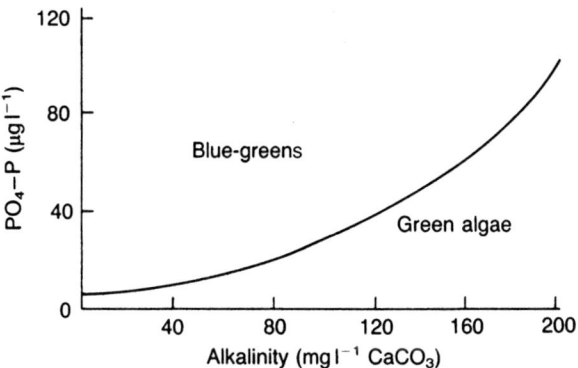

Figure 6.41 Hypothetical amount of P that must be added to extract enough CO_2 through photosynthesis to, in turn, raise pH to the level providing the critical CO_2 concentration as a function of alkalinity (King, 1972).

the water's alkalinity. Blue-greens should dominate at pH values above the line and green algae at pH values below the line. Figure 6.41 relates the amount of P required stoichiometrically to remove enough CO_2 to raise the pH to the level providing the critical concentration and a switch from green algae to blue-greens. The amount of P required would vary with the alkalinity of the water. Waters of low alkalinity theoretically need much less P enrichment to drive the pH up and, consequently, the CO_2 concentration down to the critical level than would be the case in high alkalinity waters. This is because higher alkalinity provides more buffering, i.e. less pH change for a quantity of CO_2 removed.

An experiment was performed *in situ* in 1 m deep bags in Moses Lake, Washington that tended to support King's hypothesis (Buckley, 1971). The experiment was performed to determine the effect of low-nutrient dilution water on the biomass and species composition of algae in the lake and, by adding back N and P, to determine which nutrient was responsible for the observed proportional reduction in biomass with increased dilution water. When N (as NO_3) was replaced, pH increased and CO_2 decreased as N increased (Table 6.8). Added N was expected to favour green algae, because green algae dominated in another portion of the lake where sewage effluent entered and the concentration of all macro-nutrients were high. However, the ratio of blue-greens (primarily *Aphanizomenon*) to greens increased in relation to added N (Table 6.8). Note that CO_2 was below King's critical level in all three treatments. The benefit to *Aphanizomenon* from increased N, when CO_2 is low, may be explained in later experiments by Shapiro.

Shapiro's experiments were initially conducted *in situ* in polyethylene cylinders 1 m in diameter and 1.5 m deep, utilizing the blue-green dominated phytoplankton community in eutrophic Lake Emily, Minnesota (Figure 6.41). Nutrients were added where indicated at $700\,\mu g\,N\,l^{-1}$ and $100\,\mu g\,P\,l^{-1}$. Initially, pH was lowered to 5.5, but later to various levels. Of 20 *in situ* experiments in which N, P and CO_2 were added to obtain pH 5.5, shifts to greens occurred in all except one. Shifts were observed consistently in ten other lakes and usually when pH was lowered by adding HCl. However, the shift usually did not not occur if N and P were not also added and shifts were inconsistent if pH lowering were stopped at a level higher than 5.5. The shift to greens and back to blue-greens was also accomplished to some extent by adding KOH after the shift had occurred by lowering pH (see far right, Figure 6.42).

The overall conclusion seemed to be clear. Increased eutrophication with N and P (or whichever one was limiting) could result in dominance by either green algae, which are more easily consumed in the food web, or blue-greens, which are less easily consumed and create nuisance conditions, depending on the level of CO_2 reached. The mechanism for the shift was thought to be due

Table 6.8 The ratio of blue-green cell counts to greens, maximum pH and minimum CO_2 in 1 m deep plastic-bag experiments lasting 1.5 weeks in Moses Lake, Washington

$NO_3\,N$ (mg l^{-1})	Max pH	Min CO_2 (mg l^{-1})	Blue-green:green
0.74	10.1	0.02	1.9:1
0.5	9.8	0.07	1.4:1
0.23	9.6	0.1	0.9:1

Source: Buckley (1971).

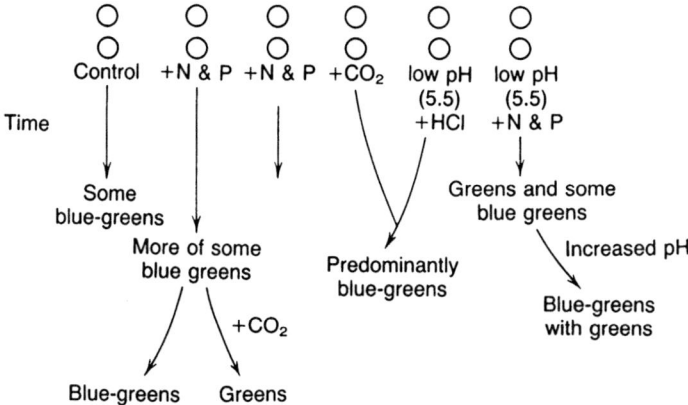

Figure 6.42 Shapiro's (1973) experiment showing the significance of CO_2 in causing a shift between green algal and blue-green dominance (schematic from personal communication).

to generally lower half-saturation uptake constants for blue-greens than green algae, and therefore, blue-greens outcompete greens as CO_2 decreases. However, Shapiro (1984) was doubtful of this, even though such uptake kinetics had been determined by Long (1976). The reason was that the shift did not occur as a gradual change in dominance, but rather the blue-greens crashed abruptly, followed by an increase in N and P in the water, possibly as a result of cell lysis due to cyanophage. Nevertheless, Shapiro (1997) demonstrates that phytoplankton assemblages dominated by blue-greens exhibited more favourable CO_2 uptake kinetics than did those dominated by green algae.

The idea that the concentration of free CO_2 is a key factor to the dominance of blue-greens or green algae was challenged by Goldman (1973), who maintained that the free CO_2 concentration is not critical to growth rate. Rather, algal growth is related to C_T. That contention was also challenged by King and Novak (1974). Although the debate was very stimulating, no consensus was reached. However, from Shapiro's experiments (Shapiro, 1973; Shapiro et al., 1975), it did appear that algal growth responded to free CO_2 because CO_2 was increased by lowering pH with acid, which in the short term should not have resulted in decreased C_T.

To test the King–Shapiro hypothesis on a large scale, an experiment was conducted during the summer of 1993 using two basins of the hypereutrophic, low-alkalinity Squaw Lake, Wisconsin (Shapiro, 1997). The south basin (9.1 ha; mean depth, 2.55 m) was artificially circulated and enriched with CO_2, while the north basin (16.8 ha; mean depth, 2.92 m) was untreated. The north basin reached a pH exceeding 10, while the enriched south basin remained around 7 all summer. In spite of the contrasting pH/CO_2 conditions, the development of large populations of *Aphanizomenon* and *Anabaena* was similar in both basins, reaching chl *a* concentrations in excess of $300\,\mu g\,l^{-1}$. Although initiation of blue-green bloom development did not depend on pH/CO_2 conditions, their more favourable CO_2 kinetics favoured their continued dominance during the summer.

Hypothesis 2 is related to the buoyancy mechanism possessed by blue-greens, but is interrelated with the $CO_2/pH/N$ and P phenomenon just described. Reynolds *et al.* (1987) have reviewed the state of knowledge on this phenomenon. Cyanobacteria have gas vacuoles, which are composed of stacks of proteinaceous gas vesicles that provide rigidity to the vacuoles. Production of gas vesicles allows blue-greens to migrate vertically in the water column, while collapse of the vesicles will result in sinking. *Oscillatoria* and *Lyngbya* are examples that tend to remain neutrally buoyant to occupy intermediate depths in deep mesotrophic lakes where nutrient content is more available than in the relatively nutrient-depleted, but lighted and mixed surface layer. Their filaments are small, providing a high 'coefficient of form resistance' to sinking. With increased nutrient content, the production of gas vesicles will increase and cells will rise into the lighted zone (Klemer *et al.*, 1982; Booker and Walsby, 1981). As photosynthesis increases the gas vesicles collapse due to increased cellular turgor pressure resulting in increased cell density, which is caused by the accumulation of carbohydrate (photosynthate) (Grant and Walsby, 1977) and potassium (Reynolds, 1975; Allison and Walsby, 1981; Reynolds, 1987). Increased cell density can also cause sinking through the accumulation of polysaccharide as ballast (Klemer *et al.*, 1988; Klemer and Konopka, 1989).

At depth, where CO_2 and/or light limit photosynthesis, but N and P are available, previously accumulated photosynthate can be assimilated into proteins and cause cell division. Turgor pressure is reduced by 'dilution' of photosynthate when the cells divide or, alternatively, by polysaccharide utilization and depletion of the ballast. Such buoyancy response to CO_2 limitation and N enrichment has been shown with *Oscillatoria* (Klemer *et al.*, 1982), with N and P enrichment in *Anabaena* (Booker and Wallsby, 1981), and with CO_2 (by pH adjustment) limitation in *Anabaena* and *Aphanizomenon* (Paerl and Ustach, 1982). The latter experiment clearly shows the effect of increasing pH (and hence reduced free CO_2 content) on buoyancy of blue-greens (Figure 6.43). More recently, Klemer and Konopka (1989) have argued that low CO_2 can promote buoyancy only if other macronutrients are limiting. In their experiments, if N and P were not limiting, then buoyancy was found to be directly related to increased inorganic C. That is, addition of N, P and C promoted more buoyancy than did N and P without C.

Large *Microcystis* colonies can sink or float through a 15 m water column in just a few hours while many days would be required for such an excursion by small *Oscillatoria* filaments (Reynolds, 1987). Thus, species like *Microcystis*

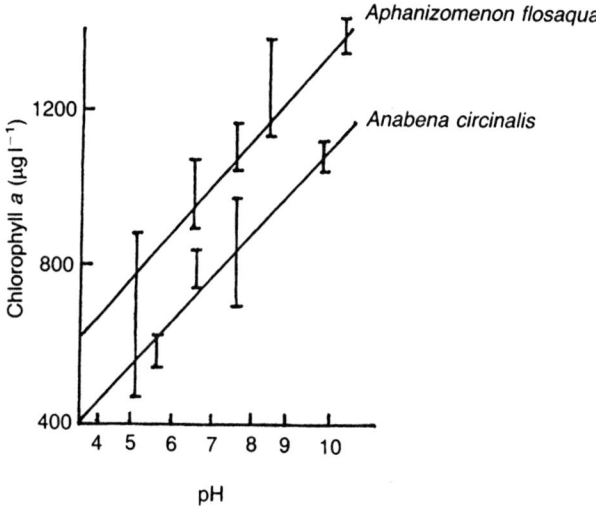

Figure 6.43 Buoyancy response of two blue-greens from exposure to increasing pH and consequent low CO_2 concentration (Paerl and Ustach, 1982). Chlorophyll *a* content is from the upper 10 ml of 100 ml graduated cylinders after 2 h.

(and apparently *Anabaena* and *Aphanizomenon*) are best adapted to rather turbid, stable (or alternately mixed/stable) water columns where they can make rather quick vertical excursions for nutrients and light, while *Oscillatoria* is adapted more to deeper, clearer waters, where even if well mixed they can maintain themselves below the mixed layer in low light and resist transport from depths where nutrients are available. As lakes become enriched and increased particulate matter restricts light penetration, the blue-greens, with their buoyancy regulation mechanisms, are favoured under conditions of strong stratification or diel partial mixing. Those conditions are not favourable to diatoms and green algae, both of which grow faster and require higher light levels than blue-greens, and lack the buoyancy mechanism (Reynolds, 1987). Spencer and King (1987) showed that even in a 1.8 m pond the light gradient was sufficient to promote strong buoyancy/sinking behaviour in *Anabaena* that led to its dominance.

Knoechel and Kalff (1975) used buoyancy regulation to explain why a blue-green (*Anabaena*) became dominant in late summer, replacing a diatom (*Tabellaria*). Because specific growth rates of the two species were not different, the authors concluded that reduced mixing as thermal stratification developed and the epilimnion deepened was the principal reason for the species shift. That is, the blue-green was favoured because its sinking rate was less than that of the diatom, which requires mixing to keep it suspended.

Reduced sinking or buoyancy of the blue-green may have been caused by changes in CO_2/pH, but the authors did not investigate that aspect.

The buoyancy mechanism is apparently unsuccessful in favouring blue-green dominance in lakes and reservoirs that are turbid from high concentrations of non-algal particulate matter. Clay particles are especially effective in attaching to the muscilaginous covering, which results in flocculating out the blue-greens and favouring greens and diatoms (Cuker et al., 1990). There are several accounts of blue-green abundance or fraction being much less than expected in turbid systems (Cuker, 1987; Søballe and Kimmel, 1988; Smith, 1990b). This was suggested as a possible explanation for green algae to continue dominating a sewage effluent-enriched arm of Moses Lake (WA) even after the effluent had been diverted (Welch et al., 1992).

Hypothesis 3 deals with zooplankton and their ability to promote the dominance of inedible colonial and filamentous blue-greens. They may do that by either grazing the small edible species and leaving the inedible ones to dominate or by recycling nutrients when the edible species are consumed, which favours the large inedible blue-greens (Porter, 1972, 1977). Toxicity of some blue-greens to animals is also a protective device against grazing loss. The literature involving the testing of these hypotheses has been reviewed recently by Burns (1987), with respect to *in situ* enclosures and by Lampert (1987) for laboratory experiments.

Even large efficient grazers (e.g. *Daphnia*) usually avoid eating most blue-greens, which are either too large to handle (*Microcystis*, *Aphanizomenon* and *Anabaena*), or if small filaments (*Oscillatoria* and *Lyngbya*) they interfere with the filter-feeding mechanism and, therefore, growth and survival of zooplankton. Also, some blue-greens may produce toxins that inhibit zooplankton (especially filter feeders) growth and reproduction (Arnold, 1971; Lampert, 1981). All of these mechanisms should lead to greater dominance by blue-greens. However, decreases in blue-greens (especially small colonies) in enclosures have been related to grazing by large zooplankton (Lynch and Shapiro, 1981; Ganf, 1983), while other results show that zooplankton have little or no direct effect on blue-green abundance (Porter, 1977; Reynolds et al., 1987). The degree to which blue-greens are consumed by large filter-feeding zooplankton depends on colony size, and whether there is an adverse effect of toxicity depends on the blue-green's growth phase. Thus, while many interactions between zooplankton and blue-greens have been observed, there is no consistent pattern of response, so this hypothesis cannot be regarded as a general case. Rather than blue-greens being favoured by zooplankton grazing, their grazing of diatoms and green algae has been shown to discourage the blooming of *Anabaena elenkinii* in eutrophic ponds. Grazing cleared the water of high densities of diatoms and greens, thus preventing the low light and low CO_2 conditions, which would normally favour blue-greens through buoyancy stimulation (Spencer and King, 1987).

Hypothesis 4 deals with the requirements for N and P. On the one hand, there is some evidence that blue-greens may be favoured by low concentrations of P, as is the case with C. The uptake rate for $^{32}PO_4$-P has been shown to be lower for blue-greens than for greens as reported by Shapiro (1973). Also, Ahlgren (1977) found that the K_s for growth of *Oscillatoria agardii* to be only $\approx 1 \,\mu g\, l^{-1}$ in continuous-flow culture experiments, with $\approx 10 \,\mu g\, l^{-1}$ being required for maximum growth rate.

On the other hand, Smith (1982, 1983, 1986) has set forth the hypothesis that blue-greens are favoured by low N:P ratios, which could mean that high P concentrations are beneficial. Smith showed in 17 lakes studied that the percentage of blue-greens in the phytoplankton was <10% in lakes with TN:TP ratios >29:1, but various percentages were observed at low ratios. The rationale is that cellular N:P required by blue-greens is lower than for other algae and as the ambient N:P ratio decreases (as eutrophication proceeds) blue-greens would be better able to exploit the available nutrient resources. The ability of some of the blue-greens to fix N could also be the reason for some of these observations. He also presented some evidence that low transparency, in addition to low TN:TP ratios, promoted the dominance by blue-greens.

The absence, or low concentrations, of trace metals (required for growth) was also suggested to affect phytoplankton dominance. In artificially fertilized Fern Lake, Washington (Olson, personal communication), 'complete medium' fertilization (including all trace elements) resulted in diatom blooms and subsequent zooplankton increase, whereas ammonium superphosphate fertilization produced an *Anabaena* bloom and relatively little zooplankton.

Hypothesis 5 deals with allelopathic excretory products from blue-greens that inhibit other algae and therefore favour the dominance of blue-greens. This has been shown by Vance (1965), who found that only *Microcystis* grew well in pond water from which that organism dominated. Keating (1977) found that summers with large blue-green blooms in Linsley Pond, Connecticut and winters with low flushing, were followed with poor diatom growth in the spring and vice versa.

Hypothesis 6 is the silica depletion concept. Increased enrichment with N and P depletes silica, which is slow to regenerate in the water column, thus limiting diatom growth, leaving more nutrients for blue-greens. Schelske and Stoermer (1971) suggested that southern Lake Michigan gradually shifted to blue-greens because silica was depleting. Schelske *et al.* (1986) have shown that the reserve of Si remaining after the stratified period is inversely proportional to the TP content.

Hypothesis 7 suggests that blue-green blooms may originate from the sediments (Fitzgerald, 1964; Silvey *et al.*, 1972; Osgood, 1988b) and rise into the water column, possibly by their buoyancy mechanism discussed earlier. Viable populations of blue-greens can exist in the sediments throughout the year (Roelofs and Ogelsby, 1970; Kappers, 1976; Fallon and Brock, 1984; Reynolds et al., 1981; Ahlgren et al., 1988a). When conditions are favourable, although initiating factors are not well known at present, a portion of the viable population rises into the water column.

Trimbee and Harris (1984) have trapped *Microcystis aeruginosa*, *Gomphosphaeria lacustris* and *Lyngbya birgei* rising from the sediments in an amount equivalent to 2–4% of the water column maximum biomass. Stimulation of *Microcystis* populations in the sediment to rise into the water column may be related to the onset of low DO and increasing light in the spring, which was indicated by the growth response of sediment populations (Reynolds et al., 1981). More directly, Trimbee and Prepas (1988) have shown that the fraction of blue-greens in the phytoplankton in 39 North-temperate lakes was inversely related to an OXYD index, which is the product of mean depth and the proportion of sediments that are oxic. This supports the hypothesis that recruitment of blue-greens from the sediments is more apt to occur in shallow lakes and is favoured by anoxic conditions. On the other hand, anoxic conditions may indicate high internal loading of P, which may be the factor favouring increased blue-greens.

Other species, such as *Aphanizomenon* and *Anabaena flos-aquae*, are not as durable as *Microcystis* in the vegetative form, but do have resting spores (akinetes) that may be important in overwintering and subsequent bloom formation. However, the role of akinetes in initiating blue-green blooms is not firmly established (Wildman et al., 1975; Rother and Fay, 1977; Jones, 1979). Barbiero (1991) and Barbiero and Welch (1992) observed that migration of sediment populations of *Gloeotrichia* (caught in traps) contributed well over 50% of the water column bloom.

In Lake Sammamish (western Washington), *Microcystis* comprised between 89 and 99% of the migrating cyanobacteria captured in traps, with *Anabaena* spp. the other migrating taxon (Johnston and Jacoby, 2003). Migration rates were more than two times higher at the shallow station (9 m) than at the deeper station (19 m). Higher migration rates from shallower sites are consistent with other studies (e.g. Lake Mendota (Wisconsin), Hansson et al., 1994; Long Lake (Michigan), Hansson, 1995; and Green Lake (Washington), Barbiero and Welch, 1992; Perakis et al., 1996).

Increased temperatures and illumination may explain elevated migration rates in the shallower regions. Increased migration rates corresponded with higher water temperatures in Green Lake (Barbiero and Welch, 1992; Perakis et al., 1996; Sonnichsen et al., 1997). Similarly, Thomas and Walsby (1986) documented lower migration of *Microcystis* with decreasing water temperatures. The stimulatory effect of light on akinete growth may also account for higher migration rates in shallower regions (Singh and Sunita, 1974). Alternatively, Hansson et al. (1994) hypothesized that a chemical substance may be associated with stratified conditions thereby hindering migration or that the pronounced thermocline associated with stratification may act as a physical barrier to migration from deeper sediments. The mechanisms that control cyanobacterial migration are poorly understood and require further study.

As the above hypotheses demonstrate, there are many factors that have been implicated in promoting dominance by blue-greens, e.g. nitrogen fixation, low soluble P concentration, low CO_2/high pH, low N:P ratio, water column stability, low transparency, buoyancy, reduced grazing, high temperature, depleted silica and allelopathy. A case can be made for each of these factors under the conditions of the particular experiments or test site. However, some of these factors are highly correlated with each other in eutrophic lakes. For example, a highly eutrophic lake during summer would usually have the following conditions: a high temperature in the epilimnion; a stable water column; low CO_2/high pH in the epilimnion; low transparency due to high algal biomass; a low N:P ratio resulting from P internal loading and N loss due to denitrification and possibly low silica due to high productivity and loss during the spring diatom bloom. The case for each of the factors is based largely through correlation and laboratory or *in situ* bag experiments, which do not prove cause and effect under whole-lake conditions where several factors are operating simultaneously. Cause and effect are most convincing through experimentation where the causitive factor is altered in a whole-lake setting and the algal dominance changes in response to the altered conditions. For example, Stockner and Shortreed (1988) promoted the bloom of *Anabaena* in one arm of a lake that was treated with a low N:P fertilizer in contrast to the other arm. Subsequently the N:P ratio was raised in the affected arm and *Ababaena* declined. We also know from Shapiro's Squaw Lake experiment that high CO_2 and low pH do not prevent blue-green blooms from developing.

Doubts can be raised about the exclusive effect of some of these factors more than others on blue-green dominance. Most blue-greens have relatively high thermal optima which have been determined experimentally, yet blooms have occurred at low temperature and many green algae have optima as high or higher than blue-greens (see Section 6.4). If blue-greens are favoured by low soluble P and high temperature, then those conditions exist in oligotrophic lakes as well as eutrophic lakes. N:P ratios are very low in Lakes Tahoe and Crater, but cyanobacteria are not dominant there. Zooplankton feeding efficiency, growth and reproduction usually decrease with an increase in large colonial and filamentous blue greens. This should contribute to blue-green dominance, yet if grazing by large zooplankton can be enhanced to clear the water of edible-size algae, which attenuate light more than large, clumped blue-greens, then the blue-greens' buoyancy mechanism is less effective.

Therefore, some combination of these factors, which may vary from time to time and place to place, would seem to be the most probable explanation for the dominance of cyanobacteria. The following general scenario is proposed as an explanation for blue-green dominance under some conditions. Increased productivity results from increased N and P and causes raised pH, lowered CO_2 and reduced light transmission in the surface water. These conditions alone would favour blue-greens because of their ability to grow at low CO_2 and access light via their buoyancy mechanism. In addition, water-column stability promotes anoxic conditions that can lead to increased P content, especially in the hypolimnion per se and shallow lakes as well. High surface temperature, which favours blue-greens directly, is also consistent with stability. Blue-greens could exploit such a physically–chemically stratified environment through buoyancy regulation. High photosynthesis and limitation of growth by N and/or P at the surface where light is not limiting would allow an accumulation of polysaccharide ballast in excess of growth demands causing loss of buoyancy. Blue-greens could then utilize the high nutrient content at depth, causing buoyancy to increase once the ballast is consumed through growth. Such an environment, with low nutrient content in the surface water and reduced mixing, would not be hospitable to non-buoyant species, which would tend to sink from the photic zone permanently. This scenario tends to fit with the concept of alternate stable states in shallow lakes (see Section 9.7).

6.6 Cyanobacterial toxicity

Some species of cyanobacteria produce toxic compounds ('cyanotoxins') that have caused livestock, wildlife and pet poisonings worldwide (see reviews by Carmichael, 1994; Chorus and Bartram, 1999; Chorus, 2001). Although these incidents have been reported for more than 100 years (e.g. Francis, 1878), investigations of the causes and consequences of cyanobacterial toxicity have only recently been conducted. As noted by Chorus (2001), two factors have triggered intensified research on cyanotoxicity in the last 20 years. One of these factors was the increased incidence of cyanobacterial blooms as a result of accelerated eutrophication of many water bodies worldwide. The other factor was the development of analytical techniques to identify the structures and concentrations of cyanotoxins. Until recently, the primary method for detection of cyanotoxins was the mouse bioassay, which is relatively insensitive and only provides a positive or negative indication of the presence of toxins. Several techniques are now available for the rapid and sensitive detection of cyanotoxins, including the enzyme-linked immunosorbent assay (ELISA) (Chu et al., 1990), the protein phosphatase inhibition assay (PPIA) and high-performance liquid chromatography (HPLC) (Chorus, 2001).

6.6.1 Occurrence of cyanotoxins

Cyanotoxins comprise a broad, diverse range of chemicals and mechanisms of toxicity (see reviews by Carmichael, 1994; Sivonen and Jones, 1999).

Major classes of cyanotoxins include the cyclic peptides, which are primarily hepatotoxins (microcystins and nodularins); alkaloids and an organophosphate, which are strong neurotoxins (anatoxin-a, anatoxin-a(S) and saxitoxins); a cyclic guanide alkaloid, which inhibits protein synthesis (cylindrospermopsin); lipopolysaccharides, which have pyrogenic properties; and dermatoxic alkaloids (aplysiatoxins and lyngbyatoxins) (Chorus, 2001).

The two groups of cyanotoxins that are of particular concern due to their prevalence and lethal effects on animals are hepatotoxins and neurotoxins. Hepatotoxins damage liver tissues, causing liver failure and even death (Carmichael, 1994). Hepatotoxins with seven amino acids are called microcystins (produced by *Microcystis* spp., *Planktothrix agardhii*, *Planktothrix rubescens* and *Anabaena* spp.) and those with five amino acids are nodularins (produced by *Nodularin spumigena*). The mechanism of toxicity involves the inhibition of the specific protein phosphatase enzymes possessed by all eukaryotic cells (Falconer, 1993). In addition, microcystins are suspected tumour promoters and teratogens (Falconer *et al.*, 1983; Falconer and Humpage, 1996).

Neurotoxins mimic the neurotransmitter acetylcholine (i.e. anatoxin-a) or inhibit acetylcholinesterase (i.e. anatoxin-a(S)) causing overstimulation of muscle cells (Carmichael, 1994). These anatoxins are primarily produced by strains of *Anabaena* spp. Saxitoxins, which are produced by some strains of *Anabaena* and *Aphanizomenon*, are also produced by several species of marine dinoflagellates that cause paralytic shellfish poisoning (PSP). Animals that have been poisoned by neurotoxins exhibit excessive salivation, staggering, muscle contractions, vomiting, diarrhoea, respiratory failure and even death.

In a comparison of surveys from different countries, Chorus (2001) found that microcystins were detected more frequently than the cyanobacterial neurotoxins. Microcystins were detected in at least 60% of all samples investigated by the Danish, German, Portuguese and Korean studies (microcystins were detected in 90% of all samples in the Czech study). Neurotoxic blooms were documented in one-quarter or less of the cases investigated in these studies (Chorus, 2001). Thus, investigations of cyanotoxicity and the development of public health guidelines have focused on microcystins due to their widespread occurrence and potential for chronic toxicity.

Microcystins are primarily found within cyanobacterial cells instead of outside the cells in dissolved form. Characteristic ranges of intracellular microcystins have been established for common cyanobacterial taxa, although there are geographic differences. For samples dominated by *Microcystis*, median values of microcystins ranged from approximately 800–900 $\mu g\,g^{-1}$ dry weight biomass, with maxima from 1500 to 5800 $\mu g\,g^{-1}$, in the data sets from Germany, the Czech Republic and Korea (Chorus, 2001). Lower ranges were found in the data set from Denmark (mean, 160 $\mu g\,g^{-1}$; maximum, 1280 $\mu g\,g^{-1}$) and higher concentrations were observed in the data from

Portugal (mean, $4100\,\mu g\,g^{-1}$; maximum, $7100\,\mu g\,g^{-1}$). Comparable levels have been measured in the USA. In lakes in western Washington, microcystins ranged from 200 to $1400\,\mu g\,g^{-1}$ in Steilacoom Lake (Jacoby et al., 2000) and from 300 to $500\,\mu g\,g^{-1}$ dry weight bloom material in Lake Sammamish (Johnston and Jacoby, 2003).

Intracellular microcystins are increasingly being measured as concentrations per litre of water to facilitate assessment of human health risks. Pelagic concentrations generally did not exceed several micrograms per litre and only occasionally reached several hundred micrograms per litre in the water bodies from different countries analyzed by Chorus (2001). However, microcystin concentrations in dense surface scums can be much higher (greater than $10\,mg\,l^{-1}$) (Chorus, 2001). Microcystins were measured in several western Washington (USA) lakes at concentrations that are comparable to these European studies (Johnston and Jacoby, 2003). Microcystin concentrations of 0.8 and $13\,\mu g\,l^{-1}$ were measured in Lake Waughop and Steilacoom Lake, respectively. In Lake Sammamish, microcystin concentrations ranged between 0.19 and $3.8\,\mu g\,l^{-1}$ throughout the lake and at all depths with the exception of a shoreline sample of a slight scum, which had a concentration of $43\,\mu g\,l^{-1}$. Microcystin concentrations as high as $32\,\mu g\,l^{-1}$ were measured in Green Lake (Seattle, Washington) (Johnston and Jacoby, 2003).

6.6.2 *Environmental factors*

As discussed in Section 6.5.6, the success of cyanobacteria in aquatic systems is attributed to many factors, including their ability to regulate buoyancy through gas vacuoles, their resistance to grazing by zooplankton due to large size or reduced palatability, their advantage at low CO_2 concentrations and high pH, and their advantage at low N:P ratios (which typically accompany eutrophication) (Reynolds, 1987; Paerl, 1988, 1996; Shapiro, 1990; Hyenstrand et al., 1998). A combination rather than any one of these factors probably accounts for cyanobacterial dominance. Once established, cyanobacteria may further alter environmental conditions to favour their growth (e.g. elevate pH due to CO_2 depletion, reduce light transparency to other phytoplankton due to surface scum formation).

Environmental conditions may affect the formation of and variation in cyanobacterial toxicity observed in natural blooms. Cyanobacterial toxicity may vary widely, between and within lakes, and during and between years, as influenced by lake morphometry, nutrient content, weather, and other environmental conditions that affect growth, accumulation and dispersal. The relative dominance of toxic and non-toxic strains within a bloom will also determine toxicity.

Few studies of the environmental factors that promote or are associated with the development of toxic cyanobacteria have been conducted. One such study occurred during summers of 1994 and 1995 in Steilacoom Lake, Washington (Jacoby et al., 2000). A toxic bloom of *Microcystis aeruginosa* occurred during the summer of 1994 but not during that of 1995. Lake characteristics that were suspected to promote the toxic bloom in 1994 were decreased water transparency, high water column stability, high surface water temperature and pH, and decreased lake flushing. Decreased water transparency during 1994 might have been due to significantly lower zooplankton abundance and thus reduced zooplankton grazing of phytoplankton. Lower zooplankton abundance was likely caused by increased planktivory by higher numbers of coho salmon fingerlings during 1994 and/or inhibition of zooplankton grazing by *Microcystis*. The success of *Microcystis* over other cyanobacteria was hypothesized by Jacoby et al. (2000) to be due to low N:P ratios and low nitrate–nitrogen with sufficient ammonium–nitrogen concentrations (Hyenstrand et al., 1998).

In a similar study, microcystins were detected in Lake Sammamish during a dense *Microcystis aeruginosa* bloom in September 1997 and during late August and early September 1999 despite low cyanobacterial abundance (Johnston and Jacoby, 2003). During the toxic episodes in 1997 and 1999, *Microcystis* was associated with a stable water column, increased surface TP concentrations ($>10\,\mu g\,l^{-1}$), surface temperatures greater than 22°C and increased water column transparency (up to $\sim$5.5 m). External loading of nutrients due to the heavy rainfall preceding the 1997 toxic episode likely provided the nutrients needed to trigger and sustain that bloom. Despite the lack of rain and subsequent external runoff, toxic *Microcystis* occurred in 1999. Migration of *Microcystis* occurred in both the deep and shallow portions of the lake, contributing to the toxic population detected in 1999 (Johnston and Jacoby, 2003).

Relationships between toxin concentrations and cyanobacterial biomass vary widely. In some studies, cyanobacterial biomass and microcystin concentrations have not been directly related (e.g. Watanabe et al., 1994; Jacoby et al., 2000). In others, microcystin concentrations were positively correlated with cyanobacterial biomass (Kotak et al., 1995, 1996). Strain composition (i.e. per cent composition of toxic and non-toxic strains) appears to be a key determinant of microcystin concentrations in natural cyanobacterial populations and likely explains the variable relationships between microcystin and biomass in most cases (Chorus, 2001).

Microcystin:chl *a* (as µg:µg) ratios measured in 55 German water bodies mostly varied between 0.1 and 0.5, with maxima of 1–2 (Fastner et al., 1999). Somewhat higher ratios (0.4–6.4) were measured in Lake Sammamish, with the highest ratios (mean = 3.2) in the hypolimnion, during the toxic episode in 1999 (Johnston and Jacoby, 2003). Thus, the toxin content of the *Microcystis* strain in Lake Sammamish appears to be relatively high, indicating an increased potential for adverse health effects during recreational uses of the lake, especially during blooms.

Although microcystins have been more commonly detected in studies of cyanotoxicity, neurotoxins have also been measured and have caused acute poisonings of animals. The first documented case of cyanotoxicity in western Washington occurred in American Lake (near Tacoma) during the winter of 1989 and caused 11 animal poisonings, including the deaths of five cats presumably due to anatoxin-a, which was isolated from the *Anabaena flos-aquae* bloom (Jacoby *et al.*, 1994). This toxic bloom was unusual in that it occurred in the winter during low light and low temperature conditions in an oligotrophic–mesotrophic lake. Jacoby *et al.* (1994) attributed the winter bloom to increased phosphorus availability following winter turnover, which remained high throughout the winter due to the lake's extremely low iron content.

6.6.3 Human health implications

Effects of cyanobacterial toxins on humans are poorly understood. Humans are typically exposed to cyanotoxins through drinking water supplies or through recreational use of water bodies with cyanobacterial blooms. Exposures to cyanotoxins have caused a variety of symptoms and illnesses including hepatotoxicity, neurotoxicity, gastrointestinal and respiratory symptoms, and allergic and dermotoxic reactions. Although outbreaks of human illness associated with exposure to cyanobacteria have been reported for years, few clinical studies have been conducted. Reviews can be found in Chorus and Batram (1999) and Chorus (2001).

Tumour-promoting activity has been found from microcystins produced by *Microcystis aeruginosa* (Falconer and Humpage, 1996). The elevated rates of liver cancer in areas of China may be linked to cyanobacterial toxins in local drinking water (Yu, 1989; Carmichael, 1994; Falconer and Humpage, 1996). Furthermore, repeated low-level exposure to such toxins could favour development of chronic gastrointestinal and liver disorders (Falconer, 1996). Exposure to cyanotoxins at a haemodialysis centre in Brazil was believed to directly or indirectly cause the deaths of 60 renal dialysis patients (Jochimsen *et al.*, 1998). To date, these are the only known human fatalities due to exposure to cyanotoxins.

Exposure to cyanotoxins can also occur during recreational water activities such as swimming and water skiing. In particular, children may ingest larger quantities of cyanobacterial cells while playing in the shallow areas of water bodies where scums accumulate. In addition to ingestion, cyanotoxin exposures may result through inhalation and dermal contact. There are numerous cases reported of illnesses due to recreational exposure to cyanotoxins that detail symptoms such as headache, nausea, diarrhoea, skin and eye irritations, sore throat, vomiting and mouth ulcers (reviewed in Chorus and Bartram, 1999; Chorus, 2001). In an epidemiological investigation of 852 people exposed to cyanotoxins during swimming, Pilotto *et al.* (1997)

demonstrated that these symptoms were associated with exposure to cyanobacteria, but surprisingly there was no direct relation to the concentrations of microcystins or neurotoxins.

The increased detection of cyanotoxins in drinking and recreational waters worldwide poses a challenge for water resource managers. The World Health Organization (WHO) has issued a guideline that microcystin concentrations should not exceed $1\,\mu g\,l^{-1}$ in drinking water (WHO, 1998; Chorus and Bartram, 1999). Several countries have also developed guidelines for recreational exposure to microcystins. For example, the Federal Environmental Agency of Germany recommends posting warning signs and conducting a remedial investigation if chl a exceeds $40\,\mu g\,l^{-1}$ and cyanobacteria are dominant (Chorus et al., 2000; Chorus, 2001). If chl a exceeds $150\,\mu g\,l^{-1}$ or if total microcystin concentrations exceed $100\,\mu g\,l^{-1}$, closure of the swimming beach is recommended until the bloom declines. WHO's guidelines differentiate three levels of hazards based on cyanobacterial abundance as measured by concentrations of chl a or cell densities (Falconer et al., 1999). However, the large variability in microcystin:chl a ratios documented in only a few studies to date (Fastner et al., 1999; Johnston and Jacoby, 2003) indicates that direct measurement of microcystins is a more accurate indicator of cyanobacterial toxicity.

6.6.4 Control of toxic cyanobacteria in lakes and reservoirs

Multiple measures have been used to kill or remove cyanobacteria in lakes and reservoirs. However, long-term control of cyanobacteria requires reducing phosphorus inputs from both external and internal sources (see Chapter 10). Public health concerns due to toxic cyanobacteria may dictate the use of immediate and aggressive control measures; however, some of these techniques may not prevent and may indeed exacerbate toxicity. In this regard, the use of copper algicides, such as copper sulfate, is not recommended for treatment of cyanobacterial blooms because these chemicals induce lysis of cyanobacterial cells, releasing the toxins to the water (Kennefick et al., 1993; Jones and Orr, 1994; Lam et al., 1995). Because extracellular, soluble toxins are more difficult to remove by conventional water treatment processes such as coagulation, chlorination and sand filtration (Hitzfeld et al., 2000), the use of algicides may actually increase human exposure to cyanotoxins in drinking water. There was a significant increase in acute liver conditions in residents of Armidale (Australia) who drank water treated with copper sulfate to control a toxic *Microcystis* bloom (Falconer et al., 1983).

However, the use of chemicals such as calcium carbonate, alum or calcium hydroxide can effectively precipitate cyanobacterial cells without inducing lysis (Kennefick et al., 1993; Lam et al., 1995; Hitzfeld et al., 2000). Activated carbon is also effective in removing microcystins in drinking water (Chu and Wedepohl, 1994). Ozonation may lyse cells and cause toxin release; however, this treatment process may be effective in destroying cyanotoxins at high doses (Hitzfeld *et al.*, 2000). The use of ozone at a high enough concentration to oxidize organic matter and toxins is essential. In addition, the creation of toxic ozonolysis by-products needs further study (Hitzfeld *et al.*, 2000).

6.7 Cyanobacteria and water supplies

In addition to producing toxins that have potential human health risks, cyanobacteria in water supplies have other undesirable effects on drinking water supplies (Cooke and Carlson, 1989; Cooke and Kennedy, 2001). Cyanobacteria are one of the primary sources of compounds that cause taste and odours, decreasing the palatability of the finished water and increasing customer complaints. Furthermore, some of the compounds that are produced by cyanobacteria in the eutrophication process also serve as precursors of disinfection by-products (DBPs), which are also associated with human health risks.

6.7.1 Taste and odours

Taste and odour problems in water supply are associated with eutrophication, and more specifically with high densities of cyanobacteria (Bierman *et al.*, 1984; Arruda and Fromm, 1989; Seligman *et al.*, 1992; Smith *et al.*, 2002). Other aquatic microorganisms (e.g. fungi, actinomycetes, green algae) also produce compounds that cause objectionable taste and odours, and become more abundant during decay of cyanobacterial blooms (Kennefick *et al.*, 1993). Volatile organic compounds responsible for taste and odours include geosmin (*trans*-1,10-dimethyl-*trans*-9-decalol), MIB (2-methyl isoborneol) and beta cyclocitral. The odour threshold concentrations of geosmin and MIB are around 4–5 and $9\,\text{ng}\,\text{l}^{-1}$, respectively, for people sensitive to smell (AWWA, 1987b). These compounds are difficult and costly to remove from water supplies; therefore, control of cyanobacteria is critical in preventing the production of taste and odour compounds.

In Cheney Reservoir, which supplies the drinking water for the City of Wichita, KS, water column concentrations of geosmin were strongly related to algal, particularly cyanobacterial, growth ($r^2 = 0.72$) (Smith *et al.*, 2002). Mid-summer blooms of *Anabaena* and *Aphanizomenon* coincided with a large peak in geosmin concentration and also with the number of taste and odour complaints received by the City of Wichita. Recommendations to reduce taste and odour problems in this reservoir were based on control measures that would achieve a TP concentration $<110\,\mu\text{g}\,\text{l}^{-1}$ throughout the system, thereby reducing phytoplankton biomass ($<10\,\mu\text{g}\,\text{l}^{-1}$ chl *a*) and thus taste and odours.

Lake Youngs is a municipal drinking water reservoir that serves over a million people in Seattle and King County, Washington. In spite of the oligotrophic state ($6\,\mu g\,l^{-1}$ TP), there were customer complaints to the Seattle Water Department regarding the earthy-musty flavour of the water in recent years (Zisette et al., 1994; Herrera Environmental Consultants, 1996). The source of the taste and odour problems in Lake Youngs water appeared to be the periphytic cyanobacteria on or near the lake sediment, as planktonic cyanobacteria had not been observed. In support of this conclusion, geosmin concentrations were elevated in bottom samples ($5-51\,ng\,l^{-1}$) relative to water column and inlet samples ($<2-3\,ng\,l^{-1}$) during 1990 (Entranco Engineers, 1993). Moreover, the water intake was at the depth where cyanobacteria were most abundant. In 1992, geosmin was detected in all bottom water samples ($4.9-17\,ng\,l^{-1}$) but not in inlet samples (Zisette et al., 1994). Actinomycete bacteria were not detected in any of the periphyton or sediment samples, indicating that they were not a major source of the taste and odours. MIB was also detected in water samples in July 1992, exceeding the odour threshold in 6 of 36 samples ($10-16\,ng\,l^{-1}$). *Oscillatoria*, a known producer of geosmin and MIB, composed more than 30% of the biovolume at the two stations where the highest geosmin concentrations were measured (Zisette et al., 1994). A strong positive relationship ($r^2 = 0.63$) between earthy-musty flavour ratings by a taste and odour panel and geosmin concentrations provided further evidence that geosmin is the primary compound causing taste and odours in Lake Youngs.

Geosmin concentrations were even higher in 1995 than in previous years (Herrera Environmental Consultants, 1996). Genera of cyanobacteria that were found in Lake Youngs' periphyton samples that are commonly associated with taste and odour problems included *Oscillatoria* (15 species), *Lyngbya* (six species) and *Phormidium* (three species). Detailed analysis of phytoplankton samples revealed the presence of *Oscillatoria limnetica*, which is a known producer of geosmin that is very small and may have been overlooked in previous phytoplankton analyses. Flavour analyses indicated that objectionable grassy and fishy flavours were moderately strong during the spring bloom of diatoms, suggesting that cyanobacteria may not be the only cause of off-flavours in Lake Youngs (Herrera Environmental Consultants, 1996). Furthermore, recent monitoring results indicate that MIB concentrations have increased in the reservoir due to changes in periphytic cyanobacteria populations (Joubert, 17 June 2003, personal communication).

6.7.2 Disinfection by-products

DBPs are produced during the disinfection of water by chlorine, chlorine dioxide, chloramines and ozone. Naturally occurring organic matter in the raw water interacts with the disinfectant to produce DBPs. Trihalomethanes (THMs) (e.g. chloroform, which is usually the most common THM) and haloacetic acids (HAA) form from chlorination, whereas disinfection with ozone produces brominated but not chlorinated by-products. DBPs are associated with various adverse health effects including spontaneous abortions (Waller et al., 1998) and stillbirths in humans (USEPA/ILSI, 1993; King et al., 2000), and cancer in animals (Krasner et al., 1994; Boorman et al., 1999). DBPs are recognized as potential health problems in water supplies throughout the world (WHO, 2000).

Because of these potential health effects, DBPs are regulated to increasingly protective levels by the USEPA under the Disinfectant/Disinfection By-Products Rule of the Safe Drinking Water Act. The new maximum allowable concentration of total THMs is $80 \mu g l^{-1}$ and for five HAAs the maximum level is $60 \mu g l^{-1}$ in finished water supplies (USEPA, 2001b). These new requirements mean that 60% of the large and 70% of the small drinking water treatment plants will be required to make some changes in their operations to meet the compliance dates and standards. These modifications will increase the cost of treatment and further strain the ability of water utilities to provide high-quality drinking water to a growing human population.

The precursors of DBPs are organic carbon molecules, especially dissolved organic carbon (DOC) species, in the raw water. Aquatic humic substances (i.e. humic and fulvic acids) compose about 50% of the naturally occurring organic matter and thus are potential precursors. There are multiple sources of these organic compounds, including aquatic plants, algae and other microorganisms, and non-living organic matter (Cooke and Kennedy, 2001). These sources may originate within the lake (i.e. autochthonous) or be derived from the watershed, i.e. allochthonous sources (Stepczuk et al., 1998b). Stream and wetland inputs of organic compounds, derived mostly from terrestrial vegetation, to reservoirs can be primary sources of DBP precursors. The organic matter from municipal wastewater treatment facilities and agricultural activities can also be THM precursors (Amy et al., 1990). In Cannonsville Reservoir (New York), the autochthonous contribution to the THM precursor pool was found to be linked to primary productivity (Stepczuk et al., 1998a). Allochthonous inputs of organic matter were also a potentially important source of THM precursors in this reservoir (Stepczuk et al., 1998b).

While alive and during decomposition, cyanobacteria and algae are significant sources of THM precursor compounds (Hoehn et al., 1980; Oliver and Shindler, 1980). The blue-green *Microcystis aeruginosa* is a significant producer of THM precursors (van Steenderen et al., 1988) as well as toxins (discussed in Section 6.6.). Other cyanobacteria including the common species *Aphanizomenon flos-aquae* also have been implicated in the release of dissolved organic compounds that are potential DBP precursors, as well as compounds that cause objectionable taste and odours in water supplies.

Thus, DBPs (as well as taste and odour problems and cyanotoxins) are strongly related to the degree of eutrophy of the reservoir (Palmstrom et al.,

1988). Water supplies that are eutrophic are more likely to have problems associated with the above compounds, higher treatment costs and more difficulty in meeting DBP standards (Cooke and Kennedy, 2001). Furthermore, identification of the relative contributions of internal versus external sources of the organic precursors is critical to the development of management measures to reduce DBPs in water supplies. In-lake and watershed measures that reduce phosphorus inputs and cycling in reservoirs will reduce algal biomass and thus the production of DBPs. The use of copper algicides should be avoided, as these treatments have been found to induce the lysis of algal and cyanobacterial cells and increase release of dissolved organic compounds (Peterson *et al.*, 1995), as well as intracellular toxins (see Section 6.6). Therefore, cost-effective controls for the cause of eutrophication, especially P, are vitally important in the water supply field.

Chapter 7

Eutrophication

7.1 Definition

Eutrophication is the process by which water bodies become more productive through increased input of inorganic nutrients. In a strict sense, the term refers to nutrients only and not necessarily to a response in production (Beeton and Edmondson, 1972). Thus, a lake could become eutrophic from increased nutrients even though productivity did not increase, as for example if algal growth were limited by light due to high suspended solids. Normally, however, increased plant productivity and biomass are considered part of the eutrophication process. Increased input of sediment can also cause eutrophication by decreasing depth, which could expand the area suitable for macrophytes, as well as encourage a more effective exchange of nutrients between sediments and water. Sediment can be allochthonous organic or inorganic and autochthonous organic.

Depending on the source of the increased nutrient supply, eutrophication can be either natural or cultural. Nutrient supply can increase naturally as a result of forest fires, earthquakes or simply increased erosion following a dramatic increase in precipitation. In these cases, the watershed and lake can stabilize when the rate of nutrient input (eutrophication) returns to earlier levels. Lakes can also age naturally and become more eutrophic. They will eventually become shallower through normal rates of sedimentation, which are typically of the order of a few millimetres per year. Assuming a sedimentation rate of 4 mm year^{-1}, 250 years would be required to reduce lake depth by 1 m. As lakes become shallower the sediment area to lake volume increases, providing more opportunity for nutrient recycling (see internal loading) from sediments, as well as greater light availability for photosynthesis. However, this ageing process could take thousands of years and many lakes may never become eutrophic, depending on nutrient inputs.

From a water quality standpoint the concern about eutrophication is usually due to cultural eutrophication. Increased nutrient input can originate from such man-caused sources as: direct discharge of treated or untreated domestic sewage (including phosphate detergents); industrial wastes, such as food-processing plants and dairies, leaching from fertilizer applied to forests, lawns, pastures or cultivated land; leaching from failing on-site sewage treatment systems; and stormwater runoff from urbanized land. Increased eutrophication in a water body occurring over a period of only 10–50 years usually indicates anthropogenic rather than natural processes.

7.2 Phosphorus mass balance models

Phosphorus was identified in Chapter 6 as usually the most limiting nutrient in freshwater, so examining some of the models used to estimate P concentration in lakes is an appropriate starting point with eutrophication. Since Vollenweider's initial work (Vollenweider, 1969a), many models have been developed that describe the mass balance of phosphorus in lakes. Notable among these are Dillon and Rigler (1974a), Vollenweider (1975, 1976), Chapra (1975), Nürnberg (1984) and Walker (1977). Walker (1987a) has adapted the P mass balance model for reservoirs. All P mass balance models are based on the kinetics of continuously stirred tank reactors (CSTR) commonly used in chemical engineering (Reckhow and Chapra, 1983). By continuously mixing the volume in such a reactor, holding that volume constant and maintaining the water inflow rate equal to the water outflow rate, the following mass balance equation applies with units of mass/time:

$$dCV/dt = C_i Q - CQ - KCV \tag{7.1}$$

where C is substance concentration in the reactor and C_i is concentration in the inflow. Q is flow rate, V is reactor volume and K is the reaction rate coefficient. If K is taken as a first-order depletion reaction and both sides are divided by Q, so that $V/Q = \tau$, the hydraulic retention time in t^{-1}, the equation becomes:

$$dC/dt = C_i - C + K\tau \tag{7.2}$$

and at steady state it becomes:

$$C = \frac{C_i}{1 + K\tau} \tag{7.3}$$

This is mathematically identical to the TP mass balance model for lakes proposed by Vollenweider (1969a):

$$dTP/dt = L/\bar{z} - \rho TP - \sigma TP \tag{7.4}$$

where L is areal loading of TP in mg m^{-2} time^{-1} ($L/\bar{z}\rho = C_i$ in Equation 7.3), $\bar{z}$ is mean depth in m, ρ is flushing rate (Q/V) in year^{-1}, TP is the lake TP concentration (assumed to equal the outflow concentration) and σ is the sedimentation rate coefficient in time^{-1}. Loading is the basic data used in mass balance models, determined by flow × concentration, and expressed in lake areal (as in Equation 7.4), volumetric (mg m^{-3} t^{-1}) or simply mass input (kg t^{-1}) units. At steady state for some time period, usually a year, the equation becomes:

$$\text{TP} = L/[\bar{z}(\rho + \sigma)] \tag{7.5}$$

This mass balance, steady-state formulation may be confusing because it implies that areal TP loading and mean lake depth help determine the predicted lake TP concentration. However, morphometry does not actually affect lake concentration, because both of the terms contain the lake surface area (SA) in their denominators, which cancel out. The formulation also contains the flow term (Q) in both its numerator and denominator. Once the surface area and flow terms are cancelled out, this equation simplifies to:

$$\text{TP} = \frac{\text{TP}_i}{1 + \sigma\tau} \tag{7.6}$$

where TP$_i$ is the average, flow-weighted input TP concentration and τ is the lake's hydraulic retention time (i.e. V/Q or the inverse of ρ). This equation is identical to Equation 7.3. Equation 7.6 predicts that in-lake TP concentrations will be higher if lakes have higher inflow concentrations, lower sedimentation rates and/or shorter hydraulic retention times.

Equation 7.6 shows that over the long term the lake will equilibrate to the given input concentrations, TP sedimentation rate and lake hydraulic retention time. If the input TP concentration is changed then a time will be required for equilibrium to the new input concentration and, if a first-order rate reaction is assumed, the times to 50 and 90% of equilibrium will be, respectively:

$$t_{50} = \frac{\ln 2}{\rho + \sigma} \quad \text{and} \quad t_{90} = \frac{\ln 10}{\rho + \sigma} \tag{7.7}$$

The difficult problem with this model is the determination of the sedimentation rate coefficient, as all other parameters can be easily determined directly. This means that for a lake with a known input concentration, known in-lake concentration and known hydraulic retention time, σ can be estimated accordingly:

$$\sigma = \frac{(\text{TP}_i/\text{TP}) - 1}{\tau} \tag{7.8}$$

This approach is useful for determining the actual sedimentation rate in individual lakes, but it provides no insight into the processes that regulate TP sedimentation losses in lakes globally. One approach that some authors have applied to conceptualize TP losses in lakes is to use a unitless retention coefficient R_{tp} (Vollenweider and Dillon, 1974; Dillon and Rigler, 1974a), which can be determined directly as

$$R_{tp} = 1 - \frac{TP}{TP_i} \qquad (7.9)$$

To develop a model that describes these losses in a large number of lakes, it is useful to have a general way of estimating sedimentation. Considerable effort has been expended by a number of limnologists to come up with a global method to estimate σ based on lake morphometrics and hydraulics. The intent with this research has been to determine whether a lake's propensity to trap phosphorus, and hence have lower in-lake concentrations relative to input concentrations, is systematically related to basic physical characteristics of the lake. From Equations 7.3 and 7.6 above it is obvious that lakes with longer hydraulic retention times should have lower relative in-lake concentrations and this has been substantiated in numerous studies. Through statistical analyses of large data sets, various authors have concluded that the sedimentation rate coefficient is positively related to the flushing rate accordingly $\sigma \approx \rho^{0.5}$ or $\tau^{-0.5}$ (Larsen and Mercer, 1976; Vollenweider, 1976) or nearly so (Canfield and Bachman, 1981). [Because this association is based on the best fit (i.e. a correlation) the units do not match.] This direct relationship between σ and ρ may seem illogical, since it was previously stated that lakes with longer retention times will trap more phosphorus (i.e. lower lake TP) and ρ is inversely related to τ. However, note that because the exponent on the water retention term (τ) in Equation 7.6 is greater than the exponent on the sedimentation term estimated as $\rho^{0.5}$ or $\tau^{-0.5}$, the lake's retention time will exert a greater impact on predicted TP concentrations than will σ estimated as $\rho^{0.5}$ (or $\tau^{-0.5}$). If $\tau^{-0.5}$ (or $\rho^{0.5}$) is used to estimate σ the Equation 7.6 can be rewritten as:

$$TP = \frac{TP_i}{1 + \tau^{0.5}} \qquad (7.10)$$

In essence, this empirical relationship tells us that lake retention time exerts the greatest effect on a lake's propensity to trap TP, but that lakes with long retention times (i.e. >10 years) will tend to trap less phosphorus than otherwise expected based purely on hydraulic retention time (if a constant sedimentation rate for all lakes is assumed). This relation also suggests that lakes with short retention times (i.e. <1 year) will tend to trap more phosphorus than expected based on their hydraulic retention time. No author has provided a mechanism for the sedimentation rate's apparent dependence on the lake retention time. However, it is conceivable that lakes with long retention times (hence low hydraulic loading rates) favour microbial assemblages (especially bacteria and phytoplankton) that are very efficient at recycling nutrients within the food web since external nutrient supplies to these lakes are limited. Similarly, lakes with short retention times will favour microbial assemblages that are not particularly efficient at recycling nutrients since external nutrient supplies to these lakes are, relatively speaking, large.

Extensive empirical research conducted on a large number of lakes suggests that flow-weighted input TP concentrations exert a first-order effect on lake TP concentrations, lake retention time exerts a second-order effect and the sedimentation rate exerts a third-order effect. Using only the input TP concentration it is possible to explain 71% of the overall variability in lake TP concentration for a database including ≈300 lakes (M.T. Brett, unpublished data). Including the lake's actual hydraulic retention time in these predictions improves the variability explained to 77%, and including the inverse association between lake hydraulic retention time and the estimated sedimentation rate improves the explained variability to 84% (M.T. Brett, unpublished data). However, because these predictions are based on log-transformed observed and predicted data, the prediction error for TP content in any given lake may be quite large. In the case described above the best model still had uncertainty in TP estimates of −41% to +68% (±1 S.D.) of the true value (see also Figure 7.1). For example, Equation 7.10 predicts

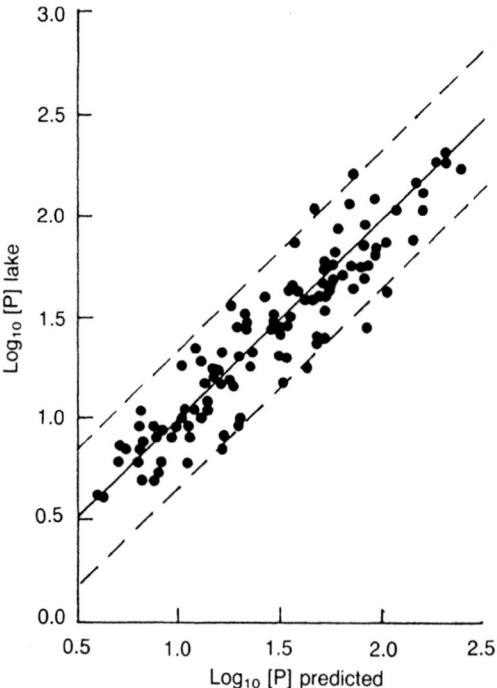

Figure 7.1 Comparison of lake TP [P] with that predicted from Vollenweider's (1976) improved loading criteria model according to a slightly modified version: $\log [P] \text{ lake} = \log [P]_i/(1 + 1.17\tau^{0.45})$ (modified from Chapra and Reckhow, 1979). Dashed line is 95% confidence interval.

pre- and post-diversion TP in Lake Washington quite well, but provides a much poorer fit for Lake Sammamish and post-diversion levels in Shagawa Lake (Table 7.1). The poor agreement for all models with post-treatment values in Shagawa Lake may be due mostly to insufficient time for internal loading to have declined. Several years were required in Lake Sammamish. The Nürnberg (1984) and OECD (1982) models show variable results, with the latter providing a better fit for Lake Sammamish (see Chapter 10 for more on the response of these lakes). This comparison suggests an alternative to calibrating a model to a particular lake data set, i.e. to compare several models and select the one that provides the best fit.

Lakes vary in nutrient content, but most of that variation is due to nutrient input, not lake morphometry such as area, shoreline length, depth and depth/area. Much of the variation in nutrient concentration among lakes, especially lakes unaffected by development, can be accounted for by external nutrient loading, or more pertinently, inflow concentration, which adjusts loading for inflow volume and lake flushing rate (see above). That is demonstrated by the most simplified version of the Vollenweider model (Equation 7.10), which has been demonstrated to explain most of the lake-to-lake variation in phosphorus content in large populations of lakes. Much of the remaining variation in TP among lakes may be explained by internal loading, which is omitted from these models, and has been demonstrated to be more a function of eutrophication than lake type. Nürnberg (1996) found that lake morphometry accounted for little, if any, variation in TP content in a group of lakes worldwide. Also, effects of nutrients are not lake-type-dependent. Thresholds for lake trophic state, regardless of the classification scheme, do not include a provision for lake type (Table 7.3). That is, a given TP concentration is expected to produce a given chl a concentration and transparency regardless of water body type. That thresholds are commonly interrelated by regression equations supports that contention (Nürnberg, 1996). Therefore, there seems little justification for considering lake type in predicting TP concentration.

Table 7.1 Annual mean TP predicted before and after P control using three models compared with observed. Data from Lake Washington and Shagawa Lake are from Edmondson and Lehman (1981) and Larsen et al. (1979), respectively

Lake	Treatment	TP(μg l^{-1})			
		Obs	Vollenweider	Nurnberg	OECD
Washington	Diversion	Pre 64	67	84	49
		Post 17	17	21	16
Sammamish	Diversion	Pre 33	43	55	34
		Post 19	29	33	25
Shagawa	P removal	Pre 51	50	54	39
		Post 30	7	8	11

Lake morphometry is, nonetheless, an important consideration in managing eutrophication of lakes. Summer TP concentrations can be 2–3 times higher in shallow lakes than in surface waters of deep lakes (Scheffer, 1998). That is due to the greater efficiency of recycling sediment P (internal loading) into the mixed layer and the concentrating effect on the internal loading flux into a shallower water column (see Chapter 4). However, without eutrophication, internal loading may not exist (sediments would always be oxic and pH near neutral), regardless of depth, so there would not be a concentrating effect of shallower water, and lake TP would be strictly due to external loading and lake flushing rate, and sedimentation as affected by ρ. However, lake flushing rate could produce differences among lakes by being less in deep than shallow lakes. That would tend to produce lower lake TP concentrations in deep (due to greater sedimentation loss) than shallow lakes, given that they had similar watershed area:lake area ratios with the same watershed yield rates for TP.

The general mass balance input/output TP models, such as Vollenweider's and others', include internal loading in the sedimentation term. That is, lakes with substantial internal loading (see Section 4.1.3) have smaller sedimentation, so a general model utilizing external loading only would underestimate lake TP. Such lakes represent some of the variations in Figure 7.1.

Whether or not internal loading is important in a lake can be determined by constructing a mass balance on a monthly or twice-monthly basis. All sources of P input from streams, groundwater, precipitation, wastewater, etc., as well as the output, must be monitored for flow and concentration on at least a twice-monthly basis with continuous gauging of major inflows recommended (Cooke *et al.*, 1993). Net internal loading can then be estimated from the following equation by solving for sedimentation (S):

$$S = I - O - \Delta(TP_l) \tag{7.11}$$

Where I is the input, O the output and $\Delta(TP_l)$ the change in whole-lake TP for the chosen time period, such as each month. If S is negative, due to an increase in output and/or lake TP that exceeds input, there is net internal loading. That is the same equation as Equation 7.4, with units in kg or mg m^{-2} per time. Internal loading is usually greatest in summer and can represent the major source (Welch and Jacoby, 2001), so time intervals should be monthly or less. The importance of internal loading as a cause for summer algal blooms would be underestimated if calculated on an annual basis, which is often the case.

If internal loading is important, as may be the case in either unstratified oxic or stratified anoxic lakes (see Section 4.1), then the model may need to be modified to account for the two sources. Nürnberg (1984) suggested using the following model to account for internal load (L_{int}):

$$TP = TP_i(1 - R_{pred}) + L_{int}/\bar{z}\rho \tag{7.12}$$

where R_{pred}, the predicted TP retention coefficient, was found in 54 oxic lakes to be best represented by:

$$R_{pred} = 15/(18 + \bar{z}\rho) \tag{7.13}$$

There are several other formulations where values refer to settling rates that range from 10 to 16 m year^{-1} (Chapra, 1975; Kirchner and Dillon, 1975; Vollenweider, 1975). Internal loading can also be added to the previous model versions (Equations 7.9 and 7.10). However, there was no attempt to treat oxic and anoxic lakes separately in the development of these other approaches of estimating sedimentation.

Using observed TP, and solving Equation 7.12 for L_{int}, makes it possible to calibrate Nürnberg's model for a particular stratified, anoxic lake. L_{int} so calculated can be compared with other estimates of internal loading for the lake in question, such as P release rates determined from sediment cores in the laboratory or observed rate of increase in hypolimnetic P concentrations. Nürnberg (1987b) has shown rather good agreement between these two methods of estimating internal loading in anoxic lakes. If there is good agreement among these different estimates of internal loading for a particular lake, the model is verified for that lake. If not, there may be an error in the estimate for sedimentation and a different modelling approach should be taken. Also, the lake may not be in equilibrium with its external loading.

Example problem 1

(a) Given a lake with a mean depth of 15 m, a flushing rate of 1.5 year^{-1} (outflow rate/lake volume) and a mean inflow TP concentration of 80 μg l^{-1}, calculate the lake's expected external TP loading in mg m^{-2} year^{-1}.
From Equations 7.5 and 7.6 we know that $L/\bar{z}\rho = TP_i$, so

$$L = TP_i \bar{z}\rho = 80 \text{ mg m}^{-3} \times 15 \text{ m} \times 1.5 \text{ year}^{-1} = 1800 \text{ mg m}^{-2} \text{ year}^{-1}$$

(b) If the lake concentration is actually 70 μg l^{-1}, calculate its expected internal loading of TP.
Using a modified Equations 7.10 and 7.12, gives

$$L_{int} = \bar{z}\rho[TP - TP_i/(1 + 1/\rho^{0.5})]$$
$$= 15 \times 1.5[70 - 80(0.55)] = 585 \text{ mg m}^{-2} \text{ year}^{-1}$$

Or using Equation 7.12, gives

$$L_{int} = 15 \times 1.5[70 - 80(1 - 15/(18 + \bar{z}\rho))] = 441 \text{ mg m}^{-2} \text{ year}^{-1}$$

Thus, the difference between internal loading with the Vollenweider and Nürnberg versions of the sedimentation term is 144 mg m^{-2} year^{-1}. Formulated in this way, L_{int} is net internal loading, because there is no loss of L_{int} due to sedimentation. Being net internal loading, it should be comparable with the observed increase in hypolimnetic TP (Nürnberg, 1984). This is reasonable, because L_{int} results in TP build-up in the hypolimnion, which is essentially net internal loading. P released into the hypolimnion should not sediment at the same rate as P in the epilimnion because being in the dark it is not as subject to uptake and sinking by algae. That is supported by the similarity between sediment P release determined in latoratory cores and field observations of hypolimnetic P build-up (Nürnberg, 1987b).

Even if one can verify a particular steady-state model for a lake, there are problems in using the steady-state version. Firstly, it is often difficult to determine an appropriate time interval (most often annual) in which the lake mean TP represents a steady state, especially if flushing rate exceeds once per year. Secondly, internal loading usually occurs during the productive period (whether stratified or unstratified) and may contribute proportionately more to growing-season TP and biomass than would external loading, especially if the latter occurs primarily during the non-productive period (e.g. winter in the Pacific Northwest). These problems can be averted by calibrating and verifying a non-steady-state version of the mass balance model (Equation 7.4), but including L_{int} as follows:

$$dTP/dt = L_{ext}/\bar{z} - \rho TP - \sigma TP + L_{int}/\bar{z} \tag{7.14}$$

Note that TP resulting from L_{ext} and L_{int} in Equation 7.14 is subject to sedimentation and, therefore, L_{int} is a gross rate. In this case, the numerator in Equation 7.5 would be $L_{ext} + L_{int}$.

No more data are usually required for the non-steady-state version, because TP loading and lake concentration data are usually collected on at least a monthly basis. While the data are, however, usually reduced to annual means for a steady-state approach (or some interval consistent with ρ), TP is computed for each time interval with the non-steady-state version. In fact, weekly time steps are recommended to obtain more realistic smooth curves even if data were collected less frequently. The model can be calibrated by determining the sedimentation rate coefficient (σ) that gives the best fit between predicted and observed data for the oxic period. Although Larsen et al. (1979) used a constant σ among years in Shagawa Lake, with good success, the model could be verified in Lake Sammamish from year to year only if σ were allowed to vary as a function of flushing rate, i.e. $\sigma = \rho^x$ (where $0 < x < 1$; $x = 0.78$, Figure 7.2, Welch et al., 1986). This is analogous to Equation 7.10 where $x = 0.5$. A formulation such as $\sigma = y\rho^x$ may be necessary where sedimentation rates are low, because as x approaches zero the sedimentation rate remains around 1.0 regardless of the flushing rate.

There still may be a problem with the non-steady-state model in stratified lakes even if there is a good fit between observed and predicted whole-lake TP. Chlorophyll a and transparency are a function of TP in the productive

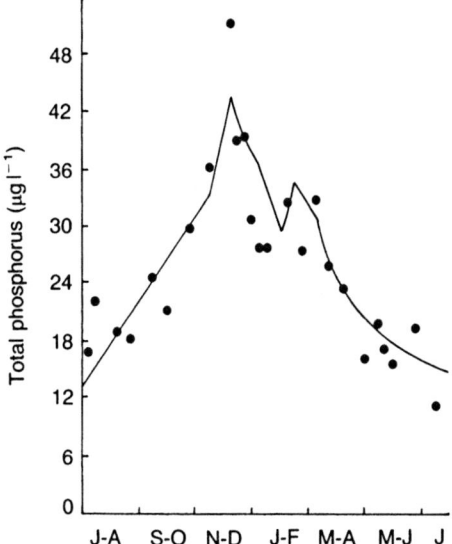

Figure 7.2 Predicted (—) versus observed TP (whole-lake) (●) in Lake Sammamish, WA.

zone, probably the epilimnion, and not a function of the whole-lake TP. Epilimnetic TP usually declines during the stratified period while hypolimnetic TP increases. Thus, either the epilimnion and hypolimnion must be modelled separately with functions included to account for exchange of TP between the two, or mean epilimnetic TP must be estimated from a relationship between epilimnetic TP and whole-lake TP. Two-layer (epilimnion and hypolimnion) TP models were developed for Lakes Onondaga and Sammamish and are routinely used in lake management (Auer *et al.*, 1997; Perkins *et al.*, 1997). Exchange between the epilimnion and hypolimnion occurs via turbulent eddy diffusion during the stratified period and by entrainment as the thermocline sinks or rises. High-TP hypolimnetic water can be captured in the epilimnion during summer storms and at turnover (Larson *et al.*, 1981). Entrainment can be estimated by observed change in volume of the two layers indicated by thermocline depth. Diffusion can be estimated from the vertical heat exchange coefficient (m per week) calculated from temperature gradients and the epilimnion and hypolimnion TP concentrations. With a two-layer model, internal loading was found to be less important to the epilimnetic TP concentration, and hence lake water quality, than previously thought from modelling the whole-lake TP concentration only.

There have been several approaches to dealing with the uncertainty in TP predictions for individual lakes. For example, in using Equation 7.14 to predict future TP concentrations in Lake Sammamish resulting from increased development in the watershed, uncertainty was included by choosing a range in TP yield coefficients (areal loss rates for different land uses) and the 5 and 95% flow probabilities for the principal inflow stream (Shuster et al., 1986). TP sedimentation was a function of ρ and increased/decreased flow resulted in dilution/concentration of the estimated TP loading. By this procedure, the prediction of 31 μg l^{-1} TP by the year 2000 had a $\pm 10\%$ error due to TP yield and a $\pm 20\%$ error due to flow. Flow was found to cause most of the year-to-year variation in loading.

Another approach is to use first-order error analysis to calculate the uncertainty in the model and loading by utilizing low, high and most probable loading estimates from yield coefficients (Reckhow and Chapra, 1983). For a model of the type of Equation 7.13, Reckhow and Chapra (1983) determined an error of $\pm 30\%$, which is added to the loading uncertainty. Once those uncertainties are summed, confidence intervals for a single model estimate for TP can be calculated. In order to evaluate small changes in TP, predicted from relatively small changes in loading, uncertainty can be applied to the TP concentration change, rather than the before and after concentration.

7.3 Trophic state criteria

Criteria exist to describe a lake's quality and its trophic state. They include the concentration and loading rate of nutrients which are the cause, as well as biological and physical indices, which are the effect. The value of numerical criteria is to allow the quality of the lake to be defined or the lake classified. Criteria can be used to chart accurately the course of a lake as it becomes more or less eutrophic or to judge if lake quality is suitable or unsuitable for recreational or water supply use and to estimate the outcomes of management alternatives to be quantified.

Some qualitative characteristics of eutrophic, compared to oligotrophic lakes, are shown in Table 7.2. Shallow lakes are more likely to be eutrophic than deep lakes, because of the greater potential of nutrient recycling from sediments, but shallow lakes can also be oligotrophic if their nutrient input has been historically low and they are naturally flushed at a fairly high rate with low nutrient water. As a corollary of depth, the relative size of the hypolimnion is apt to be less in a shallow lake and, thus, is more apt to become anoxic, facilitating internal loading of phosphorus. The amount of plankton, its productivity, frequency of blooms and resulting low water transparency are all typical of eutrophic lakes and atypical of oligotrophic lakes.

Blue-greens (cyanobacteria) frequently dominate the phytoplankton in eutrophic lakes. They can form scums on the water surface that can be very unsightly and odiferous. They can taint the flesh of fish with a musty odour

Table 7.2 Qualitative characteristics of oligotrophic and eutrophic lakes

	Oligotrophic	Eutrophic
Depth	Deep	Shallow
Hypolimnion: epilimnion	>1	<1
Primary productivity	Low	High
Rooted macrophytes	Few	Abundant
Density of plankton algae	Low	High
Number of plankton algal species	Many	Few
Frequency of plankton blooms	Rare	Common
Depletion of hypolimnetic oxygen	No	Yes
Fish species	Cold water, slow growth, restricted to hypolimnion	Warm water, fast growth, tolerate low O_2 in hypolimnion and high temperature of epilimnion
Nutrient supply	Low	High

or taste, a problem that is probably more widespread than appreciated, and also cause taste and odours in water supplies. Finally, some blue-greens are at times toxic and frequently result in deaths of domestic animals. The toxins from blue-greens ('cyanotoxins') have been grouped under neurotoxins, hepatotoxins and contact irritants (Carmichael, 1986, 1994; Chorus, 2001). Neurotoxins are primarily produced by *Anabaena* spp. while *Microcystis* spp. appear to be the primary producers of hepatotoxins worldwide. Several genera have been linked with irritants; *Gleotrichia, Anabaena, Aphanizomenon* and *Planktothrix*, formerly *Oscillatoria* (Carmichael, 1986). Surveys of lakes with blue-green blooms have shown that blue-green toxicity is probably a rather common occurrence (Chorus, 2001). Repavich *et al.* (1990) sampled 102 sites in Wisconsin, USA and found that 25% contained toxic algae as determined by the mouse bioassay. Sivonen *et al.* (1990) showed toxicity by the same method in 44% of the 188 samples from 125 sites in Finland and site results from Sweden showing >50% incidence of toxicity. The use of new, more sensitive techniques (e.g. enzyme-linked immunosorbent assay, protein phosphatase inhibition assay) to analyze cyanotoxins has permitted detection of microcystins in the majority of bloom samples analyzed in studies in Europe, Asia, Australia and North America (Chorus, 2001). Cyanotoxicity is discussed further in Section 6.6.

Rooted macrophytes are also more typical of shallow lakes, which are more apt to have larger littoral areas (<3 m) with richer sediment. Factors controlling macrophyte distribution and abundance will be discussed in Chapter 9.

Combining the higher productivity and relatively smaller hypolimnion results in greater rates of oxygen depletion and lower oxygen concentration as hypolimnetic volume declines. The kinds and abundance of fish are affected mostly by the combination of oxygen and temperature. As lakes become more eutrophic, oxygen becomes depleted in the hypolimnion (see Section 7.3.3) limiting the suitable fish habitat to the epilimnion and metalimnion. The shallower the lake, the smaller the hypolimnion and the less is its cooling effect on the epilimnion. If the epilimnion is too warm ($>20°C$) for cold-water fish, then they will disappear as the hypolimnion with more suitable temperature becomes oxygen deficient.

There are many indices that have been used to classify trophic state and lake quality. Thirty different sources for trophic state criteria were listed by Porcella *et al.* (1980) and there are still others. There are also many goals for the quality of lakes, some of which may be in conflict. The aesthetically pleasing, clear blue water of ultraoligotrophic lakes does not yield large amounts of fish. Compromises may be called for between lake quality more favourable to fish production (meso or meso-eutrophic) and that preferred for swimming, boating and viewing. However, for coldwater fish species, there may be little difference between the appropriate trophic state for fisheries and recreational uses (see Section 7.3.3).

7.3.1 Nutrients, productivity, biomass and transparency

Beginning with Sakamoto's chlorophyll *a* (chl *a*) versus TP relationship (Sakamoto, 1971) and Edmondson's observation (Edmondson, 1970) in Lake Washington's recovery from sewage diversion that chl *a* was closely tied to phosphorus concentration, TP and chl *a* have become two of the three most widely used indicators of trophic state. Transparency, determined by a black and white Secchi disc (SD), can often be empirically related to chl *a* and TP, if water colour and inorganic suspended sediment are low (see Secchi disc in Section 6.3).

There are several empirical relationships between chl *a* and TP (see Ahlgren *et al.*, 1988b), but the two earliest and often used are by Dillon and Rigler (1974b) (Equation 7.15) and Jones and Bachmann (1976) (Figure 7.3, Equation 7.16), which are respectively:

$$\log \text{ chl } a = 1.449 \log \text{ TP} - 1.136 \tag{7.15}$$
$$\log \text{ chl } a = 1.46 \log \text{ TP} - 1.09 \tag{7.16}$$

Dillon and Rigler used data from 46 lakes, mostly in eastern Canada, and Jones and Bachmann used data from 143 lakes covering a broad range in trophic state. The former data set contained TP values from spring turnover and mean summer chl *a* while the latter was composed of summer mean

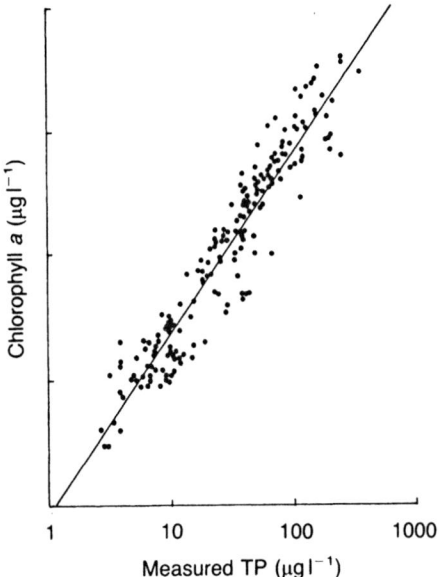

Figure 7.3 Relationship between summer levels of chl a and measured total phosphorus concentration for 143 lakes (Jones and Bachmann, 1976).

values. The equations agree rather closely, not-with-standing the difference in data-averaging times. Summer means for chl a and TP are most often used to define the trophic state of a lake. Therefore, sampling intensively throughout the non-growing season, simply to determine the lake's trophic state, may be unjustified. While TP may be higher when inflows are greater during the winter and spring, the summer mean epilimnetic concentration represents the residual after sedimentation and, therefore, should be closely tied to the P in algal cells.

Because most chl a–TP relationships using large data sets are log-log, the accuracy for prediction is not great for any single lake. In Equation 7.15, for example, errors of prediction for a chl a concentration of $5.6\,\mu g\,l^{-1}$ ($10\,\mu g\,l^{-1}$ TP) can be $\pm 60-170\%$ and $30-40\%$ for 95% and 50% confidence limits, respectively. The high correlation coefficients (0.95) tend to hide the accuracy problem which may be due to lake-to-lake variations in cellular chl a, zooplankton grazing and/or other limiting factors such as light and N (Ahlgren et al., 1988b). Sometimes it is possible to develop a relationship for the individual lake of interest that provides much greater accuracy of prediction (Smith and Shapiro, 1981). For example, nine years of data from Upper Klamath Lake, Oregon (mean depth 2 m), shows a strong relationship between summer TP and chl a and is one that could be confidentially used for management in that lake (Figure 7.4). Usually, however, this requires too much data and one must rely on the published relationship that provides the best agreement with the individual lake data.

The chl a–TP relationship was used by Carlson (1977) to develop a numerical trophic state index (TSI), which is probably the most commonly used index. However, the three indicators represent a rather narrow range within the spectrum of trophic state indicators. While Carlson's TSI and Porcella's LEI (lake evaluation index, Porcella *et al.*, 1980) reduce lake trophic state to one or more numbers, in an attempt to remove the subjectivity inherent in the terms oligotrophic, mesotrophic and eutrophic, etc., it is still necessary to use those terms in communications about lake quality. In contrast to indices in which the values are dependent on regression analyses with a particular data set, Carlson's TSIs (and LEIs) represent absolute values for chl a, TP and SD. $\log_2$ transformations were used to interrelate these three indices, so that a doubling in TP is related to a reduction by half in SD. Values of TP, chl a and SD from the following equations for TSI of 40, 50 are respectively: 12, 24; 2.6, 6.4; 4, 2.

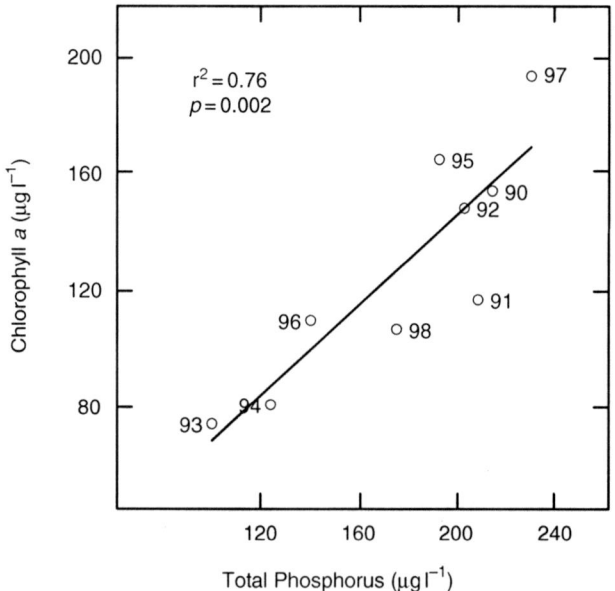

Figure 7.4 Relationship between volume-weighted lake-wide means for TP and chl a during June–September in Upper Klamath Lake, 1990–1998 (Welch and Burke, 2001).

$$\text{TSI} = 10(6 - \log_2 \text{SD}) \quad (7.17)$$
$$= 10(6 - \log_2(7.7/\text{chl } a^{0.68})) \quad (7.18)$$
$$= 10(6 - \log_2(48/\text{TP})) \quad (7.19)$$

If annual mean values are used for TP in Equation 7.19 then 64.9 is used as the numerator instead of 48. The non-linear relationship between SD and chl a is shown in Figure 7.5 (also see Figure 6.12). Note that the greatest change in SD occurs below a concentration of chl a of $\approx 30\,\mu\text{g}\,\text{l}^{-1}$. Above $30\,\mu\text{g}\,\text{l}^{-1}$, there is relatively little change in SD with increasing chl a. The significance of this will be discussed later with respect to expectations in water quality improvement relative to the costs of various reductions in TP.

Threshold values for trophic states are shown in Table 7.3. These average values are generally agreed upon as satisfactory to indicate recreational water quality and have been derived from several sources (Chapra and Tarapchak, 1976; Porcella et al., 1980; Nürnberg, 1996), but based largely on regression equations. SD thresholds are similar to values predicted from the TP and chl a thresholds using Equations 7.17–7.19 (i.e. 3.6 and 1.9 m for o–m and m–e). Absolute thresholds work well for communicating lake conditions; moreover, these values are interrelated and have recreational and water supply significance. For example, a mean chl a threshold for eutrophy of $9\,\mu\text{g}\,\text{l}^{-1}$ has been supported in work by Walker (1985), who used three TP-chl a models to show that the frequency of blooms, defined as chl $a > 30\,\mu\text{g}\,\text{l}^{-1}$, begins to increase only after the summer average reaches

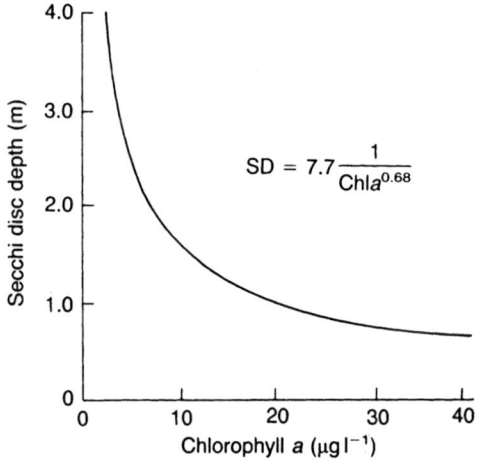

Figure 7.5 Relationship of Secchi-disc depth to chlorophyll according to equation from Carlson (1977).

Table 7.3 Trophic state values for oligotrophic–mesotrophic (o–m), mesotrophic–eutrophic (m–e) and eutrophic–hypereutrophic (e–h) boundaries (most after Nürnberg, 1996). AHOD is areal hypolimnetic oxygen deficit and AF is anoxic factor (see Section 7.3.3)

Variable	o–m	m–e	e–h
TP[a] ($\mu g\,l^{-1}$)	10	25[b]	100
Chl a[a] ($\mu g\,l^{-1}$)	3.5	9.0	25
Secchi depth[a] (m)	4.0	2.0	1.0
AHOD ($mg\,m^{-2}\,day^{-1}$)	250	400	550
Net DO ($mg\,l^{-1}$)[b]	4.5	5	–
Min DO ($mg\,l^{-1}$)[b]	7.2	6.2	–
AF (days)	20	40	60
TN[a] ($\mu g\,l^{-1}$)	350	650	1200

Notes
a Summer mean or median.
b Porcella et al. (1980).

about $10\,\mu g\,l^{-1}$ (Figure 7.6). This chl a level is related to a TP concentration of $30\,\mu g\,l^{-1}$ by Carlson's equations (Equations 7.18–7.19) and by Equation 7.15 (Equation 7.16 gives $27\,\mu g\,l^{-1}$). The absolute oligotrophic and eutrophic threshold values plotted on the LEI graph (chl a and SD slightly different from Table 7.3) show considerable consistency (Figure 7.7). The narrow

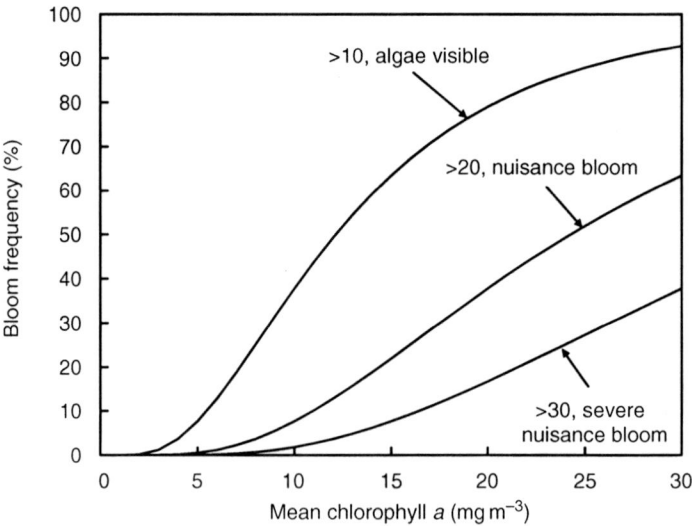

Figure 7.6 Percentage of chl a concentrations at $>10, 20$ and $30\,\mu g\,l^{-1}$ (frequency of blooms) versus mean summer chl a for three data sets (based upon Walker, 1985 and personal communication for calibration to Corps of Engineers reservoirs).

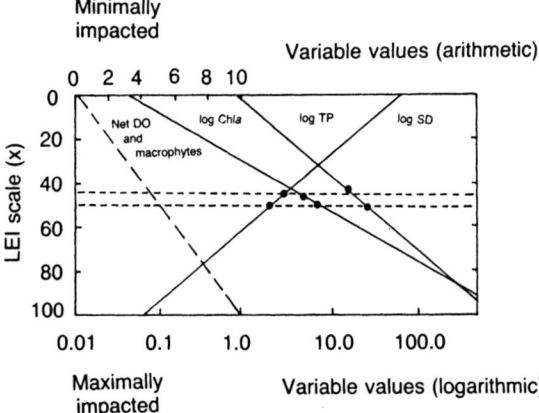

Figure 7.7 Lake evaluation index (LEI) values, where zero represents minimal impact and 100 is maximum, versus trophic state indicators (from Porcella et al., 1980). Chlorophyll a, TP and SD are plotted on the logarithmic scale, while net DO and macrophytes are on the arithmetic scale.

range of mesotrophy reflects the logarithmic nature of the index, which follows from the non-linear relation of SD with TP and chl *a* as originally derived by Carlson (Equations 7.17–7.19). Although there is logic and consistency among these threshold values, they are nonetheless averages (or medians) and may not always be consistent when applied to an individual lake.

There is some advantage to expressing lake trophic state on a probability basis (Chapra and Reckhow, 1979; OECD, 1982). This approach acknowledges that there is a high degree of uncertainty in trophic state criteria. For example, from the OECD model an annual mean TP of $40\,\mu g\,l^{-1}$ has a 38% chance of representing eutrophy, a 56% chance of mesotrophy and a 6% chance for oligotrophy. The $25\,\mu g\,l^{-1}$ threshold (Table 7.3) represents a lake with a high probability of being mesotrophic, but has a low and equal chance of being either eutrophic or oligotrophic. Also, a lake with equal chance of being either eutrophic or mesotrophic, which could represent a meso-eutrophic threshold, would have a TP concentration of almost $50\,\mu g\,l^{-1}$. While overlap and uncertainty in trophic state are realities, a threshold value of $50\,\mu g\,l^{-1}$ represents a state that is far too degraded from the standpoint of recreational use to be interpreted as mesotrophy. This would more than double chl *a* from the eutrophic threshold and, according to Figure 7.6, produce blooms ($>30\,\mu g\,l^{-1}$) more than 20% of the time.

The OECD criterion, i.e. $50\,\mu g\,l^{-1}$ TP as a mesotrophic threshold, may be more appropriate for tropical than temperate lakes. Thornton (1987a) has suggested 50–$60\,\mu g\,l^{-1}$ as the upper limit of mesotrophy in tropical lakes.

The greater tolerance for P enrichment in the reservoirs of southern Africa may result from higher temperature and consequently higher metabolic rates, as well as higher flushing rates, than temperate lakes (Thornton, 1987b). The higher metabolic rates would tend to result in lower algal biomass even though productivity was higher, associated with higher TP. The reason for greater tolerance of P is probably not that N is limiting. Thornton (1987a) cites several studies showing that although there is some variability most workers found African lakes to be P limited usually.

Example problem 2

(a) Calculate the expected average summer chl a concentration and transparency in the lake from example problem 1.

Using Equation 7.15, which yields nearly identical results as Equations 7.18 and 7.19 combined, gives

$$\log \text{chl } a = 1.449 \log 70 - 1.13$$
$$= 1.54$$
$$\text{Chl } a = 35.5 \, \mu\text{gl}^{-1}$$

and using Equations 7.17 and 7.18 combined gives

$$SD = 7.7/35.5^{0.68}$$
$$= 0.68 \text{ m}$$

(b) By what percentage must internal loading be reduced in order to improve summer transparency to 2.0 m?

Using the same equations in reverse gives

$$\log \text{chl } a = (\log 7.7/SD)/0.68$$
$$= (\log 7.7/2.0)/0.68$$
$$= 0.86$$
$$\text{Chl } a = 7.3 \, \mu\text{gl}^{-1}$$
$$\log TP = (\log \text{chl } a + 1.136)/1.449$$
$$= (\log 7.3 + 1.136)/1.449$$
$$= 1.37$$
$$TP = 24 \, \mu\text{gl}^{-1}$$
$$L_{\text{int}} = 15 \times 1.5(24 - 40 \times 0.55)$$
$$= 45 \, \text{mg m}^{-2}\text{year}^{-1}$$
$$\text{Reduction} = (585 - 45)/585 \times 100$$
$$= 92\%$$

TN (total nitrogen) is listed in Table 7.3, but has been used infrequently as an indicator. Except for unusual cases (e.g. Lake Tahoe, Goldman, 1981), TN would normally be a pertinent indicator only in highly eutrophic lakes where it could be expected to control productivity (see Section 6.5.2). Smith (1982) has presented a chl a predictive equation that includes TN and may be more useful in highly eutrophic systems than a TP–chl a relationship alone:

$$\log \text{chl } a = 0.6531 \log \text{TP} + 0.548 \log \text{TN} - 1.517 \tag{7.20}$$

Equation 7.20 predicted a chl a concentration of $21 \pm 9\,\mu\text{g}\,l^{-1}$ in Moses Lake, WA, while Equation 7.14, based only on TP, predicted $50 \pm 23\,\mu\text{g}\,l^{-1}$. The observed value in that N-limited system was $23 \pm 11\,\mu\text{g}\,l^{-1}$. Prarie et al. (1989) have presented equations for predicting chl a from TN and TP for different TN/TP ratios. This approach resulted in improved accuracy of prediction.

7.3.2 Primary productivity

Ranges in photosynthetic rate, as measured by radioactive-carbon assimilation, that are indicative of trophic states have been suggested by Rodhe (1969). These limits, with appropriate modifications, are given in Table 7.4. Similar values from two other authors are cited by Grandberg (1973). Although these ranges were determined for temperate lakes, recent comparisons of productivity at various latitudes suggest that a more general application may exist.

Although Table 7.4 shows quite a large range for eutrophy, Rodhe (1969) and others (Grandberg, 1973) have separated eutrophic from hypereutrophic at $\approx 40\%$ of the range ($250\,\text{g}\,\text{C}\,\text{m}^{-2}\,\text{year}^{-1}$ and $1000\,\text{mg}\,\text{C}\,\text{m}^{-2}\,\text{day}^{-1}$). The interval between oligotrophic and eutrophic in Table 7.4 is considered the mesotrophic state.

The main difficulty in using productivity as a trophic state indicator is that areal productivity depends as much on light availability as on nutrients. Integration over the photic zone can give rates in an oligotrophic lake that are nearly as high as those in a eutrophic lake (see Figure 7.8). Moreover, of principal importance to the quality of a lake is its transparency, which is a

Table 7.4 Ranges in rates of primary productivity as measured by total carbon uptake attributable to the trophic states of lakes

Period	Oligotrophic	Eutrophic
Mean daily rates in a growing season (mg C m^{-2} day^{-1})	30–100	300–3000
Total annual rates (g C m^{-2} year^{-1})	7–25	75–700

Source: Modified from Rodhe (1969) with permission of the National Academy of Sciences.

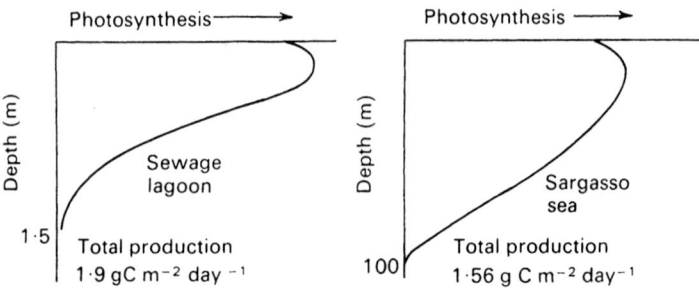

Figure 7.8 Comparison of productivity profiles in a sewage lagoon and the Sargasso Sea (modified from Ryther, 1960).

function of the concentration of particles (i.e. algal biomass) in the water and not productivity per unit area.

7.3.3 Oxygen

Algal biomass as an organic matter source is really no different from sewage in its ultimate O_2 demand or BOD. The stoichiometry of nutrients versus BOD can be hypothesized from the photosynthetic equation (Stumm, 1963). CO_2 is fixed into organic carbon with light and nutrients according to a ratio of 106:16:1, C:N:P, as previously indicated:

$$106\,CO_2 + 90\,H_2O + 16\,NO_3 + 1\,PO_4 + \text{light energy} \rightarrow$$
$$C_{106}H_{180}O_{45}N_{16}P_1 + 154.5\,O_2 \qquad (7.21)$$

The photosynthetic quotient, or PQ, $= 154/106 = 1.45$; experimentally this ratio averages 1.2. In the reverse reaction (respiration) O_2 is subsequently used to convert C_{106} to $106\,CO_2$, which results in O_2 deficits in lakes and sags (min. O_2 concentration downstream from BOD source) in streams. The potential for secondary BOD effect from treated sewage in the form of N and P can theoretically be estimated accordingly: $\approx 75\%$ of the assimilable C that is removed from sewage in secondary treatment can go back into decomposable organic matter and BOD in the form of plankton (fixed in photosynthesis) simply from the utilization of N and P that is not effectively removed by secondary treatment. In turn, the BOD potential of P, if completely utilized, can be estimated as follows:

$$\frac{154.5 \times 32}{1 \times 31} = \frac{4950}{31} = 160\,\text{mg}\,O_2\,\text{mg}^{-1}\,P$$

thus $1\,\text{mg}\,P = 160\,\text{mg}\,O_2$ if all the P remaining from treatment is converted to organic C by photosynthesis. In order for that to occur, P must be limiting the plankton growth.

To illustrate the secondary BOD effect, an experiment by Antia *et al.* (1963) is appropriate (Figure 7.9). Measurements were made during and following a phytoplankton bloom in a 6 m diameter plastic sphere submerged in the sea. The top graph shows production of carbon by the oxygen method and particulate carbon measurements. The lower graph shows decomposition of produced organic carbon. The oxgen used is expressed as carbon. Note the release of phosphorus after blackout, which caused cessation of photosynthesis and death of the plankton. The greater oxidation of C than production of particulate C was explained as the bacterial breakdown of dissolved organic matter (difference between production as measured by particulate C and O_2). The authors suggested that oxygen consumption was large as a result of the bacterial breakdown of excreted

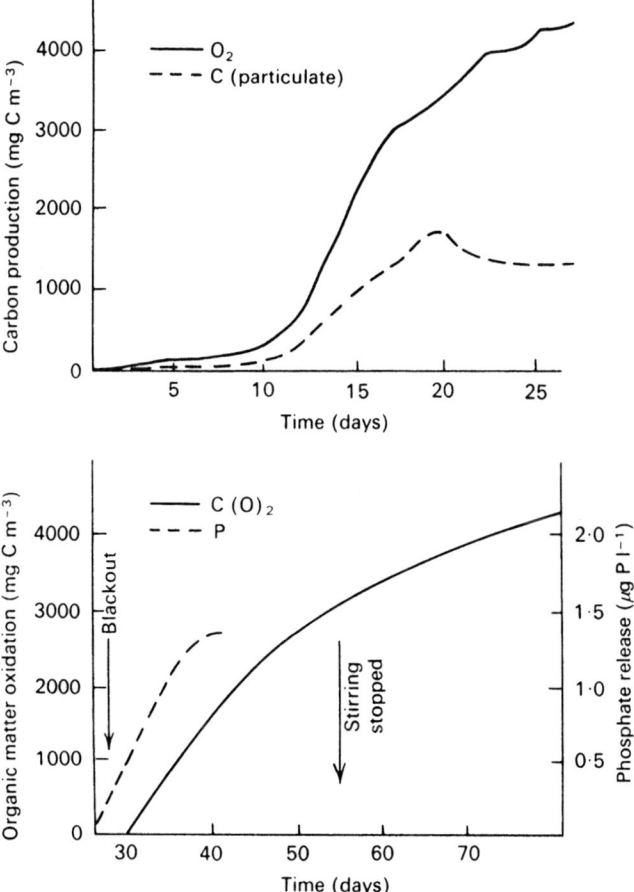

Figure 7.9 Results of a bloom and subsequent decomposition and nutrient release in a large plastic sphere (Antia *et al.*, 1963).

dissolved C and that particulate matter was used more as bacterial substratum.

This process can be observed to cause O_2 problems in overfertilized waters. For example, Baalsrud (1967) showed that the oxygen depletion problem in the Oslo Fjord, Norway, was caused largely by the fertilizing effect of sewage added to inflowing freshwater (which also furnishes needed iron). The dominant organism was *Skeletonema costatum*, a marine diatom that frequently dominates in nearshore waters. This problem can occur because for each gram of phosphorus fixed into algal tissue from treated or untreated sewage at least 50 g of oxidizable carbon is fixed from naturally occurring CO_2 (Figure 7.10).

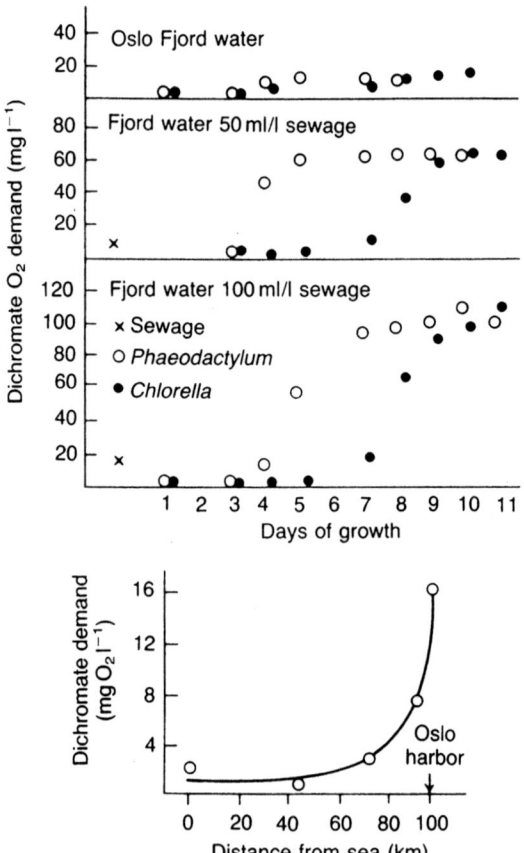

Figure 7.10 Algal growth potential experiment showing conversion of inorganic nutrients in sewage to oxygen-demanding organic matter. Top graphs show the potential of sewage in Oslo Fjord water at different distances from the sea (Baalsrud, 1967).

Indices of DO have included: oxygen deficit rate, AHOD in mg m^{-2} day^{-1} (areal hypolimnetic oxygen deficit); net DO; minimum DO; and AF in days (anoxic factor) (Table 7.3). Neither minimum DO nor net DO has been correlated with TP loading or concentration, but AHOD was related to TP retention (Cornett and Rigler, 1979), TP concentration (Nürnberg, 1996) and TP loading (Welch and Perkins, 1979):

$$\log \text{ODR} = 1.51 + 0.39 \log L/\rho \ (r = 0.73) \tag{7.22}$$

AHOD is usually calculated as the slope of the linear plot of hypolimnetic DO against time, multiplied by the hypolimnetic mean depth. DO may disappear too rapidly in some highly enriched lakes to give an accurate estimate of AHOD, even if sampling is twice per month. In that case, twice weekly sampling may be necessary.

AHOD indicated the increased eutrophication in Lake Washington (Edmondson, 1966) (Figure 7.11). AHOD reached levels of 500–800 mg m^{-2} day^{-1} early in the eutrophication of the lake, but also decreased between 1955 and 1962. The decrease in the rate was believed to be due to an increase in the quantity of blue-greens, which would tend to be buoyant and be decomposed in the epilimnion.

AF (days) = $\sum_{i=1}^{n} t_i \cdot a_i/A_o$, where t is days of detectable anoxic conditions (i.e. ≤ 1 mg l^{-1} DO), a_i is the anoxic sediment area and A_o is the lake surface

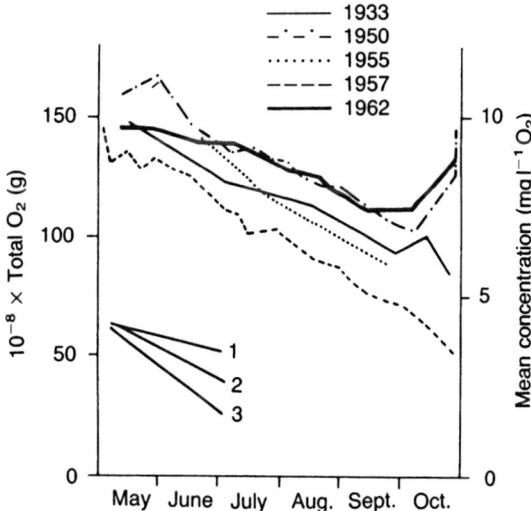

Figure 7.11 Hypolimnetic oxygen deficit below 20 m in Lake Washington in 1933, 1950, 1955, 1957 and 1962. For comparison of rates, constant slopes for deficit rates of 1, 2, and 3 mg O$_2$ cm^{-2} month^{-1} are indicated (Edmondson, 1966).

area, both in m^2 (Nürnberg, 1995a, 1995b). The index is a measure of the lake bottom area covered by $\leq 1\,mg\,l^{-1}$ DO and is more useful than AHOD in determining the extent of conditions suitable for P internal loading and bottom area unaccessible to fish. As described below, eutrophication has its greatest detrimental effect on fish populations through depletion of DO.

7.3.4 DO, eutrophication and fish

The principal effect of eutrophication on fish is one of DO depletion. An increase in plankton production to within the mesotrophic state may have a beneficial effect on the production of desirable species of fish, i.e. both cold- and warm-water sport fishes. However, once a state of eutrophy is reached and particularly if the lake stratifies in summer, there is a strong probability that DO in the hypolimnion may reach a critical level for survival toward summer's end.

As eutrophication increases, the minimum DO reached will continue to decrease and the minimum DO will occur earlier. This situation will be increasingly detrimental to particularly cold-water fishes, but may also adversely affect warm-water fish, because activity and growth will decrease progressively with decreasing DO (see Chapter 13). Both types will tend to evacuate the epilimnion when temperatures exceed their preferred level, although the problem would be faced more frequently by cold-water species. If adequate DO exists in the hypolimnion, it can be a healthy refuge during the warm summer period. If there is inadequate DO, the fish will be subjected to either stressful DO in the hypolimnion or, if excluded from the cooler but oxygenless hypolimnion, to stressful temperature in the epilimnion.

The higher the nutrient loading to the system, the greater the DO deficit and, depending upon the hypolimnetic depth, the lower the DO concentration and, consequently, the more inhospitable the entire lake will be to fish. Even shallow lakes that do not stratify will suffer from low DO in winter if they become ice covered. The problem of 'winter kill' in shallow lakes is a long-standing one in temperate areas (Halsey, 1968; Barica, 1984). Also, low DO can cause fish mortality following algal bloom die-off in summer (Ayles et al., 1976). DO concentration can decline to very low levels ($<1\,mg\,l^{-1}$) during calm periods following the collapse of an algal bloom (Barica, 1984).

The long-term effect of eutrophication will be one of changed species composition, largely due to the changed DO status (Nürnberg, 1996). Increased food supply in the form of more detritus, which tends to lead to smaller zooplankton and a less diverse worm/midge-dominated bottom fauna, would tend to favour detritus/bottom-feeding fish such as suckers and carp. Haines (1973) has shown that the growth of carp was much greater (nearly four times) and small-mouth bass much less (factor of $\approx 2-6$) in fertilized than in unfertilized experimental ponds. The fertilized ponds received phosphorus at a rather high rate of $\approx 2 \text{ g m}^{-2} \text{ year}^{-1}$. DO was probably the principal factor detrimental to the bass because the diurnal range was from 18 to $< 2 \text{ mg l}^{-1}$ in the fertilized ponds, but seldom exceeded 3.5 mg l^{-1} range in the unfertilized ponds. (See Chapter 13 for the effect of diurnal DO range.)

Regardless of which factor is more important, DO or changes in the food supply, the results are similar. In Lake Erie, for example, the populations of cisco, whitefish, walleye, sauger and blue pike drastically declined over the 40-year period that loading of nutrients to the lake was increasing, as indicated by increases in major ions (Beeton, 1965). (The principal cause for their decline was a failure to reproduce.) The total fish catch, however, did not decline; rather, the catch of the desirable species was replaced with such species as carp, perch, buffalo, drum and smelt (Table 7.5). Although other factors, such as fishing pressure and sea lamprey predation, were involved in changes in fish production in the Great Lakes, Beeton and Edmondson (1972) suggested that the changes in Lake Erie, and in particular the cisco, were closely tied to the progressive eastward movement of polluted conditions.

This pattern can be expected to occur in most lakes undergoing eutrophication, depending on the oxygen resources, which are a function of depth. Depth is important even in shallow lakes. Miranda *et al.* (2001) have shown for a 1870-ha (mean depth 3 m) lake that the probability of reaching a DO at dawn of $<1.5 \text{ mg l}^{-1}$ was 0.91 at 0.1 m water column depth, but was <0.05 at 3.5 m depth. The reason for the exponential decrease in the chance for low DO as depth increased is that more DO is available in the overlying water to satisfy the benthic demand. Collapse of an algal bloom would add another source of DO demand. Mortality of thousands of suckers in shallow (mean depth, 2 m) Upper Klamath Lake, Oregon, was coincident with collapses of blue-green (*Aphanizomenon*) blooms (Perkins *et al.*, 2000). Eutrophication is

Table 7.5 Comparative annual catch of commercial fish species in metric tons over a 40-year period in Lake Erie

Species	Early years	Recent years
Cisco	9000 (pre-1925)	3.2 (1962)
Sauger	500 (pre-1946)	0.45–1.8
Blue pike	6800	0.45 (1962)
Whitefish	1000	6.0 (1962)
Walleye	7000 (1956)	450
Total	24 300	
Drum, carp, perch, and smelt		$\approx 22\,240$ (increased catch)
Total		22 700

Source: Data from Beeton (1965).

considered to be the principal problem that has limited salmonid fisheries in European subalpine lakes (Nüman, 1972).

The general pattern of change is shown in Figure 7.12. Increased abundance and productivity of 'desirable' fish species should occur in the mesotrophic stage. As eutrophy is approached, those species tend to diminish and are replaced by a dramatic increase in production of the more DO-tolerant and/or temperature-tolerant 'undesirable' species.

There is little doubt that a eutrophic state is definitely detrimental to the production of the more DO/temperature-sensitive species. The question is how much enrichment is detrimental? Are the criteria previously described for trophic state in relation to recreation also pertinent to fish propagation? Apparently the rate of phosphorus loading that is apt to cause a eutrophic state from the standpoint of algal biomass and transparency is also similar to the loading that will cause an ODR (oxygen deficit rate) that is representative of eutrophy (Table 7.4), and which may in turn begin to strain the oxygen resources from the standpoint of the fishery. Dillon and Rigler (1975) suggested very conservative guidelines for salmonid fisheries in Ontario lakes with respect to increased P loading and plankton algae.

To evaluate the connection between eutrophic threshold criteria for recreation and effects on cold-water fish, one can compare the TP loading for a threshold TP of $25\,\mu g\,l^{-1}$ with the ultimate hypolimnetic DO concentration that loading should produce after some period of stratification.

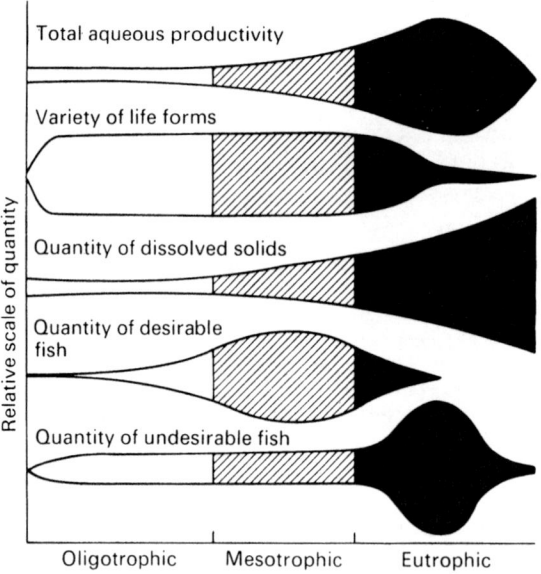

Figure 7.12 Suggested changes in various characteristics of lakes with eutrophication (modified from Lake Erie Report, 1968).

Using Equation 7.10 and an equilibrium TP of $25\,\mu g\,l^{-1}$ gives a critical (eutrophic threshold) loading of $1145\,mg\,m^{-2}\,year^{-1}$ for a lake with total and hypolimnetic mean depths of 12 m and 5 m, respectively, and a flushing rate of $2.3\,year^{-1}$ (means for 26 American OECD lakes, Welch and Perkins, 1979). After 60 days of stratification, average hypolimnetic DO content would have declined from saturation to $5.5\,mg\,l^{-1}$, using Equation 7.22, and to $4.4\,mg\,l^{-1}$, using the R_{tp}-AHOD model of Cornett and Rigler (1979), assuming 10°C and $R_{tp} = 1/(1 + \rho^{0.5})$. Those DO levels would produce stressful conditions for fish that need to reside in the cool waters of the hypolimnion during the summer, e.g. trout, striped bass and whitefish. This would be especially true if stratification lasted longer, which is usually the case. Theoretically then, an average stratified lake that had reached the threshold for eutrophy (Table 7.3), which would represent adverse conditions for purposes of recreation, would also produce adverse DO conditions for some fish.

Even non-stratified lakes with high macrophyte density may show severely depleted DO within weedbeds (Frodge *et al.*, 1987). Such DO depletion within macrophyte beds may be linked to observed die-offs of warm-water fish in shallow eutrophic lakes. Net DO may answer the dilemma of a suitable DO index for unstratified lakes, because ODR is appropriate for stratified lakes only. Porcella *et al.* (1980) developed net DO, with values ranging from 0–10, for stratified and unstratified lakes alike. Net DO is the absolute difference from an equilibrium condition (saturation) and is calculated by summing those differences (equilibrium DO − measured DO) over intervals of depth, thus incorporating the increasing tendency of supersaturation as well as deficiency in response to eutrophication.

7.3.5 Indicator organisms

While impairment of transparency (and the algal abundance that imparts it) and oxygen resources are important indicators of the suitability of lake water for recreation and water supply, there are other characteristics that are equally pertinent. The abundance of cyanobacteria is readily observable because many of them are buoyant and cause surface scums (see Section 6.5.6 for discussion of causes). Thus, the fraction of the total biovolume that are blue-greens tends to increase with eutrophication and has been demonstrated to be a very useful indicator that can have different response times to enrichment than transparency, TP and chl *a* (e.g. see Edmondson and Litt, 1982). In an attempt to develop a trophic state index using blue-greens, Smith (1986) has related the fraction of phytoplankton as blue-greens to the ratios of TN:TP and SD/Z_{mix}. For example, given a mixed depth of 10 m, a TN:TP ratio of 10 and a eutrophic threshold SD of 2 m, Smith's model predicts 27% blue-greens.

Table 7.6 Phytoplankton concentration before and after treatment indicated

Lake	Treatment	Blue-greens (%)	
		Before	After
Moses Lake Parker Horn	Dilution	96	55
Moses Lake Pelican Horn	Diversion	27	47
Lake Washington[a]	Diversion	95	20
Lake Sammamish	Diversion	68	30
Long Lake	Alum	85	20
Lake Norrviken[b]	Diversion	95	92

Notes
a Edmondson and Litt (1982).
b Ahlgren (1978).

The blue-green fraction has been known to respond sensitively to nutrient reduction as shown in Table 7.6. Prior to sewage diversion, Pelican Horn of Moses Lake, USA was dominated by small green algae representative of sewage lagoons, and the increase in % blue-greens after diversion was actually associated with an improvement in quality. Parker Horn, on the other hand, was dominated by blue-greens initially and they declined following dilution (Welch and Patmont, 1980). The blue-green fraction in Lake Norrviken remained high after treatment, because TP was still very high, but at least the lake no longer contained a monoculture of *Oscillatoria* (Ahlgren, 1978). The blue-green fraction responded slowly following diversion of sewage effluent from Lake Washington and did not reach the low level noted in Table 7.6 for about seven years (Edmondson and Litt, 1982). Relationships between the blue-green fraction and TP concentration and blue-green biomass and TP loading were presented, respectively, by Sas (1989) and Lathrop et al. (1998).

Blue-green blooms are important only in freshwater and in some instances in brackish water, such as the Baltic Sea. Diatoms and dinoflagellates are most commonly the dominants in blooms in estuaries and nearshore marine areas.

Sediment diatoms, or the ratio (A/C) of planktonic pennate (long narrow cells–Araphidinae) to centrate (round cells–Centrales) species, is also a useful index of trophic state. The advantage of sediment core data is that the history of trophic state can be determined if sediment dating is available. The disadvantage of the diatom index is that it is restricted largely to stratified lakes, because of interference from littoral pennate species (Stockner, 1972). Results from Lake Washington, for example, show a rather clear picture, where the A/C ratio rose above the eutrophic threshold of 2 during the period of maximum enrichment and then recovered promptly after diversion (Stockner, 1972). There are equally interesting results from other lakes. Benthic invertebrate indicators of eutrophication are discussed in Chapter 12.

7.3.6 Nutrient loading

Although the trophic state of a lake is dependent on the concentration of TP, chl *a*, etc., knowing the loading rate that produces that trophic state is also important. That is because: (1) nutrient inputs are assessed as mass balances to evaluate relative contributions from sources and for mass balance modelling, and (2) to improve the quality of the lake, the loading (either external or internal) must be manipulated. This was illustrated in example problem 2 above. Thus, the critical loading (L_c) to produce a mesotrophic or eutrophic state is often used as a recovery goal, compared to the current or worsening state of a lake.

There is an interesting chronology in the development of nutrient loading criteria that is worth reviewing, because it illustrates the evolution in thinking about lake behaviour. The initial attempt to define L_c involved only two variables; measured areal loading and mean depth (Vollenweider, 1968). This relationship is shown in Figure 7.13. Based on judgment about trophic state, guidelines were approximated for loading that represented thresholds or limits. Higher areal loading would be tolerated in deeper lakes due to dilution, converting areal loading to volumetric loading ($g\,m^{-3}\,year^{-1}$). The same apparent effect of depth was seen in lake response (algal biomass and productivity) to the area/volume index in the Ontario experimental lakes (Figure 7.14). While there is a dilution effect of lake volume in the index, there is also a mass loading effect that would be proportional to watershed area.

However, such loading–mean depth relationships were unreliable where flushing rate varied greatly. The flushing effect was clearly demonstrated in two Ontario lakes (Cameron and Four Mile) by Dillon (1975). Although these two lakes were approximately the same trophic state, the flushing rate of Cameron was 70 times that of Four Mile, while the TP loading to Cameron was 20 times that of Four Mile. Correction of the loading for flushing rate explained the apparent anomaly. A consideration of residence time, τ (1/flushing rate) led to refinement of the loading graph (Vollenweider and Dillon, 1974; Vollenweider, 1975), which is shown in Figure 7.15. The equation for the straight lines in the figure is:

$$L_c = 100 \quad \text{or} \quad 200\bar{z}\tau^{0.5} \tag{7.23}$$

With that correction, some most disturbing points on the loading–mean depth graph could be clarified. For example, it appeared that ultraoligotrophic Lake Tahoe was completely insensitive to increased loading due to its very great depth. However, correction for flushing rate moved Lake Tahoe to the other side of the graph where it became much more sensitive to increased loading. The reverse was true for other lakes originally located high on the left-hand side of the graph (high flushing rates and shallow).

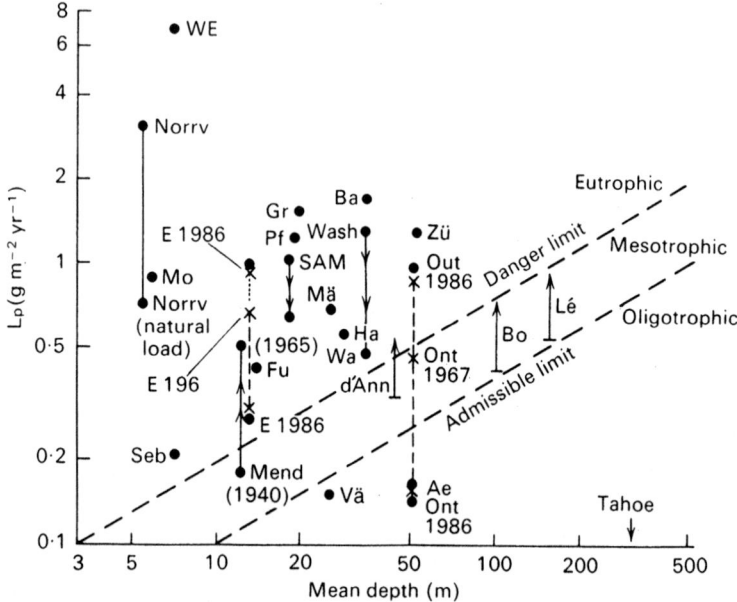

Figure 7.13 Loading graph for P in various lakes. Ae, Aegerisee (Switzerland); Ba, Baldeggersee (Switzerland); Bo, Bodensee, Lake Constance (Austria, Germany, Switzerland); d'Ann, Lake d'Annecy (France); E, Lake Erie (USA); Fu, Lake Fureso (Denmark); Gr, Greifensee (Switzerland); Ha, Hallwilersee (Switzerland); Lé, Lake Léman, Lake Geneva (France, Switzerland); Mä, Lake Mälaren (Sweden); Mend, Lake Mendota (USA); Mo, Moses Lake (USA); Norrv, Lake Norrviken (Sweden); Ont, Lake Ontario (USA); Pf, Pfäffikersee (Switzerland); Sam, Lake Sammamish (USA); Seb, Lake Sebasticook (USA); Tahoe, Lake Tahoe (USA); Wa, Lake Washington (USA); WE, Western Lake Erie (USA); Vä, Lake Vänern (Sweden); Zü, Zürichsee (Switzerland). For Lakes Erie and Ontario, dotted lines show P loading by 1986 (from 1967 estimates) without P control; dashed lines show P loading by 1986 (from 1967 estimate) with no P in detergents and with 95% P removed from all municipal and industrial wastes (Vollenweider, 1968).

Those were shifted to the right to a position more in line with their observed trophic state. The refined graph essentially converted loading to inflow concentration, which, referring back to continuous flow culture kinetics, would be expected to be more directly related to the maximum algal biomass per volume that might develop.

The next step was to include sedimentation loss in the loading relationship to produce the following critical loading equation:

$$L_c = TP_{e/m.m/o} \bar{z} (\rho + \rho^{0.5}) \qquad (7.24)$$

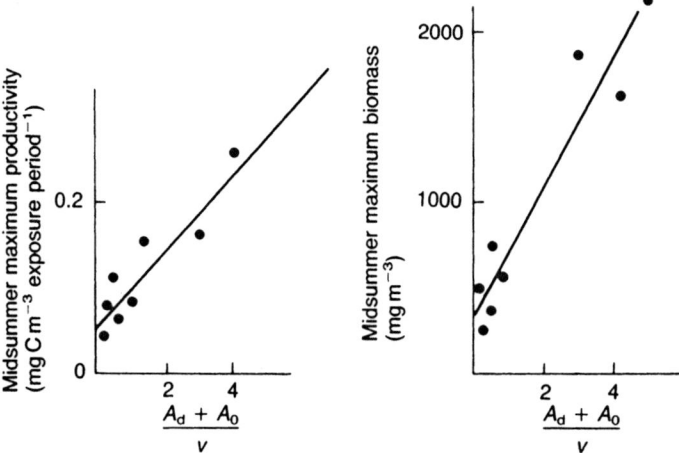

Figure 7.14 Relationships of lake response to watershed (A_d) plus lake (A_o) area divided by volume (v) (Schindler, 1971a).

where $TP_{e/m}$ is 20 or $TP_{m/o}$ is $10\,\mathrm{mg\,m^{-3}}$ suggested as the eutrophic–mesotrophic or mesotrophic–oligotrophic thresholds, respectively (Vollenweider, 1976). The effect of sedimentation is indicated in Figure 7.16 in

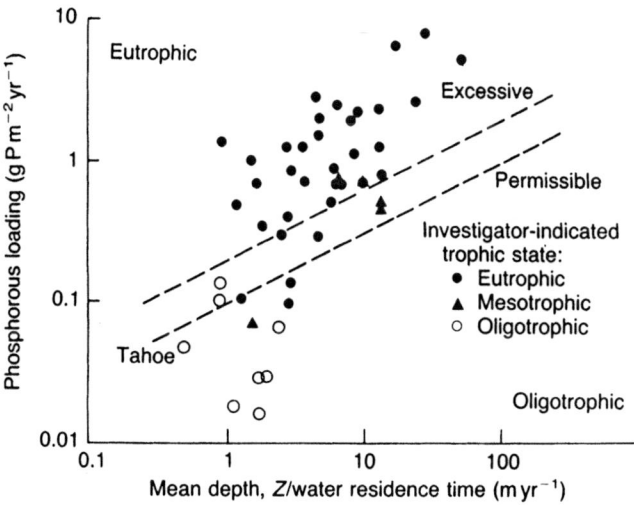

Figure 7.15 US OECD data applied to initial Vollenweider phosphorus loading and mean depth/hydraulic residence time relationship (Rast and Lee, 1978).

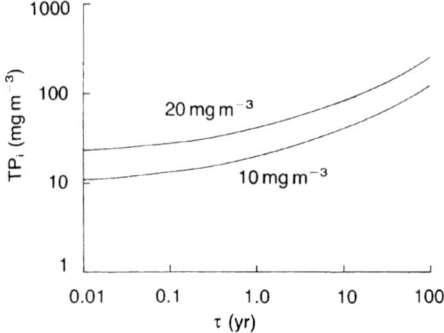

Figure 7.16 Relationship of inflow concentration (TP$_i$) with water retention time (τ) for two lake concentrations (TP = 10 and 20 mg m^{-3}) (reproduced with kind permission from Lake and Reservoir Management).

which the lines for critical lake concentrations are shown in a plot of τ, or $1/\rho$, against TP$_i$ according to Equation 7.10. If we rearrange Equation 7.10:

$$\text{TP}_i = \text{TP}(1 + \tau^{0.5}) \tag{7.25}$$

At low τ, the lines for lake concentrations of 10 and 20 mg m^{-3} become parallel with the abscissa indicating that sedimentation becomes minimal at short residence times (high flushing rates) and the lake concentration equals the inflow concentration. As τ increases, sedimentation becomes increasingly important permitting higher TP$_i$ without exceeding the critical lake concentration. That is, lakes become more tolerant of increased TP$_i$ as τ increases. See Reckhow and Chapra (1983) for a more detailed discussion of these loading relationships.

7.3.7 Sources of phosphorus

P and N enter waterways from external sources, such as: dissolution from rock; wastewater discharges, which can include a variety of waste types, e.g. treated or untreated sewage, dairies, slaughter houses, seafood canneries and other food processing plants; urban stormwater runoff; rural stormwater runoff from either fertilized or unfertilized pastures and croplands; forest; precipitation; and waterfowl. Only rock dissolution and forest runoff are considered natural background; all others are affected by man (actually forest runoff is if fertilized). The input from precipitation (even in remote areas) and waterfowl (attracted by feeding) are often affected by man. Only wastewater discharges are considered point sources, i.e. they enter directly through some conveyance system. All others are considered non-point, or diffuse sources. While point sources are usually rather easy to estimate or determine directly, non-point sources are difficult to estimate. Non-point sources are often estimated from knowing the distribution of land-use types in a watershed and applying yield coefficients that are based on a variety of data sets. For example, one set of estimates for TP (Reckhow and Chapra, 1983) is, in $mg\,m^{-2}\,year^{-1}$: forest, 2–45; precipitation, 15–60; agriculture, 10–300; urban, 50–500; and leachate from septic tank drain fields, 0.3–1.8 kg $capita^{-1}\,year^{-1}$. Non-point source nutrients were estimated to impair 51% of USA lakes and 57% of the estuaries (USEPA, 1996).

The ranges are large and it is usually difficult to select the appropriate value for a particular watershed. One approach is to select an appropriate portion of the range based on measured loading and land-use information in the watershed of interest. Assuming that ratios among the respective land-use types given above are appropriate for the watershed in question, yield coefficients can be selected by holding the ratios constant and scaling the coefficients to conform with actual loading and land-use areas. This was done for Lake Sammamish, WA, for which daily measurements of TP and continuous flow were available to calculate actual loading. Figure 7.17 shows the year-to-year agreement between lake TP concentrations observed and simulated from scaled land-use TP-yield coefficients calibrated to actual flow. The decrease in TP is due to waste water diversion in 1968 (Shuster *et al.*, 1986; Perkins *et al.*,

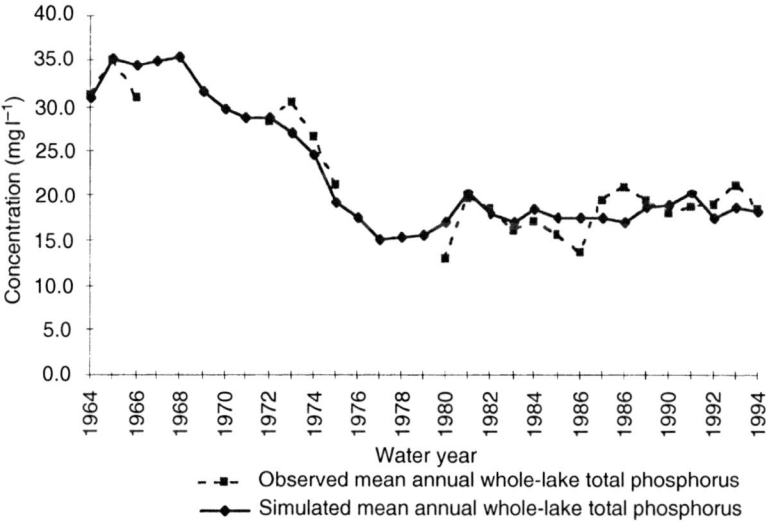

Figure 7.17. Observed annual, whole-lake TP concentration in Lake Sammamish compared to that simulated with a mass balance model using yield coefficients for TP, calibrated to land use types in the watershed (from Perkins *et al.*, 1997, reproduced with kind permission from Lake and Reservoir Management).

1997). A method of expressing uncertainty should be applied to the predictions, such as a range in probable yields (see Reckhow and Chapra, 1983).

Obviously the loading from septic tank drainfields can represent the most concentrated non-point source in a watershed. Assuming 2.5 members per family and a $4000\,m^2$ lot size, the yield would be $188-1125\,mg\,m^{-2}\,year^{-1}$ if the drainfield was not percolating adequately and removing P. Properly working drainfields ('on-site waste treatment') are actually very efficient at removing P. However, even if a few fail in the watershed the source can be significant. Moreover, P leaching from drain fields is probably in a soluble, readily available state.

Urban runoff can be a surprisingly significant source of TP as the values indicate. However, much of that TP can arrive as particulate P, sorbed to the high content of suspended matter in stormwater, and settle out shortly after entering the lake. If that happens, the contribution to productivity may not be great, even though the loading is high. This may be especially true if the loading arrives during the non-growing season, as it often does in areas with highly wet winters and dry summers, like the Pacific Northwest (Stockner and Shortreed, 1985). However, the highest fraction of TP that was bioavailable arrived during the high runoff period in winter in Lake Sammamish, a Pacific Northwest lake (Butkus et al., 1988).

7.4 Eutrophication of coastal marine waters

Eutrophication of coastal waters is also a growing global environmental problem. Over the past three decades, the ecological effects of eutrophication of coastal marine areas have become increasingly apparent. Increased nutrient loading (particularly N) to shallow coastal and estuarine waters has degraded water quality by increasing phytoplankton and macroalgae biomass, turbidity, and hypoxic (oxygen-depleted, less than $2\,mg\,l^{-1}$ DO) and anoxic areas. Changes to the phytoplankton community include an increased incidence of harmful algal blooms (HABs), including toxin-producing species. Habitat loss and altered food webs are other consequences of coastal water eutrophication.

Stratified coastal waters where exchange with ocean waters is too low for rapid dilution of nutrients in the surface and re-oxygenation of stagnant bottom waters are particularly vulnerable to the adverse effects of eutrophication. The SE Kattegat Sea on the west coast of Sweden is one such system. The effects of eutrophication were first documented in 1980. Eutrophication-enhanced primary production, increased incidence of HABs, and subsequent sedimentation and degradation of organic matter led to DO depletion in the bottom waters of this sea. Hypoxic conditions were observed in the late summer and fall every year during the 1980s (Baden et al., 1990), causing mortality of fish, lobsters and bivalves, and altered community composition and food webs, all of which has had serious repercussions on the Swedish and Danish fisheries. Eutrophication of coastal areas in the Baltic Sea also resulted in extensive hypoxic zones throughout the region.

Similar conditions have been reported in many areas of the world including the Adriatic Sea, the New York Bight, the North Sea and the Gulf of Mexico. The Gulf of Mexico mid-summer bottom-water hypoxic zone is about 20 000 km^2, which is the second largest such affected area in the world (the hypoxic zone of the Baltic basins is about 70 000 km^2) (Rabalais *et al.*, 2002). The hypoxic zone of the Gulf of Mexico typically extends throughout 20–50% of the water column. Nutrient loading from the Mississippi River, which drains 41% of the lower contiguous USA, is the primary cause of the increased phytoplankton biomass that decomposes and initiates the hypoxia. As is the case in the Baltic basins, replenishment of DO is inhibited by the physical stratification of the water column due to the density gradient between the upper, warmer, less saline waters and the higher salinity waters at lower depths, thereby promoting hypoxic conditions (Rabalais *et al.*, 2002).

Nitrogen input to the Gulf from the Mississippi River drainage basin, primarily from fertilizer application, tripled between 1955 and 1970 and again between 1980 and 1996. Sixty-one per cent of the annual average N flux of 1.6 million metric tons is in the form of nitrate. Non-point sources account for 90% of the NO_3^- inputs to the Gulf with 74% of these inputs attributed to agricultural activities (Rabalais *et al.*, 2002). Reducing hypoxia in the Gulf of Mexico will require substantial reductions in N loading particularly from non-point sources in the upper Mid-west.

Chapter 8
Zooplankton

Zooplankton are the primary consumers of phytoplankton in lakes, the ocean and in deep slow-moving rivers where the principal energy pathway is through a predator–prey-oriented food web. In highly eutrophic or organically polluted waters, where energy is diverted through detritus pathways, zooplankton may also be the transporter of that energy by consuming bacteria and detritus. Assuming that temperature is optimum and there is minimal inhibition by low DO or toxicity, the effectiveness of zooplankton consumption and production will usually be controlled by the size, quality and abundance of food (the phytoplankton), as well as by planktivory – the predation by planktivorous fishes. Control of their consumption and production effectiveness by those factors has often been related to the resulting change in their size structure. The role of planktivory can be manipulated to increase consumption of phytoplankton, improve water clarity and reduce nutrients. Results of such attempts at 'biomanipulation' are covered in Chapter 10, but the ecological principals supporting that technique will be discussed in this chapter.

Zooplankton, especially large species of *Daphnia*, e.g. *D. magna*, are sensitive to toxicants and are frequently used as test organisms in toxicity bioassays. Results of such toxicity tests are covered in Section 13.5.

8.1 Population characteristics

The zooplankton are composed primarily of populations of microscopic or near-microscopic organisms from the groups Rotifers (rotifers or 'wheeled organisms'), and the Cladocera and Copepoda (Crustacea). Rotifers and cladocerans (e.g. *Daphnia*) usually reproduce by parthenogenesis. Parthenogenesis is a form of reproduction in which populations comprised of females simply produce diploid copies of themselves. Under crowded or stressed conditions, male cladocerans develop from a diploid egg to produce a haploid (half chromosome number) sperm to unite with a haploid egg. That union produces a fertilized diploid egg as a resting stage awaiting more desirable environmental conditions for population growth. The rapid population growth that can result from these adaptations partly account for the ability of the zooplankton to control the size of the phytoplankton. Generation times in cladocerans can be as little as 2–5 days or as long as a month or more. The copepods (e.g. *Diaptomus*) have no parthenogenic stage, but can also reproduce rapidly because the female can store sperm for several fertilizations.

Adult cladocerans such as *Daphnia* (e.g. *D. magna*, *D. pulex*) can be as large as 3 mm in carapace length, while others (*Moina*, *Polyphemus* and *Ceriodaphnia*) attain a length of 1.5 mm. Small cladocerans, such as *Bosmina*, are of the order of 0.4 mm. *Daphnia* are generally large and can consume a rather wide size range of food, while small cladocerans (*Bosmina*) are more restricted in food size (Sarnelle, 1986). Size is a very important factor in zooplankton–phytoplankton interactions, because not only does it determine the size of food particles that can be consumed, but the rate at which they consume food increases with size (Burns, 1968, 1969) and planktivores select the larger size zooplankton, such as *Daphnia*.

8.2 Population dynamics

To understand zooplankton species dynamics it is helpful to separate these dynamics into demographic (and community) responses to predation and resource availability. Because cladocerans and rotifers are, as previously mentioned, parthenogenic and because these organisms are often translucent (and their eggs are therefore very easy to count), it is possible to separate their population dynamics into components of birth and death. The instantaneous birth rate (b) of a parthenogenic population of cladoceran or rotifer zooplankters can be calculated using the classic Edmondson–Paloheimo equation (Paloheimo, 1974) accordingly:

$$b = \frac{\ln(E+1)}{D} \tag{8.1}$$

where E equals the ratio of eggs to individuals (including both juveniles and adults) in the population and D is the egg development time in days. The egg development time is strictly determined by water temperature. Two days would be a typical egg development time at 20°C, while egg development could take up to 2 weeks at 4°C. The egg ratio (E) is closely related to the population's fecundity, which equals the number of eggs per adult female in the population. The fecundity of a population is a very sensitive index of that population's nutritional state. Both the quantity and quality of food available influence fecundity.

The population's instantaneous death rate (d) can be calculated as the difference between the population birth rate and the population growth rate accordingly:

$$d = b - r \qquad (8.2)$$

where r equals the population's instantaneous growth rate. This is calculated using the classic logistic growth rate equation accordingly:

$$r = \frac{\ln(N_t/N_0)}{t} \qquad (8.3)$$

where N_t equals the population size at the end of a sampling interval, N_0 equals the population size at the beginning of an interval and t equals the length of that sampling interval in days. One can also express this growth rate as the simple difference between a population's birth and death rates accordingly:

$$r = b - d \qquad (8.4)$$

Predation impacts on zooplankton community composition were first described by Hrbacek et al. (1961), who noted that zooplankton communities subject to intense fish predation were dominated by small and/or highly evasive taxa like rotifers, small cladocerans and cyclopoid copepods. In the absence of fish predation, the zooplankton assemblage of lakes is generally dominated by large taxa, and especially by *Daphnia* spp. This pattern was also noted in a classic study by Brooks and Dodson (1965). Whether a particular zooplankton species is vulnerable to fish predation depends on three key factors: visibility, evasiveness and habitat overlap. The visibility of zooplankton to zooplanktivorous fish is a function of size, pigmentation and the presence/absence of eggs. Zooplanktivorous fish are highly selective and their diets are often strongly dominated by only the largest zooplankters present at a given time. To minimize predation risk, zooplankton found in lakes containing fish are usually translucent, whereas counterparts from the same or closely related species in fishless lakes may be highly pigmented. Relative to the rest of the body, the eggs of parthenogenic cladocerans are also highly pigmented. It is not unusual to find the stomach contents of zooplanktivorous fish to be strongly dominated by egg-bearing *Daphnia* even in cases where egg-bearing *Daphnia* constitute only a small fraction of all crustacean zooplankton.

Irrespective of their size and pigmentation, zooplankton can avoid being consumed by visually oriented predators by either being evasive or by minimizing habitat overlap with predators. Zooplankton taxa differ greatly in their ability to evade fish predators. Some taxa like cyclopoid copepods have very well-developed jump responses which allow them to dart away from predators attempting to consume them, whereas other zooplankters like *Daphnia* swim quite slowly and are unable to escape from most zooplanktivorous fish (Drenner *et al.*, 1978) (Figure 8.1). These 'jump responses' are also prevalent amongst rotifers which use them to escape from predaceous zooplankton like cyclopoid copepods. One of the most fascinating aspects of zooplankton behaviour is a phenomenon called diel vertical migration (or DVM) (Lampert, 1989). Many large crustacean taxa utilize DVM to avoid fish predation by spending the daylight hours in the deeper and darker hypolimnion, and migrating only up to the epilimnion at night when visual predators cannot see them. Diel vertical migration does not occur in fishless lakes because zooplankton that do employ DVM suffer two important physiological costs. First, phytoplankton availability is usually much greater in the epilimnion, so by spending most of the day in the dark hypolimnion food consumption may be much reduced. Second, because zooplankton growth and egg development are strongly temperature-dependent, zooplankton which employ DVM will have slower growth and egg production relative to zooplankton that spend all of their time in the warmer, food-rich epilimnion. Because of these physiological constraints, smaller zooplankton taxa are also much less likely to undergo DVM.

Food availability and quality affect both zooplankton growth rates and adult female fecundity. Herbivorous zooplankton consume seston (or suspended particulate organic matter) which is a mixture of phytoplankton, bacteria and detritus. Many herbivorous zooplankton will also consume microzooplankton (i.e. rotifers and ciliated protozoans) if these organisms

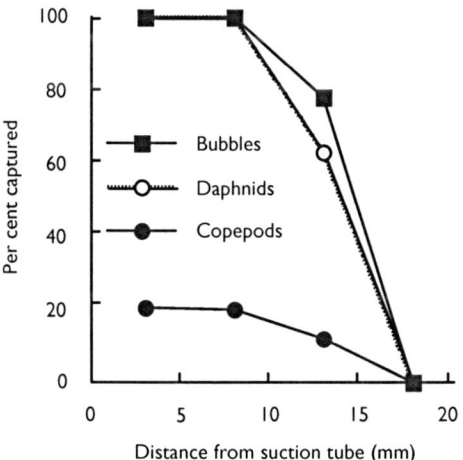

Figure 8.1 The escape response of daphnid and copepod zooplankters to a suction device simulating fish predation. Adapted from Drenner *et al.* (1978).

fall within the size range of particles they are able to ingest. For daphnids this size range is thought to be particles with diameters between 1 and 30 μm (Burns, 1968). The zooplankton's ability to utilize seston as a food resource depends on whether they are able to ingest the seston, whether they are able to digest that food they do ingest and whether the food they digest is nutritionally adequate. Very large phytoplankton cells (e.g. certain dinoflagellates) and phytoplankton colonies (especially blue-greens) may simply be too large for even the largest zooplankton taxa to ingest. Phytoplankton that are ingested may in some cases pass through the digestive system of zooplankton unscathed because they have hardened or gelatinous cell walls that make them digestion-resistant (Porter, 1977). When zooplankton are in food-rich conditions they reach their primiparous instar (the first reproductive event), earlier, at a larger individual size and with larger clutches. At 20°C, a well-fed daphnid would reach its primiparous instar in about 1 week. The growth response of zooplankton to increasing supplies of seston is identical to the classic Michaelis-Menton growth rate versus nutrient supply response for phytoplankton. At low seston availability (i.e. <0.5 mg/l), *Daphnia* growth is nearly linearly related to food supply, whereas at concentrations above this, *Daphnia* ingestion rates become saturated and their growth reaches an asymptote corresponding to their maximum potential growth rate for that food resource (Lampert and Sommer, 1997).

It has long been known that different phytoplankton groups differ greatly in their food quality for herbivorous zooplankton (Brett and Müller-Navarra, 1997). Recently, a great deal of research has explored the biochemical and elemental basis for the large observed variation in phytoplankton food quality for herbivorous zooplankton (Table 8.1). Phytoplankton taxa such as diatoms and cryptophytes were known to be of very high food quality, whereas blue-greens were known to be of very low food quality.

Table 8.1 Normalized food quality of phytoplankton monocultures. The taxa data were normalized by their C:P ratio and the C:P data were normalized by taxa. The frame of reference was Chlorophytes with C:P ratios below 300

Phytoplankton group	Normalized food quality (%)	Phytoplankton C:P ratio	Normalized food quality (%)
Bacillariophytes	124	136 (10th)*	103
Cryptophytes	122	180 (25th)*	101
Chlorophytes	100	258 (50th)*	96
Cyanophytes	56	368 (75th)*	88
		486 (90th)*	83

Source: All data adapted from Brett *et al.* (2000).

Note
*Indicates the percentile for natural lake seston C:P ratios.

Initially blue-greens were thought to be poor food resources because they often form large colonies making them difficult for many zooplankton to ingest and/or because some blue-greens produce toxic metabolites that inhibit zooplankton feeding (Lampert, 1981, 1987). While ingestibility and toxicity place clear constraints on the zooplankton's ability to consume blue-greens, more recently, it has also been noted that even small non-toxic blue-green species are very poor in food quality for herbivorous zooplankton. Ahlgren et al. (1990) attributed much of the observed differences in the food quality of common phytoplankton taxa to differences in their essential or omega-3 fatty acid content. High-food-quality phytoplankton groups like diatoms and cryptophytes tend to have very high concentrations of the essential fatty acid eicosapentaenoic acid (EPA). Blue-greens usually have very little or none of this and related essential fatty acids. Common green algae usually have very little EPA, but they often have high concentrations of related essential fatty acids such as α-linolenic and stearidonic acids, which many herbivorous zooplankton can convert to EPA. Essential fatty acids are important for all animals (including humans) because they help regulate cell membrane fluidity, lipid metabolism, gene expression and immune system functionality, and they serve as precursors for many reproductive and growth hormones (Sargent et al., 1995). Field studies have shown that EPA content of natural seston assemblages is strongly correlated with the growth rates of *Daphnia* feeding on this seston (Müller-Navarra, 1995; Müller-Navarra et al., 2000). Very recently, it has been shown that most of the food quality differences observed between mixtures of very high-food-quality cryptophytes and mixtures of low-quality blue-greens can be directly attributed to differences in their essential fatty acid content (Ravet et al., 2003). Laboratory studies looking at algal monocultures have also shown that *Daphnia* growth rates are correlated with the elemental phosphorus content of their diets. *Daphnia* are phosphorus-rich and have a nearly homoeostatic elemental composition, with a molar C:P ratio of $\approx 93:1$ (Brett et al., 2000). Because seston, and especially laboratory algal monocultures, can vary greatly in their elemental C:P ratio, it is possible to have great imbalances between the phosphorus content of *Daphnia* and their diets (Sterner and Hessen, 1994). Many studies have shown that green algae monocultures with very high C:P ratios (≥ 1000) are of quite poor food quality for *Daphnia*.

As noted in the preceding passages, much of what is known about freshwater zooplankton ecology is based on studies of *Daphnia* (Lampert and Sommer, 1997). *Daphnia* play a special role in zooplankton ecology because they are quite large and very fast growing, and because they are very effective herbivores. Because of their high particle filtering rates, the large size range of particles they are able to ingest, and because they can build up very large populations in short periods of time, *Daphnia* spp. (and especially the larger species within this genus) are the group of zooplankton most able to suppress phytoplankton biomass. *Daphnia* are also the preferred prey (because of their large size and slow swimming speed) of many zooplanktivorous fish. *Daphnia* are also the focus of many laboratory investigations of zooplankton ecology because they are easy to maintain in the lab, making them the proverbial 'white rats' of zooplankton ecology.

8.3 Filtering and grazing

Zooplankton feed by filtering the water as they move. The setae on their maxillary appendages serve to collect phytoplankton as water is circulated by swimming movements created by their antennae (see Russell-Hunter, 1970). The filtering, or clearing, rate can be determined by recording the change in food particle concentration over time according to the following equation:

$$\text{ml animal}^{-1} \text{ day}^{-1} = \text{ml animal}^{-1} (\ln C_0 - \ln C_t) \text{days}^{-1} \tag{8.5}$$

where C_0 and C_t are, respectively, the initial and ending concentration of food particles. At low food concentrations, the rate is more accurately determined over much shorter time periods by tagging the algae with a radioisotope and measuring the activity of the zooplankton themselves. Results using the copepod *Diaptomus* and the cladoceran *Daphnia* have given filtering rates ranging from 0.1–5.5 ml animal^{-1} day^{-1}, with higher rates for the larger *Daphnia* (see Jorgensen *et al.*, 1979). The grazing, or feeding rate, is in turn determined as the product of the filtering rate and the concentration of phytoplankton available:

$$L \text{ animal}^{-1} \text{ day}^{-1} \times \text{mg } C_{\text{phyto}}/L = \text{mg animal}^{-1} \text{ day}^{-1} \tag{8.6}$$

The loss rate of phytoplankton through zooplankton grazing can be compared with the rate of primary productivity by multiplying the grazing rate times the concentration of zooplankton:

$$\text{mg } C_{\text{phyto}} \text{ animal}^{-1} \text{ day}^{-1} \times \text{animal m}^{-3} = \text{mg C m}^{-3} \text{ day}^{-1} \tag{8.7}$$

which is the rate at which phytoplankton are consumed (grazed) by zooplankton and is the same unit as primary productivity. If the ratio of grazing rate to primary productivity averages around 1.0 for a growing season, it indicates that all of the phytoplankton produced is removed by grazers. Such equality of the two rates has been observed (Wright, 1958; Green and Hargrave, 1966) and is usually typical of more oligotrophic environments. Within more eutrophic environments, the ratio tends to fall below 1.0 because of the increasing presence of larger species of phytoplankton that are less easily consumed, as will be discussed later.

As discussed earlier (Figure 2.2), only a portion of food that is consumed is assimilated for growth and activity. Uhlmann (1971) cited results from several authors concluding that to produce one unit of *Daphnia* requires ≈ 5 units of phytoplankton. This represents a yield of 0.2; the values ranged from 0.15–0.4 in continuous culture of *D. magna* (Uhlmann, 1971).

Equation 8.6 suggests that given a constant filtering rate, grazing rate will continue to increase as phytoplankton concentration increases. However, at some point filtering rate will decrease and digestion will decrease as well (Uhlmann, 1971). Thus, the relationship of grazing rate to phytoplankton biomass will take a Michaelis–Menten hyperbolic shape (Figure 7.14), with grazing rate becoming saturated at some phytoplankton concentration (Uhlmann, 1971).

To comprehend the potential for zooplankton grazing to control the biomass of phytoplankton, the results from continuous culture experiments with *D. magna* and *Chlorella* (as food) are instructive. The grazing rate, G, can also be expressed in units of day^{-1}, which if multiplied by the biomass of algae would give the same units as in Equation 8.7, and:

$$GX_{\text{phyto}} = \mu_D X_D / Y \tag{8.8}$$

where X is biomass of phytoplankton and *Daphnia*, Y is the yield coefficient, taken as 0.25 (see above), and μ_D is the maximum growth rate of *Daphnia*, observed to be 0.3 in continuous culture (Uhlmann, 1971). Assuming that X_D and X_{phyto} have the same order of magnitude, the equation becomes:

$$G = \mu_D / Y \tag{8.9}$$

which is 1.2 if the values given above are used. This means that *Daphnia* is capable of consuming 1.2-times the algal biomass per day if they are of equal magnitude. Capacity of the zooplankton to remove the phytoplankton would be, of course, less with more phytoplankton and vice versa.

A similar, but less specific example, is to assume that the grazing rate of zooplankton equals the growth rate of phytoplankton. If the grazing rate is doubled from that steady-state condition, the phytoplankton biomass would decline from an arbitrary initial level of 1×10^6 cells to 27×10^3 cells in 5 days. If the grazing rate is tripled the phytoplankton biomass would be eliminated in 5 days (Sverdrup et al., 1942).

8.4 Zooplankton grazing and eutrophication

As indicated above, zooplankton grazing usually has greater control on phytoplankton biomass in oligotrophic than eutrophic waters. As a result, predator–prey conversion efficiencies would tend to decrease with eutrophication.

Table 8.2 Expected character of phytoplankton–zooplankton relationships with respect to trophic state

Variable	Oligotrophic-mesotrophic, moderately fertilized, pelagic-marine	Eutrophic, heavily fertilized ponds, neritic-marine
Ratio of zooplankton consumption/primary production	≈100%	≈30%
Efficiency of energy conversion	≈20%	≈10%
Size of zooplankton	Large	Small and do not control phytoplankton
Size of phytoplankton	Nanoplankton <50 μm	Nanoplankton not used, bacteria and detritus consumers dominate

Source: Hillbricht-Ilkowska (1972).

Table 8.3 The ratio of zooplankton to phytoplankton productivity as a measure of food-web efficiency in experimental ponds and three lakes

	Nutrient status	Zooplankton/ phytoplankton production
Ponds		
	High	0.07–0.05
	Medium	0.41–0.08
	Low	0.56–0.20
Lakes (2 years)		
Sammamish	Mesotrophic	0.04–0.04
Chester Morse	Oligotrophic	0.09–0.08
Findley	Oligotrophic	0.18–0.08

Sources: Hall et al. (1970); Pederson et al. (1976).

Observations of zooplankton–phytoplankton interrelationships across trophic gradients have shown that, along with a decrease in the ratio of zooplankton consumption and production to phytoplankton production, there is a decrease in the size of zooplankton (Hillbricht-Ilkowska, 1972; Gliwicz and Hillbricht-Ilkowska, 1973; Gliwicz, 1975; Pederson et al., 1976) (Tables 8.2 and 8.3). Associated with this change in activity and energy transfer, there was also a decrease in the size of zooplankton and an increase in the size of phytoplankton to more colonial and filamentous forms. Also, Gliwicz showed that the filtering rate increased from ultraoligotrophy through mesotrophy, but decreased in eutrophic lakes. Zooplankton biomass increased through mesotrophy, but failed to increase further in eutrophic lakes. Thus, the efficiency of phytoplankton removal was poor, with eutrophy, resulting from an increasing amount of net phytoplankton and bacteria/detritus.

There are several processes, which have been further clarified in recent years, that can account for this observation. As discussed in the chapter on phytoplankton (Chapter 6), zooplankton are relatively ineffective at grazing large colonial and filamentous species, such as the nuisance-bloom-forming blue-greens. The size and shape of colonial and filamentous blue-greens create problems for filter-feeding grazers. The filaments are either too large to consume or they interfere with their filtering mechanism resulting in added energy costs that reduce growth and survival (Porter, 1972, 1977; Porter and McDonough, 1984; Infante and Abella, 1985; Burns, 1987). Even large colonial diatoms are not favoured (Infante and Litt, 1985). Also, toxicity produced by some nuisance blue-greens has been shown to inhibit growth and reproduction of zooplankton (Arnold, 1971; Lampert, 1981). Rather than zooplankton grazing causing the increase in larger colonial and filamentous algae (especially blue-greens) as eutrophication increases, the evidence suggests that changing physical–chemical conditions (CO_2/pH, N:P, stability, etc.) as they affect buoyancy, sedimentation and light availability (see Chapter 6) are the more probable causes for the changes in phytoplankton dominance. The decreased size of zooplankton, with associated reduced capacity to graze the phytoplankton, would then be a response to the greater availability of bacteria/detritus resulting from the relatively ungrazed phytoplankton crop.

On the other hand, *Daphnia* is considered to be more representative in eutrophic than oligotrophic lakes (Patalas, 1972; McNaught, 1975). An explanation is that *Daphnia* species are large and therefore have the capacity to select a wider size range of food particles (see above), thus being more versatile than the calanoid copepods. Except in cases of blue-green toxicity and filamentous forms that interfere with filtration (Edmondson and Litt, 1982; Porter and McDonough, 1984), the large grazing-efficient *Daphnia* should be abundant and able to control most phytoplankton in eutrophic lakes. However, there is another important process that would tend to reduce the abundance of large grazing-efficient zooplankton and that is planktivory. The predatory activity of small fish, such as yellow perch (*Perca flavescens*) and bluegill sunfish (*Lepomis macrochirus*) are size-selective and can effectively reduce populations of large cladocerans (Brooks and Dodson, 1965; Hall *et al.*, 1970; Shapiro *et al.*, 1975; O'Brien, 1979; Mills and Forney, 1983; McQueen *et al.*, 1986).

There are many examples of size-selective zooplanktivory by fish. Hall *et al.* (1970) showed in a series of experimental ponds that regardless of the level of N and P enrichment, the presence of bluegill sunfish determined the zooplankton composition; an average of 53% of the biomass was contributed by the relatively large *Ceriodaphnia* in ponds without fish, but it made up only 3% in ponds with fish. In Oneida Lake, New York, year-to-year and seasonal fluctuations in *Daphnia pulex* were controlled largely by the abundance of young yellow perch (Mills and Forney, 1983). Fish removal or exclusion experiments have also clearly demonstrated the adverse effect of planktivores on large cladoceran zooplankton (Lynch and Shapiro, 1981; Shapiro and Wright, 1984; Spencer and King, 1984; Raess and Maly, 1986; Post and McQueen, 1987). Preserving the large, grazing efficient zooplankton has also been accomplished by stocking piscivorous fish, such as large trout or bass, which is another way of removing the planktivores (Benndorf *et al.*, 1984; Wagner, 1986; Carpenter *et al.*, 1987). Without exception, such manipulations have resulted in increased abundance and size of cladocerans.

8.5 Trophic cascades

With an increase in the abundance of large zooplankton, one would expect an increased loss rate, and consequently a decreased biomass, in the phytoplankton. In most of the above cited experiments, that is what happened. The positive effect of planktivorous fish on phytoplankton biomass was initially documented by Hrbacek *et al.* (1961), but the deliberate alteration of fish populations to control phytoplankton by preserving large, grazing efficient zooplankton, was originally termed 'biomanipulation' by Shapiro *et al.* (1975). Later the process was referred to as 'cascading trophic interactions' by Carpenter *et al.* (1985). It is also known as the 'top-down effect', as opposed to the 'bottom-up effect' (i.e. through nutrients), of controlling phytoplankton. The essence of the process can be illustrated with three relationships: indirect between piscivorous fish and planktivorous fish; direct between piscivorous fish biomass and the biomass of large zooplankton; and a resulting inverse relationship between piscivorous fish biomass and phytoplankton biomass (Figure 8.2). As an example, Shapiro *et al.* (1975) showed that phytoplankton biomass in enclosures were ultimately (50 days) reduced to low levels by the grazing of *Daphnia* regardless of whether the water had been enriched as long as perch were absent. The trophic cascade, however, does not always carry down to the phytoplankton, especially in highly enriched waters (Post and McQueen, 1987), and the reason is not well understood. It could be due to the reduced filtering capabilities encountered by *Daphnia* feeding on filamentous blue-greens, because blue-greens are promoted by physical–chemical conditions associated with enrichment, or by stimulation from increased nutrient recycling (Post and McQueen, 1987), or by the inability of the process to reduce phosphorus to controlling levels in highly enriched lakes (Benndorf, 1990).

The nature and magnitude of food web impacts on phytoplankton biomass has been an intensely debated topic (DeMelo *et al.*, 1992; Harris, 1994; Sarnelle, 1996). In a quantitative review of 54 trophic cascade experiments, Brett and Goldman (1996) showed that zooplanktivorous fish treatments

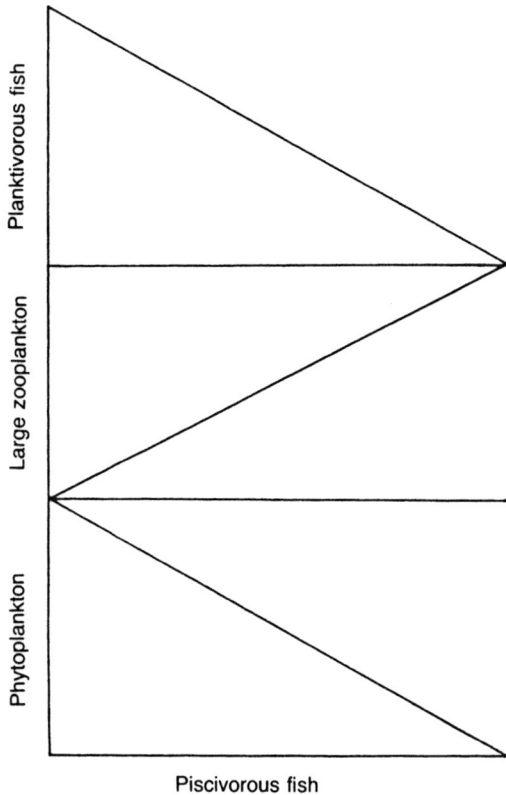

Figure 8.2 Hypothetical relationships between the biomass of piscivorous fish, planktivorous fish and phytoplankton.

on average reduced zooplankton biomass to only 25% of fishless control treatment levels (Figure 8.3). However, there was great variability in the phytoplankton community's response to these treatments. In about two-thirds of these experiments phytoplankton biomass increases in response to fish treatments (and reductions in zooplankton biomass) were quite small ($\approx 40\%$ increase in algal biomass), while in the other cases responses to the fish treatments were quite large ($\approx 700\%$ increase in algal biomass). Interestingly, Brett and Goldman (1996) found that the response of the zooplankton community to fish treatments was not correlated with the phytoplankton's response to these treatments. These results suggest that reducing or removing zooplanktivorous fish from most lakes would only result in a small (and probably imperceptible) improvement in water clarity, whereas in some cases removing zooplanktivores might lead to dramatic improvements in clarity. This result supports the hypothesis that trophic cascades

224 Effects of pollutants in standing water

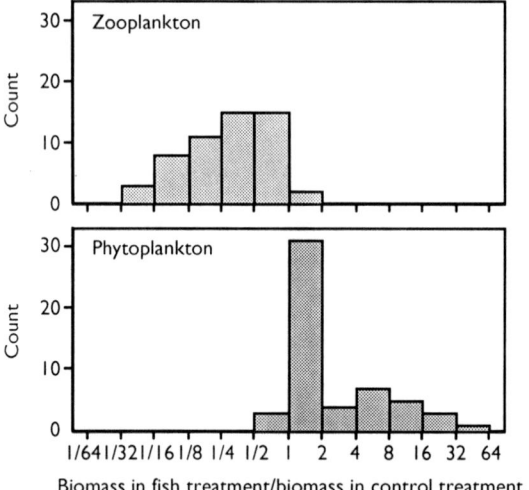

Figure 8.3 The response of zooplankton and phytoplankton biomass to zooplanktivorous fish treatments in 54 mesocosm experiments. Adapted from Brett and Goldman (1996).

are a common occurrence in lakes, but lends much weaker support to the idea that these food web processes can be consistently applied to suppress phytoplankton biomass in overly productive lakes. Another area of debate in the food web interactions literature is the relative importance of top-down (i.e. fish control) and bottom-up (nutrient control) processes for regulation of zooplankton and especially phytoplankton biomass in lakes. By examining the results of experiments that utilized both with/without fish and with/without nutrient addition treatments, Brett and Goldman (1997) showed that zooplankton biomass was strongly regulated by fish predation but was, in these experiments, unrelated to nutrient additions. This analysis also showed that phytoplankton biomass increased on average by $\approx 100\%$ in fish addition treatments and by $\approx 200\%$ in nutrient addition treatments. These results are quite similar to McQueen et al.'s (1986) top-down versus bottom-up hypothesis, where these authors suggested that zooplankton biomass will be most strongly regulated by fish predation and phytoplankton biomass will be most strongly regulated by nutrient supplies.

McQueen et al. (1986) also suggested that 'trophic-decoupling' may be a common feature of planktonic food web interactions. Trophic-decoupling means interactions at the zooplankton–phytoplankton interface are either weak or highly variable and difficult to predict. Evidence of trophic-decoupling could be seen in Brett and Goldman's (1996) study, where they did not find a correlation between zooplankton and fish responses to fish treatments, and in Brett and Goldman's (1997) study, where these authors noted that zooplankton biomass did not have a statistically significant response to nutrient addition treatments despite the fact that these treatments caused algal biomass to triple. Similar trophic-decoupling was noted in a study of coastal and marine planktonic food webs (Micheli, 1999). Some field studies have suggested that this decoupling may be due to physical feeding interference and the very low food quality of blue-green-dominated phytoplankton assemblages (Müller-Navarra et al., 2000). Recently, Danielsdottir and Brett (unpublished data) developed a mechanistic model to test the potential impact of phytoplankton food quality, nutrient availability and fish predation on planktonic food web dynamics. According to the results of this model, when phytoplankton food quality for herbivorous zooplankton was high, strong trophic cascades suppressed phytoplankton biomass, the zooplankton withstood intense zooplanktivory, and energy was efficiently transferred through the food web sustaining higher trophic level production. Danielsdottir and Brett also found that low food quality resulted in trophic-decoupling at the plant–animal interface, with phytoplankton biomass determined primarily by nutrient availability, zooplankton easily driven to extinction by fish predation and poor energy transfer through the food web. Interestingly, these model results are quite similar to Vollenweider's (1976) prediction that 'the phytoplankton-zooplankton interrelationship appears to be particularly dependent on the species composition of the phytoplankton. If the phytoplankton is composed primarily of species edible for zooplankton, one may find a relatively low phytoplankton standing crop'. The feasibility of biomanipulation for lake restoration will also be discussed in Chapter 10.

8.6 Temperature and oxygen

The lethal temperature (LT_{50}) for most crustacean zooplankton is rather high, between 28 and 35°C, and acclimation temperature seems to have little influence in adjusting the lethal temperature (Welch and Wojtalik, 1968). For example, *Eurytemora affinis*, an estuarine copepod, had a lethal temperature between 28 and 32°C whether the acclimation temperature was 15, 20 or 25°C (Heinle, 1969a). The acclimation time in that experiment was two months or about one generation time. The rather absolute nature of the lethal temperature for zooplankton will be illustrated again in Chapter 10 with the discussion of invertebrate survival during passage through power-plant cooling systems. There are stenothermal species, however, such as *Limnocalanus* discussed below.

Considering their apparent high thermal tolerance, compared to typical maximum temperatures in lakes, there would seem to be some capacity for heat assimilation with respect to zooplankton in most temperate lakes. One, therefore, might expect that moderate increases in temperature from the discharge of heated water would increase zooplankton production. That is because metabolic activity increases with temperature resulting in increased growth so long as the organism's optimum is not surpassed.

Such an increase in production occurred in a Polish lake (Lichenskie) that received heated water from a power plant. The temperature of that lake reached an average during the warm part of the summer of 27.1°C compared to 21.6°C in an unheated control lake (Mikorzynskie). Zooplankton production and turnover rate (P:B) doubled in the heated lake compared to the unheated lake (Table 8.4). Both lakes were eutrophic and had similar biomass and species composition, although *Daphnia* were less important in the heated lake. So the effect of an average difference of 5–6°C resulted in a doubling in herbivorous zooplankton production. That was in spite of the dominance of the phytoplankton being *Melosira granulata* and *Microcystis aeruginosa*, which are not easily consumed (see Chapter 6). Although zooplankton production and turnover rate increased through the trophic series, with even a greater increase when increased temperature was combined with a eutrophic state, there still may have been a shift in size and a decrease in grazing efficiency as discussed above. *Daphnia* did decrease in importance in the heated lake and copepods were the principal producers in both lakes. Such increases in production could result even if the detritus/bacteria pathway were increasingly utilized by smaller zooplankton throughout the trophic series shown in Table 8.4. However, phytoplankton production also increased with added heat from 3.7 to 7.3 g O_2 m^{-2} day^{-1}, which suggests that a change in energy pathway was not associated with the temperature effect. However, the quality of the lakes may have declined with eutrophication (e.g. the dominance of *Microcystis*) and heat, in spite of the increased production. Moreover, other species such as many fish are less tolerant of temperature increases than are zooplankton and would take precedence when judging whether or not eutrophication and/or heated water input were benefits.

Zooplankton is adversely affected by low dissolved oxygen. The large copepod, *Limnocalanus*, was nearly eliminated in Lake Erie as eutrophication increased. The reason was thought to be related more to planktivory

Table 8.4 Zooplankton (herbivores) production and turnover rate (P:B) over a trophic series of lakes and with heat added

Lake	Trophic state	Production (kcal m^{-3} day^{-1})	P:B
Naroch	Mesotrophic	0.12	0.06
Miastro	Mesotrophic-eutrophic	0.30	0.08
Batorin	Eutrophic	0.71	0.18
Mikorzynskie	Eutrophic, unheated	0.62	0.11
Lichenskie	Eutrophic, heated	1.38	0.24

Source: Patalas (1970).

than to enrichment *per se*. Gannon and Beeton (1971) suggested that decreased hypolimnetic DO forced the ordinarily cold-loving organism from the hypolimnion into the epilimnion where DO was plentiful, but was more susceptible to grazing by increasing populations of yellow perch. Reversing the process has been suggested as a way to enhance the preservation of large-bodied zooplankton subjected to excessive planktivory. By oxygenating anoxic, relatively dark hypolimnia, refuges could be created into which zooplankton could migrate during the day when they are most susceptible to sight-feeding planktivores. This has been shown to work in enclosure experiments (McQueen and Post, 1988).

Cladocerans living in low-DO waters often become red in colour as a result of producing haemoglobin (Fox, 1950). Red *Daphnia* survived longer, fed on *Chlorella* at a higher rate and produced more eggs than did pale organisms under low DO. Red patches at the surface of highly enriched environments, such as sewage lagoons, are often the result of swarms of cladocerans with high haemoglobin content.

8.7 Invasive, non-native zooplankton

Examples of non-native zooplankton species that have been introduced to the USA with significant ecological effects include the opossum shrimp (*Mysis relicata*) (Spencer *et al.*, 1991) in Montana, *Daphnia lumboltzi* in the south-east and south-central regions (Havel and Hebert, 1993; Kolar and Wahl, 1998), and *Bythrotrephes cederstroemi* in the north-east (National Research Council, 1996). The latter two species are probably more resistant to predation by planktivorous fish than native zooplankton, giving them a competitive advantage. *B. cederstroemi* was introduced to the USA via ship ballast water from Europe, whereas the source of *D. lumboltzi* is unknown but was likely human-caused.

The introduction of opossum shrimp to Flathead Lake (Montana) was intentional. Between 1968 and 1975, fish managers introduced opossum shrimp to more than 100 lakes in western North America to provide additional food for kokanee salmon (*Oncoryhychus nerka*) (Spencer *et al.*, 1991). The shrimp reached Flathead Lake in 1981 resulting in dramatic changes in the native zooplankton populations with far-reaching effects on the food web. Opossum shrimp removed the native zooplankton, which were a major component of the salmon's diet, and escaped from fish predation by migrating to the dark lake bottom during the day. The establishment of opossum shrimp in Flathead Lake led to a sharp decline in kokanee, which in turn deprived bald eagles (*Haliaeetus leucocephalus*) and grizzly bears (*Ursus horribilus*) from eating the kokanee as they spawned in the lake's tributaries (Spencer *et al.*, 1991).

Chapter 9
Macrophytes

The higher plants in aquatic environments are usually rooted with rigid cell walls and asexual (e.g. fragmentation, tubers) and sexual (flowers and seeds) reproduction. Asexual, or vegetative, reproduction is usually highly effective in expanding a plant's distribution within a water system. Plants emphasized here are the submersed species, which are usually rooted and occupy the littoral zones of lakes and often the slow-moving or stagnant reaches of rivers. Emergent species dominate in the very shallow edges of the littoral and in marshy wetland areas, but they usually do not present the nuisance problems provoked by some submersed species.

Of interest are the effects of macrophytes on water quality, such as nutrient recycling through senescense, and controls on growth and distribution of especially the nuisance submersed species, such as water milfoil and hydrilla. Much has been learned about these aspects in the past 30 years. Rooted submersed macrophytes are known to depend largely on the sediments for their nutrition. They release very little nutrient to the surrounding water via excretion, but some contribute substantially to internal nutrient loading through senescense and decay. Although sediment texture, and organic and nutrient content have been shown to affect the distribution and growth of rooted macrophytes, sediment characteristics have not been effective in predicting such attributes. However, given an adequate substratum, the maximum depth of colonization can be predicted.

9.1 Habitats

9.1.1 Running water

Most rooted macrophytes are not able to withstand too great a current and an adequate sediment deposit must exist for their rooting. Light is generally satisfactory, because rivers are usually shallow, but turbidity may still limit in some instances. Butcher has listed five types of waters, based on current and sediment type, that influence nutrient availability and in which the associated turbidity affects light penetration (Butcher, 1933; Hynes, 1960). The types with representative plants are as follows.

1   Torrential on rock or shingle – this habitat contains mostly mosses (*Fontinalis*); larger macrophytes are not stable in the high velocities.
2   Non-silted on shingle – a 'sluggish' stream; milfoil (*Myriophyllum*) and water crowfoot (*Ranunculus*).
3   Partly silted on gravel and sand – this habitat, because of bottom type and turbidity, has, for example, *Ranunculus*, pondweeds (*Potamogeton*), and arrowweed (*Sagittaria*).
4   Silted on silt – the sedimented bottom and increased turbidity favours pondweeds (*Potamogeton*), water lilies (*Nuphar*), and the aquarium plant (*Elodea, Egeria*).
5   Littoral – very little current with mud substrate. Emergent vegetation is common, for example, reeds and bulrushes.

9.1.2 Standing water

The distribution of macrophyte development in standing water is dependent on slope in the littoral region. The more gradual the slope, the greater the potential to collect sediment, as well as the more area available to light penetration and thus optimal growth. Spence (1967) has observed in Scottish lochs that macrophyte diversity and abundance were favoured by brown muds and an alkalinity >50 mg l^{-1} as $CaCO_3$. Vegetation was usually absent from rocky or sandy beaches.

The characteristic zones in the littoral region are as follows.

1   Emergent vegetation zone: grasses, rushes, and sedges, which depend on lake sediments for nutrients but otherwise carry on photosynthesis in the atmosphere, which supplies CO_2.
2   Floating leaf plant zone: water lilies, some pondweeds, and duckweed (*Lemna*), most of which obtain their nutrients primarily from sediments but carry on some photosynthesis in direct contact with the atmosphere.
3   Submerged vegetation zone: pondweeds, hornwort (*Ceratophyllum*), naiads (*Najas*), milfoil, arrowweed, and stonewort (*Chara*), which obtain nutrients from both soil and water and are dependent on CO_2 in water for photosynthesis.

9.2   Significance of macrophytes

Rooted macrophytes are significant in lakes in at least four ways. With respect primarily to the submersed growth forms in the littoral zone, macrophytes have the following effects: contribute productivity themselves and provide a substratum for periphyton and insects to attach; contribute to lake ageing by recycling nutrients and accumulating sediment; serve to protect juvenile fishes from predation; and create nuisance biomass levels, especially if they are exotic species, that interfere with water recreation.

Because the colonizable area for submersed macrophytes is limited by light penetration to the lake bottom, macrophytes normally supply a relatively small fraction of primary productivity in large deep lakes. That is, littoral production is much less than pelagic production. However, Wetzel and Hough (1973) argue that most of the lakes in the world are small and shallow and consequently have a large fraction of their productivity supplied by macrophytes and attached periphyton. It follows, therefore, that macrophytes must provide a substantial portion of primary productivity in freshwater although there are few cases where they actually dominate productivity (Wetzel, 1975).

Macrophytes also recycle nutrients from the sediment that contributes to phytoplankton production, which in turn contributes additional sediment (together with the organic matter from macrophytes themselves that is trapped by macrophytes further extending the littoral zone. In that manner, macrophytes accelerate lake ageing. Carpenter (1981) quantitatively analysed this process for Lake Wingra, Wisconsin and concluded that ageing was more sensitive to internal phosphorus recycling via rooted macrophytes (milfoil) than to external sediment and nutrient inputs.

Submersed macrophytes provide a surface for fish egg incubation, e.g. in the yellow perch. They also protect juvenile fish from predation. That function has a positive effect in the case of northern pike (*Esox lucious*) which are very cannibalistic. Without substantial macrophyte cover their survival is low. On the other hand, if macrophyte stands are too dense and widespread, protection from predation, especially in the centrachid (sun-fishes) and yellow perch (*Perca flavens*), leads to stunting. Therefore, numbers increase but not biomass (Nichols and Shaw, 1986). Reducing the area colonized by macrophytes (and concentrating the fish for a summer) by drawing down the water level led to decreased numbers and increased size of black crappie and, consequently, increased growth of large-mouth bass in Long Lake (Kitsap), WA (Gross, 1983). Thus, there seems to be an optimum density and distribution of macrophytes with respect to fish production and individual size. Critical limits for surface coverage of macrophytes have been suggested, but would no doubt vary from lake to lake depending on fish species and macrophyte density (i.e. g or stems m^{-2}).

Excessive macrophyte abundance can represent a nuisance to water recreation, such as swimming, fishing, boating and water skiing. Increased density and distribution of macrophytes are often perceived by lake-shore residents and water users to occur over a period of 10–20 years. Such increases may be caused by sediment accretion due to increased nutrient input and recycling as suggested by Carpenter (1981) and involve native species. However, many problems with submersed macrophytes are due to exotic species, such as Eurasian water milfoil (*Myriophyllum spicatum*), curlyleaf pondweed (*Potamogeton crispus*), elodea (*Elodea canadensis* and *Egeria densa* (=*Elodea Elodea densa*)) and hydrilla (*Hydrilla verticillata*). Milfoil, *E. densa*, hydrilla and curlyleaf pondweed are exotic to the USA and *E. canadensis* is exotic to Europe. It is of interest to review briefly the case of milfoil in the USA.

Milfoil had been present in the Chesapeake Bay region since 1881, but did not reach nuisance proportions until the 1950s and 1960s, possibly related to hurricanes accompanied by increased salinity and input of sediment (Nichols and Shaw, 1986). Subsequently, it spread to the midwestern USA and the Northwest where it reached nuisance proportions in the 1960s and 1970s, respectively. In some areas it has declined inexplicably to 10–20% of earlier, explosive levels (Elser, 1967; Wile, 1975; Carpenter and Adams, 1977, 1980b). In those areas it has begun to fluctuate from year to year in a fashion similar to native species (Nichols and Shaw, 1986).

Much of the success of milfoil is related to an effective method of vegetative reproduction. The plant easily fragments with parts floating for long distances before settling to the sediment to colonize a new area. Although the plant is exotic to the USA, environmental reasons have been suggested to account for its success. Coffey and McNabb (1974) observed that milfoil grew better under the ice at low light intensities than other species. It therefore had a larger biomass at the beginning of the growing season than other species, which it could shade. However, milfoil rose to dominance in parts of Lake Washington, especially Union Bay, in spite of increasing clarity due to decreased phytoplankton biomass (recovery from eutrophy to oligotrophy) and no ice cover. While this case does not support the low-light explanation, its invasion into many rather turbid Tennessee Valley reservoirs where macrophytes were previously scarce is consistent with the idea. Its ability to reach the surface in water as deep as 3 m and then extend horizontally creating a tangled mass of stems and leaves on the surface allows it to compete effectively for light among other species and in turbid environments.

Milfoil may also be favoured by water hardness and pH (Hutchinson, 1970b). Results from Swedish lakes showed that it seemed to prefer higher pH and Ca content compared to two other species of *Myriophyllum* (Figure 9.1). Hutchinson hypothesized that its success may be related to its ability to use HCO_3. Spence (1967) also noted the presence of *M. spicatum* in lakes with $HCO_3 \geq 60$ mg l^{-1}. However, Nichols and Shaw (1986) suggest that explanation may be over-simplistic even though they cite evidence that it tolerates salinity up to 2% and grows vigorously at salinities to 1%. Its explosion and dominance in Lake Washington and Lake Sammamish, WA, indicates that it can be successful in soft-water lakes as well. Nevertheless, eutrophication, which tends to increase pH and shift the carbon equilibrium to more HCO_3 and less CO_2, due to greater demand on CO_2 by phytoplankton, may contribute to the explosive success of milfoil. On

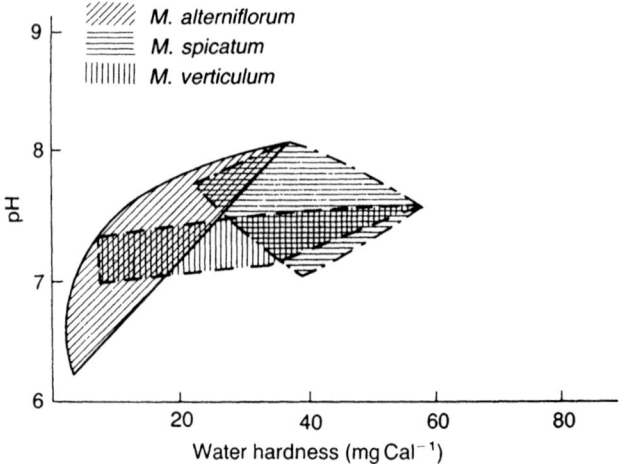

Figure 9.1 Distribution of three species of *Myriophyllum* in Sweden as a function of pH and water hardness (modified from Hutchinson, 1970b).

the other hand it has risen to dominance in oligotrophic systems as well (Newroth, 1975).

Alternatives to control macrophytes will be discussed in Chapter 10.

9.3 Effects of light

Submersed macrophytes are adapted to high light intensities. This is understandable considering that they can grow to great abundance in water less than 1 m deep. For example, *E. canadensis* was shown to have an optimum between 75 and 100% of full sunlight, based on *in situ* measurements of ^{14}C assimilation (Figure 9.2). Note that rates were very low for plants exposed to 12.5% of sunlight. Nichols and Shaw (1986) cite studies showing the optimum for elodea to occur between 15% and full sunlight, as well as it being found deeper than other species in some instances. Thus, elodea is tolerant of a wide range of light, but of special interest is its utilization of high light intensities. In contrast, phytoplankton are generally considered to be saturated at between 30 and 50% of full sunlight (see Chapter 6).

The apparent benefit of high light intensities to submersed macrophytes is consistent with regression models developed by Canfield *et al.* (1985) for 108 lakes and Chambers and Kalff (1985) for 90 lakes in which the maximum depth of plant colonization (MDC) was a function of the Secchi disc depth, which disappears at $\approx 10-15\%$ of surface intensity. The MDC can vary depending on the species present. From the following models, the actual depth of colonization is slightly more than the Secchi depth: log

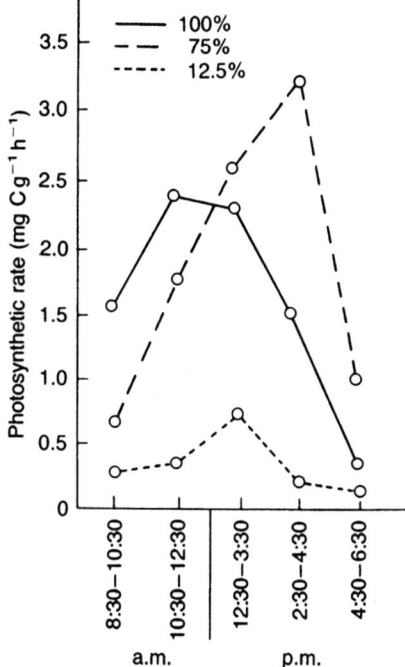

Figure 9.2 Photosynthetic rates of *Elodea canadensis* suspended in pickle jars at the indicated percentages of incident light in a shallow pond (modified from Hartman and Brown, 1967).

$MDC = 0.62 \log SD + 0.26$ (Canfield *et al.*, 1985); $Z_c^{0.5} = 1.33 \log SD + 1.4$ (Chambers and Kalff, 1985).

The great sensitivity of colonization depth to relatively high light intensities can have marked year-to-year effects on plant abundance and distribution in reservoirs subjected to variations in turbidity. For example, the year-to-year variation in plant abundance in Pickwick Reservoir, Alabama was tied closely to light and rainfall during the spring growth period (Figure 9.3). The two years in which the plants (mostly *Najas*) reached nuisance proportions had sunny springs with low rainfall. Nuisance proportions refer to a coverage of 2400 ha, of the 20 000 ha reservoir, with submersed macrophytes that reached the surface, allowing mats of periphyton to develop. The two sunny, dry springs at Pickwick Reservoir with more incident light and less rainfall resulted in greater water clarity, because less rainfall produced less runoff and less turbidity. As a result, the depth of 10% surface intensity ranged from 1 to 2 m, which would represent a large difference in reservoir area colonized. Robel (1961) has also shown the effect of turbidity on the abundance of *Potamogeton pectinatus* (sago pondweed) in a Utah marsh (Figure 9.4).

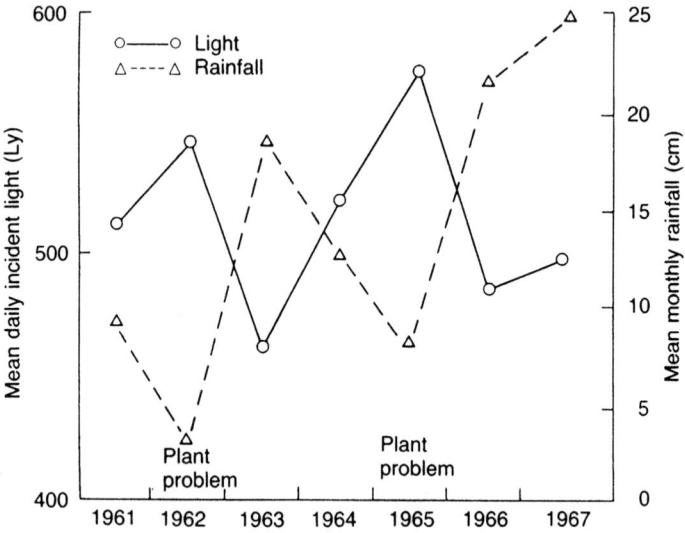

Figure 9.3 Mean daily incident light and mean monthly rainfall recorded at Florence, Alabama, during the 'critical' period of plant growth in Pickwick Reservoir from 15 April to 31 May during 1961–7 (Peltier and Welch, 1970).

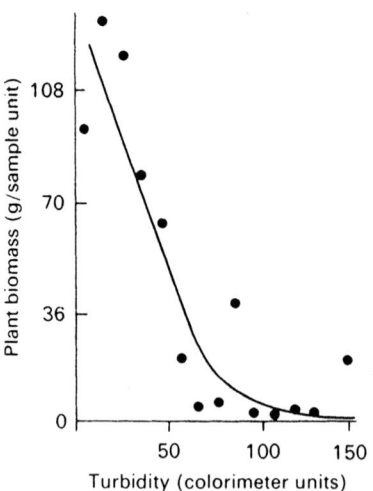

Figure 9.4 Relation between *P. pectinatus* biomass and turbidity in a Utah marsh (Robel, 1961).

The lake area colonized can similarly vary greatly with the concentration of phytoplankton and in turn with the level of enrichment. Secchi disc depth is a function of both phytoplankton and inorganic sediment (as in Pickwick Reservoir). The potential inverse relationship between phytoplankton and macrophytes must be considered when choosing restoration alternatives for lakes. For example, for Green Lake, Seattle (WA), it was shown that halving the summer average TP content would decrease the chl a content threefold and increase the area colonizable by macrophytes by threefold as well. The effect of water clarity on area colonizable by macrophytes is an important consideration where manipulating the nutrient, and hence phytoplankton, concentration over the long term. However, sediment type and nutritional content are also important as will be shown.

9.4 Effects of temperature

Temperature acts as a control on light-saturated rates of photosynthesis and could be expected to approximate the Q_{10} rule, over a portion of its temperature range at least, as was shown with plankton and periphytic algae. Field observations of *P. pectinatus* have shown the maximum growth rate to occur in the spring, coincident with greatest increase in light while temperature remained rather low (10°C). Hodgson and Otto (1963) state that the pondweed can develop at temperatures from 10–28°C. The interpretation must be that the greatest rate of growth occurred in young plants, but that this rate would increase as temperature increased up to its thermal optimum. This is because light-saturated rates of photosynthesis increase with temperature. Thus, if the river had been heated during the spring when light increased, growth would have been even greater than observed.

A temperature increase from heated-water discharges would probably select species with temperature optima matching the new conditions, as was the case with algae. Hynes (1960) observed that *Vallisneria spiralis* (wild celery) is usually a warm-water aquarium plant in the UK and has been found in the natural environment only where heated effluents occur. However, nuisance macrophyte growth related to heated-water discharges has not been a recognized problem.

9.5 Effects of nutrients

Although distribution and growth of submersed macrophytes is dependent on year-to-year and seasonal variations in water clarity, the more long-term increases (and decreases) are usually attributed to increases in enrichment. While N, P and C (and possibly K) are the important macronutrients for submersed macrophytes, as they are for algae, experimental work shows that macrophytes may not be affected by relatively high concentrations of N and P (500 and 50 µg l^{-1}) in the surrounding water as are plankton algae. For example, experiments with *P. pectinatus* (sago pondweed) conducted in flow-through aquaria to simulate river growth conditions, showed no significant difference in growth over a tenfold gradient in SRP, which should have been the limiting nutrient (Table 9.1). Growth appeared to be enhanced slightly if plants were rooted in river sediment. This suggests that growth was either saturated at low water concentrations ($<20\,\mu g\,l^{-1}$, which was the lowest concentration tested) or that water concentrations were relatively unimportant and the plants were more dependent on sediments for nutrition. Plants with roots in sand sealed from overlying water showed poor growth indicating the roots to be more important in nutrient uptake than shoots.

The dominant effect of sediment nutrition and minor effect of high nutrient concentrations in the water was also shown for milfoil (Table 9.2). There was some limitation at very low concentrations ($5\,\mu g\,l^{-1}$ SRP) with roots in sand, but growth in full sediment (undiluted with sand) was nearly five times greater than growth in full sand and at very high water concentrations of nutrients. Likewise, *Najas* growth was found to be proportional to the fraction of lake sediment (mixed with sand), irrespective of water concentrations, although there appeared to be some biomass dependent depletion of K (Figure 9.5). Gessner (1959) cited similar results with elodea and milfoil and Chambers *et al.* (1989) have shown that *P. crispus* grew better in sediments enriched by sewage effluent regardless of the surrounding water nutrient content.

Table 9.1 Growth (stem elongation) of *P. pectinatus* in aquaria with continuous flow

	NO_3-N(mg l^{-1})	PO_4-P(mg l^{-1})	22-day growth (mm)	
			Sediment	Sand
Full nutrient	1.71	0.26	359	310
Half nutrient	0.89	0.11	303	264
Tap water	0.44	0.03	351	288
Tap water with sealed roots	0.58	0.02	303	84

Source: Peltier and Welch (1969).

Table 9.2 Growth (biomass dry weight) of *Myriophyllum spicatum* under different treatments of sediment type and water nutrient concentration

Water nutrient (mg l^{-1})	Sand/sediment (%/%)			
	100/0	67/33	33/67	0/100
0.5 P; 5.0 N	1.43	–	–	–
0.05 P; 0.5 N	1.26	–	–	–
0.005 P; 0.05 N	0.68	3.78	4.51	5.63

Source: Modified after Mulligan and Barnowski (1969).

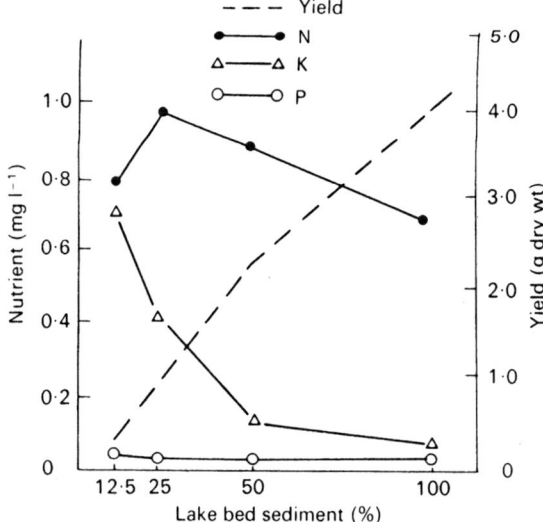

Figure 9.5 Effect of increasing fractions of lake-bed sediment (mixed with sand) on the yield of *Najas* and the measured concentrations of nutrients in experimental containers (Martin et al., 1969).

The more definitive experiments evaluating the relative role of sediments versus surrounding water as sources of nutrition, are those using ^{32}P. Carignan and Kalff (1980) showed that most of the P taken up by milfoil, *Heteranthera* and *Vallisneria* came from the sediments; 95–103% in oligotrophic, 86–96% in mesotrophic and 70–74% in eutrophic lakes. Gabrielson et al. (1984) found that 85% of the foliar demand of *E. densa* for P originated from the sediment and also found, as have others (see Nichols and Shaw, 1986), that there was no net excretion of P to the surrounding water. Other researchers to demonstrate that sediments are the principal source of the plant's nutrition when surrounding water concentrations are relatively low, are: Bristow and Whitcombe (1971) for P with *M. brasiliense*, *M. spicatum* and *E. densa*; Nichols and Keeney (1976a,b) for N with milfoil; Bole and Allan (1978) for P with milfoil and hydrilla; and Best and Mantai (1978) for N and P with *M. exalbescens*. Schultz and Malueg (1971) did show, however, that elodea, *P. amplifolias* and *V. americana* took up more P from the water than from sediments. The preponderance of results, however, demonstrate that the roots of several species of submersed macrophytes supply most of the plant's nutritional needs from the sediments. Foliar uptake may supply most needs if surrounding waters are relatively high in dissolved nutrients.

While the above experimental results show that sediments are a more important source for plant nutrition than the surrounding water, they do not explain the factor(s) in sediments responsible for improved growth. If sediment were the principal source of P in P-limited lakes, then one might expect a relationship between sediment P content (e.g. pore-water SRP + hydroxide exchangeable P) and plant growth, within and among lakes. However, such a relationship has not been found, which suggests that the availability of P in most lake sediments is large and some other factor(s) in sediment is more critical to growth and distribution. In support of this, critical tissue concentrations have been determined experimentally for several species by relating tissue content to growth (yield) in plants exposed to complete media of elements (Figure 9.6). Tissue analyses of naturally occurring populations have shown very few to be nutrient-limited (Gerloff and Krombholz, 1966). Exceptions were in Lake Wingra, Wisconsin where Gerloff (1975) showed K to limit milfoil growth (yield) and Schmitt and Adams (1981) found that net photosynthesis of milfoil was reduced at tissue levels of P (0.28%) much higher than those (0.13%) that limit short-term yield.

Organic content has been considered as a critical factor and favours plant growth when at intermediate levels, e.g. 10–25% of dry matter (Nichols and

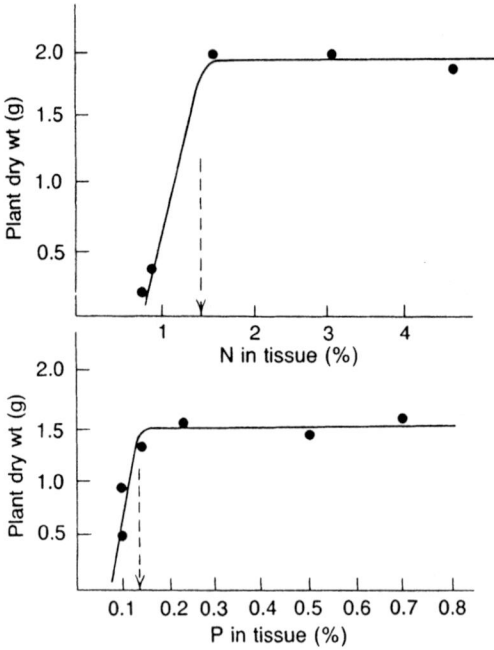

Figure 9.6 The critical tissue concentrations of N and P (concentrations allowing maximum growth, dashed line) in algae-free cultures of several species of macrophytes were consistently around 1.3% N and 0.13% P. The above results are with *Vallisneria americana* (modified from Gerloff and Krombholz, 1966).

Shaw, 1986). In support of this, Misra (1938) observed a species selection by sediments of different humus content in Lake Windemere, UK following work by Pearsall (1920, 1929) that plant distribution was controlled by the substratum. Three types of sediment were identified in the lake and each had a characteristic plant assemblage (Table 9.3). Laboratory experiments showed that *P. perfoliatus* grew best on the sediment type (intermediate humus content) in which it was found in the lake. Subsequent experiments showed that each dominant species preferred the sediment type in which it naturally occurred. Spencer (1990) recently found that *P. pectinatus* biomass and tuber production increased with increased humus addition to sediment and suggested that plants may prefer intermediate sediment density, not too flocculant and not too dense (sandy). The role of humus was one of reducing sediment density. On the other hand, additions of labile organic matter such as glucose (Spencer, 1990) or plant remains (Barko, 1983) have shown, respectively, either neutral or negative effects on plant growth. Apparently too much organic matter may in some cases create conditions of organic acid accumulation, low pH, increased metal availability and inhibiting gases that together are inhibitory to plant growth. At intermediate levels, however, organic matter may create a sediment density most favourable for growth because of greater nutrient availability (Barko and Smart, 1986).

The increased growth and distribution of rooted submersed macrophytes resulting from eutrophication is thus apparently an indirect effect of lake-water enrichment. Enrichment of the lake water will result in increased phytoplankton production, which will in turn result in increased sedimentation of organic matter. Enrichment of the water will probably not contribute directly to increases and decreases in macrophyte biomass. If there is ample shallow area in the lake with a gradual enough slope to accumulate the increased production of phytoplankton-derived low density organic sediment, conditions should improve for macrophytes to root and maintain themselves. The importance of slope in this regard has been demonstrated by Duarte and Kalff (1986), who showed that for 17 lakes in which turbidity and light did not limit growth ($SD > 2\,m$), 72% of the variation in macrophyte

Table 9.3 Association of plant and sediment types in Lake Windermere, England, and 1 month's growth of *P. perfoliatus*

Mud type	Dominant plant species	Humus (%)	Dry weight (mg)
Inorganic coarse brown silt	*Isoetes* (quill wort)	8.04	467
Moderately organic black flocculent mud	*Potamogeton perfoliatus*	12.26	778
Highly organic brown mud	*Sparganium minimum* and *Potamogeton alpinus*	24.00	298

Source: Misra (1938) with permission of Blackwell Scientific Publications Ltd.

biomass was explained by slope. The relationship was steeply inverse below a critical slope of 2.24% and if organic content were included, the variance explained increased to 88%. They suggest that at greater slope, high-water content, low density sediment tends to be unstable.

As enrichment of lake water increases, the maximum depth of colonization should decrease due to reduced clarity of the water restricting the availability of light. Thus, slope, which controls the distribution of suitable sediment, and available light are apparently the principal factors controlling the growth and distribution of submersed macrophytes. The increase in macrophyte biomass with eutrophication, until light begins to limit, is probably related to the increased deposition of organic matter that creates a low-density sediment more suitable for rooting. Wetzel and Hough (1973) have suggested a generalized sequence among the various types of primary producers and increased eutrophication (Figure 9.7). Phytoplankton and periphyton (attached algae) dominate as enrichment begins to increase, followed by submersed macrophytes. Periphyton are probably the first indicator of increased enrichment because its attachment near nutrient sources allow it to respond to increased nutrient concentrations, while dilution would prevent pelagic phytoplankton from experiencing such increased concentrations. This has been shown in ultra-oligotrophic Lakes Tahoe (Loeb, 1986) and Chelan (Jacoby et al., 1990). The subsequent greater increase in periphyton than phytoplankton as enrichment proceeds, as depicted in Figure 9.7, is probably related to the increased surface area for attachment provided by macrophytes. The dominance of periphyton and macrophytes in this example suggests that the lake is relatively shallow with a sufficiently small bottom slope to accumulate suitable sediment for macrophyte growth over much of its area. In deeper lakes with greater littoral bottom slope, a greater time separation

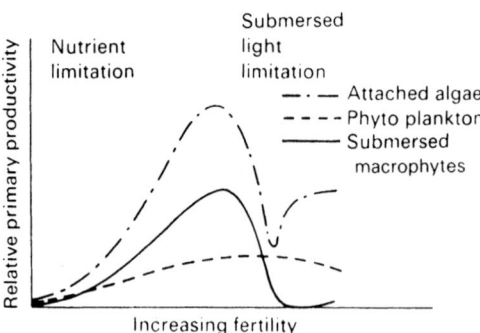

Figure 9.7 A generalized scheme for the changes in submersed macrophytes, attached algae and phytoplankton in lentic environments (modified from Wetzel and Hough, 1973).

may occur in the sequence of phytoplankton/periphyton and macrophytes in order to develop suitable sediment. The sediment and macrophyte development process, as Carpenter (1981) has suggested, includes a feedback loop from macrophytes to phytoplankton through P recycling.

Shading of macrophytes by periphyton may also account for the greater rate of increase by periphyton and, as Phillips *et al.* (1978) have demonstrated for the Norfolk Broads, UK, it is the most important reason for the decline in submersed macrophytes depicted in Figure 9.7.

9.6 Nutrient recycling

Because rooted macrophytes obtain most of their nutrient demand from the sediment, the fate of their tissue P and N content may be important to the nutrient economy of a lake if there is substantial coverage by plants. As indicated above, most research has shown that plants do not excrete significant amounts of dissolved P or N into the surrounding water (Nichols and Shaw, 1986). However, they can contribute significant amounts of nutrients through senescense and decay, which takes place throughout the summer growing season in some species. Milfoil continually sloughs leaves during the summer; the plants can senesce and break apart as they remain in the water column with much of the plant material near the water surface. A similar pattern has been observed for *P. praelongus* with most of its die-back complete by late July.

Landers (1982) showed that enclosures with milfoil in an Indiana reservoir contained more SRP than enclosures without milfoil or in the open water. Enclosures with milfoil also contained over four times more chl *a* than those without. Although this source of internal loading of P is considered important, realistic estimates of whole-lake contributions of P from macrophytes exist only for milfoil in Lake Wingra, Wisconsin. Smith and Adams (1986) determined shoot and root uptake of ^{32}P by plants in laboratory experiments and tissue content of P was determined in plants in the lake throughout the year. By combining laboratory and lake data, they constructed the following budget for P in the lake's weedbeds: in g P m^{-2} year^{-1}, uptake by roots (from sediment) 2.2 plus uptake by shoots (from water) $0.8 = 3.0$; loss to water from healthy shoots 0.0 plus loss to water by senenscence/decay 2.8 minus uptake from water $0.8 =$ gain to water (internal loading) 2.0 or in mg P m^{-2} day^{-1}, whole year on daily basis $= 5.5$ and summer on daily basis $= 17.0$.

These results are from 1977, after milfoil had started its decline to ultimately $\approx 10\%$ of its previous level in the 1960s and 1970s. Carpenter (1980a) estimated a contribution of 3.0 g m^{-2} year^{-1} from weed beds in 1975 when the plants were more abundant. As an indication of the significance of milfoil senescense as a source of P, the summer weedbed release of 17 mg m^{-2} day^{-1} is intermediate in the range of measured release rates from sediments in anoxic hypolimnia: 1.5–34 mg m^{-2} day^{-1} (Nürnberg, 1984).

In contrast, invasion of a Norwegian lake (Steinfjord) by *E. canadensis* in 1976 resulted in increased phytoplankton on only a few occasions in the 1980s following summer die-back (Rorslett *et al.*, 1986). While the plant's establishment raised the lake P storage (water + plants) from 2 to 6 tonnes, there was no significant change in water P content and the lake remained mesotrophic. Large year-to-year fluctuations in *E. densa* in Long Lake (Kitsap), WA have also produced opposite effects on water P content (Welch and Kelly, 1990). A virtual crash in the plant in 1985 and 1986, resulting in a 75% reduction in overall lake macrophyte biomass, produced the marked increases in lake water P content. The plant's normal presence as a thick carpet was hypothesized to protect the sediment–water interface from wind mixing and, thus, reduce internal loading rather than increase it through senescence and decay. The importance of macrophytes in influencing nutrient cycling and productivity in shallow lakes is further discussed in the following section.

9.7 Alternate stable states

The interaction and influence of aquatic macrophytes on nutrient cycling and phytoplankton abundance in shallow lakes is increasingly recognized (Jeppesen *et al.*, 1998; Scheffer, 1998). The pristine condition of most shallow lakes is probably clear water with rich aquatic vegetation. Increased nutrient loading typically causes a shift from clear to turbid conditions. The alternate stable-state theory explains how shallow lakes can change from a clear, macrophyte-dominated state to turbid, algal-dominated state without an increase in nutrient loading (Scheffer *et al.*, 1993). This theory provides a framework for understanding the dynamic relationship between macrophytes and phytoplankton in shallow systems. At low nutrient loading a lake will likely be dominated by macrophytes and at high nutrient loading, the lake will be algal-dominated because the high turbidity (from phytoplankton) will reduce light penetration below that is needed to support macrophytes. At intermediate loading, the lake can be in either state depending on the lake's history and environmental setting (Moss *et al.*, 1996; Scheffer, 1998) (Figure 9.8).

Macrophytes can be lost from a macrophyte-dominated lake through severe storms, herbicide applications, or grazing by herbivorous fish or waterfowl. A case study that illustrates the alternate stable-state concept is Lake Apopka, which is a large (124 km^2), shallow (mean depth, 1.7 m) and hypereutrophic lake in central Florida. The lake supported a dense submersed and floating-leaved macrophyte community until a hurricane initiated the switch to a turbid algal state in 1947. The macrophytes disappeared in only a few years following the storm and have not yet returned (Bachmann *et al.*, 1999). Following the loss of the macrophytes, sediments were resuspended by wind-driven waves on a regular basis. Prior to the

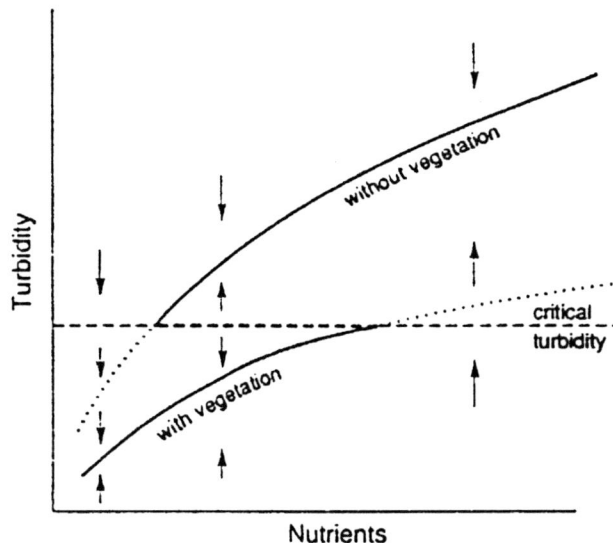

Figure 9.8 Relationship between nutrient content and turbidity for a shallow lake in an algal-dominated state and in a macrophyte-dominated state (from Scheffer et al., 1993). Turbidity increases with increasing nutrient concentration, but as vegetation (macrophytes) reduces turbidity, two different relationships result depending on the presence or absence of vegetation. The critical turbidity represents the point at which the macrophytes are lost due to insufficient light. The dashed parts of the two equilibrium lines at low and high nutrients indicate non-stable states. Alternate stable states can exist in the intermediate nutrient range. Arrows indicate the change in turbidity under non-equilibrium conditions. (Reprinted from Scheffer et al., Trends in Ecological Evolution, Vol. 8, Alternative equilibria in shallow lakes, 1993, pp. 275–279, with permission from Elsevier).

hurricane, sediments were protected from water movements by the dense-rooted macrophyte beds (Bachmann et al., 2000). Massive fish kills occurred during periods of high wind and sediment resuspension, likely due to the oxidation of sediment organic matter derived from the dead macrophytes (Bachmann et al., 2000).

As illustrated by the Lake Apopka case, shallow lakes may exist in one of two alternate stable states over a range of nutrient concentrations: a clear state dominated by aquatic macrophytes and a turbid state characterized by high algal biomass (Scheffer et al., 1993). The cornerstone of this phenomenon is the interaction between macrophytes and turbidity. High phytoplankton biomass, and thus high turbidity, suppresses macrophytes by limiting light penetration. Alternatively, lakes with high submersed macrophyte coverage tend to have lower turbidity and higher water clarity than lakes with similar nutrient content and sparse macrophytes. Several ecological mechanisms may explain the positive effect of macrophytes on water clarity, including macrophyte stabilization of bottom sediments (resisting resuspension), protection of phytoplankton-grazing zooplankton, uptake and subsequent reduction of available P, and release of allelopathic substances toxic to phytoplankton (Scheffer et al., 1993). Thus, despite the general transition from benthic to pelagic domination of primary productivity as nutrients increase, alternate stable states may exist in shallow lakes, particularly at intermediate nutrient concentrations. The positive feedback mechanisms above may promote clear water conditions so that once macrophytes are established, the water becomes clear and continued improved light conditions permit the persistence of the macrophytes (Scheffer et al., 1993, Scheffer, 1998). However, clear (little algae), shallow lakes without substantial macrophyte coverage are ecologically impossible without intense management, usually involving use of toxic chemicals.

Inverse relationships between macrophyte biomass and phytoplankton abundance have been documented in many shallow lakes. In Long Lake, a shallow lake in western Washington, mean summer TP and whole-lake macrophyte (primarily *Egeria densa*) biomass were inversely related (Welch and Kelly, 1990; Jacoby et al., 2001). *E. densa* was hypothesized to reduce internal loading of P, the primary cause of phytoplankton blooms in the lake, by stabilizing and protecting the sediment from wind-caused resuspension. In support of this hypothesis, internal P loading was found to be greater during years with low versus high wind speeds (Welch and Kelly, 1990). Furthermore, mean summer Secchi disk depth and late summer macrophyte biomass were positively correlated over a 19-year period ($r = 0.53$, $p < 0.05$) (Figure 9.9) (Jacoby et al., 2001). The large year-to-year shift between a clearer vegetated state with low TP and a more turbid state with high TP and low macrophyte biomass indicates that Long Lake is in a transitional stage between alternate stable states. Other shallow lakes have shown this bimodality of states (Mitchell, 1989; Moss et al., 1990; Blindow et al., 1993; McKinnon and Mitchell, 1994).

Although the negative effect of macrophytes on phytoplankton biomass has been documented in many shallow lakes, the specific mechanisms responsible for this effect have been difficult to document for specific lakes. Furthermore, some of these mechanisms may be direct and associated with the macrophytes themselves (e.g. decreased resuspension of sediment, shading, secretion of allelopathic substances) and others may be indirect (e.g. provision of grazing zooplankton refuges, modification of nutrient cycling by plant metabolism, hydrodynamic effects) (Søndergaard and Moss, 1998). The interaction of all these mechanisms also complicates the elucidation of macrophyte–phytoplankton interactions in a given lake.

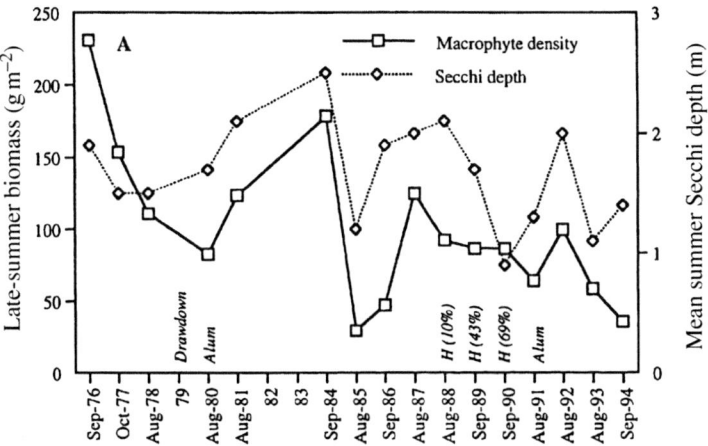

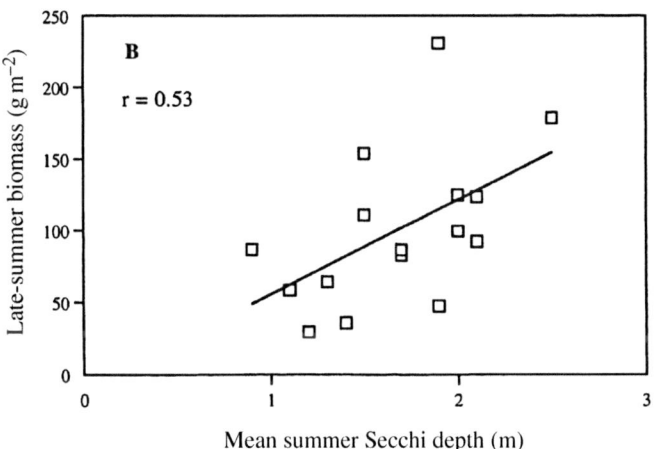

Figure 9.9 (A) Whole-lake, late-summer macrophyte biomass and mean summer (June–August) Secchi disk depth from 1976 to 1994 in Long Lake, Washington. (B) Relationship between whole-lake, late-summer macrophyte biomass and mean summer Secchi disk depth (from Jacoby et al., 2001, reproduced with kind permission from E. Schweizerbart'sche Verlagbuchhandlung. http://www.schweizerbart.de).

Several studies also indicated that macrophyte density must attain a threshold level of 15–30 per cent volume infestation (PVI) before they have a significant effect on phytoplankton (Canfield et al., 1984; Schriver et al., 1995; Jeppesen et al., 1998; Søndergaard and Moss, 1998). Furthermore, this threshold effect may be modified depending on fish abundance (e.g. at very high planktivorous fish levels, a high macrophyte biomass may not provide sufficient refuge for grazing invertebrates). This non-linear, threshold relationship may be important from a management viewpoint.

The clear-water, macrophyte-dominated stable state is generally recognized as the preferred condition over the alternate state of high algae, turbid water and sparse macrophytes (Scheffer et al., 1993). However, switching from a turbid, algal-dominated lake to a clear, macrophyte-dominated lake is extremely difficult (Moss et al., 1996). Reducing external nutrient loading is rarely effective in restoring a clear state in shallow lakes, primarily due to the importance of the internal loading from one or more of the mechanisms described above. One key to switching from one state to the other is altering the types and abundance of fish in the lake (Scheffer et al., 1993; Moss et al., 1996). Reducing the abundance of planktivorous fish to decrease predation pressure on the large-bodied zooplankton, which then can increase and graze upon the algal biomass, may force a switch to clearer water (Meijer et al., 1990; Søndergaard et al., 1990; Van Donk et al., 1990). This approach is based upon the 'trophic cascades effect' and has been used with varying levels of success in lake management (Shapiro and Wright, 1984; Gulati et al., 1990; Carpenter and Kitchell, 1992). The increased transparency following successful biomanipulation in shallow lakes has resulted in the development of submerged macrophytes that have persisted for a number of years and sustained clear-water conditions (Meijer et al., 1994). Herbivory of re-established macrophytes by waterfowl may cause the lake to return to a turbid, algal-dominated state following initial success of biomanipulation of the fish community (van Donk and Gulati, 1995). Removal of benthivorous fish such as carp (*Cyprinius carpio*), which resuspend sediments and increase turbidity, may also increase transparency and macrophyte growth (Meijer et al., 1990). Lowering the water level so that the critical light extinction coefficient is higher may also increase light availability to macrophytes and thus promote their return. Lake management tools that target internal loading mechanisms (e.g. alum application, artificial circulation) (Welch and Cooke, 1995) can supplement biomanipulation efforts.

9.8 Invasive, non-native macrophytes

More than 20 invasive aquatic plant species occur in the USA, including Eurasian water milfoil (*Myriophyllum spicatum*), which was previously discussed in Section 9.2, alligator weed (*Alternanthera philoxeroides*), water hyacinth (*Eichornia crassipes*) and hydrilla (*Hydrilla verticillata*) (Benson, 2000). Hydrilla is native to Asia and Africa, and was introduced to North America as a common aquarium plant. First found growing wild in Florida in 1960, it has spread as far north as Connecticut and as far west as Washington and California. Hydrilla tolerates a wide range of environmental conditions and persists at low light levels. Typical of many invasive aquatic plants, hydrilla reproduces through a variety of methods including fragmentation, turions (buds that form in the leaf axils) and tubers (subterranean turions). Similar to Eurasian water milfoil, hydrilla branches into dense mats near the water surface, creating inhospitable habitat for other plants.

Invasive wetland plants of concern in the USA include purple loosestrife (*Lythrum salicaria*) and melaleuca (*Melaleuca quinquenervia*). *Melaleuca* was deliberately introduced to Florida in 1906 to help drain the Everglades. This plant is highly invasive, grows rapidly and has a high water need, which has lowered the water table. Its introduction has resulted in 19 000 ha of monoculture stands with no wildlife value and up to 600 000 ha of lightly to moderately infested areas (Bush, 2003). Attempts to control this plant have been unsuccessful.

Exotic plants of concern in Europe include *Elodea canadensis* (a native of North America), floating pennywort (*Hydrocotyle ranunculoides*; also from North America) and swamp stonewort (*Crassula helmsii*) (Leach and Dawson, 1999). Water hyacinth causes water management problems in at least 50 countries on five continents and kariba weed (*Salvinia molesta*) is a problem in tropical regions worldwide. *Lagrasiphon* (oxygenweed) has invaded many lakes and rivers in New Zealand.

Excessive biomass levels of non-native macrophytes occur because the natural predators and pathogens that control the plants in their native water are usually absent in the introduced system. The consequences of excessive macrophyte abundance not only affect water recreation but also can clog irrigation canals, hydroelectric systems and navigational waterways, and cause flooding by obstructing drainage (Barrett, 1989). Dense macrophyte beds alter water chemistry (e.g. dissolved oxygen, pH, nutrients) causing changes to invertebrate and fish communities (Frodge *et al.*, 1990). During their decomposition, macrophytes are also a source of nutrients and organic matter to the water column and sediment, thus affecting nutrient cycling as described in Section 9.6. Food chain effects also occur due to the low nutritive value of some exotic plants such as purple loosestrife and water hyacinth.

Chapter 10

Lake and reservoir restoration

The control of eutrophication by restoring degraded lakes and reservoirs is a relatively new science. While some activity began in the 1960s, organization of knowledge into an assortment of techniques to improve the quality of standing bodies of freshwater did not occur until the 1970s. The US Environmental Protection Agency (USEPA) began funding lake restoration activities in 1976 and instituted a Clean Lakes Program in 1980. As a result, activity profoundly increased. Dunst *et al.* (1974) reported a total of 81 restoration projects in the USA, employing one or more techniques, that were either started or finished prior to 1974. Between 1976 and 1987, USEPA had funded 362 projects through the Clean Lakes Program (USEPA, 1987). Additional projects were also funded solely through state 'Clean Lake Programs'. Lake improvement declined in the mid-1990s when funding to the Clean Lakes Programs at the federal and state levels markedly declined.

The need for increased restoration activities was evident from several surveys. The North American Lake Management Society reported in 1983 that 12 000 lakes in 38 states were adversely affected by macrophytes and algae. The American Water Works Association (AWWA, 1987a) reported that 61% of surface drinking water supplies in the USA and Canada had algae-related (primarily planktonic blue-greens) taste and odour problems that affected a total of 38 million people. In response to the need, books and manuals have been published on techniques for restoring lakes and reservoirs (Cooke *et al.*, 1986, 1993; Moore and Thornton, 1988; Cooke and Carlson, 1989; NALMS, 2001). Although funding of lake restoration projects has greatly declined in the USA, the need is still there as evidenced by reports that cultural eutrophication is still one of the primary impairments to surface waters (USEPA, 1996). Also, an effort was initiated in the mid-1990s to set numerical nutrient criteria for lakes, streams, estuaries and wetlands (USEPA, 1998b).

The water quality problems caused by eutrophication were briefly mentioned in Chapter 7, but will be discussed further here. Increased enrichment of lakes and reservoirs can have the following direct and indirect effects: increased biomass of planktonic algae, with usually a dominance of blue-greens, several of which produce unsightly surface scums; taste and odour problems, caused largely by excretory products of cyanobacteria, which can taint drinking water supplies and fish flesh; oxygen depletion, caused by decomposition of sinking phytoplankton and organic sediment, that may render the hypolimnion anoxic or low enough in DO to limit the production and even survival of cold-water fish as well as causing episodic fish kills in unstratified lakes; reduced transparency of the water; increased incidence of swimmer's itch, as a result of increased production of algae and detritus eating snails, the intermediate host for the parasite; increased trihalomethane precursors resulting from the increased organic matter; and increased incidence of toxic cyanobacterial blooms.

Lake quality can degrade from either increased abundance of phytoplankton or macrophytes, or both. Increased algal or cyanobacterial abundance is a direct result of increased concentration of limiting nutrient, usually P, and would be expected to respond to reductions in nutrient concentration. However, increased macrophyte abundance is an indirect effect of enrichment, as discussed in Chapter 9, and therefore would not be expected to respond quickly to nutrient controls. Therefore, most external and in-lake controls on P are expected to reduce biomass and change species composition of phytoplankton. In many cases, however, controls have not resulted in significant reductions in P concentrations, so reductions in phytoplankton and changes in species composition were often less than expected. As a result, restoration results have ranged from dramatic success to little or no change in water quality. While reduction of external nutrient loading is usually considered the primary objective to improving lake quality, in-lake controls on P that are designed to reduce internal loading may also be necessary to achieve expected improvements.

10.1 Pre-restoration data

To estimate the degree of success of a restoration measure(s) to reduce lake P content, there are some important data needs and some alternative ways of using those data to estimate lake response and, therefore benefits versus costs of treatment. Most important are the budgets for water and P. A reasonably accurate budget on sources and losses of water volume is necessary to develop a sufficiently accurate budget on P. Point and various non-point sources of P are described in Chapter 5. Probably the most difficult source to assess is groundwater. If sources of P via groundwater are important, then more direct estimates of groundwater than simply a calculation of the unknown residual in the budget may be necessary (Winter, 1981). Even with estimates of flow, data on the nutrient content in groundwater are equally important. From measured inputs and outputs of P, a residual accounting for exchange across the sediment–water interface can be calculated. A negative residual is an estimate of net internal loading (see Chapter 4). N loading is usually not computed, because of the following: there are very few cases where N control has been attempted or was successful; there is one more gain (fixation) and one more loss (denitrification) than in the P budget, so a N budget is more difficult to construct; and P is usually easier to control, because of more sinks and fewer sources than N (see Chapter 4).

The next step is to develop a P model that is calibrated and verified for the lake in question. There are several alternatives and a Vollenweider-type, mass-balance model is often chosen (see Chapter 7). However, a dynamic P model may be more useful than a steady-state one. With a dynamic model, seasonal effects of inputs from external and/or internal sources can be separated, which will allow alternatives for selection of average P values to relate to algal biomass and more specifically define the source of problems (Welch *et al.*, 1986; Ahlgren *et al.*, 1988). A two-layered model for stratified lakes would further define the relative importance of internal loading (see Chapter 7). Algal biomass, as chl *a*, and subsequently Secchi transparency, are usually predicted from predicted P concentration based on choices of several regression models available (see Chapter 7 and Ahlgren *et al.*, 1988). However, development of predictive equations based on data from the lake in question would be preferred (Smith and Shapiro, 1981). Although predictions of species composition (e.g. percentage of blue-greens) are not yet highly reliable, relationships with P have been observed (see Chapter 7).

Sediment cores can be used to estimate the net sedimentation rate of P. This can be compared to the current net annual retention of P, as well as to past rates of sedimentation or retention. The sum of the annual P net sedimentation rate and the measured outflow rate of P can represent an additional estimate of the current external loading rate. Sedimentation rate can be determined either by stable lead measurements, assuming that the lead increase was due to the use of leaded gasoline, starting about 1930, or by lead-210 dating of the core (Schell, 1976). The latter gives recent rates as opposed to post-1930 average rates that are obtained from using stable lead. Because of the decrease in use of leaded gasoline starting in about 1972, stable lead may also give an estimate of recent sedimentation rates. Net sedimentation rate of P can also be determined by calibrating a dynamic mass balance model, i.e. solving for the unknown σ (see Chapter 7).

There are nearly 20 different techniques that have been used to restore and/or improve/preserve the quality of lakes and reservoirs. These can be divided into external and in-lake controls. External controls are largely designed to control P and include diversion of point sources of stormwater or wastewater (sewage effluent), advanced treatment of wastewater (alum, lime or ferric chloride) and detention/retention, infiltration or polishing (with alum) of stormwater. In-lake treatments include P inactivation (with alum, iron, lime or Riplox), dilution/flushing, dredging, artificial circulation, hypolimnetic aeration, hypolimnetic withdrawl, N addition, biomanipulation (planktivore predation/removal), mechanical harvesting (including rotavating), bottom covering, biological harvesting (grass carp) and draw-down (lake-level lowering). Several techniques have been predictable and cost effective, while others have been less so. However, with most there has been insufficient controlled, whole-lake demonstration, with adequate data collection, with which to establish a completely reliable understanding of cost effectiveness.

10.2 External controls on P

10.2.1 Diversion and wastewater treatment

These two techniques have been employed most to restore and/or preserve the water quality in lakes and reservoirs. Diversion of treated sewage usually involves the construction of interceptor sewers and other piping systems through which treated sewage is transported away from the degraded lake to a body of water with greater assimilative capacity. Advanced wastewater treatment is usually synonymous with P removal with either alum (aluminium sulphate), lime (calcium carbonate) or iron (ferric chloride). Removals usually exceed 90% with residuals as low as $0.1 \, mg \, P \, l^{-1}$, but more typically $0.5-1.0 \, mg \, l^{-1}$. A large volume sludge is created with advanced treatment that requires disposal, usually on land. P may be leached from the sludge over time (Ryding, 1996).

Recovery following external P reduction has not always been prompt or complete. Cullen and Forsberg (1988) have reviewed the response in 46 lakes resulting from decreases in external P loading by either diversion of point sources or advanced treatment. The responses to P input reduction varied from 'sufficient to change the trophic state' (13) to 'no response' (6), with the remaining 27 responding by 'either P and alga reduction with no trophic state change' or 'P reduction with no change in algae content'. The failure of lakes to recover promptly (in proportion to flushing rate) following external P control is usually due to high rates of recycling of nutrients from sediments, termed internal loading (see Chapter 4). Shallow lakes are the most difficult to improve (Ryding and Forsberg, 1976). In fact, internal loading was observed to increase on the average following external P load reduction in a study of the recovery of 18 European lakes (Sas, 1989).

In most cases lake P concentration will decline in spite of internal loading, which has also been found eventually decline (Ahlgren, 1977; Welch *et al.*, 1986; Sas, 1989). While internal loading continued to supply P in several of the nine shallow and nine deep European lakes, the in-lake TP decreased 73% on average in response to an average decrease in inflow TP concentration of 81% (Sas, 1989). TP decreased as predicted, based on the TP lake: TP inflow ratio observed by Sas (1989) in three large Swedish lakes over a 20-year period, but remained high in the fourth lake, which was shallow, due to internal loading (Wilander and Persson, 2001). Unfortunately, there has been little success at predicting the persistence of internal loading. Five contrasting case histories of nutrient input reduction will be briefly described.

Diversion of secondary treated sewage from ten treatment plants that discharged effluent to Lake Washington is one of the most frequently cited cases of recovery in the world. The lake had begun to recover promptly even before the three-year construction project, which diverted 88% of the P load, was complete (Edmondson, 1970, 1972; Edmondson and Lehman, 1981; Edmondson, 1994). The diversion from the lake was accomplished over the period 1964–1967. TP declined rather quickly from a mean annual $64\,\mu g\,l^{-1}$ prior to diversion to about 25 µg by 1969 and to a near-equilibrium value of about $21\,\mu g\,l^{-1}$ by 1972, five years after diversion was complete (Figure 10.1). The lake should have reached 90% of its total decrease in 2.2 years, based on a first-order decline [$\ln 10/(\rho + \rho^{0.5})$], where $\rho = 0.4$ year^{-1}. Lake TP has remained rather stable with a 1990–2001 mean of $15\,\mu g\,l^{-1}$ in large part because the main tributary supplying 50% of the water input has a flow-weighted mean TP of $17\,\mu g\,l^{-1}$ (King County, 2003). Thus, Lake Washington responded precisely as the Vollenweider model predicted while predictions by the Nürnberg (1984) and OECD (1982) models are less accurate (see Chapter 7).

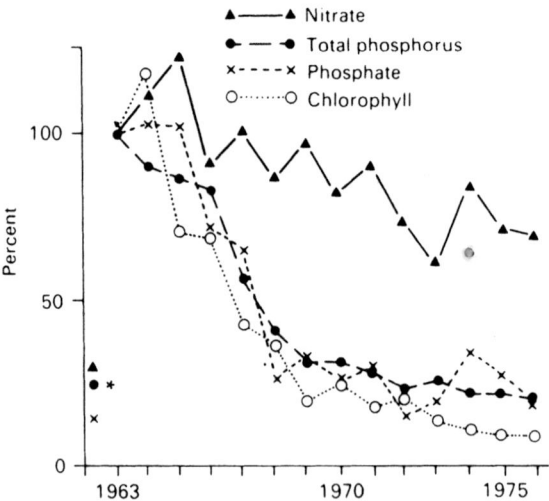

Figure 10.1 Relative values in surface water (upper 10 m) of phosphate-phosphorus and nitrate-nitrogen (January to March), total P (whole year), and summer (July and August) chl *a* in Lake Washington. The 1963 mean values, plotted as 100%, were in $\mu g\,l^{-1}$; total P, 65.7; phosphate-P, 55.3, nitrate-N, 425; and chl *a*, 34.8 *1933 data (Edmondson, 1978).

Algal biomass as chl *a* declined from a four-year summer mean of 36 µg l^{-1} before diversion to a seven-year mean of 6 µg l^{-1} after, in direct proportion to the decline in P (Figure 10.1). This was some of the first direct evidence of the singular importance of P to algal control. However, the fraction of blue-greens in the algal biomass did not decline until the mid-1970s, which was promptly followed by a dramatic increase in *Daphnia* in 1976 (Edmondson and Litt, 1982). Transparency had improved from 1 m, as a summer average, prior to diversion, to 3.1 m in the mid-1970s, but the advent of *Daphnia* reduced the chl *a* by one-half and approximately doubled the transparency. Thus, the lake condition in the late 1970s of about 17 µg l^{-1} TP, 3 µg l^{-1} chl *a* and a Secchi transparency of nearly 7 m was the result of both chemical and biological recovery. Chlorophyll *a* and transparency have continued to remain at similar levels during the 1990s with means of 2.7 µg l^{-1} and 7.1 m, respectively (King County, 2003).

Unfortunately, the Lake Washington case is not typical as indicated by the analysis of Cullen and Forsberg (1988), Sas (1989) and others. The principal reasons that Lake Washington recovered so promptly and completely are its relatively great depth (64 m maximum, 37 m mean), fast renewal rate (0.4 year^{-1}), its oxic hypolimnion and its relatively short history of enrichment. The lake's large hypolimnion and short period of enrichment (first signs were observed in the early 1950s, Edmondson *et al.*, 1956) prevented the hypolimnion from reaching an anoxic state. Thus, internal loading was insignificant and demonstrates the advantage of treating a lake before reaching an advanced state of eutrophy.

In contrast to Lake Washington, the response of Lake Sammamish, which lies about 12 km east of Lake Washington, is half as deep (18 m mean depth) and has about the same renewal rate (0.55 year^{-1}), was substantially delayed following diversion of sewage and dairy wastewater in 1968 (Welch *et al.*, 1980, 1986). There was a slight decrease in whole-lake TP, from 33 to 27 µg l^{-1}, following diversion of ≈35% of its P load, but not until seven years after diversion did TP approach the ultimate equilibrium level of 18 µg l^{-1} (Figure 10.2). The latter value (± 2 µg l^{-1}) has persisted during the 1980s. The response of TP was predicted most accurately by the OECD model, but all models overpredicted the post-diversion level (Table 10.1). Change in chl *a* and Secchi transparency paralleled that of TP. Chlorophyll *a* decreased by 50% and transparency, which had remained at a summer mean of 3.3 m during the early 1970s after diversion, increased to a mean of 4.9 m.

The delay in recovery of Lake Sammamish was due to internal loading from anoxic sediments. Although the lake had not exceeded a state of mesotrophy during its enrichment, it nonetheless had an oxygen deficit rate typical of eutrophy and reached anoxia during late summer, due to its relatively small hypolimnion. The ultimate improvement in the lake was related to a reduction in the sediment release rate of P, which was in turn related to a decline in oxygen deficit rate (Figure 10.3). Sediment P

254 Effects of pollutants in standing water

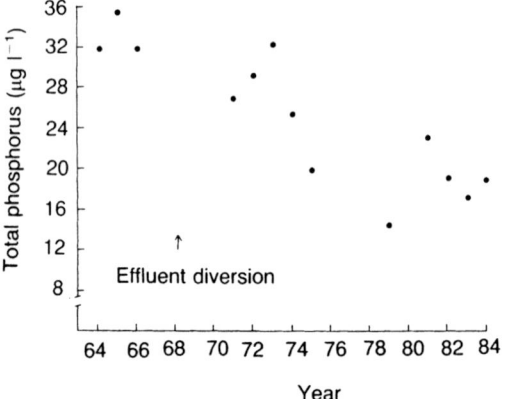

Figure 10.2 Whole-lake TP in Lake Sammamish, WA before and after wastewater diversion (Welch et al., 1986).

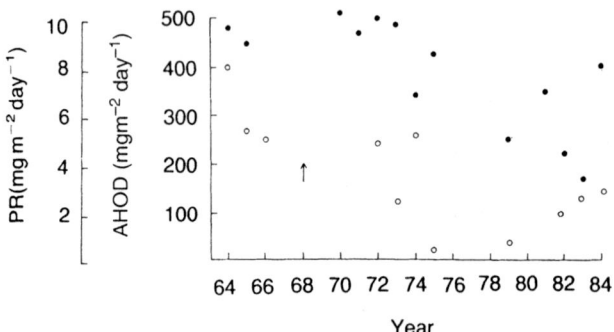

Figure 10.3 Sediment P release (PR) and areal oxygen deficit (AHOD) in Lake Sammamish before and after wastewater diversion (Welch et al., 1986). o, PR; •, AHOD.

release has remained low at about 20% of the pre-diversion rates through the mid-1990s (Perkins et al., 1997).

A similar long-term response to nutrient diversion has occurred in Lake Norrviken, Sweden. However, in contrast to Lakes Sammamish and Washington, Lake Norrviken was hypereutrophic prior to diversion having received industrial and domestic wastewater for nearly 100 years (Ahlgren, I., 1977). A large fraction of incoming P was diverted (87%) and the lake P content decreased predictably (see Chapter 7) as high-P lake water was replaced with the new low-P inflow water (Ahlgren, I., 1977, 1978, 1979). Initially, from 1969 to 1975, average annual lake TP content decreased from 450 to $150\,\mu g\,l^{-1}$, which was still well above the hypereutrophic threshold.

Summer TP dropped from 263 µg l^{-1} before to values of generally 100 µg l^{-1} or less after 1974. Internal loading showed a decrease through 1977, but increased again as indicated in Figure 10.4. Chlorophyll *a* steadily decreased from summer concentrations in excess of 100 µg l^{-1} to 36 µg l^{-1} in 1980. Inorganic N also decreased in the lake from levels >1000 µg l^{-1} to usually 100 µg l^{-1} or less and the decrease in algal biomass was more closely correlated with the N decrease, which is considered the more probable cause (Ahlgren, I., 1978). Transparency approximately doubled but the depth of visibility was still only 1.2 m (Sas, 1989). The phytoplankton in summer was no longer a monospecific crop of *Oscillatoria* (Ahlgren, G., 1978).

In one of the earliest efforts to reduce external nutrients and alleviate algal blooms, sewage effluent from the city of Madison, Wisconsin, USA was diverted in 1958 from Lakes Waubesa and Kegonsa, the last two lakes in the Madison chain of lakes. Effluent contributed 88% of the P to Lake Waubesa and the four- to seven-fold increase in loading to the lower lakes is shown in Table 10.1. Following diversion, winter SRP decreased from >500 µg l^{-1} to ≈100 µg l^{-1} as predicted from a simple hydraulic washout model (Sonzogni and Lee, 1974). Although there was little effect on algal biomass, due to the continued high P, dominance changed from 99% *Microcystis* to a more mixed assemblage (Fitzgerald, 1964). Through the early 1970s, Stewart (1976) found no significant change in either oxygen or transparency that was detectable above normal year-to-year fluctuations. Since 1975, SRP during spring in the lower lakes has decreased to near undetectable levels from values between 50 and 90 µg l^{-1} (Lathrop, 1988, 1990). This further decrease is attributed to interception of sewage effluents from communities

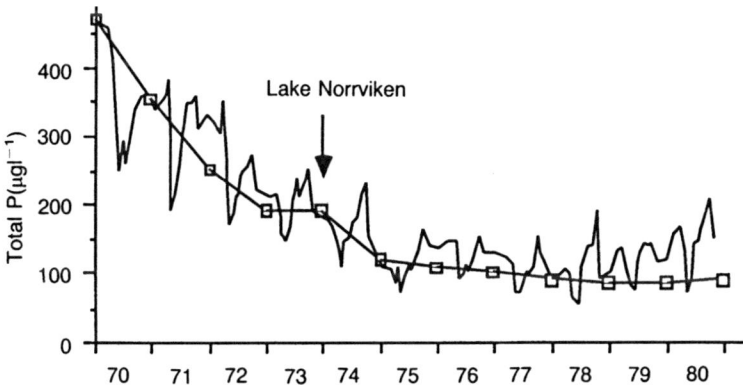

Figure 10.4 Lake mean concentrations of total phosphorus in Lake Norrviken. □—□, Theoretical dilution curve without sedimentation. Arrows, three water retention times after onset of restoration (modified from Ahlgren, 1988).

Table 10.1 Nutrient loading to the Madison lakes

Lake	Loading ($g\,m^{-2}\,year^{-1}$)	
	N	P
Mendota	2.2	0.07
Monona	8.8	0.90
Waubesa	47.0	7.0
Kegonsa	6.8	3.9
Koshkonong	9.8	4.3

Source: Lawton (1961).

upstream from Madison that drained to Lake Mendota, resulting in lower P in the inflow to the lower lakes, as well as less spring runoff since 1977. Transparency and chl *a* have improved accordingly (Lathrop, 1988, 1990).

Finally, there is the well-studied case of lake response to advanced wastewater treatment (AWT) in Shagawa Lake, Minnesota, USA. Treatment of sewage effluent with lime beginning in 1973 removed 85% of the incoming P to the lake (Larsen *et al.*, 1975). The lake responded quickly with average annual TP dropping from $51\,\mu g\,l^{-1}$ to $30\,\mu g\,l^{-1}$ and spring chl *a* also dropping by 50%. However, the lake remained eutrophic during summer due to internal loading from an anoxic hypolimnion and the partial entrainment of that high-P water into the epilimnion during summer storm events (see Chapter 4). Although summer levels of TP also declined following treatment, they were still too high to allow much of a decrease in summer chl *a*. This is shown in Figure 10.5 by the decreasing trend in TP between 1973 and 1976 compared to the predicted pretreatment level (Larsen *et al.*, 1979). Shagawa Lake is expected to continue to recover, in spite of continued internal loading, as has been observed in other lakes, but the rate of recovery will be slower than if internal loading did not exist (Sas, 1989).

Recent results from Shagawa Lake show that the increase in summer TP has decreased from $35-50\,\mu g\,l^{-1}$ before input reduction (Figure 10.5) to $20-30\,\mu g\,l^{-1}$, indicating that internal loading was still occurring 16 years after treatment (Wilson and Musick, 1989). The persistence of internal loading was indicated by outflow TP still exceeding inflow TP with internal loading accounting for $\approx 30\%$ of the annual loading. Peak chl *a* concentrations during summer were less in 1988 than in the early 1970s; $30-45\,\mu g\,l^{-1}$ versus $50-60\,\mu g\,l^{-1}$, respectively. However, recovery to an oligotrophic state will apparently be long term, as the gradual burial of P results in reduced internal loading, possibly as long as 80 years to reach equilibrium (Chapra and Canale, 1991). Thus far, annual mean TP has remained at about $35\,\mu g\,l^{-1}$,

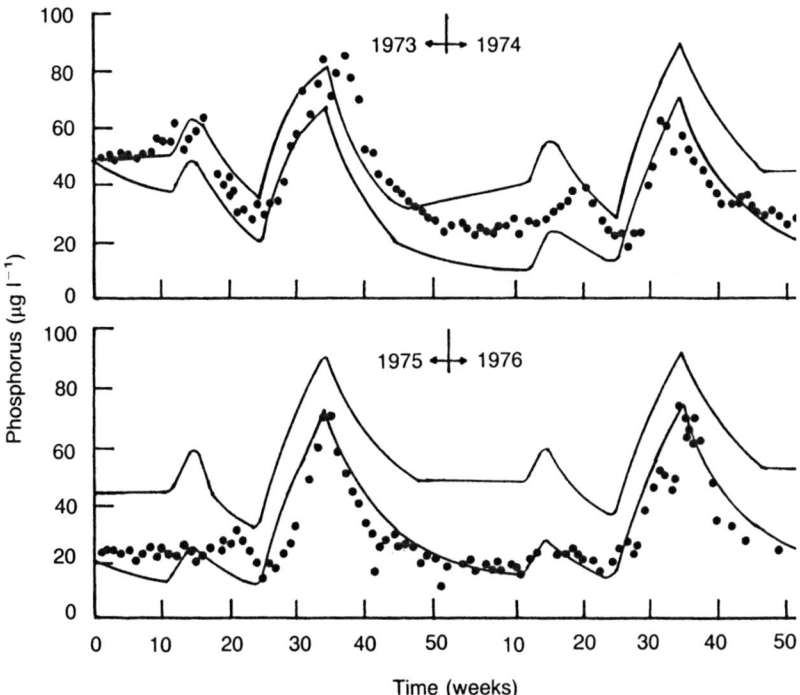

Figure 10.5 Comparison of predicted (–) and observed (•) TP concentrations in Shagawa Lake, MN before (upper line) and after (bottom line) institution of phosphorus removal from sewage effluent (modified from Larsen et al., 1979).

well above the ultimate equilibrium concentration of $12\,\mu g\,l^{-1}$ (Wilson, 1994, personal communication).

These case histories of stratified lakes illustrate the effect of internal loading on the rate and extent of recovery. While all lakes responded to TP load reduction, recovery to equilibrium levels was prolonged for those lakes enriched long enough to develop a substantial internal loading. Although still present, internal loading declined relatively soon in Lake Sammamish, but has continued at levels that maintain lake TP concentrations well above the eutrophic threshold in Lakes Norrviken and Shagawa. The recovery delay in the latter two lakes has been exacerbated by the greater availability of hypolimnetic P to the epilimnion than is the case for Lake Sammamish.

Unstratified lakes with high internal loading are often slower yet to show improvements, due to the greater availability of sediment-released P to algae than in stratified lakes. While only a fraction of the very high $(100\,s\,\mu g\,l^{-1})$

TP concentrations in the anoxic hypolimnia of stratified lakes is available to the epilimnion, all P released from sediments in unstratified lakes immediately enters the mixed water column and is available to algae. Given that internal loading rate has been related to eutrophication and not depth or degree of stratification (Nürnberg, 1996), surface waters of eutrophic shallow lakes are usually much higher in TP than eutrophic stratified lakes (Scheffer, 1998). This means that shallow lakes are not only more sensitive to enrichment, but also are more difficult to recover from a eutrophic state.

10.2.2 Watershed treatment

After any sewage effluents have been diverted or given advanced treatment, the next most important sources of enrichment causing eutrophication are stormwater runoff and failing individual septic systems. Although the P content of stormwater runoff is not as high, by a factor of ten or more, as that of sewage effluent and the fraction that is SRP is usually much less, it is nonetheless an important source of enrichment to lakes (see Section 7.37). There are several techniques to remove P from stormwater, but their effectiveness varies greatly and there are few cases of lake improvement from treatment of stormwater. Detention of stormwater in ponds results in effective settlement of P if time is sufficient, but it is mostly the particulate fraction. Directing stormwater through a grassy swale can remove a significant fraction of P if contact time between water and the soil/plant surface is sufficient. If areas of high percolation capacity are available, infiltration of stormwater is a possibility. Infiltration can result in a long and effective contact between water and soil. The use of wetlands for P retention is also effective (removals as high as 90%), especially in constructed wetlands, although they require much greater area than other techniques (Cooke et al., 1993; Kadlec and Knight, 1996). There is a maximum P loading of about $1 g m^{-2}$ year^{-1} beyond which P retention declines and output TP rises above an average of $40 \mu g l^{-1}$ (Richardson and Qian, 1999). Wet detention ponds with polishing (removal of P with alum or iron) are more consistently effective, but are costly as well.

The removal efficiency of P in wet detention basins varies from 35 to 60% over a water detention time range of 2–15 days (Walker, 1987b; Benndorf and Putz, 1987). A longer detention means more pond area and/or volume to achieve the desired removal. However, constraints on available area in urbanized watersheds can limit detention basin size so that adequate removal efficiency is not possible to protect a lake or reservoir sufficiently. As a result, P interception has been a successful alternative. Water from 'pre-dam' detention basins, at flows up to $5 m^3 s^{-1}$, has been treated with ferric chloride to precipitate P (Bernhardt, 1981). Wahnbach Reservoir and Lake Schlachten in the Federal Republic of Germany are examples of eutrophic water bodies that have recovered to a state of high quality following such treatment (Sas, 1989). Others have employed P interception to protect water bodies (Cooke and Carlson, 1986; Walker et al., 1989; van Duin et al., 1998).

Leachate from inadequately percolating or failing septic tank (on-site) drainfields may be another important source of enrichment, although if the drainfield is working properly, P is effectively sorbed in the soil and significant amounts may not reach the lake. P retention is especially effective if there is an unsaturated zone above the groundwater level. As groundwater level raises, the effectiveness of drainfields decreases.

10.2.3 Dilution/flushing

The nutrient content of lakes can be reduced, or diluted, by adding low-nutrient water, the availability of which is the greatest limitation to this technique. Where such low-nutrient sources were used, marked improvement in lake quality resulted; e.g. Green Lake and Moses Lake in Washington state, USA (Welch and Patmont, 1980; Welch et al., 1992; Cooke et al., 1993). If low-nutrient water is not available, algal biomass may be reduced by increasing the flushing rate to a level that significantly increases the loss rate of cells, relative to their growth rate. A critical level in that regard seems to be $\approx 10\%$ day^{-1} which, for anything but a small lake, would require a large daily supply of water. Where increased flushing rate with high nutrient water was used, the results were unsatisfactory; e.g. Rotsee near Lucerne, Switzerland, where flushing rate was increased from 0.1 to 0.7% day^{-1} in 1921–1922 (Stadelman, 1980; see Cooke et al., 1993).

The most extensive dilution project is Moses Lake, where low-nutrient Columbia River water (25 µg l^{-1} TP) has been routed through the 2700 ha lake at various intervals during spring–summer since 1977. On a time-weighted average, dilution water inflows entered the lake at an average of 8.2 m^3 s^{-1} (128 × 10^6 m^3) during the spring–summer period of the subsequent 12 years, replacing water in the arm of the lake proximal to the inflow at an average rate of 5.8% day^{-1}, but in the whole lake at only 0.27% day^{-1}. TP and chl a were reduced by >50% and transparency doubled in the proximal arm during the first years of treatment (Welch et al., 1992). Similar improvements were also evident in other parts of the lake accessed by wind-enhanced transport of the water. Algal biomass control initially was due to decreased NO$_3$ in the inflow, even though little change in that variable was noted in the lake (Welch et al., 1984).

Further improvements in the quality of Moses Lake have resulted from diversion of sewage effluent in 1984 and a progressive decrease in P concentration in 50% of the lake's undiluted inflow water. P has become the most limiting nutrient and further control hinges on the rather large and variable ($\pm 100\%$) internal load of P, which averaged about 30% of the total load (Jones and Welch, 1990). Moreover, while dilution effectively reduced in-lake P content, the benefit diminished as water input increases, due to enhanced internal loading of P, apparently resulting from the reduced lake P causing an increased diffusion gradient between sediment and overlying water. While simple dilution provided a two-thirds reduction in lake TP, the effect of internal loading resulted in only a one-third reduction (Welch et al., 1989a; Welch and Jones, 1990).

Dilution water addition continued at an even higher rate during the 1990s, with inputs averaging 204×10^6 m^3 over the 11-year period or an increase of one-third over the previous 12 years. Although monitoring was infrequent during the 1990s, water quality was considered good except for 1997, which was a low-input year. Intensive monitoring during 2001, a high dilution water year, showed a very low mean TP ($20\,\mu g\,l^{-1}$), low chl a and blue-greens were largely unrepresented (Carroll, 2003). Moreover, there was no net internal loading during the spring-summer period.

10.3 In-lake controls on P

Techniques that are most appropriate at controlling internal loading of P are P inactivation, Riplox, hypolimnetic aeration, hypolimnetic withdrawal, and dredging. Three of these five techniques are aimed primarily at controlling mobile P in sediments that are affected by the iron redox mechanism (see Chapter 4). Alum [$Al_2(SO_4)_3 \cdot (H_2O)_{18}$], when added to lake water, forms a hydroxide floc that rather quickly settles to the lake bottom creating a barrier to further releases of sediment P by binding mobile P as Al–P. The continued effectiveness of Al is due to its solubility being independent of redox, which, in contrast, controls Fe. Hypolimnetic aeration produces an oxic hypolimnion, which should increase the control of P by Fe and reduce sediment release. The Riplox technique is designed to reduce interstitial SRP, and thus diffusion of P from sediment, by oxidizing the organic matter in sediments. Oxidation is accomplished by denitrification through the addition of $Ca(NO_3)_2$ under alkaline conditions. Fe is added if naturally low or if the S/Fe ratio is high. The idea is to keep Fe in the oxidized ferric state and P complexed to the ferric hyroxy complex.

Alum has been used extensively in the USA with well over 100 treated lakes, some being supported by USEPA and additional projects receiving sole support from individual states or local entities. Hypolimnetic aeration using primarily compressed air has also been popular, but while its record is good for increasing hypolimnetic oxygen content without destratifying the lake, it has been much less effective than alum in controlling sediment-P release. Low Fe/P ratios in sediment may be the cause for some ineffectiveness. Riplox is an effective technique, but has been used in only three lakes, two in Sweden and one in Minnesota, USA (Cooke et al., 1993).

Representative examples of treatment effectiveness for alum, Riplox and hypolimnetic aeration are given in Table 10.2. Of these three techniques to

Table 10.2 Representative results from in-lake restoration techniques to reduce internal loading from eutrophic lake sediments. Values are summer means and are averaged if more than one year's data were available (from Welch, 1988; Welch and Cooke, 1999)

Treatment	Lake	Whole-lake TP ($\mu g\,l^{-1}$) Pre	Whole-lake TP ($\mu g\,l^{-1}$) Post	Reduction (%)	Internal Load Reduction (%)	Reference
Alum	Dollar, OH	240	52	78	–	Cooke et al. (1986)
	W. Twin, OH	97	32	67	62	Cooke et al. (1978)
	Annabessacook, ME	72	34	53	65	Dominie (1980)
	Long, WA	65	30	54	100	Jacoby et al. (1982); Welch et al. (1982)
	Six unstratified[a] lakes/basins-means	55	26	48	68	Welch and Cooke (1999)
	Seven stratified[b,c] lakes-means	37	19	42	80	Welch and Cooke (1999)
Riplox	Lillesjon, Sweden	3000	50[c]	97	–	Ripl (1986)
	Long, Mn	87	85[c]	2	80	Noonan (1986)
Hypolimnetic aeration	Sodra Horken, Sweden	400	25[c]	91	–	Bjork (1985)
	Kolbotvatn, Norway	900	600[d]	33	–	Holton (1981)
Hypolimnetic withdrawal	Ballinger, WA	–	–	–	86	USEPA (1987)
	Mauensee, Switzerland	100	40[c]	–	98	Gachter (1976)
	Reither See, Austria	41	21	49	–	Pechlaner (1978)
Dredging	Lilly, WI	40	18	55	–	Dunst (1981)
	Trummen, Sweden	1000	100	90	–	Bjork (1985)

Notes
a Long Lake included.
b W. Twin, Dollar, Anabessacook lakes included.
c epilimnion only.
d hypolimnion.

retard sediment-P release, alum was most effective and consistent and these results are typical of other treated lakes (Cooke *et al.*, 1993; Welch and Cooke, 1999). The first three lakes listed are thermally stratified in summer. Long Lake is shallow (mean depth 2 m) and unstratified and there were doubts that treatment of a shallow lake would be successful, because three previous shallow-lake treatments were unsuccessful (Welch *et al.*, 1988b). However, the Long Lake treatment was highly effective for four years (Table 10.2). Moreover, lake TP has averaged 40% less than the pretreatment level for 8 of 9 years following the application. Two of the previous three shallow-lake failures were due to continued high external P loading. Treatment of stratified lakes has been, without exception, highly effective and long-lasting. However, whether stratified or unstratified, unless external P loading is controlled first, an alum treatment may fail to produce the expected improvement in quality.

Alum treatment effectiveness and longevity were evaluated in 19 USA lakes with sufficient data, for which at least 3 years had elapsed since treatment (Welch and Cooke, 1999). Alum effectiveness could not be evaluated for five lakes because external loading remained high in one, so there was no improvement and the effects of wastewater diversion could not be separated from alum for the other four. Of the remaining 14, seven were stratified and seven were unstratified (two had two basins). TP and internal loading were reduced on average by about one-half and two-thirds, respectively, in six of the nine unstratified lakes/basins and the effect persisted for at least 8 years (Table 10.2). Treatments were ineffective or short-lived in three unstratified lakes/basins with dense beds of macrophytes, which apparently interfered with alum effectiveness via poor floc distribution of P release from their decay. TP and internal loading were similarly reduced in the stratified lakes with effectiveness lasting at least 13 years (Table 10.2).

The effect of treatment in stratified lakes is most pertinent to lake quality in epilimnetic waters, because hypolimnetic P may be largely unavailable to epilimnetic waters and algae (see Chapter 4). That was the case in West Twin Lake in which its alum treatment did not reduce epilimnetic TP any more than in the untreated control, East Twin Lake (Cooke *et al.*, 1993; Welch and Cooke, 1999), even though alum greatly reduced internal loading and whole-lake TP (Table 10.3). Chlorophyll *a* decreased to about the same extent as TP following treatment in both the stratified lake epilimnia and the unstratified lakes (Welch and Cooke, 1999).

Iron and calcium have also been used to inactivate P in lake sediments. Iron would be expected to have a beneficial effect in lakes with low sediment Fe/P ratios. Jensen *et al.* (1992) found that internal loading was controlled by iron if TFe:TP ratios were <15 by weight. However, Fe is redox-sensitive and its effect has not been proven to be as effective or long lasting as that of alum (Cooke *et al.*, 1993). Ca, added as slaked lime, has been shown to be effective at reducing P in a hardwater lake (Babine *et al.*, 1994), but while Ca

Table 10.3 Cost and possible frequencies of in-lake treatments for a hypothetical 100 ha lake over 20 years

Control of	Treatment	Whole lake	Hypolimnion only	Number of treatments	Total cost ($ × 10⁻³)
Phosphorus	Alum	X (unstratified)		4	280
			X	1	68
	Riplox	X		1	530[a]
				1	265
	Hypolimnetic aeration		X	6 months/yr	350
	Hypolimnetic withdrawal		X	6 months/yr	320[a]
	Dredging	X		1	2240
	Complete circulation	X		6 months/yr	280[a]
Biomass	Biomanipulation				
	Fish removal				
	Netting[b]	30% (littoral)		2	68[a]
	Poisoning[c]	X		2	38
	Piscivore stocking[c]	X		4	198
	Macrophyte harvesting	30% (littoral)		Annual	350
	Macrophyte grazing[d]	30% (littoral)		2	60[a]

Notes
a Costs are median values (or single examples) in 2002 dollars converted from Cooke et al. (1986) except as indicated (from Welch, 1988).
b Wagner, personal communication.
c Mongello, personal communication (stocking 0.5 kg trout at 50 kg ha⁻¹).
d Devils L., Oregon; Bonar, personal communication.
X = whole lake or hypolimnion only.

treatments have improved lake quality, the longevity of single treatments has not been evaluated (Cooke *et al.*, 1993).

Riplox has a greater longevity potential than alum. While the alum floc layer is covered with new sediment and dispersed downward diminishing its effectiveness, Riplox removes the organic matter and the cause for Fe reduction. Sediment oxygen demand in Lake Lillesjon has continued to decline even after ten years and hypolimnetic P has remained low (Ripl, 1986). Unfortunately, external P loading in the other two lakes treated (Long Lake, MN and Lake Trekanten, Sweden) remained great enough to offset any improvement in water quality even though sediment-P release was controlled (see Long Lake, Table 10.2). Internal loading recurred in Trekanten, apparently because newly deposited algal organic matter reduced sulphate to sulphide, which tied up iron allowing P release. Excess iron was not added to Trekanten. A comparison of alum and Riplox treatments both showed well over 90% reduction in P release from Green Lake (Seattle, USA) sediment in laboratory incubated cores even after 64 days (Cooke *et al.*, 1993; DeGasperi *et al.*, 1993).

Hypolimnetic aeration has been frequently used but with variable effectiveness in controlling sediment-P release (Cooke *et al.*, 1993). The technique was effective in controlling hypolimnetic P in Sodra Horken, but the unimpressive results from Kolbotvatn often occur (Table 10.2). One reason for poor effectiveness has been low Fe/P ratios, such that internal loading was not controlled by Fe (Walker *et al.*, 1989). The technique does not seem to have any long-term effect. Sediment-P release is usually partially controlled when aeration is in progress, but rather quickly resumes once aeration is curtailed.

The advantage of alum is that the hydroxide floc forms a physical–chemical barrier that traps dissolved P that would diffuse from sediments to overlying water even if the water is anoxic. Furthermore, it may also be a barrier to other mechanisms of release, such as microbial mineralization, excretion from microbial cells and blue-green migration. On the other hand, Riplox and hypolimnetic aeration will work only if oxic conditions exist. However, Riplox does remove organic matter, which is the cause for anoxia in sediments, so in the longterm, it can restore oxic conditions.

Hypolimnetic withdrawal has been successful in several lakes (Table 10.2). Nürenburg (1987b) reviewed data from 17 cases, and showed that hypolimnetic TP decreased in 10 of 11 lakes where data were available as well as in the epilimnion in 8 of 11. Epilimnetic decrease in TP was correlated with the magnitude of TP export from the lake. However, the technique has not been successful in improving water quality in lakes that are meromictic in which the monomolimnion remained anoxid (Pechlaner, 1978) or where external loading remained high, e.g. Lake Ballinger.

Lake Trummen represents the classic case for dredging, which can provide near permanent control on P internal loading by simply removing the P-rich layer in the sediments (Table 10.2). The P-rich layer extended to 0.5 m, so 1 m of sediment was removed. Lilly Lake was dredged to greater depths, ranging from 1.8 to 6 m, because the principal objective was to control macrophyte growth and increase fish habitat. However, the improved transparency, to some extent related to P control, allowed macrophytes to colonize to a greater depth even though macrophyte biomass was 70–85% less (Dunst, 1981). Internal loading was also reduced in Lake Braband, Denmark, following removal of 0.4×10^5 m^3 of P-rich sediment (Søndergaard et al., 2000).

Comparative costs for the above mentioned in-lake treatments are shown in Table 10.3. Only application costs are considered. Retreatment over a 20-year period is assumed necessary only for alum in an unstratified lake. Clearly, alum is probably the most cost-effective of the five treatments for controlling sediment-P release. Riplox should be equally effective and more long-lasting, although the practical evidence is not yet convincing, and the cost is over fourfold greater. While hypolimnetic aeration is usually less cost-effective than alum for sediment P control, there may be other reasons for aeration, such as creating an aerobic environment for fish and benthos. Dredging is clearly the most costly, however, in return there is the added benefit of having a deeper lake with possibly less noticeable problems with macrophytes.

10.4 In-lake controls on biomass

Techniques to control algal and aquatic plant biomass may be possible in cases where external sources of nutrients cannot be controlled. The causes for nuisance blooms of blue-greens have become better understood in recent years and some of the important factors can be affected by artificial circulation. One of the most frequently used techniques to improve lake quality is artificial circulation (whole-lake aeration), but the goals for that technique are several. As a result, blue-green biomass has been reduced in some cases but increased in others (Pastorak et al., 1982). Control of algal biomass has been achieved by manipulating grazer food chains, both in small-scale experiments and in whole-lake treatments (Shapiro et al., 1975; Andersson et al., 1978; Shapiro, 1979; Lynch and Shapiro, 1981; Benndorf et al., 1984; Shapiro and Wright, 1984; Wagner, 1986; Søndergaard et al., 2000). A highly effective and popular (in the USA) biological technique is the use of triploid grass carp (*Ctenopharyngodon idella*) to control aquatic macrophytes. However, there are unsolved problems, such as appropriate stocking rates, nutrient recycling and interactions with other fish populations (Cooke et al., 1993).

The understanding of blue-green blooms, in relation to their buoyancy regulation mechanism (Reynolds et al., 1987; Spencer and King, 1987) and changes in CO_2/pH (King, 1970; Shapiro, 1973; Paerl and Ustach, 1982; Shapiro, 1984, 1990) directly affects the feasibility of controlling such blooms with artificial circulation. Such blue-greens as *Aphanizomenon*, *Microcystis* and *Anabaena* can conceivably rise through a 10 m water column in only a few hours if turgor pressure in their gas vacuoles is high, which occurs if CO_2 or light limitation curtail the cellular accumulation of carbohydrate. Previously accumulated carbohydrate can be assimilated into proteins at depth, where N and P are available for growth (and carbohydrate dilution) and photosynthesis is limited by light. This results in gas vesicle production and turgor rise in the vacuoles. Cell photosynthate (carbohydrate) accumulates when cells enter the surface water with greater light and CO_2 availability, but with N and P limitation. Either vacuole collapse or increased carbohydrate ballast causes the cells to sink. Such a buoyancy response to CO_2 limitation and N addition was shown for *Oscillatoria* (Klemer, 1973; Klemer et al., 1982), for *Anabaena* with N and P addition (Booker and Walsby, 1981) and light (Spencer and King, 1987), and with pH adjustment (CO_2 limitation) for *Aphanizomenon* and *Anabaena* (Paerl and Ustach, 1982).

The dominance by blue-greens may occur under conditions of strong stratification or infrequent diel mixing, because their buoyancy mechanism allows them to exploit nutrients and light from a water column that becomes increasingly turbid as enrichment proceeds. Diatoms and green algae would be less favoured under such relatively static, light limiting conditions, even though they generally grow faster than blue-greens (Reynolds et al., 1987). Increased circulation could induce the algal community to revert to a dominance by diatoms and greens in spite of high nutrient concentrations. This could occur because increased circulation would lower pH and raise CO_2, causing gas vacuole collapse/carbohydrate dilution and increased light availability to mixed plankton cells. Experiments in 7-m deep enclosures in a stratified lake have shown that rapid deep mixing was more successful than slow deep mixing in lowering pH, raising CO_2 and decreasing blue-green dominance (Shapiro et al., 1982). Observations following artificial circulation have shown that a shift occurred from blue-greens to greens in about one-half the cases examined (Pastorak et al., 1982). In the majority of cases where a shift occurred, pH also decreased. However, blue-greens increased in the other half of the cases where there was not a shift. Thus, while there is some evidence that changes in CO_2/pH as a result of circulation will cause a shift, there is also a moderate possibility a shift will not occur and blue-green abundance may even increase. To isolate the effect of CO_2/pH on shifting from blue-green dominance, an experiment was conducted on a two-basin lake with one basin enriched with CO_2 while the other basin was untreated (Shapiro, 1997). Blue-green dominance continued in both basins despite higher CO_2 concentrations in the treated lake. While blue-greens showed low half-saturation constants for CO_2 in the experiment, the results indicate that some other factor with circulation than CO_2/pH must be operating to switch dominance. In the enclosure experiments, as Shapiro et al. (1982)

showed, whether a shift occurs depends on mixing rate. To achieve adequate mixing, a critical air flow rate of $9.2\,\text{m}^3\,\text{km}^{-2}\,\text{min}^{-1}$ has been recommended (Lorenzen and Fast, 1977), although results suggest that the critical air flow may also depend on other factors, such as velocity of mixing, rate of CO_2 introduction, buffering capacity of the water and rate of nutrient input to the photic zone (Shapiro, 1984). Nevertheless, the critical air flow rate was demonstrated in Lake Nieuwe Meer (132 ha, 18 m mean depth) in the Netherlands (Visser et al., 1996). With an air flow rate of $9.9\,\text{m}^3\,\text{km}^{-2}\,\text{min}^{-1}$, direct measurement of the buoyancy state (% sinking colonies) of *Microcystis* showed that the mixing rate with that air flow was sufficient to largely neutralize its buoyancy mechanism and substantially reduce algal biomass and blooms and shift dominance to diatoms and greens. Inadequate air flow has been a principal cause for poor results of artificial circulation (Cooke et al., 1993).

Control of algal biomass may also be attained by increasing the grazing rate by zooplankton, and, thus, the loss rate of algae. While the effect of planktivorous fish on algal abundance has been known for some time (Hrbacek et al., 1961), attempts to refine biomanipulation as a restoration technique began in the 1970s (Shapiro et al., 1975; Shapiro, 1979). There has been much effort in Europe to improve water quality in shallow lakes by removing benthivorous and planktivorous (roach and bream) or stocking piscivorous fish (pike). Søndergaard et al. (2000) reported on 30 whole-lake treatments in Denmark, while manipulations have also been extensive in the Netherlands (Meijer et al., 1999). Grazing loss rates of algae can be increased by preserving the large-bodied zooplankton, especially *Daphnia*. This may be accomplished by: complete removal of fish by poisoning or partial removal by netting and replacement with a favourable planktivore:piscivore ratio; improving that ratio by planting piscivores; or installing hypolimnetic aeration to provide a refuge for *Daphnia* from predation. Positive results have been reported from planktivorous fish removal without planting piscivorous fish (Søndergaard et al., 2000). There was success with the technique early on in experimental enclosures (e.g. Andersson et al., 1978; Lynch and Shapiro, 1981; McQueen et al., 1986), as well as in whole-lake treatments (Stensen et al., 1978; Shapiro and Wright, 1984; Benndorf et al., 1984; Wagner, 1986). The work in Europe in the 1990s has developed biomanipulation into a dependable technique and important tool in switching lakes to a clear-water state (see Section 9.7), although its long-term sustainability is still uncertain (Søndergaard et al., 2000). Benndorf (1990) reviewed 25 whole-lake biomanipulations and found that the changed fish stock had remained stable for 3–5 years in only nine. In most instances, the presence of planktivores resulted in removal of the large-bodied zooplankton. And without planktivores, algal biomass was often reduced, which of course is the desired effect (Figure 8.2). This was even true of blue-greens except where *Aphanizomenon* formed flakes (Lynch and Shapiro, 1981). Some experiments, however, have shown that large-bodied zooplankton are preserved, but without the desired reductions in algal biomass (McQueen and Post, 1988).

Procedures for and results from removing zooplanktivorous fish have varied. The complete removal of fish from Round Lake, Minnesota, USA, followed by restocking with a reduced planktivore:piscivore ratio (165:1 to 2.2:1), was successful for two years (Shapiro and Wright, 1984). *Daphnia* quickly became dominant causing a reduction in chl *a* of >50% and an increase in transparency from 2 m to >4 m. However, the effect did not persist. Planktivores may be effectively removed by netting, as was done in a small New Jersey pond. There, the planktivore:piscivore ratio was changed from 6:1 to 2.4:1. An increase in the existing daphnid species doubled the overall zooplankton size, but large-bodied daphnids did not appear, presumably because planktivore density was still too high, although planktivores remained controlled for four years (Wagner, 1986; Wagner, personal communication). Piscivores (rainbow trout) were introduced to a small quarry in the Federal Republic of Germany at a rate of $117 \, \mathrm{kg \, ha^{-1}}$, matching the abundance of planktivores. The planktivores were decimated after one year. The planktivore reduction was apparently responsible for the post-stocking dominance (95%) of the zooplankton biomass by large-bodied daphnids (Benndorf *et al.*, 1984). Results in Denmark indicate that about 80% of the zooplanktivorous fish should be removed over 1–2 years to achieve improved lake quality. Also external P loading should have been reduced to $0.5–1.0 \, \mathrm{g \, m^{-2} \, year^{-1}}$ (Søndergaard *et al.*, 2000). Overall results indicate that large-bodied zooplankton usually increase in abundance and that a reduction in algae often occurs but is less certain.

The use of triploid (sterile) grass carp to control macrophytes has become very popular in the USA. Stocking densities have ranged from 27 to $185 \, \mathrm{ha^{-1}}$ and in most cases 90% or more plant removal effectiveness was achieved within two or three years (Cooke *et al.*, 1993). The use of grass carp will probably increase because control is relatively quick and lasts for several years. However, grass carp can potentially increase nutrient recycling, which can cause algal blooms. They could also have adverse effects on other fish populations through removal of habitat, competition for food and introduction of disease. These potential problems must be recognized when using this technique.

There are several techniques designed to physically remove rooted macrophytes, including harvesting (mowing), rotavating, bottom covering and dredging (Cooke *et al.*, 1993). Harvesting is the most common mechanical technique and although relatively slow compared to, e.g. herbicides, harvesting has several advantages: nutrients are removed at a rate that could significantly reduce the lake-P content; risk of toxicity to animals and public objection to a toxicant are avoided; and some plant habitat will remain. Neither herbicides nor harvesting are very long-lasting, however, Although the beneficial effects of nutrient removal by harvesting are logical (Wile, 1975; Carpenter and Adams, 1977; Welch *et al.*, 1979), they have not been adequately demonstrated.

Example costs for in-lake treatments are compared in Table 10.3. Mechanical harvesting and complete circulation are relatively more expensive than the biological techniques of macrophyte and algal control. Also, fewer treatments may be necessary for the biological than physical techniques, although frequency, as well as costs, are not well established. While costs for grass carp may be relatively consistent, because the sources for triploid individuals are limited, sources are not so limited for piscivores and their costs may vary widely. Greater use of biological techniques can be expected in the future because of their relatively low cost, minimal side effects and effectiveness in waters where nutrients cannot be sufficiently controlled.

Part III

Effects of pollutants in running water

Running waters, or lotic environments, contrast sharply in many respects to their lentic counterparts. The fauna and flora are distinct from that in lentic systems. Productivity in running water is dependent primarily on attached algae, which is part of the periphyton. That is because plankton algae adapted to lentic waters are usually intolerant of the short residence time and turbulence that characterize running water. However, a modest population abundance of plankton algae can be found in reasonably clear, slow-moving rivers. The dominance of periphytic (attached) producers in running water explains their reported low abundance of algae (in the water) per unit phosphorus (Soballe and Kimmel, 1987). The effect of residence time on dominant producer organisms is illustrated in Chapter 11.

The ecological effects of pollution have been described more for running than standing water (e.g. Hynes, 1960; Klein, 1962). The majority of the effort to control domestic and industrial wastes has been directed to protecting running water environments, e.g. restoring oxygen to acceptable levels in the Thames (UK), Willamette (USA) and Cuyahoga (USA) Rivers. That organic and toxic wastes would have a greater effect in lotic than in lentic waters is largely because populations of organisms downstream from the waste source are exposed to high concentrations of waste with little or no opportunity for avoidance. The plug-flow conditions of wastewater transported downstream from a point source would explain the greater exposure than if the same volume of waste were dispersed into a lake from a point source. This is consistent with pollution-caused fish kills being observed more often in steams than lakes. For these reasons, effects of organic and toxic waste are discussed under running water, primarily in Chapters 12 and 13. Eutrophication in running water is usually not included in discussions on effects of pollution and there has been little effort until recently at setting criteria or controlling nuisance periphytic algal biomass in streams (USEPA, 1998b, 2000). Eutrophication is often mistakenly thought to be a problem in lakes only. The effect of stream enrichment and difficulty in setting nutrient criteria are treated extensively in this part.

As in lakes, there is substantial variability in organism populations in streams due to natural factors, such as light, temperature, current velocity, depth and substratum character. The effect of these factors must be considered in separating the effects of pollutant inputs. A given stream's response to pollutants or other disturbances will depend on its natural chemical and physical conditions. For example, streams of the Pacific coastal ecoregion are naturally cool, clear, oligotrophic and low in alkalinity. As such, these streams and their biota are highly sensitive to temperature increases, sediment inputs, nutrient enrichment and acidic precipitation (Welch et al., 1998). Establishing acceptable experimental controls to define variations in populations due to natural factors is often difficult. While upstream sections from the source may represent acceptable controls, one or more of the above-mentioned factors are often substantially different between upstream and downstream sections. This problem will be discussed primarily in Chapter 12.

Chapter 11

Periphyton

Periphyton is the slimy material attached to the surface of rocks but also other substrata (e.g. wood, sediment, etc.) in streams and in the littoral zone of lakes. It is composed primarily of algae, but can be dominated by bacteria and fungi if there is a source of dissolved organic matter and can include invertebrates (e.g. protozoans, insects, etc.). The periphyton mat is usually rather thin, but very slippery in fast-flowing streams with moderate enrichment. In that case, the community is usually dominated by diatoms, such as *Navicula*, *Gomphonema*, *Diatoma*, *Cymbella*, *Cocconeis* and *Synedra*. Most of these genera were not mentioned in the discussion on plankton because they are peculiar to the periphyton.

In streams enriched with inorganic nutrients (N and P), long waving strands of filamentous green algae can develop, especially if current velocity is relatively low. Typical representatives of the filamentous green algae are *Cladophora* and *Stigeoclonium*. Cyanobacteria (blue-greens) such as *Phormidium* and *Oscillatoria* may also be important, but green algae are usually more abundant in running water. These mats can represent an aesthetic and water quality nuisance as great as that from planktonic blue-greens in lakes. Once biomass reaches a level that is unstable under the existing current velocity, the mats slough and the dislodged biomass may create a secondary BOD downstream. Diatoms can also develop a thick brownish mat and can even appear filamentous in character under enriched conditions.

Filamentous bacteria, such as *Sphaerotilus*, may develop to nuisance levels in streams receiving organic enrichment. The filaments of bacteria are greyish or colourless, but they may become covered with diatoms (e.g. *Gomphoneis*) and be mistaken for a diatom mat and, conversely, mats of filamentous-appearing diatoms may be mistaken for *Sphaerotilus*. Fungi and ciliated protozoans, such as *Stentor*, *Carchesium* and *Vorticella* may also be abundant in the periphyton mat in organically enriched streams. Insects, such as chironomid larvae (midges), also occur in the thick mat and are considered part of the periphyton, whether it is composed primarily of algae or bacteria. Even in unenriched streams, other insects, such as mayfly nymphs, blackfly (*Simulium*) larvae and even rotifers may be abundant in the mat.

While the emphasis in this chapter is on the character and cause of nuisance levels of periphyton in running water, filamentous green algae can reach very visible nuisance levels in standing water as well. As an example of nuisance conditions in lakes, Wezernak *et al.* (1974) mapped the extent of mats of the filamentous green alga *Cladophora* (blanket weed) along the Lake Ontario shoreline using remote sensing. A nearshore strip 350 m wide was 66% covered with *Cladophora* from Niagara to Rochester, New York, amounting to about 15 700 kg km^{-1}. Waves break the strands loose from their substrata and deposit windrows of decaying plant material on the beach in such situations. Such nuisance conditions may be much more obvious to lake shore users than those caused by phytoplankton in the open water. In this chapter, results from studies of filamentous green algae in lakes and estuaries will be integrated with those from streams to understand better the causes for nuisance levels.

11.1 Significance to productivity

The turbulence and short residence time in streams is intolerable to planktonic organisms and, therefore, attached algae are the dominant primary producers. However, plankton may represent a significant contribution in deep, slow rivers where periphyton may be limited by light and substrate instability. Even plankton productivity may be limited by light in such environments if turbidity is too high. Rickert *et al.* (1977) illustrated the distribution of algal biomass per unit photic zone with detention time, showing the shift from periphyton to plankton dominance in running water as detention time increases (Figure 11.1). However, they omitted filamentous green algae, which are often important in moderate velocity (and even high velocity), shallow rivers.

The river continuum concept of Vannote *et al.* (1980) is useful in considering community shifts from shallow, fast-flowing streams (usually low-order tributaries) to slow-flowing rivers (usually high-order). Heterotrophic fungi and bacteria usually dominate the shaded, fast-flowing, shallow first- to third-order streams in the forest, which is the principal source of primary production and organic matter to the stream, i.e. allochthonous sources. The organic matter from leaves, twigs and other large particulate material, along with attached bacteria and fungi, is degraded by the detritivore insects, as well as by the fungi, protozoans and bacteria. As a result, respiration exceeds primary production in such environments ($P/R < 1$). Farther downstream in the fourth- to sixth-order range, shade by the forest canopy and the terrestrial source of organic matter decrease, or that source becomes well diluted by the increased stream volume. In such reaches, in-stream primary production by periphytic algae becomes dominant, grazing invertebrates increase and the stream is autotrophic with $P/R > 1$. Still farther downstream (seventh order), the P/R ratio may again drop below 1 if turbidity and depth create light-limited conditions.

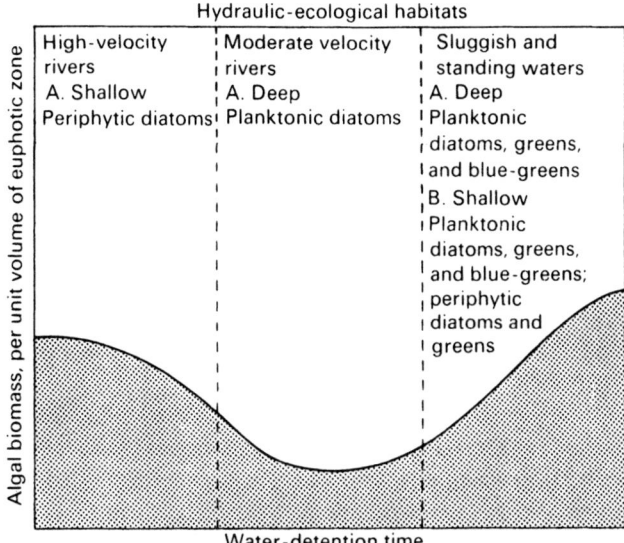

Figure 11.1 Conceptual diagram relating algal biomass and dominant algal types to water-detention time and light penetration. Shallow implies that the euphotic zone extends to the bottom; deep implies the euphotic zone does not reach the bottom (Rickert et al., 1977).

Periphyton comprises a relatively small fraction of the primary production in deep lakes, whose area of shallow littoral is relatively small. The contribution to production depends on the percentage of area where shallowness is sufficient to allow enough light to reach the bottom. If macrophytes are abundant, the zone of periphyton growth can be larger due to the large lighted substratum supplied by the rooted plants. Periphyton production per unit area can be great in oligotrophic lakes or shallow lakes with large littoral areas (Loeb et al., 1983) and in the intertidal areas of estuaries. In such cases, the contribution to total production is significant. While rooted plants, logs and rocks are suitable substrata, shifting sand beaches are not stable substrata and, therefore, support little periphyton.

Wetzel (1964) compared the three sources of primary production in a large (44 ha) shallow saline lake in California (Table 11.1). In the shallow area, periphyton production per unit area exceeded phytoplankton and macrophyte productivity. But overall, phytoplankton was greatest because of the larger pelagic area. This is probably typical of most shallow lakes. Although not the largest in most cases, the periphyton contribution may still be significant. There are other examples showing that the contribution of periphyton to production in lentic waters can range from 1% in an oligotrophic lake without macrophytes to 62% (the dominant producer) in a shallow,

Table 11.1 Comparison of the three sources of productivity in Borax Lake, California

Source	Annual mean (mg C m^{-2} day^{-1})	Total annual mean (kg C lake^{-1} yr^{-1})
Phytoplankton	249.3	101.0
Periphyton	731.5	75.5
Macrophytes	76.5	1.4
Total	1057.3	177.9

Source: Wetzel (1964).

rapidly flushed lake where phytoplankton production is controlled by washout (Wetzel, 1975, p. 414).

Streams have often been considered less productive than lakes. However, McConnell and Sigler (1959) studied a 32-ha area in the upper canyon of the Logan River, Utah, USA which is a beautiful, fast-moving, productive trout stream. They measured periphyton productivity at 1020 mg Cm^{-2} day^{-1} for that year. Considering that $\approx$ 200 mg Cm^{-2} day^{-1} annually is the lower limit for the meso-eutrophic threshold of a lake (Rodhe, 1969), this stream exceeded that rate by 5.0 times in the upper canyon and 20 times in the lower river. Thus, relatively shallow (less than the photic-zone depth), swift streams can produce large quantities of organic matter per unit area and still not take on the nuisance conditions of many eutrophic waters. However, cyanobacteria dominated in the canyon with an average chl a of 300 mg m^{-2} and *Cladophora* was prevalent downstream. Effects of enrichment in streams will be discussed later.

11.2 Methods of measurement

Periphytic biomass can be collected by scraping an area from natural or artificial substrata and analysing for wet weight, dry weight, ash-free dry weight (AFDW), pigment content, or cell and/or species counts. The available time and specific purpose usually dictate the extent of analysis. The most convenient determinations for biomass are chl a and AFDW. The ratio of AFDW to chl a is often used as a heterotrophic–autotrophic index for water quality.

Accumulation or accrual of biomass on artificial substrata, such as Plexiglas slides, glass microscope slides, wood shingles and concrete blocks, can be conveniently collected and analysed for any of the above indices. The general procedure for using artificial substrata is to stagger incubation times for pairs of slides in order to develop a curve of biomass accrual over periods of 2, 4, 6 and 8 weeks. The purpose is to determine the time to maximum accrual rate and maximum biomass. The time for maximum accrual depends on growth and loss rates. Once determined for a particular stream and season, the incubation period of maximum accrual is often used exclusively to minimize sample analysis.

The disadvantages of artificial substrata are the following: the species that normally occur on the stream bottom may not be selected, diatoms are usually the first to colonize and more time is required for filamentous green algae, thus, maximum biomass may be underestimated; and productivity as indicated by net accrual rate is probably over-estimated because accrual starts from a bare area in contrast to that proceeding on natural stream substrata over the same time period.

Advantages of artificial substrata are as follows: they provide standardized, readily comparable substrata of known area on which organisms at each station have an equal chance for attachment and growth; a precise, comparable rate of accrual can be determined; data collection is easy; and analysis of material can provide sensitive indices of water quality (e.g. AFDW/chl *a*) and effects are integrated over time.

Productivity and respiration rates by periphyton can also be determined by light/dark 'bell jars' or a light/dark flow-through box using O_2 evolution/uptake, or ^{14}C uptake (Vollenweider, 1969b) or the O_2 curve method of Odum (1956). These methods are more time-consuming, but in some cases may be preferred due to their greater sensitivity and desire to understand the metabolic dynamics of the stream.

11.3 Factors affecting growth of periphytic algae

11.3.1 Temperature

The same principles pertaining to the effect of temperature on phytoplankton growth (Chapter 6) also apply to periphyton, but some added examples may be useful. McIntire and Phinney (1965) showed that the short-term metabolism of periphyton in an artificial stream was greatly affected by temperature. The following changes in respiration rates were in response to abrupt temperature changes over a period of 5 h: a change from 6.5 to 16.5°C produced a change in metabolism from 41 to 132 mg O_2 m^{-2} h^{-1} and change from 17.5 to 9.4°C a change in metabolism from 105 to 63 mg O_2 m^{-2} h^{-1}.

The photosynthetic rate increased accordingly in response to a temperature rise over an 8-h period at a light intensity of 20 000 lux (1966 footcandles, fC): a change from 11.9 to 20°C produced a change in photosynthesis from 335 to 447 mg O_2 m^{-2} h^{-1}.

There was no effect of temperature on photosynthesis below 11 000 lux (1000 fC), which is $\approx 28\%$ of photosynthetically active radiation (PAR) in average full sunlight (14% of total). In this case, the Q_{10} for photosynthesis was ≈1.7 and that for respiration was ≈2.0 (Figure 11.2). The respiratory Q_{10} can be calculated as follows:

$$Q_{10} = \left(\frac{K_2}{K_1}\right)^{10/t_1-t_2} = \left(\frac{110 \text{ mg O}_2 \text{ m}^{-2} \text{ hr}^{-1}}{56 \text{ mg O}_2 \text{ m}^{-2} \text{ hr}^{-1}}\right)^{10/10} = 1.96$$

Periphyton can thus be expected to respond rather quickly, at $Q_{10} \approx 2$, over a similar range in temperature. However, Q_{10} may not be the most appropriate model to represent temperature response (see Chapter 6). At low light intensities an increase in temperature would cause a disproportionate increase in respiration over that of photosynthesis and net production would decrease in the short term.

What can be expected in the long term from a temperature increase? As with the plankton, periphyton community composition would probably change gradually towards a dominance by blue-greens and lowered diversity. Coutant (1966) showed such a trend in some of the first results on the effects from a heated (power-plant) discharge. The study was done in the Delaware River where temperature at times reached 40°C (104°F). When the temperature exceeded about 30°C (86°F), blue-greens were favoured.

The shifts in community composition and diversity in response to heated water are shown in Figures 11.3 and 11.4. The increased temperature would also likely result in greater light-saturated rates of photosynthesis and net production. In support of that, Coutant (1966) noted that periphytic biomass was 8-times greater and chl a was 2.5-times greater in a heated channel (38–40°C) compared to ambient levels in the Columbia River, where the maximum daily mean reached 21°C.

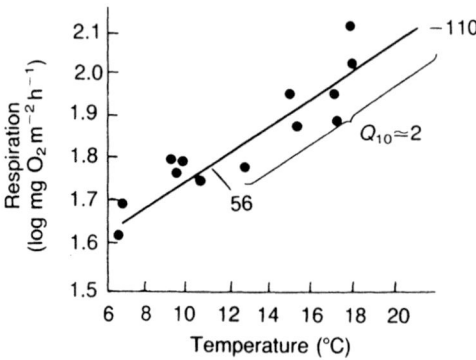

Figure 11.2 Relation between respiration rate and temperature in stream periphyton, with rates indicated for an interval of 10°C (Phinney and McIntire, 1965).

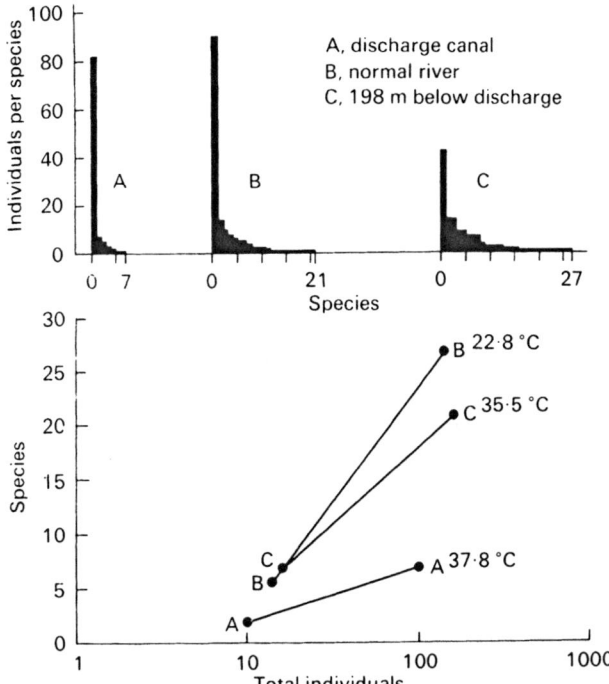

Figure 11.3 Results of periphyton collection from artificial substrata over a 7-day period ending 4 June 1959, in the Delaware River at Martin's Creek Plant (after Trembly, 1960; Coutant, 1966).

11.3.2 Light

The periphytic algal community can also be expected to respond to light in a fashion similar to the phytoplankton (see Chapter 6). However, depth of the photic zone and depth of mixing are not critical factors in shallow, fast-flowing streams. Rather, there is the question of too much light and whether periphyton can adapt to nearly full sunlight conditions. Adaptation in the periphytic algae does occur and the direct effect of light is modified by high and low light-adapted communities ('sun' and 'shade').

This is important because shaded conditions along streams are common and the extent to which adaptation can occur determines how much of a possible increase in enrichment could effectively be utilized. Moreover, shade is often removed and the effect of enrichment in low/high light streams is of interest.

McIntire and Phinney (1965) showed very interesting physiological and compositional shifts in their artificial stream community in response to changing light intensities (Figure 11.5). The principal effect of shade adaptation

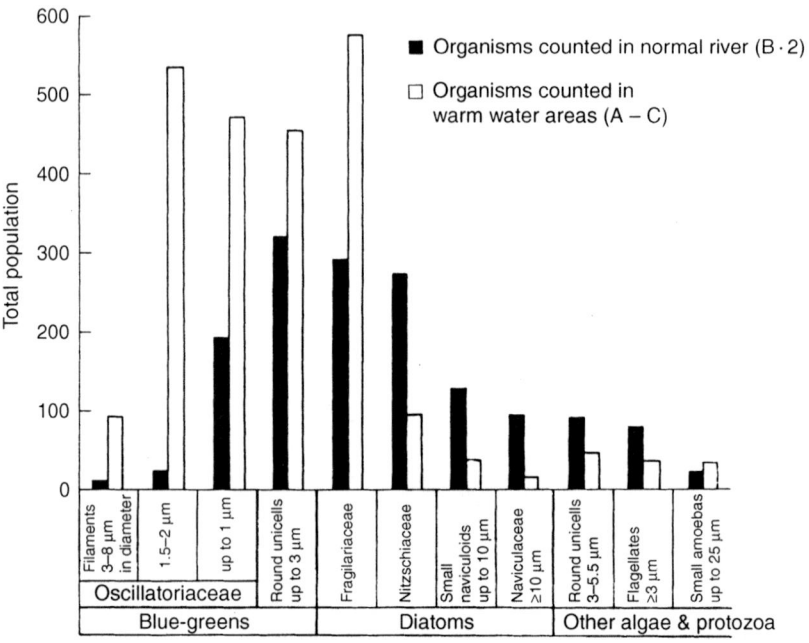

Figure 11.4 Comparison of microorganisms in warm water and normal Delaware River water at Martin's Creek Plant (after Trembly, 1960; Coutant, 1966).

was a slightly higher rate of photosynthesis, with increasing light, at low-light intensity, but a lower light-saturated rate for the shade-adapted community. Sun and shade adaptation was performed at ≈7% and 3% of full sunlight (14 and 6% PAR).

The light-adapted periphyton appeared to saturate ($\approx I_c$) at ≈800 fC (Figure 11.5), which is ≈11% of full sunlight or ≈22% of full PAR (McIntire and Phinney, 1965). Furthermore, the periphyton showed no inhibition up to ≈27% full sunlight (54% PAR), suggesting tolerance over a rather wide range of intensity. Jasper and Bothwell (1986) showed that a periphytic diatom community exposed to natural light was saturated at $500\,\mu E\,m^{-2}\,s^{-1}$ from April to September and at $150\,\mu E\,m^{-2}\,s^{-1}$ from January to March; these levels correspond to 25 and 7.5% of full sunlight. Furthermore, that community was not inhibited during summer unless intensities exceeded that of full sunlight by factors of 2–3. Thus, stream periphyton appear to be highly adapted to wide ranges in light intensity. However, as shall be discussed later, mat thickness may be subject to light limitation.

In spite of the overall slower growth in the McIntire and Phinney shade-adapted periphyton, biomass accrual attained levels similar to that at high

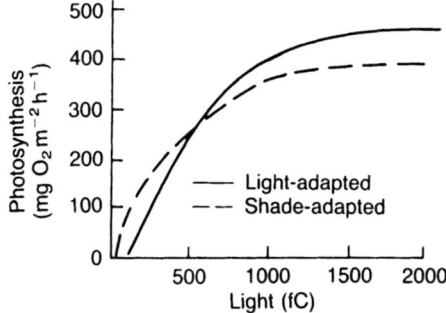

Figure 11.5 Effect of light on shade-adapted (200 fC) and light-adapted (550 fC) communities composed of diatoms (D) blue-greens (B/G), and green algae (G). The light-adapted community consisted of 46% D, 42% B/G, and 12% G. The shade-adapted community consisted of 67% D, 26% B/G, and 7% G (McIntire and Phinney, 1965).

light but required more time (Table 11.2). Biomass in the lighted stream reached an apparent saturated biomass level in two-thirds the time necessary for the shaded stream.

The interaction of low light and nutrient enrichment on the McIntire and Phinney periphyton is worth mentioning at this time although effects of enrichment will be discussed more fully later. The effect of shade adaption was that nutrient enrichment did not increase the photosynthetic rate, but there was a positive response to CO_2 increase in the light-adapted community (Figure 11.6). The explanation for this phenomenon seems to be that if photosynthesis-caused CO_2 removal is great enough and the CO_2 level (or any other nutrient) becomes low enough to limit growth, the light-adapted community will remove more CO_2 and cause self-limitation. As a result, increases in CO_2 will stimulate further increases in photosythesis. Because the pH change was only 1.45, it was discounted as directly influential in that case. However, as will be shown later, situations may exist in which nutrient

Table 11.2 Biomass accumulated in artificial stream communities held in light (550 fC) and shade (200 fC) conditions (values are maxima)

Duration	Biomass (mg slide^{-1})	
	Light	Shade
180 h	140	5.4
2 wk	120	89
6.5 mo	593	–
9 mo	–	565

Source: McIntire and Phinney (1965).

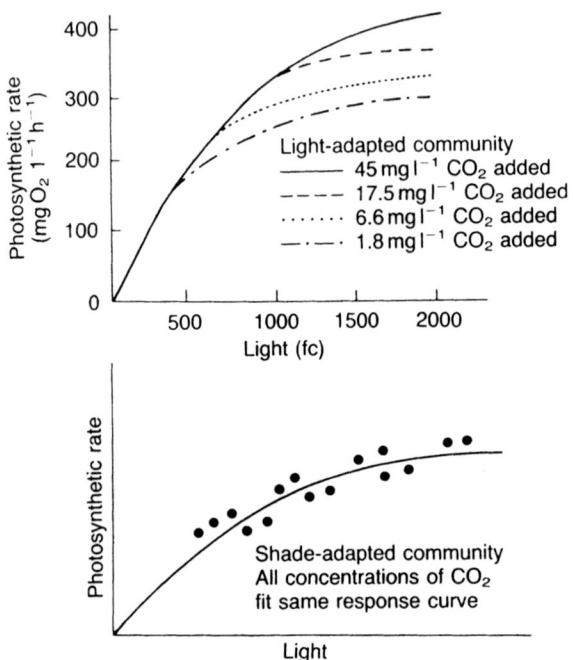

Figure 11.6 Light- and shade-adapted periphyton community response to increased light over a gradient of CO_2 content (McIntire and Phinney, 1965).

content is low enough to provoke limitation even in shade-adapted communities so that increased enrichment would have a stimulatory effect.

The interaction of light and temperature can have a dominant influence on the seasonal occurrence of any given species. Alteration of either factor can shift species composition due to tolerance, as was shown with limits to both factors. For example, there is interest in the extent to which light and temperature cause seasonal changes in the abundance of *Cladophora glomerata*, a nuisance filamentous green algae in Lake Erie. For temperature, Storr and Sweeney (1971) found that its optimum growth occurred at 18°C, while growth stopped at 25°C. Predictions of growth rate from natural photoperiod and temperature alone, based on laboratory growth related to photoperiod and temperature, compared rather closely with observed biomass (Figure 11.7). In spite of sub-optimum temperature and photoperiod in October, which resulted in low predicted biomass, the observed biomass increased, probably in response to increased nutrient content following lake overturn. A similar increase in periphyton biomass in the autumn following lake turnover and SRP increase was also observed in Lake Sammamish (Porath, 1976). This was in spite of decreasing light conditions. Thus, a close

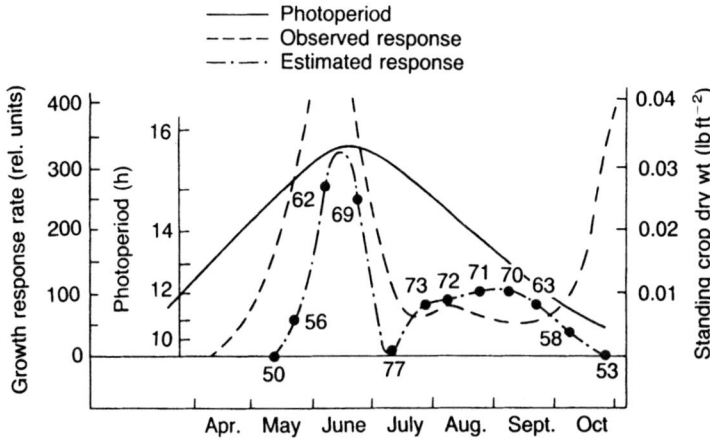

Figure 11.7 Observed biomass of Cladophora in Lake Erie and estimated levels based on experimental light and temperature response. Observed values for temperature given in degrees Fahrenheit (Storr and Sweeney, 1971).

fit between periphyton growth and the physical factors light and temperature may hold for a rather stable nutrient regime, but where limiting conditions occur, fluctuations in nutrient levels can dominate growth response.

11.3.3 Current velocity

Current velocity is often the most important factor controlling the biomass of stream periphyton. The frictional shear stress created on the periphyton mat at high velocity may substantially erode or scour the attached organisms from the substratum (Horner and Welch, 1981; Biggs et al., 1998). Thus, the biomass of periphyton observed represents a balance between growth and loss due to scouring (Biggs, 1995). As a result, the frequency of flood events has been shown to be an important determinant of periphytic biomass in streams (Tett et al., 1978; Biggs, 1988; Biggs and Close, 1989; Biggs et al., 1999). Several researchers have observed early that $\approx 50 \text{ cm s}^{-1}$ represents a reasonable estimate of a maximum tolerable velocity beyond which loss rates substantially reduce biomass (see Horner and Welch, 1981).

Field observation in western Washington streams tends to agree with such a velocity limit to biomass. However, community growth form has been shown to determine, to a large extent, the effect of current velocity on periphytic biomass (Biggs et al., 1998). Muscilagenous diatom communities had the greatest resistance to velocity, stalked/short filamentous diatoms showed low resistance above 20 cm s^{-1}, while long filamentous green algal communities showed progressively decreased biomass over a range of velocity from 15 to 80 cm s^{-1}, with very low biomass occurring at rates >50 cm s^{-1}.

The above observations are for coarse, gravel/cobble bed streams at relatively low summer flow. Periphyton on unstable substrata, such as sand or silt, is highly susceptible to disturbance and biomass loss, regardless of the growth form. During flood or freshet events, scouring/sloughing is accelerated due to not only velocity, but also suspended sediment. Observations in artificial channels have shown that periphyton communities, diatom and even filamentous algae, adapt with low loss rates even at velocities as high as 70 cm s^{-1} (Horner et al., 1983). Adaptation to high velocities (>80 cm s^{-1}) have been observed by others (Traaen and Lindstrom, 1983). That suggests that another factor may interact with velocity to enhance scouring of even relatively resistant growth forms in natural streams. That other factor may be suspended matter, because the addition of sediment (glacial flour) to those same artificial channels resulted in enhanced scouring (Horner et al., 1990) (Figure 11.8). Thus, a synergistic effect of velocity and suspended sediment is usually a normal occurrence in natural streams since sediment transport increases with velocity. In addition, sudden increases from relatively low to high velocity (e.g. 20 to 60–70 cm s^{-1}) resulted in increased scouring, even without suspended sediment, if the mat had developed under stable flow (Horner et al., 1990). Also, high relative changes in velocity are

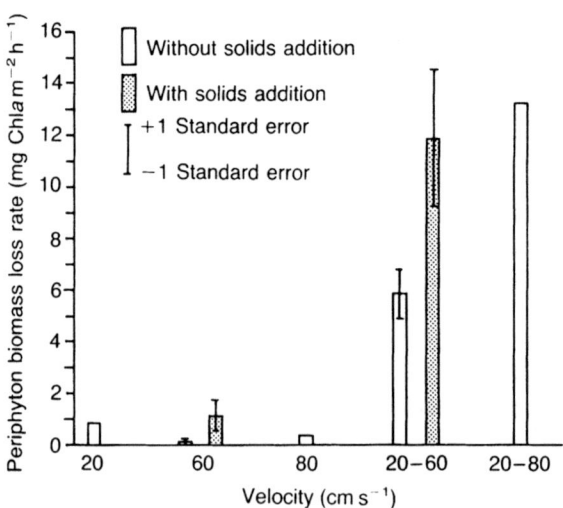

Figure 11.8 Periphyton biomass loss rates before and immediately (15 min) after velocity increase and addition of solids (to 25 mg l^{-1} TISS (total inorganic suspended solids)) in channels that receive 25 µg SRPl^{-1}. Velocities of 60 and 20–60 cm s^{-1} refer to channels with and without addition of solids (from Horner et al., 1990).

required to slough low biomass, but high biomass is susceptible to low relative changes (Biggs and Close, 1989). Limitation by light may also be a consequence of increased suspended sediment, but the enhanced scouring effect is probably more important in shallow streams where light is usually supersaturated.

Velocity increases over the range below the 50–60 cm s^{-1} level have been shown to enhance nutrient uptake and growth of periphyton. The stimulatory effect of velocity is shown by the growth of *Sphaerotilus* in artificial channels (Figure 11.9). Velocities above ≈ 15 cm s^{-1} greatly increased accrual rates. Such a critical level has been shown for filamentous green algae in artificial channels (Horner *et al.*, 1983, 1990) and was suggested earlier by Whitford (1960) and Whitford and Schumacher (1961). Similar to the response for *Sphaerotilus*, a filamentous bacterium (Figure 11.9), velocities above about 15 cm s^{-1} seemed to maximize accrual of filamentous algae (*Mougeotia*) in artificial channels.

An explanation for the stimulatory effect of velocity was given by Schumacher and Whitford (1965), following observation that ^{32}P uptake by *Spirogyra*, a filamentous green alga, increased with velocity increases over a range of 0–4 cm s^{-1} (Table 11.3). They concluded that the velocity effect was one of maximizing turbulent diffusion. The turbulence across the mat surface by current velocity maintained a maximum gradient of nutrient concentration between the ambient water and the surface of periphytic algal cells. Others have described this process as well (Lock and John, 1979; Dodds, 1989; Biggs and Hickey, 1994; Borchardt *et al.*, 1994). Logically, as mat thickness increases, diffusion of nutrients (and light penetration as well)

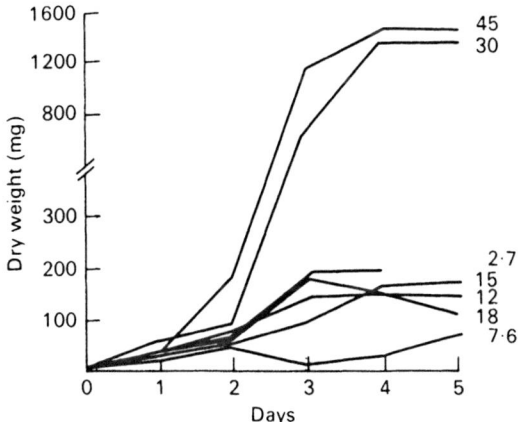

Figure 11.9 Effect of current velocity on *Sphaerotilus* growth. In all cases sucrose was added at 5 mg l^{-1}; and velocity is indicated in cm s^{-1} (modified from Phaup and Gannon, 1967).

Table 11.3 Effect of velocity over a range of low rates on the uptake of ^{32}P by Spirogyra

Velocity (cm s^{-1})	Uptake (counts min^{-1} g^{-1} dry weight)
0	78 424
1	92 748
2	98 602
4	158 861

Source: Schumacher and Whitford (1965).

into the mat decreases at any particular velocity. Uptake of SRP per unit biomass of filamentous green algae (Horner et al., 1990) and respiration rate per unit biomass of heterotrophytic slime bacteria (Quinn and McFarlane, 1989) have both been shown to decrease with increased mat thickness. Thus, a velocity increase should tend to increase diffusion and enhance greater mat thickness and, as a result, greater areal biomass.

The two opposing effects of increased current velocity on periphyton mats in flowing water are: (1) scouring/sloughing due to increased frictional shear that apparently varies with community growth form and (2) enhanced mass transfer of nutrients (Biggs et al., 1998). These authors argue that stimulation is more important than scouring loss in mucilagenous 'dense and coherent gelatinous' diatom (and presumably filamentous bacteria and cyanobacteria) communities and vice versa in open and loosely woven filamentous (green algae) communities.

This phenomenon was also shown by McIntire (1966). A felt-like community, composed largely of diatoms, developed in unenriched artificial channels at a velocity of 38 cm s^{-1}. However, at a velocity of 9 cm s^{-1}, long oscillating filaments of the green algae *Stigeoclonium*, *Oedogonium* and *Tribonema* developed. Seasonal changes in natural streams can be partly explained by this phenomenon where mats of filamentous green algae become abundant at low summer flow-rates where at earlier higher flows, brown mats of diatoms occurred. As will be shown subsequently, taxonomic composition is also affected by enrichment and as Stanford and Ward (1988) and Biggs (1985) have suggested, the low-flow proliferation of filamentous green algae during summer low flow, especially along the edges of streams, may be partly related to infiltration of higher-nutrient groundwater.

11.3.4 Inorganic nutrients

N and P are the primary macronutrients that enrich streams and rivers and cause nuisance biomass levels of algae. As stated above, conditions that allow biomass to accumulate, i.e. adequate light, optimum current velocity for periphyton, sufficient detention time for plankton, as well as low loss to grazing, will not cause high biomass without nutrients. Nutrients, especially P, are the key stimulus to increased and high maximum/mean algal biomass.

While P is the key nutrient controlling productivity and causing excess algal biomass in most lakes and reservoirs worldwide, N may have more importance as a limiting element of biomass in streams. Lohman *et al.* (1991) reported low NO_3–N causing N limitation at 16 sites in 10 Ozark Mountain streams and cited sources for N limitation in northern California and the Pacific Northwest. N was clearly the limiting nutrient in the upper Spokane River, Washington (Welch *et al.*, 1989b). Chessman *et al.* (1992) observed N to limit more than P in Australian streams.

Nevertheless, P has been the principal cause for increased periphyton accrual in running water. Stockner and Shortreed (1978) increased periphytic diatom accrual in stream-side channels (with ambient light) tenfold by increasing SRP from <1 to $9 \mu g \, l^{-1}$. Perrin *et al.* (1987) increased periphyton biomass from <10 to >100 mg chl *a* m^{-2} by increasing SRP from a background level of $1-20 \mu g \, l^{-1}$ in another oligotrophic British Columbia stream. A lesser increase was observed at SRP addition to $15 \mu g \, l^{-1}$. They also showed that the filamentous green algae *Ulothrix* and *Spirogyra* replaced diatoms in the enriched areas during the late-summer, low-flow period, indicating that dominance by filamentous green algae during low flow may be related to increased nutrients as well as low velocity. Elwood *et al.* (1981) also increased periphytic accrual and produced mats of filamentous green algae and blue-greens by raising in stream SRP content above background of $4 \mu g \, l^{-1}$. Also, Chetelat *et al.* (1999) have shown for a group of Canadian streams that *Cladophora* begins to dominate the periphyton at a TP concentration of about $20 \mu g \, l^{-1}$. That filamentous green algae and cyanobacteria can reach biomass levels exceeding 500 mg chl *a* m^{-2}, in a week or so in response to modest increases in P, has been shown in artificial channels by raising SRP from ≈ 2 to $10-15 \mu g \, l^{-1}$ (Horner *et al.*, 1983, 1990; Walton *et al.*, 1995; Anderson *et al.*, 1999).

Clearly, eutrophication of running water can produce large masses of periphyton, especially filamentous green algae, that can represent a nuisance condition to water quality and recreational water use (Biggs, 1985; Freeman, 1986; Biggs and Price, 1987). Mats and/or filamentous strands can break loose and clog water supply intakes, cause low DO and high pH, alter the substratum and habitat for bottom fauna, interfere with angling and generally degrade the aesthetic environment. For example, water quality violations due to large mats of *Cladophora*, *Ulothrix*, and *Spirogyra* occur along much of a 100 km stretch of the South Umpquah River, Oregon, where pH and DO extremes reach 9.3 and $1 \, mg \, l^{-1}$, respectively (Anderson *et al.*, 1994). Similar violations due to excessive *Cladophora* biomass occur over a 150 km stretch of the Clark Fork River, Montana (Watson *et al.*, 1990). Also, DO ranged from 3 to $25 \, mg \, l^{-1}$ in seven Ontario streams with excessive *Cladophora* biomass (Wang and Clark, 1976). However, these detrimental effects have not been quantitatively related to biomass level. From 19 cases of stream enrichment reported in the literature and results of a multi-stream survey in which the percentage stream area covered by filamentous forms increased with total biomass, a lower threshold for possible nuisance conditions of 100–150 mg chl a m^{-2} was suggested (Horner $et\,al.$, 1983; Welch $et\,al.$, 1988c).

As indicated, the filamentous green alga, $Cladophora$, known as blanketweed, has special significance with respect to nuisance conditions. It is especially obnoxious because of the long streamers that can develop, tear loose after collecting sediment and drift downstream. Some of the highest periphyton biomass levels, 1200 mg chl a m^{-2}, have been determined for $Cladophora$ growth in streams (Welch $et\,al.$, 1992).

$Cladophora$ and other filamentous greens can also proliferate and cause nuisances in the littoral zone of lakes proximal to sources of wastewater or agricultural runoff, as is the case in Lakes Erie and Huron. Although $Cladophora$ could accumulate at greater depths in areas not affected by wastewater, probably due to greater light penetration, experimental results $in\ situ$ showed that enrichment with P was the principal causative factor producing nuisance proliferations (Neil and Owen, 1964). The experiment consisted of adding N, P and organic fertilizer, and combined N, P and K, to nearshore areas in Lake Huron with no existing $Cladophora$ growth. All experimental areas were seeded with rocks covered with the alga. The experimental period was from 24 June to 25 September and final results are illustrated in Figure 11.10. Inorganic nutrient enrichment produced the greatest proliferation of biomass with P alone producing nearly as much as N, P and K together. N was obviously not the nutrient limiting the growth of $Cladophora$ in the Lake Huron shoreline. The rate of 2.7 kg day^{-1} of P input represents a daily concentration renewal in the affected area (based on crude assumptions of depth/area) of $\approx 250\,\mu g\,l^{-1}$ and a wasteload from ≈ 1000 people. In other work on $Cladophora$ problems in the Great Lakes, the biomass near waste inputs were found to be longer, greener, denser and have higher cellular P content (Lin, 1971). More recently, Auer and Canale (1982) effectively modelled nearshore periphyton biomass in Lakes Michigan and Huron with P as the controlling nutrient.

Higher densities of littoral periphyton (i.e. 100–150 mg chl a m^{-2}) have been observed adjacent to developed areas of the Lake Tahoe watershed (Loeb, 1986) and in the vicinity of nutrient-rich tributaries to Lake Chelan (Jacoby $et\,al.$, 1990) relative to periphyton biomass (<10–$20\,\mu g$ chl a m^{-2}) near undistributed areas of the lakes' watersheds.

The problem of determining the critical concentration of P, beyond which nuisance biomass levels can be expected, has received considerable attention. Although there are 'critical' concentrations or guidelines for P that can be related to the trophic state and expected phytoplankton biomass in lakes, no such relationships have existed for streams until recently. Fundamentally,

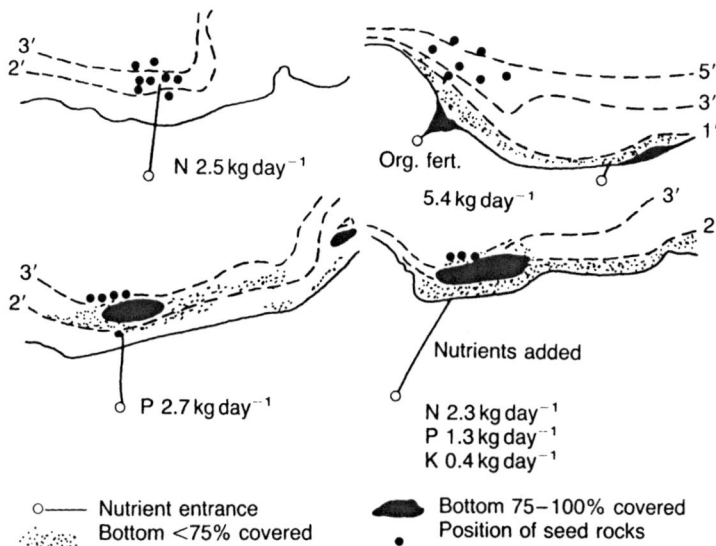

Figure 11.10 Effect of inorganic nutrients on *Cladophora* growth in Lake Huron. Experimental period was 24 June to 25 September (Neil and Owen, 1964).

it is not surprising that development of guidelines for streams has been much slower. The basis for guidelines for P in lakes is analogous to conditions in continuous culture systems where the steady-state biomass of algae per unit volume of water is proportional to the inflowing nutrient concentration (or initial concentration prior to growth). Biomass per unit volume increases as the soluble nutrient per volume is converted into biomass. Thus, direct correlations between total limiting nutrient (e.g. TP) and algal biomass (e.g. chl *a*) such as those developed by Dillon and Rigler (1974b) and Jones and Bachmann (1976) are logically expected. On the other hand, periphyton biomass in running water (or even in littoral areas of lakes) accrues on a substratum surface and extracts its nutrients from the overlying water continually supplied from upstream (or groundwater sources). Therefore, the limitation for maximum biomass per area of periphyton is not restricted to the mass of nutrient per volume of water. Rather, in that sense, one would expect that total biomass developed over a stream reach may be related to the total mass of limiting nutrient originating upstream. However, in the case of biomass per unit area at one point, it can be easily demonstrated that mass of nutrient from upstream (i.e. loading or concentration × flow rate) is many orders of magnitude greater in a very short time than is needed to produce periphytic algae per unit area at the maximum observed rate. Therefore, the pertinent unit for nutrient uptake, as well as biomass accrual, is nutrient concentration, not supply rate. Growth rate, and ultimately biomass, of periphyton is controlled by the concentration of soluble, limiting nutrient and that unit is the basis for modelling (McIntire et al., 1975; Newbold et al., 1981; Auer and Canale, 1982; Horner et al., 1983). However, determination of available nutrient concentrations in streams is complicated by algal uptake, and thus removal from the overlying water, and contributions from diffuse groundwater sources that are difficult to measure.

If biomass is in effect saturated above some nutrient concentration, then it follows that biomass should not be enhanced if nutrient concentration is raised above the growth-saturating level, which could represent the 'critical' level for running water. That is consistent with the general failure to develop meaningful causal relations between periphytic biomass and nutrient concentrations in streams; high biomass is often coincident with low soluble (and also total) nutrient concentrations, because the periphyton mat has absorbed it from the water (Jones et al., 1984; Welch et al., 1988c). That problem was avoided to some extent by Biggs and Close (1989), who correlated ($r^2 = 0.53$) mean annual periphyton biomass with average SRP content over a 13-month period.

Growth-rate saturating concentrations have been found to be very low for thin films of diatoms in artificial channels ($1-4\,\mu g\,l^{-1}$ SRP) (Bothwell, 1985, 1988) as well as for *Cladophora* ($5\,\mu g\,l^{-1}$ SRP) (Freeman, 1986). Such low growth-saturating levels make control of biomass by manipulation of in-stream nutrient concentration virtually impossible and cast doubt on experimental observation of biomass increase in response to increased SRP concentrations above background levels, which were similar to the above growth-saturating levels (see earlier results of enrichment).

The apparent explanation for increased biomass accrual from enrichment of waters with low, but rate-saturating concentrations, is due to mat-thickness limitation. In order to develop a thick mat, and hence a high periphyton biomass, increased nutrient concentration is apparently necessary to offset the diffusion limitation created by increased mat thickness. In this regard, Bothwell (1989) has shown that while growth rate is saturated at $1-2\,\mu g\,l^{-1}$ SRP, maximum biomass of diatoms (150 mg chl *a* m^{-2}) after some accrual time period, continues to increase up to $25\,\mu g\,l^{-1}$. A maximum biomass of diatoms and blue-greens (*Phormidium*) of almost 1000 mg chl *a* m^{-2} in three weeks required an in-channel SRP of nearly $20\,\mu g\,l^{-1}$ (Walton, 1990) (Figure 11.11). In the same channel system, a mat dominated by a filamentous green alga (*Mougeotia*) required SRP of $\approx 7\,\mu g\,l^{-1}$ to reach a maximum 2-week biomass of 350 mg chl *a* m^{-2} (Horner et al., 1990). Thus, the nutrient concentration that limits maximum biomass, because of mat-diffusion limitation, would seem to be the most appropriate for management of stream quality. Such a maximum-biomass limiting value for filamentous forms may, therefore, range from about $7-20\,\mu g\,l^{-1}$ SRP, probably determined as an annual mean.

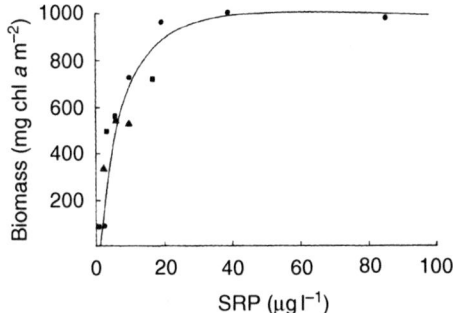

Figure 11.11 Maximum periphyton biomass on natural rocks in artificial channels during three separate experimental periods (symbols) of about three weeks each, at an average light intensity of 194 µE m^{-2} s^{-1}, an average temperature of 19°C and current velocities of 20 cm s^{-1} (■, ▲) and 40 cm s^{-1} (●). The line is calculated from the steady-state model with $K_s = 5$ µg l^{-1} SRP and $K_1 = 1.46 + 0.276$ ($r^2 = 0.85$) (Walton, 1990).

Biggs (2000) recently showed that three factors determined periphytic biomass levels in gravel-bed streams: days of accrual, dissolved N and SRP (as annual means). Biomass reached about 100 mg m^{-2} chl a after 30 days accrual following a scouring disturbance. Incorporating separate relationships between nutrients and biomass predicted that with 30-day accrual time, biomass would reach the 100 mg m^{-2} level with a dissolved N of 100 µg l^{-1} and SRP of 10 µg l^{-1} (Figure 11.12).

Given that P is the most important nutrient to control and chl a is the appropriate biomass indicator, what should the critical levels be for those variables to prevent habitat and water quality degradation in streams?

11.3.5 Recommended limits for biomass and nutrients

Criteria have been suggested from several sources for levels of biomass that present a nuisance condition in streams. There is general agreement for a criterion of about 150 mg m^{-2} chl a (Table 11.4). For example, control strategies were developed for the Clark Fork River, Montana, using a maximum biomass of 100–150 mg m^{-2} chl a as a criterion (Watson and Gestring, 1996; Dodds et al., 1997). That is consistent with a maximum of 100 mg m^{-2} chl a and a 40% coverage of filamentous forms that were proposed for New Zealand streams to 'protect contact recreation'. However, there was insufficient evidence for protection of other water quality characteristics, such as DO and pH, which would vary due to atmospheric exchange and buffering capacity (Quinn, 1991). In support of the biomass criterion, per cent coverage of filamentous forms was less than 20, but

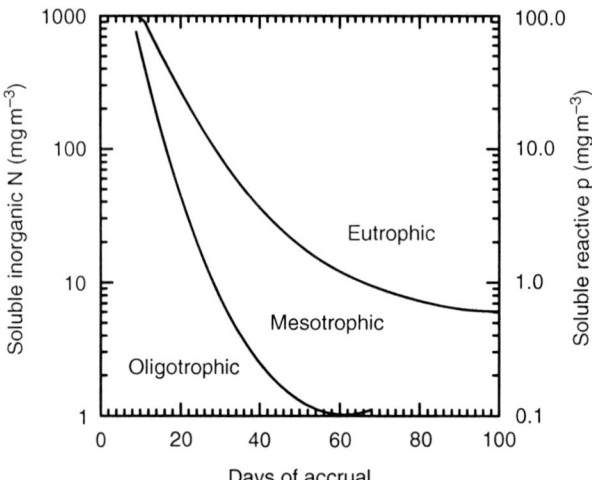

Figure 11.12 Suggested boundaries for trophic state in gravel-bed streams for 100 and 200 mg m^{-2} chl a. Note that for 30 days accrual time, 200 mg m^{-2} is related to 100 µg l^{-1} soluble N and 10 µg l^{-1} P (from Biggs et al., 2000, reproduced with kind permission from the Journal of the North American Benthological Society).

increased with increased biomass, and aesthetic quality was noticeably affected (Welch et al., 1988c). However, there were no apparent effects on DO, pH or benthic invertebrates at that biomass level. Such effects are more likely to occur at higher biomass levels, as previously discussed. Other observations show an association between this biomass level and enrichment. As shown from 19 cases surveyed by Horner et al. (1983), biomass levels higher than 150 mg m^{-2} tended to occur with enrichment, and filamentous forms were more prevalent. Also, Lohman et al. (1992) observed that biomass rapidly recovered following flood-scour events in 12 Ozark streams when biomass exceeded the 150 mg m^{-2} level at sites with moderate to high enrichment; biomass did not recover as rapidly at unenriched sites where initial levels did not exceed about 75 mg m^{-2}.

While the 150 mg m^{-2} level cannot be supported as an absolute threshold above which adverse effects on water quality and benthic habitat readily occur, it nonetheless is a level below which aesthetic quality will probably not be appreciably degraded by unsightly filamentous mats or other adverse effects attributed to dense mats of filamentous algae, such as taste and odours in water supplies and fish flesh, impediment of water movement, clogging of water intakes, restriction of intra-gravel water flow and DO replenishment or the above cited adverse effects on DO/pH in the water column, or degradation of benthic habitat (see Section 11.8). Avoidance of

Table 11.4 Nutrient ($\mu g\,l^{-1}$) and algal biomass criteria limits for periphyton and phytoplankton recommended to prevent nuisance conditions and water quality degradation in streams based either on nutrient–chl a relationships or preventing risks to stream impairment as indicated (modified from USEPA, 2000)

TN	DIN	TP	SRP	Chl a[a]	Risk	Source
Periphyton						
				100–200	Nuisance	Welch et al. (1988c, 1989b)
1500		75		200	Eutrophy	Dodds et al. (1998)
275–650		38–90		100–200	Nuisance	Dodds et al. (1997)
300		20		150	Nuisance	C.F. Tri-St Coun.MT
		20			Cladophora	Chetelat et al. (1999)
		10–20			Cladophora	Stevenson, personal communication
	430		60		Eutrophy	UK Environmental Agency (1998)
	100[b]		10[b]	200	Eutrophy	Biggs (2000)
	25		3	100	Invertebrates	Nordin (1985)
		15		100	Nuisance	Quinn (1991)
	1000	18	6[c]	150	Nuisance	Sosiak (2002)
Phytoplankton mean in $\mu g\,l^{-1}$						
300[d]		42		8	Eutrophy	Van Nieuwenhuyse and Jones (1996)
		70		15	Chl standard	Tualatin R. OR (2000)
250[d]		35		8	Eutrophy	OECD (1982) (for lakes)

Notes
a Maximum biomass in $mg\,m^{-2}$ chl a for periphyton and mean biomass in $\mu g\,l^{-1}$ chl a for phytoplankton.
b 30-day biomass accrual time.
c Total dissolved P.
d Based on Redfield ratio of 7.2N:1P.

these problems will probably be achieved with the criterion for maximum biomass of $150\,mg\,m^{-2}$ chl a.

Development of critical concentrations of P (and N) that produce a nuisance level of biomass requires predictive relationships between the two variables such as exist for lakes. Relationships between biomass and soluble nutrients have produced good results in experimental channels and with field samples in some instances (Horner et al., 1990; Biggs, 2000). However, use of soluble nutrients poses some problems. Relationships developed with periphyton biomass and ambient total nutrient in gravel/cobble-bed streams have shown more variability compared to those for lakes (Lohman et al., 1992; Dodds et al., 1997).

An explanation for the greater variability and lower chl:TP ratios in streams than lakes may be that sloughed detritus from the substrata is part of total nutrient determined in stream water, but excludes the living biomass in the periphyton mat. In lakes and reservoirs, and to some extent plankton in deep, slow-flowing rivers, total nutrient concentration includes living plankton cells. The high detritus level in streams is indicated by TP–chl a relationships in streams with low chl a:TP ratios ranging from 0.08 to 0.22 (Van Nieuwenhuyse and Jones, 1996). Chlorophyll a:TP ratios from lakes are typically 0.5–1.0 (Ahlgren et al., 1988b).

Total nutrient–periphyton chl a relationships using field-site data from streams predict lower biomass levels per nutrient than do relationships developed with data from controlled channel experiments. A mean chl a of $100\,\text{mg}\,\text{m}^{-2}$ would require about $100-200\,\mu\text{g}\,\text{l}^{-1}$ TP based on regression models for 12 Ozark streams (Lohman et al., 1992) and for a 200-site database that included streams from North America, Europe and New Zealand (Dodds et al., 1997). Using the maximum:mean biomass ratio of about 4.5, the maximum chl a biomass for the $100-200\,\text{mg}\,\text{m}^{-2}$ mean would be around $450\,\text{mg}\,\text{m}^{-2}$. However, mixed filamentous green algae, cyanobacteria and diatoms well in excess of that maximum have resulted from in-channel SRP and TP concentrations of 10–15 and $20-50\,\mu\text{g}\,\text{l}^{-1}$, respectively (Horner et al., 1983; Horner et al., 1990; Walton et al., 1995; Anderson et al., 1999, unpublished data).

This discrepancy may result from the nutrient demand by heterotrophic organisms in detritus, which would produce a lower chl a:TP ratio in streams than in lakes. Residence time in the above-cited channel experiments was short (16 min or less), nutrient input was controlled to low levels and velocity was usually constant with minimal sloughing during the growth period (Horner et al., 1990). Such conditions produced little detritus and higher chl a:TP ratios than in natural streams sampled throughout the year. Nevertheless, regression models, developed from large data sets, may be more appropriate in recommending nutrient criteria in spite of the low chl a:TP ratios for streams with high turbidity.

There is a case for using soluble nutrient concentrations to develop criteria, however. Uptake and maximum biomass are clearly saturated at very low ($<10\,\mu\text{g}\,\text{l}^{-1}$ SRP) concentrations (Bothwell, 1985, 1989; Walton et al., 1995) independent of TP concentrations. However, as mentioned earlier, soluble nutrient is usually lowest when biomass is highest, due to uptake, which is similar in lakes. Estimates of inflow concentration or in-stream concentration during non-growth periods would be necessary to use soluble nutrients to set criteria for maximum biomass. For example, averaging SRP and periphytic biomass over a 13-month period in New Zealand streams took account of growth and non-growth periods and produced good results (Biggs and Close, 1989).

The Clark Fork River, Montana, is the site of the most comprehensive effort to set nutrient criteria in streams to prevent nuisance periphytic biomass (i.e. $150\,\text{mg}\,\text{m}^{-2}$ chl a). Biomass dominated by *Cladophora* has frequently reached levels of around $600\,\text{mg}\,\text{m}^{-2}$ chl a (Watson et al., 1990; Watson and Gestring, 1996). Criteria of $350\,\mu\text{g}\,\text{l}^{-1}$ TN and $30\,\mu\text{g}\,\text{l}^{-1}$ TP were recommended using regression models from the large 200-site data set, probabilistic estimates for frequency of biomass exceedence levels for respective ranges of TP and TN, and comparison of biomass and nutrients at low-biomass reference sites (Dodds et al., 1997). Although periphytic biomass, including filamentous greens, much greater than 150 mg m^{-2} chl a has been consistently produced in channels from lower TP concentrations (as discussed above), 30 µg l^{-1} is reasonable for natural streams given their lower chl a:TP ratios. The Tristate Implementation Council for the Clark Fork River emphasized the reference site method and set more conservative criteria of 300 µg l^{-1} TN and 20 µg l^{-1} TP. Interestingly enough, these values for TP encompass mesotrophic–eutrophic boundaries recommended for lakes and reservoirs (Vollenweider, 1976; Porcella et al., 1980; OECD, 1982).

These criteria are much lower than mesotrophic–eutrophic boundaries of 75 µg l^{-1} TP and 1500 µg l^{-1} TN suggested by Dodds et al. (1998), which were based on the upper one-third of the respective distributions. The TN and TP data sets used to delineate trophic state categories may have included highly turbid streams with low periphyton biomass in which nutrients were not intended to be linked with algal biomass.

Existing nutrient and biomass criteria established for some streams and agencies are listed in Table 11.4. Periphytic biomass levels considered to represent a nuisance threshold are listed along with the nutrient concentrations that cause those levels. There is a general agreement that >100–200 mg m^{-2} chl a represents a nuisance condition. In four cases, a nuisance condition for biomass and the appearance of *Cladophora* would be expected at a TP concentration around 20 µg l^{-1}. TP–chl a relations developed from a large data set, however, predict much higher TP to produce such nuisance levels (Dodds et al., 1997).

The Bow River example deserves some explanation because it represents the most thoroughly investigated case of periphyton response to decreased nutrient input in a gravel-bed river (Sosiak, 2002). The river was monitored for over 16 years to evaluate the effect of a reduction in, first, P (80%) and, later, N ($\sim$50%) in Calgary wastewater. Both periphyton and macrophytes decreased downstream in response to nutrient reduction, but the distribution and timing of the decreases were to some extent unexpected.

Periphyton, consisting mostly of diatoms, although filamentous greens including *Cladophora* were present, had reached a summer maximum of around 600 mg chl a m^{-2} downstream prior to P reduction. Such maximums have persisted within 50 km of the effluent input since P reduction in 1983, but decreased markedly farther downstream over $\sim$90–250 km reach. However, the decrease in periphyton occurred rather gradually over 13 years following P reduction, as total dissolved P (TDP) declined to very low levels downstream (median <5 µg l^{-1}). Within $\sim$50 km downstream of the effluent input, however, where TDP declined initially from a median from $\sim$90 to $\sim$25 µg l^{-1}, a further decrease in TDP has not occurred and periphyton biomass showed no change from the high pre-treatment levels (Sosiak, 2002).

What these data show is that: (1) periphyton biomass in streams and rivers does respond to nutrient reduction, (2) biomass levels below nuisance levels ($\sim 150\,\text{mg chl}\,a\,\text{m}^{-2}$) can be attained if P can be sufficiently reduced, (3) 'sufficiently' here apparently meant to levels around $6\,\mu\text{g}\,\text{l}^{-1}$ TDP, and (4) as demonstrated with a regression relationship, the response may not occur quickly even in rivers where water exchange is immediate (Sosiak, 2002). The gradual reduction in river TDP suggests that bottom sediments, even in rubble-bottom rivers, may have a memory. Macrophytes (mostly pond weeds) also had reached biomass levels $>2000\,\text{g}\,\text{m}^{-2}$ within 30 km downstream prior to effluent treatment, but declined soon after N reduction (1987), reaching levels $<\sim 200\,\text{mg}\,\text{m}^{-2}$ in 1995–1996 (Sosiak, 2002).

11.4 Grazing

The removal of periphyton biomass by macroinvertebrate grazing may be the most important control on periphyton in streams of moderate velocity even in enriched streams. Several researchers have demonstrated that snails and caddis flies are capable of holding periphyton to 5–50% of the ungrazed biomass (Lamberti and Resh, 1983; Jacoby, 1985, 1987; Lamberti et al., 1987; McCormick and Stevenson, 1989; Walton, 1990; Anderson et al., 1999; Welch et al., 2000). A large-cased caddis fly, *Dicosmoecus gilvipes* (2–4 cm) has been used experimentally by several investigators and found to be a highly effective grazer, producing some of the highest removal rates observed, even under enriched conditions. The organism reduced periphyton (diatoms) biomass by 80% *in situ* at natural densities ($41\pm 8\,\text{m}^{-2}$) in an unenriched Cascade Mountain foothill stream (Jacoby, 1987). In laboratory channels, biomass (composed of diatoms and filamentous forms) was reduced to 5% of the unenriched level at a density of 200 organisms m^{-2} ($4.8\,\text{g}$ dry weight m^{-2}) (Lamberti et al., 1987) while at $100\,\text{m}^{-2}$ ($2.4\,\text{g}\,\text{m}^{-2}$), the caddis fly reduced biomass of primarily a filamentous blue-green and diatom to 7% of the ungrazed level in unenriched and 12% in enriched (SRP $25\,\mu\text{g}\,\text{l}^{-1}$) treatments (Walton, 1990).

In these same channels, *Neophylax*, another relatively large caddis fly, held biomass (chl a), dominated by a filamentous green (*Stigeoclonium*), at 2–14% of ungrazed and 6% of ungrazed for *Dicosmoecus* (Anderson et al., 1999). Both taxa were stocked at 1.5 and $3.0\,\text{g}\,\text{m}^{-2}$ for those effects. Reductions were substantial even at a stocking density less than $1.0\,\text{g}\,\text{m}^{-2}$. Grazing rates for *Neophylax* and *Dicosmoecus* were 6.1 and 8.3 mg chl a (g animal)$^{-1}$ day^{-1}, respectively, at high SRP level, but decreased as SRP decreased (Table 11.5). However, SRP level did not compensate for grazer effect on biomass. That is, biomass in grazed channels were still <50% of ungrazed, regardless of SRP level. Some results showed grazing rates for these two taxa at fivefold those cited above (Anderson et al., 1999).

Table 11.5 Periphyton growth rates and invertebrate grazing rates for both experiments with SRP and grazing combined. Grazing rates ((g invertebrate)$^{-1}$ day^{-1}) are expressed as gram dry weight present in the channels. L=lowest level of SRP enrichment (2 µg l^{-1}), M=middle level of SRP enrichment (15 µg l^{-1}) and H=highest level of SRP enrichment (25 µg l^{-1}). Grazing rates for *Neophylax* were calculated using day 29 data while those for *Dicosmoecus* were calculated using day 18 data (Anderson et al., 1999)

Invertebrate grazer	SRP level	Periphyton growth rate (mg chl a m^{-2} day^{-1})	Grazing rate (mg chl a (g invertebrate)$^{-1}$ day^{-1})
Neophylax	L	1.98	0.79
	M	7.18	4.85
	H	14.38	6.14
Dicosmoecus	L	0.75	0.44
	M	19.91	7.17
	H	31.75	8.34

Smaller grazers were not as effective in some experiments. Small mayflies, such as *Nixe rosea* (Jacoby, 1987) and *Centroptilum elsa* (Lamberti et al., 1987), consumed periphyton but were not effective in reducing periphyton biomass of either diatoms or filamentous forms. Snails, on the other hand, have been shown to be effective grazers (Jacoby, 1985), but less so than *Dicosmoecus* (Lamberti et al., 1987). On the other hand, grazing on a filamentous green (*Ulothrix*) by relatively small New Zealand taxa (*Deleatidium*, mayfly at 3 mg dry weight; *Potamopyrgus*, snail at 1–2 mg; and caddis fly *Pcynocentroides* at 1–2 mg) in outdoor channels was effective (Welch et al., 2000). Grazed biomass was 50–60% of that in ungrazed channel sections with rates averaging about 3 mg (g animal^{-1}) day^{-1} and grazer densities up to 6 g m^{-2} (Welch et al., 2000). The effect of the mayfly on *Ulothrix* at three grazer densities is shown in Figure 11.13.

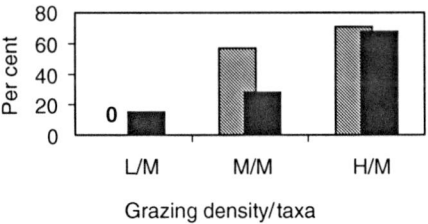

Figure 11.13 Per cent removal of filamentous green algae (FGA) (*Ulothrix*) by low, medium and high density (4.5, 3.0 and 1.5 g m^{-2}, respectively) of mayflies (*Deleatidium*) after 15 days of grazing (data from Welch et al., 2000).

The growing evidence on the effectiveness of grazers in reducing periphytic biomass of both diatoms and filamentous green algae and blue-greens, may help to explain the frequently observed poor relationship between biomass and nutrient concentration. Considering the relatively low, biomass-saturating-SRP concentration (in P-limited streams), the occurrence of periphytic biomass levels of the order of 1000 mg chl a m^{-2} may be more related to poor habitat conditions for macroinvertebrate grazers than to enrichment, regardless of how nutrient availability is expressed.

Results from seven New Zealand streams receiving point-source enrichment suggest that grazing was a more important determinant of periphyton biomass than were nutrients (Welch *et al.*, 1992). All streams received treated sewage effluent; five from oxidation ponds, one from a trickling filter plant and one from a combination of primary and secondary treatment of sewage, dairy and slaughterhouse wastewaters. SRP concentrations only doubled or tripled from low upstream levels (5 µg l^{-1}) in two streams, but increased to well over 100 µg l^{-1} in all others. From results in channel experiments (see above), most observed biomass levels were much below that expected from SRP, temperature and velocity. Rather, the data suggested a grazer-threshold effect, grazer densities >3000 m^{-2}(3–9 g m^{-2} for average dry weight of 1–3 mg) being associated with periphyton biomass <50 mg chl a m^{-2}. The highest periphyton biomasses ($\approx$1200 mg chl a m^{-2}, *Cladophora* dominated) occurred where grazer densities were very low. In one case a thick mat of primarily *Cladophora* developed on flat bedrock throughout the stream in a virtual absence of grazers due to the limited habitat. A study of ten Quebec streams showed that periphyton biomass was held below nuisance levels at grazer levels as low as 1 g m^{-2} (Bourassa and Cattaneo, 1998).

Evaluating the effect of enrichment on stream periphytic algae should probably begin with an assessment of substratum. Rubble bottom streams with relatively little fine sediment clogging up interstices may be able to assimilate greater inputs of enrichment than streams with relatively little clean and secluded surface area as macroinvertebrate habitat. This may point to management strategies of preventing flow variation and sediment input to preserve a grazer-rich substratum. Such an approach may be more effective than putting the effort into controlling nutrient inputs, especially since biomass-saturation levels of SRP are so low (<20 µg l^{-1}) and SRP levels in some cases are >1000 µg l^{-1}.

11.5 Organic nutrients

Dissolved organic nutrients favour bacterial slimes dominated by filamentous bacteria (*Sphaerotilus*) and fungi. Bacterial slimes are aggregates of *Sphaerotilus*, *Zooglea* and fungi. *Sphaerotilus* may have a different appearance morphologically under varying environmental conditions. For example,

S. natans may appear similar to *Cladothrix dichotoma* and in turn to *Leptothrix ochracea*. All these organisms are obligate organotrophs. They can create nuisance slime conditions, usually in running water, downstream from a source of dissolved organic substances. They have created nuisance problems near untreated waste discharges from pulp mills, dairy plants, landfills (leachate) and slaughterhouses as well as domestic sewage.

The culture-determined nutrient requirements for *Sphaerotilus* are monosaccharide and disaccharide sugars at a concentration of 20–50 mg l^{-1} as well as organic acids, amino acids and inorganic N and P (Phaup and Gannon, 1967). The presence or absence of *Sphaerotilus*, however, seems definitely related to the presence of sugars and low molecular weight BOD (Curtis and Harrington, 1971; Quinn and McFarlane, 1989).

As with algae, the response of heterotrophic slimes has also been studied in experimental channels. Phaup and Gannon (1967) determined the effects of sucrose and velocity on biomass accrual and community type in experimental, 210-m long, outdoor channels. The periphyton that developed on strings held in the running river water without additional enrichment included *S. natans*, the algae *Melosira*, *Navicula*, *Cosmarium*, *Euglena*, the protozoans *Tetrahymena*, *Colpedium*, amoeba, and some insects: chironomids (midges) and simulids (blackflies). Sucrose enrichment increased biomass accumulation on the string substrata. Growth was maximum within 30 h with 5 mg l^{-1} sucrose and increased with current velocity from 18 to 45 cm s^{-1}, at a temperature of 20–28°C as noted earlier (Figure 11.9). The river water contained 500 µg l^{-1} N and 1 µg l^{-1} SRP. The maximum biomass accrued was not obtained at temperatures below 17°C, because *S. natans* was replaced by another filamentous bacterium that was much less prolific at higher temperature. This finding has a bearing on the winter phenomenon in which *S. natans* often covers more stream distance than in summer. The cause for this is related to temperature and metabolic rate. At high temperature, utilization of nutrient and growth are also high, resulting in the quick depletion of nutrients downstream. Consequently, the minimum required organic content occurs over a shorter distance than in winter when temperatures are low.

In a similar experimental channel study, Ormerod *et al.* (1966) showed that a succession of species within the bacterial slime community was affected by a combination of pulp-mill wastewater and inorganic nutrients. Bacterial slimes replaced algae when spent sulphite liquor (SSL) was added, but algae was only partly replaced with sewage addition, because apparently the additional inorganic nutrients favoured algae. Two streams were studied with the same concentration of SSL (0.1 ml l^{-1}). The following organisms dominated under two different stream conditions: soft-water stream, *Fusarium aquaeductum*; hard-water stream, *Sphaerotilus natans*. Under variable temperature with velocity of 5 cm s^{-1}, the following change in the periphyton occurred: SSL addition at 0.1 ml l^{-1}, change from algae to *Sphaerotilus*;

SSL + sewage at $0.5-1.0 \text{ ml l}^{-1}$, algae increased with some *Fusarium*. At a constant temperature of 12 °C and a water velocity of 15 cm s^{-1}, the following community changes were observed (Figure 11.14): sucrose addition at 10 mg l^{-1}, change from algae to *Sphaerotilus* and *Zooglea*; sucrose (10 mg l^{-1}) + SRP at 2 μg l^{-1} and up to 1000 μg l^{-1}, change from algae to *Sphaerotilus* and then to *Fusarium*.

Although an increase in temperature resulted in an increase in growth rate (Figure 11.15), the pattern of ultimate dominance did not change – only the rate of change. Regardless of temperature, the addition of P favoured dominance by *Fusarium*. This is most interesting because it shows that although heterotrophic microorganisms are favoured by dissolved organic matter, the species actually dominating can be strongly influenced by inorganic nutrients. The addition of other nutrients, such as N, K, Na, Mg, Ca, Fe and vitamin B_{12}, had no effect upon succession.

Nuisance biomass levels of *Sphaerotilus* were also produced in a natural stream (Berry Creek, Oregon, USA) by adding sucrose at $1-4 \text{ mg l}^{-1}$ (Warren *et al.*, 1964). Effects of those nuisance biomass levels on fish and invertebrates will be discussed later.

Although 'outbreaks' of nuisance biomass levels of heterotrophic slimes have been clearly associated with the discharge of organic wastewater from a variety of activities (Gray and Hunter, 1985), establishing a critical concentration of organic matter has been rather difficult. The usual measures of organic matter, such as BOD_5, COD and DOC (dissolved organic carbon), have not always been useful. Curtis and Harrington (1971) stressed that heavy slime outbreaks have been noted where very low BOD concentrations

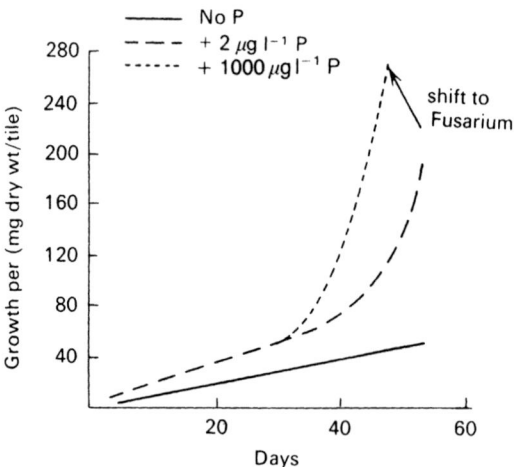

Figure 11.14 Effect of phosphate on slime community growth. Channels contained 10 mg l^{-1} sucrose and 1.9 mg l^{-1} NH_4Cl (Ormerod *et al.*, 1966).

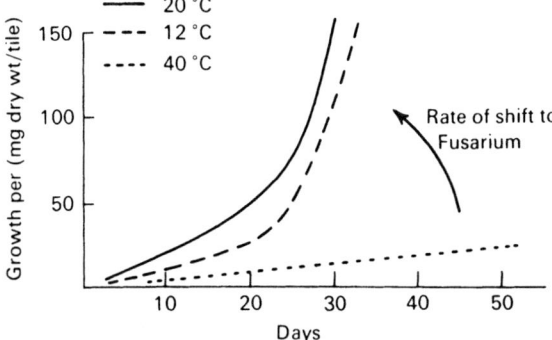

Figure 11.15 Effect of temperature on slime community growth. Channels contained 10 mg l^{-1} sucrose, 0.19 mg NH_4Cl l^{-1} and 0.34 mg K_2HPO_4 l^{-1} (Ormerod et al., 1966).

occurred, as well as no slime development in spite of existing high BODs. They suggested that the former could have been due to the presence of rapidly degradable substances, such as sugars, whereas the latter may have been due to inhibitors.

Quinn and McFarlane (1989) have recently provided some practical insight to the problem. They showed that low molecular weight compounds ($M_r < 1000$) determined by COD and BOD_5 in water passed through an ultra-filtration membrane were the key factors determining respiration rates of heterotrophic slimes in artificial channels. They further showed that restricting BOD_5 to 5 mg l^{-1} eliminated nuisance biomass levels downstream of sources if the wastewater were from domestic sewage or a slaughterhouse. However, dairy wastewater contained a higher proportion of low M_r organics and if that source represented half the BOD load entering the river, the critical BOD_5 needed to eliminate nuisance biomass was 2–3 mg l^{-1} (Quinn and McFarlane, 1989).

The concentration of total dissolved organic matter is, therefore, of less importance than its composition. If the DOC is diverse, with few compounds of low M_r, nuisance conditions may not develop. For example, Cummins et al. (1972) added leaf leachate, a natural source of allochthonous organic enrichment, to an artificial stream at concentrations up to 30 mg l^{-1} DOC, which is 10-times the natural level and similar to that received by heavily polluted streams. In 10 days, 85% of the DOC was utilized and no change in the diversity of macroinvertebrates resulted. Moreover, a diverse fungal flora developed without dominance by *Sphaerotilus*, which probably would have occurred if the 30 mg l^{-1} had been DOC as sucrose or low-M_r organics.

Other indications that a diverse DOC source will not lead to a dominance by nuisance heterotrophic slimes was apparent in experiments by Ehrlich

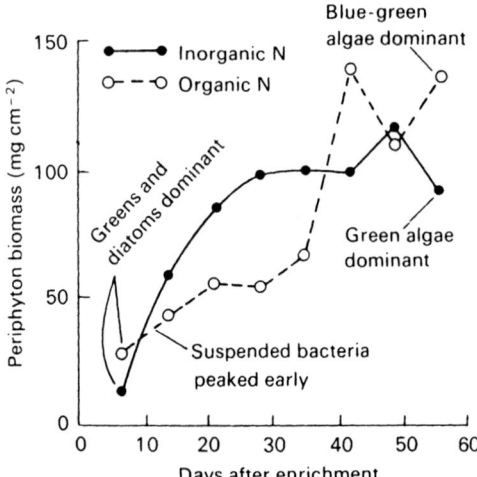

Figure 11.16 Biomass of periphyton in two artificial streams, one receiving inorganic and the other organic (yeast extract) N. Diatoms and green algae were dominant at the start of the experiment in both streams, bacteria peaked early in stream with organic N, but at the end, green algae dominated the inorganic stream and blue-greens dominated the organic stream (Ehrlich and Slack, 1969).

and Slack (1969). Figure 11.16 shows the accumulation of periphyton biomass in two artificial streams; one received inorganic N and the other a yeast extract of $50\,mg\,l^{-1}$. The rate of increase of algal biomass was greatest with the addition of inorganic N because it was readily available and mineralization was unnecessary. With the yeast extract, an intermediate step was required for bacterial mineralization, which then led to algae. Although cyanobacteria dominated in the stream receiving organic N, filamentous heterotrophic slime bacteria like *Sphaerotilus* did not.

Based on the above evidence from artificial-stream studies, one can conclude that the principal cause for nuisance levels of heterotrophic slime growth (mainly *Sphaerotilus*) is the presence of low-M_r, highly degradable, dissolved organic matter, particularly sugars.

1. Wastes containing sugar in the range of $1-30\,mg\,l^{-1}$ should lead to nuisance heterotrophic slimes dominated by *Sphaerotilus*. BOD_5 concentrations causing nuisance slimes could be as low as $2-3\,mg\,l^{-1}$ if a high proportion of the wastewater is low M_r organics.

2. Sugars and/or low M_r organics plus inorganic nutrient (mainly P) enrichment should lead first to *Sphaerotilus* then to a domination by fungi.

3   Diverse DOC compounds even up to $30\,\mathrm{mg\,l^{-1}}$ (such as natural leaf leachate) should lead to a variety of non-filamentous (and therefore non-nuisance) heterotrophs and algae.

11.6   Periphyton community change as an index of waste type

Although organic wastewater, such as sewage, can result in an increase in periphyton biomass with mono-species dominance (e.g. *Sphaerotilus* or *Cladophora*), certain species are typically characteristic of various levels of organic matter, resulting in a downstream community change or succession during natural degradation. Fjerdingstad (1964) has described such a successional response of the periphyton community to a gradient of sewage enrichment (Table 11.6). This successional response is similar to the Saprobien System first described by Kolkwitz and Marsson (1908) and has

Table 11.6 Water quality characteristics and the associated response of the microorganism community to organic waste

Zone	Chemical	Biological
Oligosaprobity (clean water)	$BOD < 3\,mg\,l^{-1}$ O_2 high Mineralization of organic matter is complete	Diatoms diverse Filamentous green algae present Filamentous bacteria scarce Ciliated protozoans scarce
Polysaprobity (septic)	H_2S high O_2 low NH_3 high	Algae present but not abundant Protozoa absent Bacteria abundant–faecal, saprobic
α Mesosaprobity (polluted)	Amino acids high H_2S low–none $O_2 < 50\%$ saturated $BOD > 10\,mg\,l^{-1}$	Algae scarce–some tolerant forms[a] Filamentous bacteria abundant[b] Ciliated protozoans abundant[c] Few species–biomass great
β Mesosaprobity (recovery)	$NO_3 > NO_2 > NH_3$ $O_2 > 50\%$ saturated $BOD < 10\,mg\,l^{-1}$	Diatoms not diverse–biomass great[d] Ciliated protozoans only present[e] Cyanobacteria abundant[f] Filamentous green algae abundant[g]
Oligosaprobity (clean water)	Stream recovered or 'purified'	

Sources: Fjerdingstad (1964); Sládečková and Sládeček (1963).

Notes
a *Gomphonema, Nitzschia, Oscillatoria, Phormidium, Stigeoclonium* often dominate.
b *Sphaerotilus, Zooglea, Beggiatoa.*
c *Colpidium, Glaucoma, Paramecium, Carchesium, Vorticella.*
d *Melosira, Gomphonema, Nitzschia, Cocconeis.*
e *Stentor.*
f *Phormidium, Oscillatoria.*
g *Cladophora, Stigeoclonium, Ulothrix.*

often been characterized as representing the process of 'stream purification'. However, the chemical concentrations are only relative. For example, according to the previous discussion on BOD and filamentous bacteria, problems could still be present in the recovery zone even if BOD is $< 10\,\mathrm{mg\,l^{-1}}$.

The length or presence of these zones in a stream depends upon flow rate (dilution), waste concentration and temperature. These factors, in turn, determine the dispersed concentration, its contact time and, therefore, the uptake rate of organic matter (i.e. BOD). A high flow rate and low temperature results in a low concentration and low uptake rate, so that a lesser effect (but possibly still a nuisance biomass) is usually stretched over a longer stream distance. With a low flow and high temperature, the uptake rate is greater so the BOD is depleted faster and the adverse effect occurs over a shorter stream distance.

Sládečková and Sládeček (1963) have described the succession of periphyton in a series of Czechoslovakian reservoirs polluted with sugar-beet waste and sewage. The combined effect of these wastes showed typical zones, which are illustrated (Figure 11.17). These zones are analogous to those in streams shown in Table 11.6. Three representative periphyton communities in the reservoirs, receiving various amounts and types of waste effluents (sugar beet and sewage), were collected on glass slides. The communities indicate different degrees of water quality change: community 51 = relatively 'clean' water conditions (oligosaprobity); community 52 = 'recovery' conditions (β-mesosaprobity); community 53 = 'polluted' conditions (α-mesosaprobity).

Note the lack of diatoms and preponderance of ciliated protozoans and filamentous bacteria in the 'polluted' water community (community 53). According to this system of evaluating water-quality change, the various species are grouped according to their position with respect to distance or travel time from a waste outfall and, therefore, to the amount of 'self-purification' that has occurred. The occurrence of these groups of species or communities is considered indicative of various degrees of organic 'pollution'.

Another useful method of evaluating the effects of organic wastewater is the heterotrophic or autotrophic index, HI or AI, respectively. AI is more appropriate from the standpoint of unit values and has been the more accepted terminology (Collins and Weber, 1978). The index is determined as the ratio of AFDW:chl a in the periphyton collected on artificial substrata and indicates the degree of effect of organic waste. AI increases in proportion to the concentration of organic matter (or BOD), because heterotrophs occupy a greater portion of the biomass as organic waste increases.

Collins and Weber (1978) suggested a value of 400 as the upper limit indicating clean water conditions. An interesting example of the effectiveness of the index was shown by Biggs (1989) for a spring-fed New Zealand river. Upstream and downstream stations were monitored before and after the diversion of slaughterhouse wastewater. Figure 11.18 illustrates the quick

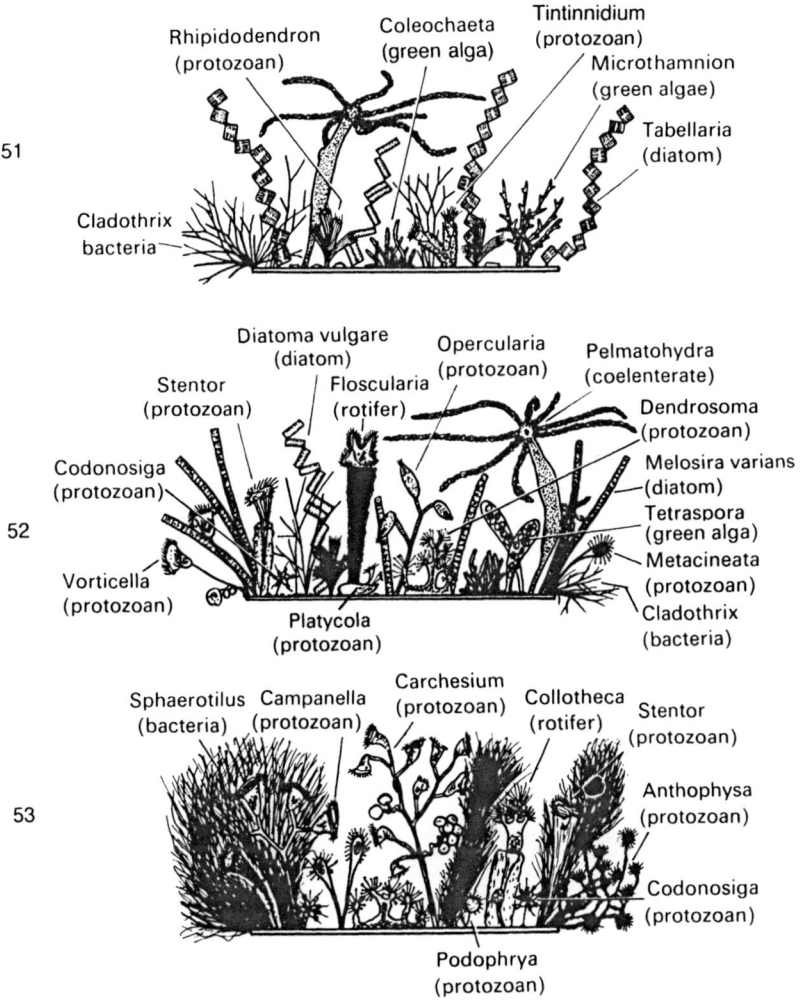

Figure 11.17 Illustrated pattern of periphyton response to waste in three Czechoslovakian reservoirs. See text for explanation (from Sládečková and Sládeček, 1963).

recovery (to below 400) of the periphyton community following the beginning of volume reduction of the discharge in May 1986 and complete diversion in August 1986. Prior to diversion, the AI downstream was >10-fold higher than the upstream control, but shortly after diversion values were similar. Biggs also found AI to be highly correlated with BOD_5 ($r^2 = 0.86$). Heterotrophs were dominated by *Sphaerotilus* and *Zooglea*.

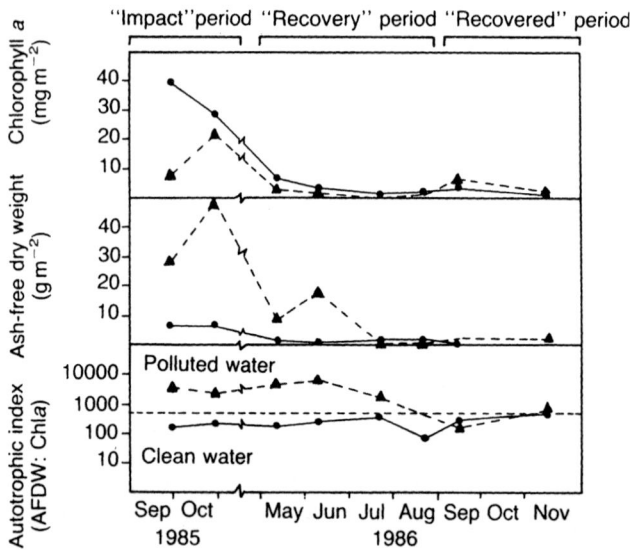

Figure 11.18 Changes in periphyton biomass (chl a, AFDW) before and after diversion of slaughterhouse wastewater from South Branch, Canterbury, New Zealand, Diversion was complete before August. Dashed line is recommended upper limit of AI for clean water (see text). ●, Control; ▲, affected section (Biggs, 1989).

11.7 Effects of toxicants

Toxicants may affect periphyton both positively and negatively, but periphyton is not usually investigated as an indicator of toxicity. Total biomass would normally be expected to decline downstream from the input of an inhibitor, but the tolerant species would become dominant and possibly increase in abundance. The increase in some species may result from nutrients released from dead animals. Grazers may be eliminated, which should allow a rapid biomass increase in algal periphyton as a result of no grazing once the toxicant has dispersed. Insecticides would probably produce such an effect, because their negative effect on algae would be minimal. For example, about a week after spraying with DDT to control forest insects, forest streams have become intensely green. The positive effect of insecticide on periphytic biomass has also been shown in artificial channel experiments (Eichenberger and Schlatter, 1978).

Metal toxicants have a more detrimental effect on periphyton. The addition of Zn^{2+} to artificial streams showed an increased reduction in the number of algal species over a concentration range of $0-9\, mg\, l^{-1}$ (Williams and Mount, 1965). Noteworthy in that experiment was the elimination of

Table 11.7 The effects of Cu waste on stream periphyton collected on artificial slides

Upstream station Cu^{2+} low	Periphyton density $1000\,mm^{-2}(3\,wk)^{-1}$ *Stigeoclonium, Nitzschia, Gomphonema,* *Chaemosiphon,* and *Cocconeis* were most abundant
Downstream from waste $Cu^{2+}\,1.0\,mg\,l^{-1}$	Density $150-200\,mm^{-2}(3\,wk)^{-1}$ *Chlorococcum, Achnanthes*

Source: Hynes (1960).

Cladophora at the lowest concentration tested, $1\,mg\,l^{-1}$. In this instance water was taken from a holding pond that also supplied a high concentration of particulate food; consequently, *Sphaerotilus* and fungal slime organisms increased as Zn^{2+} increased and algal species decreased. Without the dead organisms, little increase in decomposer biomass would have occurred.

Cu^{2+} is another heavy metal to which algae are very sensitive. $CuSO_4$ is a common algicide. Hynes (1960, p. 81) cited work by Butcher showing the effects of Cu^{2+} in industrial waste in an English stream (Table 11.7). Recovery of species diversity and biomass ($33\,000\,mm^{-2}$) occurred 8 km downstream from the point of discharge.

Diatom diversity was shown to be a sensitive indicator of toxic effects of Zn and Cu in a Montana stream (Prickly Pear Creek) which received leachate from abandoned mine tailing ponds (Kinney, personal communication; Miller *et al.*, 1983). Figure 11.19 shows the marked reduction in diatom taxa at the two downstream stations where Zu and Cu concentrations were elevated most above USEPA criteria determined for fish. Further downstream,

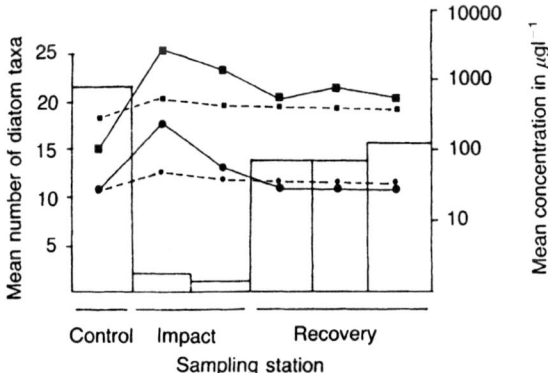

Figure 11.19 Mean diatom taxa related to stream concentrations of total Zn and Cu and EPA acute criteria for those metals in Prickly Pear Creek, Montana in 1980 (data from Miller *et al.*, 1983). ■—■, Zn concentration; ■---■, Zn criteria; ●—●, Cu concentration; ●---●, Cu criteria.

308 Effects of pollutants in running water

as metal concentrations dropped, taxa increased to near the level observed at the upstream control station.

11.8 Nuisance

Under proper conditions of light and temperature, which usually sets limits on nutrient-saturated growth rate, enrichment with either inorganic or organic nutrients promotes the dominance of certain prolific periphyton species and thus results in an increase in production and possibly a nuisance biomass. Whether such a biomass is considered to be an undesirable nuisance condition may depend on how the users of the stream are affected. Some of the adverse effects that may be caused by high periphyton biomass are as follows.

1 A secondary BOD is created that can deplete O_2 downstream as the filaments break off, float away and decompose.
2 Water intakes may be clogged with floating clumps of filaments.
3 Undesirable taste and odours can be created if the affected stream is used for a water supply.
4 Dense mats can cover the bottom, restricting intragravel water flow, which can inhibit fish reproduction.
5 Dense mats and long stringy filaments can interfere with fishing, recreation and aesthetic enjoyment.
6 Dense mats and filamentous clumps reduce the habitat for benthic animals and can cause direct physical damage.

The relative importance of these and other effects were determined for 103 cases of heterotrophic slime outbreaks in Ireland (Gray and Hunter, 1985) (Table 11.8). Although these effects on stream uses seem to prove that excessive biomass levels do in fact represent a nuisance, the increased production could be interpreted as a possible benefit under some circumstance if it leads to increased production of fish food (invertebrates) and fish. Examples of benefits were illustrated in Berry Creek, Oregon, where sucrose added at concentrations from $1-4\,\mathrm{mg\,l^{-1}}$ produced mats of *Sphaerotilus* and dense populations of midges (Warren *et al.*, 1964). The diversity of both periphytic microscopic and macroscopic organisms decreased, while mats of heterotrophic slimes, which may not be aesthetically pleasing, and fish production nonetheless increased. Table 11.9 shows the results of annual biomass levels of benthos, fish-food consumption and fish production in energy units in enriched and unenriched sections of the stream.

The production of cut-throat trout appears to have been enhanced in this case from the enrichment, in spite of other adverse effects of the mats of *Sphaerotilus* on macroinvertebrate diversity. The insect biomass was composed mostly of chironomids. From these results one could argue that the

Table 11.8 The relative importance of various adverse effects of slime growths in Irish rivers ($n = 103$)

Effect	%
Appearance and amenity	86.4
Smell and deoxygenation	46.6
Damage to fish	37.9
Sloughed flocs	24.3
None	12.6
Smell only	11.7
Deoxygenation only	9.7
Habitat destroyed	3.9
Spawning ground smothered	2.9
Possible public health risk	2.9
Blockage of flow	1.0
Agricultural supplies affected	1.0
Public supplies affected	1.0
Slime growth on fauna	1.0

Note
From Gray and Hunter (1985).

Table 11.9 Insect biomass, fish-food consumption, and fish (cut-throat trout) production in artificially enriched ($1-4\,\mathrm{mg\,l^{-1}}$ sucrose) and unenriched sections of Berry Creek, Oregon

	Annual values (kcal m^{-2})				
	Unenriched (U) dark	Unenriched (U) light	Enriched (E) dark	Enriched (E) light	E:U
Insect biomass					
1960–61	2.19	5.03	20.4	12.2	4.5
Fish-food consumption					
1961	8.38	6.06	20.15	15.64	
1962	9.46	8.09	19.07	21.55	
1963	7.45	8.36	13.05	9.00	
Mean	8.43	7.50	17.42	15.40	2.1
Fish production					
1961	−0.21	0.01	2.13	2.20	
1962	0.49	−0.07	3.70	4.80	
1963	0.58	0.99	2.51	1.65	
Mean	0.29	0.31	2.78	2.88	6.3

Source: Modified from Warren et al. (1964).

Note
Values are totals for a portion of each year.

benefit of increased fish production met the goal for sound water-quality management; the ecosystem was used but not abused. However, the following points must be considered in similar cases.

1. The knowledge base necessary for 'fine tuning' the discharge of wastewater to allow 'benefits' without 'detriments' for a variety of ecosystems receiving a variety of wastes.
2. The costs and problems of monitoring stream conditions to provide data for fine tuning.
3. The instability risks of intense management.

Considering the last point, even if a fine-tuned system is achieved, it may be highly susceptible to environmental changes. For example, trout are known to grow very fast in highly productive areas, but with associated temperature, pH, NH_3 and O_2 conditions that may become marginal at times. For Berry Creek, slight changes in weather patterns, such as cloudiness and/or high temperature, could cause total elimination of the highly prized, fat trout. Moreover, the trout were planted as adults and whether or not reproduction was successful in the enriched portions of the stream was not determined. For example, Smith *et al.* (1965) showed interference from *Sphaerotilus* on reproduction by walleye pike. Thus, total cost may be less if waste inputs were minimized rather than attempting the objective of use without abuse.

Chapter 12
Benthic macroinvertebrates

The benthic macroinvertebrates of running waters include many orders of organisms that represent an interesting, usually unappreciated part of the aquatic environment. Benthic fauna in standing waters, which will be briefly discussed in this chapter, are relatively unnoticed as well. The mere existence of large populations of strange-looking nymphs, naiads and larvae of insects within the rubble and gravelly substrata of streams is fascinating. That so few people know of their existence is a pity. They have an important function in the aquatic food web and, as such, the structural composition of the macroinvertebrates will reflect changes in the quality or quantity of energy input.

The macroinvertebrates include grazers, detrital feeders and predators. They process and utilize the energy entering streams from either autochthonous periphyton production or from allochthonous sources, such as leaves, needles and other particulate matter from the forest, or organic wastes from humans (e.g. sewage) or other animals in the watershed. Macroinvertebrate species can be separated into feeding types, such as large-particle detritivores (shredders), small-particle detritivores (collectors and selectors), grazers (periphyton scrapers) and predators. Furthermore, the representation of the community among these groups is distinctive, as streams change naturally from upstream, low-order and usually heterotrophic, to autotrophic in mid-regions and ultimately back to heterotrophic, high-order lower downstream (Cummins, 1974). In lakes or the sea they are largely dependent upon the autochthonous or allochthonous production that is sedimented to the bottom. A diverse community of macroinvertebrates is instrumental in stream purification – the processsing of organic matter from either human or natural sources that leads ultimately to CO_2, water and heat. As noted in the preface, the efficiency with which that process proceeds depends to a large extent upon the diversity of the community – the more individual niches that are occupied, the quicker and more completely will be the conversion process. Overloading streams with organic enrichment or the introduction of toxic substances will tend to decrease the diversity and thus the efficiency of 'purification'.

The macroinvertebrates of freshwater are dominated by the insects and their diversity is greatest in running water. The orders Ephemeroptera (mayflies), Plecoptera (stone flies), Trichoptera (caddis flies), Diptera (true flies) and Odonata (dragonflies and damselflies) usually make up the majority of the biomass. Because of the abundance of species – 50–100 is not unusual in a running water environment – their identification to the species level is difficult. Other important groups in freshwater are the Mollusca (snails and clams), the Annelida (worms and leeches) and the Crustacea (scuds, sow bugs and crayfish). In the marine environment the latter groups represent the community to the practical exclusion of the insects. In brackish water environments there is considerable overlap and some of the more tolerant freshwater species of insects can be relatively important. See Appendix A for a classification system for freshwater macroinvertebrates.

The principal purpose in this chapter is to treat the macroinvertebrates as indicators of the kind and relative magnitude of waste entering aquatic systems. As such, they offer advantages not available from any other group. They are sedentary and relatively easy to sample compared to the highly motile fishes, and because their longevity is greater, fluctuations in biomass and species composition are less pronounced than in plankton. Therefore, the macroinvertebrate fauna can readily reflect changes in the input load or type of waste even when sampled as infrequently as monthly or bimonthly. Semiannual sampling is often adequate to detect significant changes.

The life history of the macroinvertebrates includes either three or four stages in the case of the insects; egg, naiad (or nymph) and adult, or egg, larvae, pupa and adult. Although completion of more than one life cycle in a year is not uncommon, particularly with the midges, many require one or more years for completion (Usinger, 1956). The Oligochaeta, Mollusca and Crustacae have two- or three-stage life histories and have one or more generations per year. Most of the insect's life is spent in the immature stage, with the adult terrestrial portion usually devoted to the purpose of reproduction. Therefore, bottom substratum samples largely contain the naiads and larvae of insects, and major changes occur in the species composition through the spring and summer because emergence time varies among the populations with photoperiod and temperature. Natural community changes can be identified apart from the effect of wastes by previous study or with a representative control area. The mere presence of a variety of long-lived species that remains relatively stable in diversity and biomass from one month to another means that water quality has remained relatively unchanged. Any alteration in water quality, even though brief, can readily be reflected in that community. If decimated in the interim, the only replenishment source for the remainder of that life cycle is from tributaries or upstream. Repopulation can be relatively fast as a result of drift. As indicators of water quality, together with their inherent interspecies variability in tolerance to the variety of wastes that enter aquatic ecosystems, the macroinvertebrates are extremely valuable.

12.1 Sampling for benthic macroinvertebrates

12.1.1 Quantitative estimates

The existing standing crop and species representation of stream invertebrates can be sampled with several types of units, including the following.

1. Surber square foot sampler is most often used in shallow areas.
2. Standard area enclosure (heavy metal frame) with a screen is often used in riffle areas of deeper streams.
3. Hess sampler is a cylindrical unit used in deep or shallow riffles and is more efficient than the above two units.
4. Peterson and Ekman grabs are used either in deep pools in rivers or in lakes and the Ekman is restricted to small-particle substrates.

None of these samplers is representative for all habitats (riffles, pools, shorelines, etc.). The size of the area sampled must be determined by the density of organisms so that a minimum sample size is possible with minimum sorting effort. Selectivity of certain samplers for different organisms and size must be considered. For example, the Surber selects for those organisms that wash from rocks easily and are large enough to be caught with a net. Small and newly hatched larvae and naiads are easily lost.

Artificial substrata have been used with considerable success. Multiplate samplers, which are similar to slides for periphyton, are one type (Hester and Dendy, 1962). Wire baskets filled with natural rubble substrata and buried at each station have been successful (Anderson and Mason, 1968; Mason et al., 1970). Some species are probably missed by these methods, particularly less mobile ones. However, comparisons have shown good agreement with dredge collections (Anderson and Mason, 1968). Also, more individuals are usually collected with the artificial units. Of distinct advantage is the comparison among stations by using the same type substratum. Nevertheless, some analysis of the naturally occurring community structure should be included to determine what portion of the community is sampled with artificial substrata.

12.1.2 Qualitative estimates

Hand nets can be used in weedy areas, along mudbanks and in debris or in places where it is difficult to use any of the above quantitative samplers. A visual search on protruding substrata such as logs, large rocks, sticks etc. is often useful.

Although these procedures give occurrence and therefore contribute to a species list, they cannot give quantitative data on relative abundance and thus are not as useful for communicating findings and statistical analysis.

12.2 Natural factors effecting community change

Substratum and current are important factors in selecting kinds and quantity of macroinvertebrates. Macon (1974), with respect to the effect of current on macroinvertebrate community type, grouped velocities as very swift ($>100\,\text{cm s}^{-1}$), swift ($50-100\,\text{cm s}^{-1}$), moderate ($25-50\,\text{cm s}^{-1}$), slight ($10-25\,\text{cm s}^{-1}$), and very slight ($<10\,\text{cm s}^{-1}$). Current erodes substrata composed of either rocks, stones or gravel, which interact to determine the kinds of organisms in the community (Hynes, 1960). Current determines sediment transport. The greater the current, the larger the substratum type (gravel to boulders) and the freer the interstices within the substratum of fine sediments; as current decreases, deposition increases and the interstices tend to fill up. Physical characteristics of the two types of environments, running and standing water, and associated community types are compared.

12.2.1 Running water

Eroding substrata is the rule in running water, particularly in the upper reaches of streams (low order). Species are usually numerous and of many groups, including worms, leeches, snails, clams, insects, and some crustaceans (scuds), but insects usually dominate the community. In such an environment the substratum is usually clean, interstices are free of sediment and animals are adapted in characteristic ways: limpets adhere to smooth stone surfaces, insects have sharp claws (stone flies, caddis flies, and mayflies) to cling to stone surfaces. Some are dorsoventrally flattened so that the whole body fits firmly against stones (Heptageniidae). Many species, particularly stone flies, do not tolerate sediment because it fills interstices within the substratum, which hinders attachment for herbivorous (or predaceous) feeding and restricts hiding space. Dense algal growth on rocks tends to clog interstices, thus producing a surface unfavourable for flattened forms but favourable for midges and small mayflies. Snails and leeches attach to relatively clean rocks. Many of these forms can tolerate velocities from $100-200\,\text{cm s}^{-1}$ (Macon, 1974, p. 136).

An example of how the type of eroding substrata affects density, particularly as the substratum becomes more fixed, is shown in some results from English streams (Table 12.1). The number of taxa varied from 11 to 25 but showed no special trend with substratum type. However, obvious species selection did occur within these types of eroding substrata. For

Table 12.1 Fauna abundance in English streams versus substratum type; fauna abundance in an unpolluted African river in summer

Substratum type	Abundance (No. m^{-2})	
English streams		
Loose stones	3316	
Stones embedded in the bottom	4600	
Small stones with fine gravel	3375	
Blanket-weed on stones	44 383	
Loose moss	79 782	
Thick moss	441 941	
Pondweed on stone	243 979	
African river	Coarse sediment	Stones
Eroding substrata	6710	4730
Stable depositing	12 590	7570
Unstable depositing	4450	6660

Source: English streams: Hynes (1960), Table 2 data from Percival and Whitehead (1929); African river: Chutter (1969), Tables II and VI.

example, *Rhithrogena* (mayfly) preferred stones free of vegetation, whereas *Hydropsyche* (net spinning caddis) was mostly found on periphyton-covered stones. Midges comprised from 40 to 54% of the fauna on the four vegetation-covered substratum types, but only 5–17% on the other three types. Diversity (ratio of species number to individuals) decreased considerably as density increased. Typical representatives of the depositing and eroding substrata environments were clearly illustrated by Hynes (1960, pp. 30–31).

Depositing substrata, where interstices are filled because of reduced current, is typical in lower reaches of streams where sediment accumulates. This can also occur from watershed or in-channel erosion or sediment containing waste. In such an environment burrowers become dominant, e.g. worms which live under the sediment surface, build tubes and feed selectively on detritus, as well as insects that burrow (midges and ephemerid mayflies). Without a clean hard substratum to adhere to and living space among sediment-free rubble, clingers disappear. Midges, most of which are tube builders and net spinners, become abundant. Also detrital-feeding clams occur. However, abundance of individuals may not be greatly reduced in depositing substrata compared to eroding substrata.

Stability of substrata may have more influence on organism density than sedimentation (Table 12.1). For example, the hydraulic effects of urban stormwater runoff may have a greater effect on macroinvertebrate abundance than do pollutants in the stormwater including sediment. The increased area of impervious surface in the watershed (streets, houses, etc.)

results in much higher peak flows than the stream had experienced resulting in substantial substrata movement.

Biggs et al. (1999) found that invertebrate abundance was fairly uniform until an intermediate frequency of substrata disturbance occurred (approximately 10 bed-moving events per year); then densities decreased similar to that of periphyton biomass (Figure 12.1). Furthermore, invertebrates controlled periphyton biomass in streams with low frequency of bed movement but had little or no effect in highly disturbed streams.

Substrata stability and invertebrate species richness also appear to be strongly correlated, with highest richness in intermediately disturbed habitats and lowest in undisturbed and severely disturbed habitats (Townsend et al., 1997). These findings have been explained by the 'intermediate disturbance' hypothesis, which proposes that ecological communities (such as stream benthos) seldom reach equilibrium because disturbances continually re-set the process of competitive elimination by opening space for colonization by less-competitive species, preventing domination by competitively superior species. Some velocity-sensitive species may find flow refugia during a disturbance event and thus move back to substrata surfaces, re-joining the velocity-resistant taxa following the disturbance (Townsend et al., 1997). If a severe disturbance destroys substrata refugia, most taxa may be eliminated

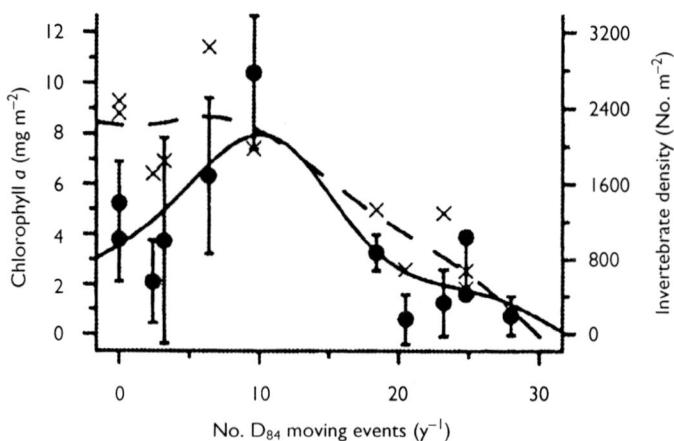

Figure 12.1 Mean monthly densities of invertebrates as a function of the frequency of bed sediment movement (up to D_{84} size fraction). Also shown are the mean monthly chl a concentrations (± 1 SE) from the stones on which the invertebrates resided (●, periphyton biomass; ×, invertebrate density) (from Biggs et al., 1999, reproduced with kind permission from the Journal of the North American Benthological Society).

resulting in low-diversity communities dominated by taxa with high growth rates (e.g. some mayflies). Thus, invertebrate community structure in streams could be largely controlled by sediment stability and availability of flow refugia (Townsend and Hildrew, 1994).

12.2.2 Standing water

Lakes are represented in general by two types of substrata, which relate to trophic state. An oligotrophic lake without an accumulation of sediment in the littoral region has clean rocks and fauna similar in some respects to that of eroded substrata in streams, e.g. stone flies and case-building caddis, but few of the burrowing types. Where fine sediment prevails at depth, the community is composed of largely midges, clams and worms. Eutrophic lakes and reservoirs with macrovegetation and sedimented littoral regions harbour a varied fauna. In the depositing substrata community in such an environment, whether at depth or in shallows, midges usually dominate and some species of clams and worms occur, and densities are much higher than in oligotrophic lakes. A varied fauna occurs in and on the vegetation, including insects represented by stone flies, mayflies, dragonflies, beetles and midges.

12.2.3 Experimental controls

To monitor and detect effects in receiving waters resulting from waste discharges, a 'control' must be selected to account for the natural variations caused by current, substrata, elevation and temperature. Control stations must be located in areas unaffected by the waste but comparable in substratum and current to the test (waste-affected) stations. Otherwise, a significant difference in mean densities, species abundance or diversity between test and control cannot be interpreted as clearly an effect of only the waste. If control stations are physically comparable with test stations, then interpretation of effect can be rather straightforward. If controls are not comparable, conclusions must be qualified.

If control stations are unavailable, i.e. current, substratum and elevation at stations unaffected by waste are not similar to test stations (or if there is no unaffected upstream section), a control period may be required before waste is introduced to receiving waters. Such a period must be sufficiently long to include normal variability. One year is usually a minimum, but two years of data are much preferred to allow for variations in temperature and flow.

Because elevation difference usually causes emergence differences within insect groups, the interpretation of the presence or absence of such a group must not be mistaken for an elevation-emergence effect.

12.3 Oxygen as a factor affecting community change

Certain species and even whole groups of macroinvertebrates show various levels of tolerance to DO (dissolved oxygen). The groups in which most species are usually intolerant of low DO are the stone flies, mayflies and caddis flies (Gaufin and Tarzwell, 1956). Thus, a diverse representation among these orders indicates clean water, especially with respect to organic waste.

However, even within these groups many species can survive for a limited period of time at concentrations as low as 1 mg l^{-1} (Macon, 1974). Resistance to low DO can be partly due to morphological adaptations for maintaining a current over the animal's gills, such as with most mayflies (Hynes, 1960).

At the other extreme, there are a few organisms whose special physiological and morphological capabilities allow them to tolerate or resist low DO for a considerable period. These include the air breathers: the moth fly (*Psychoda*), mosquito (*Culex*), rat-tailed maggot (*Eristalis*), pulmonate snails (*Physa* and *Limnaea*), beetle, and adult water boatman and back swimmers. Some tubificid worms (*Tubifex* and *Limnodrus*) and chironomid midges (*Chironomus*) have haemoglobin, which can increase in oxygen-poor conditions. A high glycogen content and reduced activity allow these organisms to withstand prolonged oxygen minima (even anoxia) by meeting reduced metabolism with an aerobic glycolysis. Haemoglobin has a greater affinity for O_2 and probably loads up more easily under lower O_2 tension than haemocyanin. The success of tubificids, particularly *Tubifex tubifex* and *Limnodrilus hoffmeisteri*, in environments that become anaerobic for extended periods, partly results from their rapid reproductive rate. They can quickly repopulate the area once the anaerobic period has passed. The same can be said for the tolerant chironomids (Brinkhurst, 1965).

Those species with an intermediate tolerance to low DO include the crustaceans, *Asellus* and *Gammarus*, blackflies (*Simulium*), some leeches (Hirudinea), and craneflies (*Tipula*), and *Sphaerium* and *Pisidium* clams may be abundant in low-oxygen environments. Some mayflies are adapted to low tensions with the consumption of O_2 remaining rather constant down to low levels. Figure 12.2 shows curves for *Cloean* and *Leptophlebia* (Fox *et al.*, 1935; Macon, 1974). Thus, these two species can show normal activity down to very low O_2 levels, whereas one *Baetis* species failed to survive at low concentrations.

As indicated earlier, some mayflies tolerate sedimented conditions under which they can maintain adequate circulation of water over gills. *Cloean* (above) has such a morphological advantage. Because some caddis flies (*Hydropsyche*) can undulate and thereby increase circulation, they are found in areas of intermediate DO. Dragonflies and damselflies (Oonata) are often found in silty conditions. They have highly developed rectal and caudal gills, respectively, as well as other structures and thus are able to tolerate low DO for short periods (Gaufin and Tarzwell, 1956).

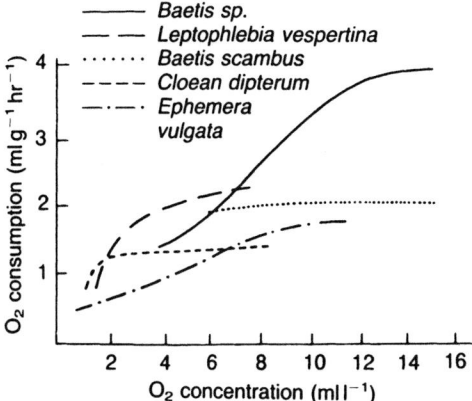

Figure 12.2 Respiration rate (O_2 consumption) versus O_2 concentration in four ephemeropterans (Fox et al., 1935).

The demand for DO has been found to vary greatly within the so-called intolerant groups, the stone flies, mayflies and caddis flies (Figure 12.3). Thus the concept of group tolerance is only very generally valid. The species variation among the five mayflies shown in Figure 12.2 also illustrates the difficulty in generalizing about group tolerance. On the other hand, oxygen consumption rates of the individual species, according to Olson and Rueger (1968), correspond well with their observed distribution in organically polluted streams. In order to use the occurrence or absence of aquatic macroinvertebrates as an indication of oxygen availability, at least the genera, if not the species, level must be identified because of the variability among species. This point will be taken up later.

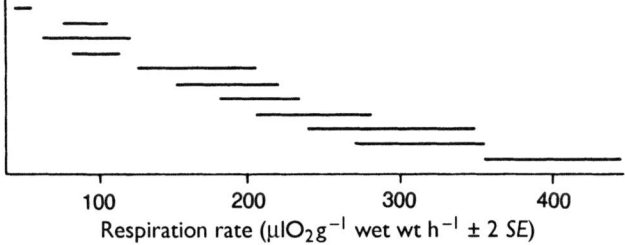

Figure 12.3 Respiration rates of aquatic insects determined at 20°C. Insects indicated in figure in descending order from left to right are *Tipula* (crane fly), *Caloteryx* (dragonfly), *Limnephilus* (caddis fly), *Pteronarcys* (stone fly), *Hetaerina* (dragonfly), *Paragnetina* (stone fly), *Macronemum* (caddis fly), *Ephemera* (mayfly), *Potamanthus* (mayfly), *Baetisca* (mayfly), *Leptophlebia* (mayfly) (modified from Olson and Rueger, 1968).

Experimental evidence indicates that no single critical level of DO exists for macroinvertebrates in streams (Macon, 1974). The incipient limiting levels vary even within a family from rather low to quite high concentrations and vary even more with current velocity. For organisms adapted to high velocities (*Rhithrogena*), the lethal limit for DO increased greatly at low velocity (from 3 to 6 mg l^{-1}); but for *Ephemerella*, which is morphologically capable of increased circulation at low DO and/or velocity (Ambühl, 1959), the limit remained rather low and constant (1–2 mg l^{-1}) over a wide range of velocity.

12.4 Temperature

The range of tolerance for each critical stage in an organism's lifetime can be quite different for each species. Thus, an alteration in environmental temperature can result in changed community composition even if the ultimate lethal temperature in summer is not exceeded for any species. That is, a species can be eliminated from the community without mortality of adults. The replacement may result from an altered ability to utilize food resources or to reproduce and thereby compete for living space. A general tolerance pattern for a hypothetical species is shown in Figure 12.4.

Although the range for survival of adults is quite large, the optimum range for growth usually occurs at a lower temperature. The reproductive stage and the early free-living stage in the development of the young are therefore critical to population survival. Gradual shifts in species dominance can thus occur as the seasonal pattern of daily mean temperature changes from normal. This has been shown for quite small increases in temperature. Tolerance is apparently great for relatively wide diurnal fluctuations.

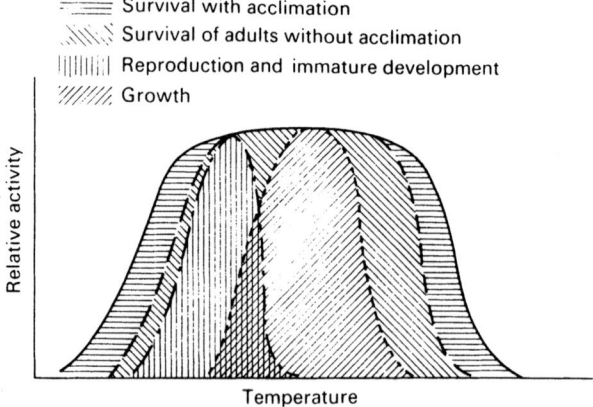

Figure 12.4 Generalized temperature tolerance graph for the important life stages of an invertebrate.

Although most of these principles have been demonstrated more thoroughly for fish, they probably also hold for most macroinvertebrates. In the determination of effect, the important factor associated with temperature change is time. A very high temperature can often be tolerated for a short time, whereas long-term survival can be guaranteed only at usually much lower temperature. Thus, when stating tolerance limits for any life stage of a given species the time of exposure must be included. Other factors that are almost as important include time and temperature for acclimation, food availability, DO, level of activity, age or developmental stage and presence of inhibitory substances.

12.4.1 Temperature criteria – rate of rise

Aquatic organisms that are entrained and pass through power plants may experience a relatively large and rapid rise in temperature (i.e. without cooling towers/ponds); the entrained water can undergo the increase in temperature in a matter of minutes. Some evidence suggests that invertebrates can tolerate a relatively high $\Delta T/t$ (increment rise above ambient per unit time). Direct observations of the condition of entrained zooplankton show that rate of rise is not detrimental unless the final temperature reached equals or exceeds the acute lethal level. This is indicated by results summarized in Table 12.2. The lethal temperature for zooplankton seems to be the same whether subjected to abrupt or gradual temperature increase (Heinle, 1969a; Welch, 1969b). There was little difference observed in the 24-h TL_{50} (temperature at 50% survival) for *Eurytemora affinis* whether previously acclimated for months at 15, 20 or 25°C. A similar result was found with *Mysis relicta* (Smith, 1970), although the lethal temperature for that cold-water species was 16°C.

12.4.2 Temperature criteria – increment rise above ambient

The total increment rise above ambient temperature (ΔT) can be important because reproduction in benthic animals in temperate areas is triggered

Table 12.2 Zooplankton mortality compared to the maximum temperature and the ΔT

	Mortality (%)	Maximum	ΔT (°C)
Chalk Point Plant, Chesapeake Bay[a]	90	37	6.4
British plant on Thames[b]	0	24.4	7
Paradise Plant, KY[c]	100	35.5	9

Notes
a Heinle (1969b).
b Markowski (1959).
c Welch (1969b).

primarily when an appropriate temperature level is reached rather than by a daily rate of rise. For the most part, temperature level seems to be more important than photoperiod. Moreover, wide diurnal fluctuations in temperature can apparently be tolerated, whereas small changes in daily means cannot. As an example, a 7-year study in a British pond showed that the dominant mayfly emerged when the temperature in the spring reached 10–11°C, which occurred within a 2-week period in April in spite of what seemed to be a wide variation in weather conditions from year to year. The slow response of water bodies to climate change evidently makes water temperature a safer cue for aquatic animals than for terrestrial animals.

The seasonal changes (low temperature in winter and high in summer) are thus very important. The life cycles of aquatic animals are geared to seasonal temperature changes as well as to light. Crustaceans, such as the crayfish, must go through a low-temperature stage in which they do not moult but rather put their energy into reproductive cell development. Crayfish held in a cooling pond that was warm all year did not stop moulting and growing and did not reproduce until the winter temperature was reduced. The same response has been suggested for insects that responded with lower adult longevity and poorer emergence if held at constant but sublethal temperatures (Nebeker, 1971a,b). Also, emergence may occur as much as five months early in heated stream sections. This would be detrimental to adults in temperate climates (Nebeker, 1971b). A similar increment rise above ambient may not have as important an effect on life cycles in tropical climates, because natural temperature changes are relatively small.

12.4.3 Temperature criteria – daily maximum

During the period of high temperature, most damage results if the maximum reached exceeds the tolerable level regardless of the increment rise above ambient (i.e. the increment added to the normal river temperature). This is illustrated in Figure 12.5 for a hypothetical constant flow stream with a heated-water addition. There would be an additional variation about the daily mean that is not shown.

Heat addition to the Thames River, UK resulted in a 12°C rise above the ambient daily mean with a maximum of 28°C. No change in the number of species present was observed, but the abundance of leeches, scuds (*Gammarus*) and midges decreased, whereas snails and clams increased (Mann, 1965). The same rise above ambient in the Delaware River, USA (12°C), but with a maximum of 32–35°C, resulted in an extensive reduction in the number of species as well as total abundance (Trembly, 1960). Although the area repopulated in winter, the high temperature reached in summer was nevertheless the damaging factor.

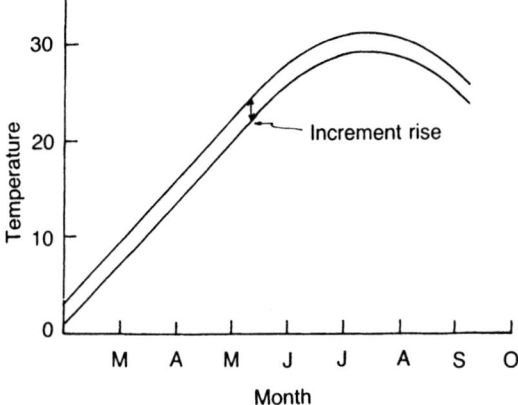

Figure 12.5 Increment rise above ambient temperature in a hypothetical constant flow river receiving a constant flow and heated effluent.

12.4.4 Community shift versus temperature level

Evidence suggests that moderate tolerance by aquatic freshwater invertebrates in general exists up to about 30°C. Figure 12.6 shows that the mode for the lethal temperatures of a large number of invertebrate species occurs at ≈35–40°C using available data. Allowing that a plotting of the preferred temperature (if available) would put the mode somewhat lower, say ≈30–35°C, means that a temperature increase to ≈30°C could result in an increase in species diversity in some environments, but that increases to levels >30°C would probably result in decreases in diversity. This can be

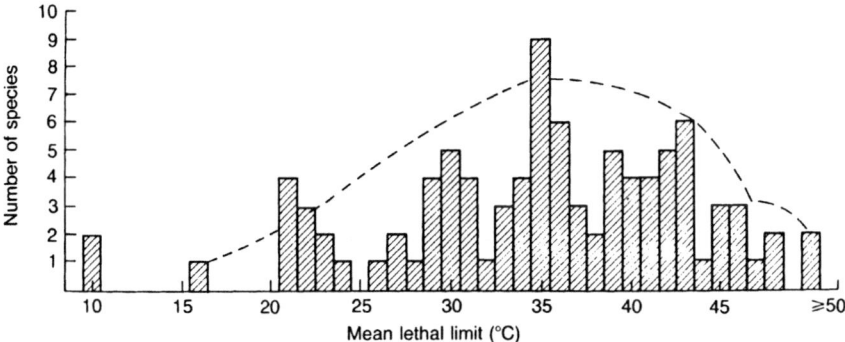

Figure 12.6 Distribution of freshwater invertebrate species tolerance to temperature. Dashed line shows approximate mode. (Bush et al., 1974 with permission of the American Chemical Society).

Table 12.3 Observed tolerances of macroinvertebrates

	Limit for normal diversity and abundance (°C)	Upper limit for highly tolerant forms (°C)
Delaware River[a]	32	37
Chironomid midges[b] ($n = 8$)	28	34
Caddis flies[c]	28	35

Notes
a Trembly (1960).
b see text.
c Mihursky and Kennedy (1967).

supported by observations summarized in Table 12.3. The Delaware River case was cited previously where the decrease in species and individual abundances was greatest above 32°C (Trembly, 1960). TL_{50} values for seven species of chironomids ranged from 29 to 39°C (Walshe, 1948). The more tolerant species were from lentic habitats indicating adaptation to warmer conditions. The optimum for normal activity, such as feeding, development and emergence, was 28°C for *Tanytarus dissimilis*, which has a TL_{50} of 33°C. This represents the median for the total of eight species tested. If the 5°C difference between optimum and the TL_{50} were typical, the upper limit for the most tolerant species of chironomids ($TL_{50} = 39$) may be ≈ 34°C.

The above limited sources of information suggest that to maintain normal diversity and abundance of most aquatic invertebrates the mean daily temperature should not exceed 30°C. However, this does not mean that the same species would necessarily be present up to 30°C.

For additional evidence to support the 30°C level, a summary of temperature tolerances of 13 invertebrates and vertebrates in Chesapeake Bay is shown in Table 12.4. In this instance damage in terms of species reduction would presumably occur at temperatures >30°C.

In the marine environment, the maximum tolerable limit of mean daily temperature is no doubt lower than in freshwater because oceanic temperature is more stable and the ambient temperature does not reach the high

Table 12.4 Predicted species in the suboptimal range and those lost with increases in temperature in Chesapeake Bay

Temp (°C)	Species in suboptimal (%)	Species lost (%)
26.7	0	0
29.4	8	0
32.2	61	8
35.0	16	69
37.8	15	85

Source: Mihursky (1969).

levels, at least in temperate oceans, that occur in freshwater. However, studies of thermal effects are scarce in the marine environment. Adams (1969) cites results from a study of a power plant on Morrow Bay, California, USA that showed predictable shifts from cold-water to warm-water invertebrate species with a temperature increase. A rise above ambient of 5.5°C was observed at the surface as far away from the discharge as 150 m. At that point 54 species were present, 39% of which were warm-water types, which was considered normal for the area. At 90 m, of the 34 species observed, 67% were warm-water types; near the heated-water discharge only 21 species were observed, 95% of which were warm-water types.

Compared with the freshwater and estuarine examples of heated-water effects cited earlier, a 5.5°C increase seems rather small. Nevertheless, the above examples suggest that marine communities may be very sensitive to temperature change.

The importance of such shifts in species as indicated for freshwater and marine environments depends upon the nuisance aspect of the invading species and the economic importance of the species affected. A shift to warm-water species may be undesirable because generally warm-water forms grow fast and mature early but live short lives, whereas cold-water forms grow relatively slowly and mature late but live long lives. Nebeker (1971a) has shown for several insect species that even though feeding activity was high over a wide range (15–30°C) emergence was increasingly less successful and the longevity of adults was shortened at temperatures much above 15°C. Thus animals adapted to lower temperatures may be expected to reach larger size because they live longer. Although productivity would tend to be greater for warm-water species, the efficiency for energy transfer in the system would possibly be reduced and its overall stability impaired.

The potential effects of temperature increase due to global warming on stream benthic invertebrates were investigated by Hogg and Williams (1996) in a first-order stream near Toronto, Ontario. They divided the stream longitudinally into two channels and raised the ambient water temperature by 2–3.5°C in one of the channels consistent with global warming predictions for that region. Decreased densities of animals (especially chironomids), increased growth rates and earlier onset of emergence were found in the elevated-temperature channel relative to the control. In addition, smaller size at maturity and precocious breeding in the amphipod *Hyalella azteca* and altered sex ratios in the caddis fly *Lepidostoma vernale* were documented with increased temperature. Hogg and Williams (1996) concluded that invertebrate life history characteristics may be more sensitive indicators of small, gradual temperature change than are metrics such as community composition, biomass and species richness. Prediction of the consequences of elevated temperature on aquatic invertebrates is difficult due to our limited understanding of the genotypic and phenotypic variability within species, which allows them to adapt to altered conditions (Sweeney et al., 1992).

12.5 Effect of food supply on macroinvertebrates

Benthic community composition depends upon the type and availability of food as well as on the chemical/thermal conditions. If the type and/or availability of food changes, the community will respond even if pronounced changes in chemical characteristics or temperature do not occur. Although difficult to generalize, the following indicates how changes in food supply could cause community shifts:

1 Detrital feeders can include the net-spinning caddis flies, the aquatic sow bug, chironomids, clams, snails, most mayflies, some stone flies, and blackflies. These are largely collectors of fine particulate matter. Most of the mayflies, caddis flies and stone flies eat both detritus and algae, but usually detritus feeding is more typical (Grafius and Anderson, 1972).
2 Grazers of periphyton include some mayflies, case-building caddis flies, some beetles and snails.
3 Predators are represented by the dragonflies, some leeches, a few stone flies, some beetles, some midges, alder flies and a few caddis flies.

For example, a change in the food supply could affect stream invertebrate composition in the following way. In a stream of low productivity the amount of detritus resulting from primary producers is low, thus grazers utilizing periphyton on substratum surfaces are expected to be most abundant, with fewer collectors designed to utilize detritus accumulating on surfaces from autochthonous production. Now, if the amount of detritus is increased through inorganic nutrient or organic waste input, detrital feeders would be expected to flourish downstream of those inputs at the expense of grazers. However, grazers would be expected to respond in the immediate area of enrichment in response to increased autotrophic production. Such an increase in detritivores would be particularly true in the case of fine particulate inputs such as sewage. If poor water quality resulted from the increased production, such as low DO, and thus limited predators, which ordinarily kept prey populations low, the tolerant detritivore prey species could reach a large biomass in the predator's absence. Such an interaction has been noted with leeches and tubificid worms.

A community dominated by shredders and collectors (large and small particulate detritivores, respectively) in low-order streams with forest cover (with $P/R<1$) will normally shift to one dominated by grazers and collectors in higher-order streams with less shade and more autotrophic production (with $P/R>1$). This has been described as a probable outcome for streams not receiving cultural sources of enrichment (Cummins, 1974; Vannote et al., 1980).

12.6 Effect of organic matter

12.6.1 Natural inputs

Natural inputs of allochthonous organic matter can change community structure and actually result in species increases if the physical–chemical changes are not severe and diversity was previously limited by a scarcity of food. Terrestrial sources of organic matter are leaves, needles and other forms of detritus which furnish the food supply to low-order streams as indicated above. However, moderate influxes of other types of organic matter will also change communities by increasing species richness. This was shown in results from a Californian mountain stream that received particulate organic matter input (algal detritus) from a lake. At the point where particulate matter (seston) peaked, due to enrichment from the lake, the number of species increased. Whereas 5 species were detected upstream from the lake, the number increased to 11 and 14 at two points downstream. Here the detrital-feeding blackflies (Simulidae) were very abundant and midges (chironomids) increased in abundance. These are the 'collectors' of detritus. Simulids removed an estimated 60% of the suspended plankton algae in 0.4 km of stream distance from the lake (Maciolek and Maciolek, 1968). The species number of oligochaetes (worms) can also be low if organic enrichment is low and increase with further enrichment (Milbrink, 1980).

Similar circumstances can occur in the tail waters from reservoirs without degrading stream quality. Productive stream fisheries have resulted from the increased plankton detritus produced in the reservoir that subsequently stimulates the collector insect community downstream. On the other hand, dense mats of periphytic algae may occur downstream from dams responding to the release of deep, hypolimnetic water with high dissolved nutrient content.

12.6.2 Cultural inputs

The effect of cultural wastewaters is usually more severe than the natural sources of allochthonous organic matter just described. The most obvious effect of organic wastewater is a demand on DO. In addition, the associated input of suspended solids alters the bottom substratum to one that is more depositing than eroding. Rather than an increase in species as well as biomass, the increase in biomass and productivity is usually accompanied by a reduction in species diversity.

The principal reason for the reduction in species abundance is that the increasing severity of the physical and chemical factors eliminates the intolerant species. The remaining tolerant species flourish because of increased survival, a result of the reduction of predators and a more favourable food supply, and, consequently, increased biomass results from increased food supply. But if food supply is too great for flow and DO resources, even biomass of tolerant species could decrease in septic zones. As in the example with natural organic matter, if the physical and chemical changes are not severe and the system were highly oligotrophic initially, increases in biomass as well as diversity could result from mild inputs of cultural organic matter. Thus, the degree of community shift is a function of the type of waste, the waste load, the dilution volume or rate of river flow, and turbulence (reaeration potential). Sewage effluent has produced these predictable effects in numerous cases. Results from one of the early classic, but still pertinent studies are shown in Figure 12.7. The associated BOD and DO results are shown in Figure 12.8.

Figure 12.7 shows the seasonal abundance of taxa upstream and downstream from a sewage outfall in Lytle Creek, Ohio, USA. Note the relative difference in the effect of the effluent depending on the season considered. The effect throughout the stream was much more severe in late summer and winter when flow was low. The higher BOD in August and December that resulted from the low flow are shown in Figure 12.8. Consequently maximum and minimum DO were lower during summer low flow. Also note the extension of the zone of minimum taxa abundance during the winter (Figure 12.7). This probably resulted from a reduced rate of waste decomposition during the period of low temperature and low dilution, although flow data were not present for those months.

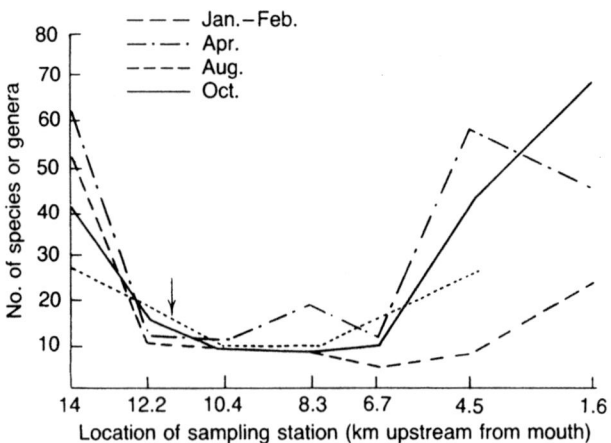

Figure 12.7 Distribution of species abundance in Lytle Creek, Ohio, upstream and downstream from a sewage outfall (arrow) at various seasons (Gaufin and Tarzwell, 1956).

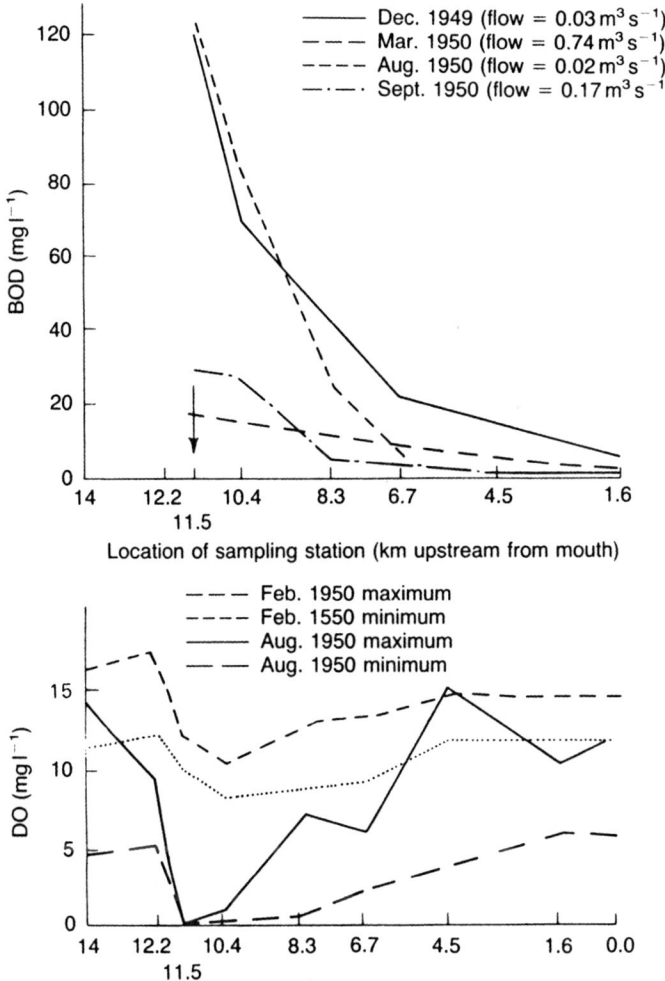

Figure 12.8 BOD and DO in Lytle Creek, Ohio, upstream and downstream from a sewage outfall (arrow) at various seasons (Gaufin and Tarzwell, 1956).

To illustrate the effect of flow, distance and time to the maximum DO deficit (DO sag) can be calculated for a hypothetical river using the Streeter–Phelps equation:

$$\frac{dD}{dt} = K_d L_t - K_a D_t \qquad (12.1)$$

where K_d and L_t are the decomposition rate and BOD at time t (see waste characterization), respectively, and they represent the deoxygenation reaction, while K_a and D_t are the reaeration rate and DO deficit at time t, respectively, and represent the reaeration reaction. For an influent BOD of 150 mg l^{-1} and river temperature of 25°C, the time and distance to the maximum DO deficit (minimum DO) would be 1 day and 1.5 km if river flow is 5000 m^3 day^{-1}. The time to maximum deficit is still 1 day and if river flow is 10 000 m^3 day^{-1} but distance is 4.1 km and actual deficit is more than halved from 6.5 to 3.0 mg l^{-1} DO.

Hawkes (in Klein, 1962) has generalized the community response to organic waste in the form of sewage in terms of the genera that are typical for stony, rapid streams as well as for streams with primarily depositing substrata.

Decreases occur in intolerant species of stony rapids with the addition of organic waste according to the following transition:

Rithrogena → *Ephemerella* (mayfly) → *Gammarus* (scud)

Decreased competition and increased food supply results in the following sequence of elimination:

Baetis (mayfly) → *Simulium* (blackfly) → *Hydropsyche* (caddis fly) → *Limnaea* (snail) → *Herpobdella* (leech)

Further increase in waste and substratum change favours dominance according to the following sequence:

Nais (worm) → *Asellus* (sow bug) → *Sialis* (alderfly) → *Chironomus* (midge) → *Tubifex* (worm)

Even in depositing substrata the end-point is the same with *Sialis*, *Chironomus* and *Tubifex*, but sediment-tolerant organisms such as *Ephemera* and *Caenis* (mayflies) are characteristic of the initial depositing substrata community. The general pattern of abundance for species and individuals in response to organic waste is shown in Figure 12.9. In Figure 12.9(a) the typical effect of a single organic waste is seen. Such a pronounced effect would necessarily result from a heavy loading, particularly when a sludge deposit occurs. Abundance increases, whereas variety decreases.

Two oligochaete worms that dominate environments grossly polluted with organic waste are *Tubifex tubifex* and *Limnodrilus hoffmeisteri*. Heavy loading of organic detritus and the accompanying low DO (they can survive to 0.5 mg l^{-1}) contribute to the success of these species over all others in grossly polluted streams. Brinkhurst (1965) suggested that a principal reason for

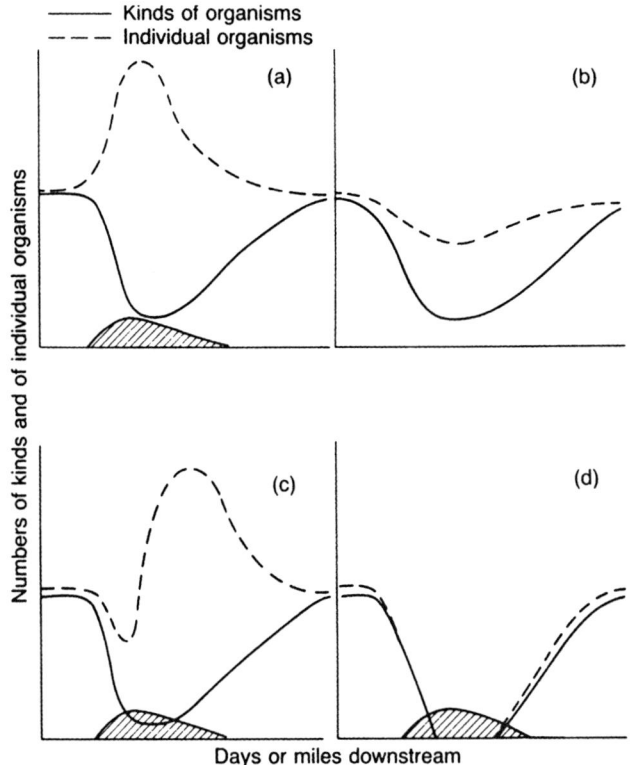

Figure 12.9 Generalized response of macroinvertebrates to (a) organic waste, (b) low levels of toxic waste, (c) organic and toxic waste, (d) high levels of organic and toxic waste. Sludge deposit is indicated by cross-hatched area (Keup, 1966).

their dominance in large numbers can be largely ascribed to reproductive habits. In direct contrast to many of the less tolerant inhabitants of running water, these two species were found to breed at all times of the year. Brinkhurst suggested that the worms too are affected adversely by low DO, but are able to maintain a large biomass in response to the large food supply in spite of low DO because of their frequent breeding. The same concept of life history trait was used to explain the dominance of the polychaete worm *Capitella capitata* in organically polluted marine environments (Grassle and Grassle, 1974).

The success of tubificids in utilizing and at the same time stabilizing organic waste loads can be illustrated by the potential sludge recycling rate of a high density population. *Tubifex* has been reported to produce 170 cm of faecal pellets per day (Bartsch and Ingram, 1959). They orient with the head down in the sediment and the posterior end in the overlying water producing a mound of faecal pellets, which are often lighter in colour than the surrounding sediment. If a density of $100\,000\,m^{-2}$ is assumed, and *Tubifex* has been observed at two- to three-times that amount (Gaufin and Tarzwell, 1956; Hynes, 1960), then the animals would recycle the top 3 cm of sediment per day. *Tubifex* has been observed to move to a depth of 15 cm in lakes (Milbrink, 1980). The computation considers that a *Tubifex* is about 4-cm long with a posterior diameter of 0.06 cm.

Cairns and Dickson (1971a) have generalized the tolerance to organic waste into three principal groups. The intolerant group includes mayflies, stone flies, caddis flies, riffle beetles and hellgrammites; the tolerant group includes sludge worms, certain midges, leeches and certain snails; and the moderately tolerant group includes most snails, sow bugs, scuds, blackflies, crane flies, fingernail clams, dragonflies and some midges.

Although this grouping may appear rather crude, particularly when the overlap in tolerance among genera within families and species within genera is relatively great, conditions of gross pollution are nonetheless often separable from no, or only moderate, pollution with these groupings. In order to detect subtle changes in community response, a more detailed approach is necessary.

12.6.3 Eutrophication in lakes

Oligochaetes increase in lakes relative to other organisms, particularly Chironomidae, as a result of cultural eutrophication. The worms also show greater abundance with depth, inversely to the trend of midges. This can be associated with oxygen depletion and sedimentation with depth, but the degree to which species competition causes the change is unknown (Thut, 1969). Note in Table 12.5 that Lake Washington compared closely with western Lake Erie and also the degree to which Lake Erie was degraded

Table 12.5 Relative composition of profundal benthic biomass in various lakes

Lake	Oligochaetes (%)	Chironomids (%)	Sphaerids (%)
Lake Washington	51	43	3
Lake Erie (1929–30)[a]	1	10	2
Lake Erie (1958)[b]	60	27	5
Cultus Lake (B.C.)	34	65	
Convict Lake (CA)	31	65	
Lake Constance (CA)	20	57	20
Lake Dorothy (CA)	23	69	3

Source: After Thut (1969) with permission of the Ecological Society of America.

Notes
a *Hexagenia* abundant.
b *Hexagenia* absent.

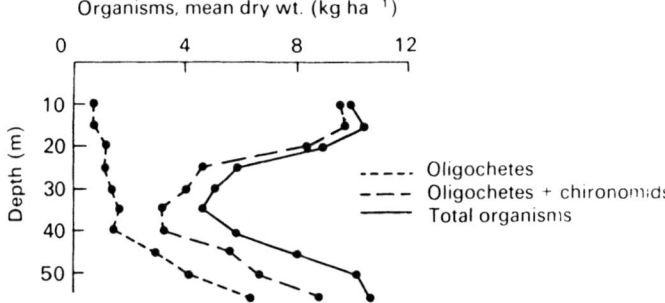

Figure 12.10 The distribution of oligochaets and chironomids with depth in Lake Washington (Thut, 1969).

from early surveys. The density of oligochaetes can reach tens of thousands per square metre. Carr and Hiltunen (1965) found the greatest densities in Lake Erie, which averaged $6000\,m^{-2}$ in 1961, near the mouths of the Maumee, Raisin, and Detroit rivers. Densities in 1930 were $<2000\,m^{-2}$.

The distribution with depth in Lake Washington is shown in Figure 12.10. Results from oligotrophic Cree Lake in Saskatchewan, Canada (Figure 12.11), did not show the proportional increase in oligochaetes with depth as occurred in Lake Washington. Thus, the proportional biomass change is an important indicator in lake eutrophication.

Species of chironomids have been associated with the trophic state of lakes. In the large lakes of Central Sweden, Wiederholm (1974) has suggested

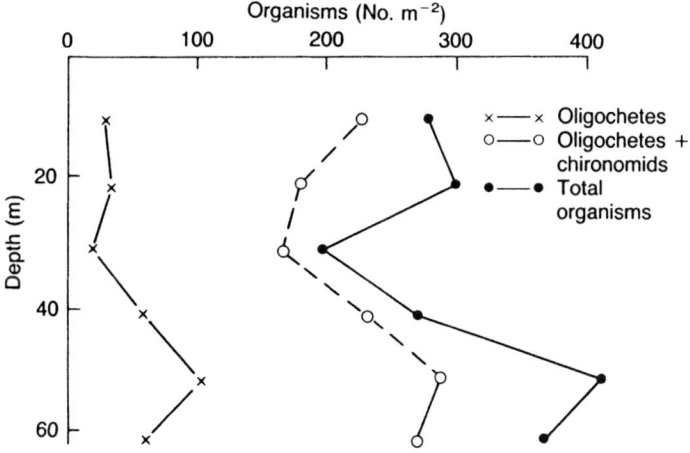

Figure 12.11 Distribution of chironomids and oligochaets in Cree Lake (data from Rawson, 1959).

Table 12.6 Relationship between four chironomid species and trophic state indicators

	Heterotrissocladius subpilosus	Micropsectra spp.	Chironomus anthracinus	Chironomus plumosus
Chl a (μg l^{-1})	<3	3–10	10–20	>20
Total P (μg l^{-1})	<15	15–30	30–60	>60
P load (g m^{-2} yr^{-1})	<0.5	0.5–1.0	1.0–2.0	>2.0

Source: After Wiederholm (1974).

that each of four species were found to represent oligotrophy, mesotrophy, eutrophy and possibly hypereutrophy, respectively, which were in turn related to the levels of P and chl *a* (Table 12.6). Warwick (1980) has further separated lakes into trophic indices according 15 chironomid assemblages recognized by Saether (1979). Wiederholm (1980) has further related their chironomid trophic indices to P and chl *a* per unit mean depth. Mean depth was argued as an important factor in the response of chironomids to increased eutrophication because more P and algal biomass per epilimnetic volume is required in a deep lake for a given change in chironomids than in a shallow lake. The profundal invertebrates depend on organic matter reaching the bottom and less reaches the bottom undecomposed as depth increases. Thus shallow lakes have a more eutrophic assemblage of chironomids than deep lakes under similar enrichment levels.

12.7 Effects of toxic wastes

The response of macroinvertebrates to toxicity is very different from that to organic waste. The number of species decreases, as with organic waste, but biomass remains the same or increases slightly following the mortality of other species that could be predators or represent a significant food source. The initial biomass can be maintained in the presence of only a very low level of toxicity. However, in most cases biomass will decrease as well as the number of species, and the effect of increased food supply would be a short-term effect and probably too minimal to notice.

The degree of change in species number and biomass depends on the resulting concentration of toxicant in relation to the range in tolerance of the species present. Considering the foreign nature of most toxicants, particularly the synthetic organics, an increase in species number as predicted from temperature and organic matter increase, in relatively cold and oligotrophic environments, respectively, would not be expected to occur with toxicant input.

A combination of both toxicants and organic wastes result in the predicted general community response shown in Figure 12.9. In Figure 12.9(b) the decrease in variety is similar to that which occurs with organic waste input, but the primarily toxic waste usually has little food value and, consequently, the tolerant forms have no extra energy with which to increase their numbers. In Figure 12.9(c) an initial toxicity exists, which reduces variety but then toxicity diminishes (complexed physically or biologically or diluted) and the full effect of the organic matter is observed. In Figure 12.9(d) the level of the toxic substance(s) does not 'diminish' and the combined effect is worse than with either waste alone.

A typical response to toxicity is shown by results from a stream receiving drainage from mine tailings ponds and spoil piles that remain from mining activities in the late 1800s and early 1900s. Six stations were established on Prickly Pear Creek, Montana, USA extending from an upstream control to the farthest (11 km) downstream sample point from the waste source (Miller et al., 1983; LaPoint et al., 1984). Zn and Cu were the principal toxicants being leached from the mine waste and increased to mean concentrations exceeding 1000 and 100 μg l^{-1}; respectively, immediately downstream of the source (Figure 12.12). These levels were well above USEPA criteria, which varied due to hardness (Chapter 13). Cu was also above the criteria level in the impact zone. Total abundance of individuals and numbers of species decreased substantially in the maximum impact zone within 2 km downstream of the source and in direct proportion to the increase in metals. The invertebrates had still not recovered to that of the upstream control at the most downstream station.

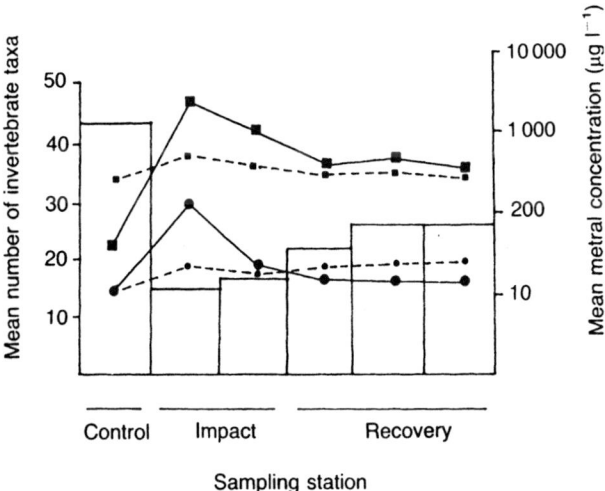

Figure 12.12 Mean number of invertebrate taxa related to stream concentrations of total Zn and Cu and EPA acute criteria for those metals in Prickly Pear Creek, Montana in 1980 (data from Miller et al., 1983). ■—■, Zn concentration; ■---■, Zn criteria; ●—●, Cu concentration; ●---●, Cu criteria.

As with DO, a wide range of tolerance among species in the different orders can be expected. However, there is an indication that midges may be more tolerant to heavy metals than are caddis flies, and mayflies may be least tolerant (Savage and Rabe, 1973; Winner *et al.*, 1975, 1980). However, relative tolerance, as judged from species presence in waste-affected streams, may be related to the number of species in the respective families. Given the same range of tolerance within a family, there is a greater chance of encountering more tolerant species in the family with the greatest number of species.

The numbers of species of macroinvertebrates, representing a wide range in tolerance, along with a paucity of information of individual species tolerance, make the development of specific toxicant criteria largely impossible. However, it may be useful to contrast the tolerance of some invertebrate representatives with that of fish, for which there are specific criteria. With regard to metals, there is an indication that, in general, aquatic insects may be more tolerant to acute levels than fish (Warnick and Bell, 1969; Clubb *et al.*, 1975). However, when exposure time was longer, some insects were as sensitive as, or more sensitive than, fish (Spehar *et al.*, 1978). There are also sublethal effects on aquatic insect life stages that are analogous to those in fish. For example, moulting and emergence were reduced in several species as a result of Cd concentrations considerably below lethal levels and those of smaller size were more sensitive to acute levels than larger ones (Clubb *et al.*, 1975). Greater sensitivity of younger, smaller specimens than older, larger ones has also been shown with pesticides (Jensen and Gaufin, 1964).

While it is important to know the concentrations of toxicants that cause damage to critical life stages of aquatic invertebrates, it is probably even more important to determine the pertinent chemical form of the toxicant and the role of interacting factors. That is particularly true for metals. The relative toxicity of a metal usually varies with pH (due to solubility), DO and temperature (due to rate of metabolism), hardness (due to competitive inhibition of Ca and Mg with the toxic metal(s)) and the amount of organic matter present. Due to less organic matter available for complexation, the effectiveness of metals may be greater in oligotrophic than eutrophic waters. The specific nature of these interactions will be discussed with regard to fish (Chapter 13).

12.8 Suspended sediment

The sediment carried by running water can have an adverse effect on bottom-dwelling organisms. As pointed out previously, the distribution of organism types and abundance is determined to a large extent by substrata type. Excessive erosion from abuse of the watershed often results from overgrazing, deforestation and urban development. By removing vegetative cover, the rate of runoff increases, producing more force for dislodging soil particles and transporting them to water courses. The peak stream flow increases as the fraction of watershed impervious surface increases. Such flow increases, beyond extremes previously experienced by a stream, will result in greater within-channel erosion and sediment deposition creating depositing substrata. The greatest effect (change) on stream macroinvertebrates can be expected where increased deposition occurs on a previously eroding substratum.

Suspended sediment is an insidious and widespread pollutant. The deterioration of a stream is gradual with no dramatic kill of fish or invertebrates, as is often the case with a toxicant. Nevertheless, the community can be altered just as significantly. The principal effect of sediment is to decrease habitat of the eroding substratum community, especially by filling the interstices in rubble/gravel bottom streams. As a result, the number of species as well as the abundance of organisms should decrease because there would usually be little organic matter in the eroded material that could add to the food supply of deposit feeders, which will dominate the sedimented environment. However, a moderate increase in suspended sediment to an eroding substrata could actually result in no appreciable change or in an increase in abundance, whereas the number of species could increase if the substratum were stable. Chutter (1969) found the latter effect in an African stream where in summer the number of species in a stony stream increased from 30 to 42 and then decreased to 22 as the substratum progressed from eroding to stable depositing to unstable depositing. Abundance for the three environmental types showed the same trend (Table 12.1). Effects from logging usually result in reduced diversity and decreased abundance of some taxa (e.g. stoneflies), but increased abundance of others, e.g. worms, midges and mayflies, which are more tolerant of sediment (Graynoth, 1979; Newbold et al., 1980).

An adverse effect on bottom fauna can result from rather modest increases in sediment concentration. Such an effect was observed in Bluewater Creek, Montana, USA, an eroding substratum stream, as a result of irrigation return flows. The effect was noticed at an average level of $\approx 100 \, \text{mg} \, l^{-1}$. Although such a concentration of suspended sediment is not readily visible when viewed through a beaker of the stream water, a reduction in transparency would be quite visible in a natural water course where the light path is greater. There would have been no toxicity result from such a concentration in the stream, yet the abundance of organisms was depleted tenfold (Figure 12.13).

Developing criteria for sediment concentration is difficult because the effect results from the amount deposited and retained in the substratum, not on the concentration left in the water. The fraction of load deposited in turn depends upon river flow. Therefore, low-velocity streams will usually show a greater effect from a given average sediment concentration than high velocity streams.

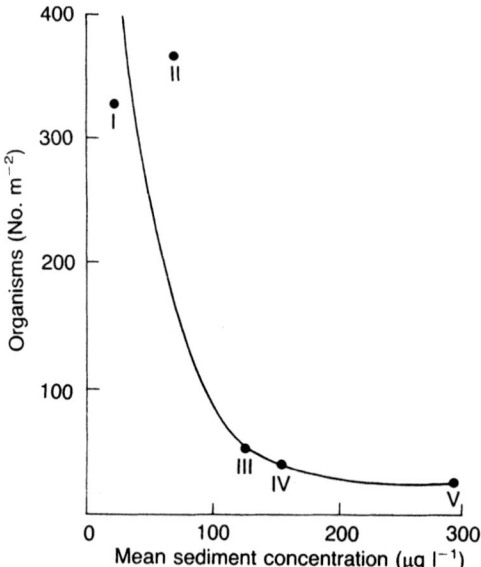

Figure 12.13 The relationship between annual mean suspended sediment concentration and the abundance of benthic organisms at five stations in Bluewater Creek, Montana. Downstream (I–V) increase in sediment was caused by irrigation return flow (data from Peters, 1960).

Effects from urban runoff are often related more to physical scouring, shifting and replacement of the eroding substratum in streams than to sediment deposition. This is due to peak flows well above those experienced during predevelopment. As a result, populations may be dislodged and transported downstream with decreased diversity and only the opportunistic species surviving. Effects are difficult to predict so evaluation of stream receiving water quality following corrective action should be an integral part of every project as is now the case for lake restoration in the USA.

12.9 Recovery

The recovery of stream systems is relatively rapid because in a running-water system the contaminated water is moved downstream relatively quickly following curtailment of waste input. After one hydrologic cycle, which includes high flow that would transport much of the contaminated sediment downstream (or deposit new uncontaminated sediment), the physical environment could be largely recovered. All that is necessary is repopulation, which requires recolonization of the organism-depleted area and time for regrowth to pretreatment abundance. Because most taxa have 1-year life cycles, it is not surprising that 1- to 2-year recoveries are common. Recovery could require a longer time in the lower reaches of rivers where, due to low velocity, sediment is not displaced readily, or where most of the tributaries are depleted of fauna, and successful recolonization would thus be delayed.

Table 12.7 lists some examples of recovery in streams from various types of wastes. Most of the recovery times are of the order of 2 years. With most of these examples it is difficult to be anything but qualitative about changes in water quality and/or the bottom fauna composition. They nevertheless convey the impression that recovery is relatively 'fast' if sources for recolonization exist and the waste input is greatly reduced. The Clinch River fish kill in 1967, following a fly-ash pond leakage, resulted in a spill of water of pH 12 moving down river killing nearly everything in its path. Opinions, at the time of the kill, of a 10-year recovery were obviously quite incorrect.

12.10 Assessment of water quality

The state of water quality can be effectively assessed by use of the benthic macrofauna. The advantages of this community were stressed earlier. An important question, however, is what methodology or technique should be used. Although the presence or absence of certain species has qualitative meaning to someone with experience in the relative tolerance of benthic organisms, such descriptive information is difficult to communicate or treat statistically. Thus, a knowledge of the relative tolerance of organisms present (and absent) is usually combined with a numerical expression of community structure or indicator populations to assess water quality. Numerical expressions may or may not treat the relative tolerances of one species versus another.

12.10.1 Indicator species and taxonomy

Indicator species are often employed to judge the relative quality of an environment, i.e. when a certain species is present a particular water quality condition is indicated. The absence of a particular species is also important and although this is not implied in the approach, it is nonetheless often considered.

The principal drawbacks in this approach are taxonomic and the lack of definitive information on comparative tolerance to given types of waste. On the taxonomic problem, Resh and Unzicker (1975) have stressed the importance of identification to species. They found that of 89 genera of macroinvertebrates that had known tolerances to organic waste, i.e. intolerant (I), facultative (F) or tolerant (T), 65 have species in more than one general tolerance category (I/F, I/F/T, or F/T). Although the genera of invertebrates are relatively easy to determine in most cases, there apparently exists such an overlap in tolerance that a definitive analysis is not really possible unless

Table 12.7 Recovery of macroinvertebrate and fish communities affected by different types of wastes following waste controls

Stream ecosystem	Waste	State of recovery	Waste removed	Elapsed time
Clark Fork River, Montana[a]	Acid mine	Full recovery of trout and bottom fauna	100% of acid mine water after 6 months	2 yr
Maine, several streams[b]	DDT	Full recovery of bottom fauna	100% (discontinuous applications)	2–3 yr
Thames River estuary, England[c]	Sewage	Partial recovery; from 0 to 61 fish species, from 0 to 10% minimum O_2 saturation	60% sewage through secondary treatment	7 yr
Clinch River, Virginia[d]	Acid waste spill	Complete recovery of bottom invertebrate diversity, but some mollusks still absent	100%	2 yr
Clinch River, Virginia[d]	Alkaline fly-ash spill	Complete recovery of bottom invertebrate diversity	100%	6 mo
Blue Water Creek, Montana[e]	Silt	Marked improvement in trout/rough fish from 0.64 to 3.5 and 0.14 to 1.0 at two stations, respectively	52–44% of sediment load at two stations, respectively	2–3 yr
Plum Creek (Clink River)[f]	Fuel-oil spill	Complete recovery of bottom fauna abundance	100%	5 mo

Notes
a Averett (1961).
b Dimond (1967).
c Gameson et al. (1973).
d Cairns et al. (1971b).
e Marcuson (1970).
f Hoehn et al. (1974).

identification is to the species level. With many groups this is difficult and time consuming because the state of taxonomy is relatively poor with North American fauna, for example, and opinions of experts are often required.

Equally important is the known tolerance of a particular species to the various types of wastes that can be discharged into streams. Only with organic wastes, such as sewage, is there any general understanding of tolerance among species. An excellent example of the use of indicator species is that of profundal chironomids and lake trophic state described earlier.

This is not to say that nothing can be gained from a description of the organisms present, identified to some level short of species, along with their relative abundances. As has already been shown, Hawkes (1962) described a typical succession of genera within English streams of differing substrata in response to organic waste. Certainly the presence of *Tubifex* in large numbers is quickly identified as gross organic pollution. However, the problem of detecting the effect of small incremental changes in waste input is probably difficult to solve by the qualitative indicator organism approach alone.

12.10.2 Numerical indices

Formulae to express the 'state of stream quality' or 'degree of pollution' using the macroinvertebrates are numerous. They are generally of three types: biotic, diversity and similarity (Washington, 1984). The biotic ('pollution') indices require a judgment as to the relative tolerance (or saprobic valency) of the various taxa identified, usually tolerant, intolerant or facultative, and the distribution among those groups is weighted. Thus, they require considerable experience by the investigator. A major problem with these formulae is that they are largely restricted to the assessment of organic waste effects and thus lack a basis to categorize the tolerance of the taxa to the myriad of toxic wastes or physical factors (e.g. sediment, flow) that could be present (Sládeček, 1973; Washington, 1984). However they are considered by many to be more biologically relevant than the other indices because they consider the physiological requirements of the taxa.

Washington (1984) has reviewed 19 biotic indices. He suggested that Beck's was the first true biotic index (Beck, 1955), which compares the abundance of species that are tolerant and intolerant to organic pollution. It includes only species abundance and not individual abundance or any other attributes besides tolerance to organic pollution. Beaks' river index (Beak, 1965), on the other hand, includes judgments regarding trophic character (feeding habits), sensitivity to organic pollution and relative abundance of individual species organized into a table. There are other indices that include one or more of these factors, such as the Trent Biotic Index, Graham's index, and Chandler's Biotic Score (see Washington, 1984). These indices generally depend on a gradient in tolerance of certain taxa to organic pollution and require identification to species in some cases, but only to order or family in others.

Hilsenhoff (1977) adapted Chutter's (1972) biotic index, developed for South African streams, to North American streams. The index uses quality (Q) values and abundance of each taxon for which a Q value is chosen. An average Q value, weighted to taxa abundance is calculated according to:

$$\text{Index} = \frac{\sum_{i=1}^{K}(n_i Q_i)}{n} \qquad (12.2)$$

where n_i is the number of individuals in each of K taxa and n is the total number of individuals. Hilsenhoff's Q values range from 0–5, which is the range of final index values with <1.75 representing a clean undisturbed stream and 3.75 representing gross enrichment or disturbance.

USEPA has developed a bioassessment approach that uses indices of biotic integrity to assess the existence and magnitude of impairment to aquatic resources. Because biological communities reflect the overall ecological integrity of an ecosystem, monitoring these communities is critical to determining the effects of pollution or other disturbances. To this end, USEPA has developed guidance for conducting biosurveys, such as the Rapid Bioassessment Protocols (RBPs), for three aquatic assemblages: periphyton, fish and benthic invertebrates (Plafkin et al., 1989; Barbour et al., 1999). These protocols have been tested in different regions of the country and have been adopted by most of the states in the USA for water resource assessment and management.

The biotic indices presented in RBPs are based on the Index of Biotic Integrity (IBI), which integrates various functional and structural/compositional metrics (Karr, 1981). The metrics are then combined into an overall single number. For example, the fish IBI includes measures of abundance, total species richness, the numbers of various fish groups (e.g. suckers, darters, sunfish), the numbers of sensitive and tolerant species, trophic composition and a measure of fish condition (Karr, 1991). The resulting values for the 12 metrics are added together to form the IBI (range: 12–60), which provides a measure of the overall integrity of the fish community at that site.

Metrics have also been defined and evaluated for benthic macroinvertebrates (Barbour et al., 1999). The most effective metrics are those that show a response across a range of human influence (Fore et al., 1996; Karr and Chu, 1999). The best benthic metrics for streams include measures of richness, species composition, pollution tolerance or sensitivity, feeding group and habitat type (Table 12.8). Different bioassessment protocols and metrics are used for lakes (USEPA, 1998a) and wetlands (Danielson, 1998).

Table 12.8 Best candidate benthic metrics and predicted direction of metric response to increasing perturbation (from Barbour et al., 1999)

Category	Metric	Definition	Predicted response to increasing perturbation
Richness measures	Total number of taxa	Measures the overall variety of the macroinvertebrate assemblage	Decrease
	Number of EPT taxa	Number of taxa in the insect orders Ephemeroptera (mayflies), Plecoptera (stoneflies) and Trichoptera (caddisflies)	Decrease
	Number of Ephemeroptera taxa	Number of mayfly taxa (usually genus or species level)	Decrease
	Number of Plecoptera taxa	Number of stonefly taxa (usually genus or species level)	Decrease
	Number of Trichoptera taxa	Number of caddisfly taxa (usually genus or species level)	Decrease
Composition measures	%EPT	Per cent of the composite of mayfly, stonefly and caddisfly larvae	Decrease
	%Ephemeroptera	Per cent of mayfly nymphs	Decrease
Tolerance/ intolerance measures	Number of intolerant taxa	Taxa richness of those organisms considered to be sensitive to perturbation	Decrease
	%Tolerant organisms	Per cent of macrobenthos considered to be tolerant to various types of perturbation	Increase
	%Dominant taxon	Measures the dominance of the single most abundant taxon. Can be calculated as dominant 2, 3, 4 or 5 taxa	Increase
Feeding measures	%Filterers	Per cent of the macrobenthos that filter FPOM from either the water column or sediment	Variable
	%Grazers and scrapers	Per cent of the macrobenthos that scrape or graze upon periphyton	Decrease
Habit measures	Number of clinger taxa	Number of taxa of insects	Decrease
	%Clingers	Per cent of insects having fixed retreats or adaptations for attachment to surfaces in flowing water	Decrease

IBI has been used to discern the effects of various types of human disturbances including organic pollution, logging, livestock grazing, recreation and urbanization in different regions of the USA (e.g. Fore et al., 1996; Karr, 1998). In a 2-year study of 19 Puget Sound lowland streams, a nine-metric IBI was used to assess the macroinvertebrate community across a range of urbanization (Kleindl, 1995; Karr, 1998). The nine metrics (mayfly, stonefly, caddisfly, intolerant, long-lived, and total taxa richness, relative abundance of planaria and amphipods, tolerant taxa, and predator taxa) changed in proportion to urbanization, which was measured as per cent total impervious area (%TIA). Biological integrity was found to be linearly and inversely related to %TIA ($r^2 = -0.76$). The rural, least degraded streams had IBIs between 35 and 45 (out of 45), and the most degraded urban streams had IBIs below 15.

Diversity indices do not consider the tolerances of individual species, only the number of species and number of individuals and, thus, relate more to community structure (Washington, 1984). There are many formulations for diversity (Washington lists 19), but only four will be reviewed here. A frequently used index that considers only species richness is as follows (after Margalef, 1958):

$$d = \frac{S-1}{\log_2 N} \tag{12.3}$$

where S is the number of species and N is the number of individuals. The natural log of N is also used. Practically, this formula illustrates that if one searches an environment and collects individuals, the number of species encountered increases linearly as the logarithm of the number of individuals. At some point the finite number of species is approached and the curve flattens out. This formula does not, however, account for the relative abundance of each species or evenness in distribution.

One of the most popular formulae used to express diversity, which includes richness (number of species) and evenness, is the Shannon approximate index from information theory (Shannon and Weaver, 1948):

$$H'' = -\sum_{i=1}^{s} \frac{N_i}{N} \log_2 \frac{N_i}{N} \tag{12.4}$$

where s is the total number of species, N_i is the number of individuals in the ith species, and N is the total number of individuals. The unit for H'' is bits per individual if $\log_2$ is used, which is customary in information theory because binary systems are of interest (Zand, 1976). Approximate refers to the proportion N_i/N in the sample and not the infinite population as p_i represents in the formula for H' as is often given for the Shannon index

(Kaesler *et al.*, 1978). The symbol $\bar{d}$ is also used (Wilhm, 1972) and is equivalent to H''. The maximum diversity H''_{max} from this formula is defined as $\log_2 S$. For example, $H''_{max} = 5$ if 32 species are present.

Another formula, that describes diversity per individual is that of Brillouin (1962):

$$H = \frac{1}{N} \log \frac{N}{N_1! N_2! \ldots N_s!} \tag{12.5}$$

According to Zand (1976) and Kaesler *et al.* (1978), the Brillouin index describes diversity when the number of individuals in each species is low more accurately than does the Shannon index. Kaesler *et al.* (1978) gives a theoretical comparison of the two indices.

Both of the above formulae treat species richness or abundance and the evenness or equability of those species, which are the two ingredients of diversity. Thus, the environment with a more even distribution of its members or individuals, for the numbers of species present, could be thought of as being more varied and in a sense more 'balanced'. As indicated in Chapter 1, this does not necessarily mean the community is more stable and resistant to disturbance. If an unpolluted environment receives a diversity of food materials and has a diverse bottom substratum, it should also have a more equitable distribution of the abundance among species with differing requirements for food type and space.

Theoretically, one would expect the index to have more sensitivity to physical–chemical changes with evenness considered as well as richness. A hypothetical example is shown in Table 12.9. Nevertheless, some diversity formulations have been widely criticized because they essentially define the improbability of an event and are considered to have little biological relevance (Hulbert, 1971; Washington, 1984). For example, Hulbert suggested a formula (Hulbert's PIE) that describes the probability of encounters and the relative importance of interspecific competition. Washington (1984) suggests that Hulbert's PIE has promise as a biologically relevant diversity index.

Table 12.9 Diversity formula comparison for three hypothetical communities

Communities	N_1	N_2	N_3	N_4	N_5	ΣN	S	$\frac{S-1}{\ln N}$	$\bar{d}$
A	20	20	20	20	20	100	5	0.87	2.32
B	40	30	15	10	5	100	5	0.87	1.67
C	96	1	1	1	1	100	5	0.87	0.12

Source: After Wilhm and Dorris (1968).

Notes
N = number of individuals in species 1–5; S = number of species.

A convenient diversity index that requires little or no taxonomic knowledge is the Sequential Comparison Index (SCI) that is based on the 'theory of runs' (Cairns and Dickson, 1971). With this index, runs are enumerated on a sample and a diversity index, DI, determined:

$$\text{DI} = \frac{\Sigma \text{no. runs/no. specimens}}{\text{no. trials}} \tag{12.6}$$

The number of taxa is then used to calculate the SCI or DI_{total}:

$$\text{SCI} = \text{DI} \times \text{no. taxa} \tag{12.7}$$

Cairns and Dixon suggested that values >12 indicate a healthy condition and <8 would indicate pollution.

As mentioned previously, diversity indices do not consider the specific taxa present among the sites being compared. Therefore, the numerical diversity index may be the same but the specific taxa representing the two communities could be completely different (and with different tolerances to the waste). Similarity, or comparative, indices are measures of community structure but compare the abundance of particular species (Washington, 1984).

Gaufin (1958) illustrated the need to keep track of the abundance of particular species in the Mad River, Ohio, USA where the pollution-tolerant snail *Physa*, the leech *Macrobdella* and the worm *Limnodrilus* were all present at the clean-water sites, but they became more abundant downstream from the waste sources. Hence, their abundance, rather than their presence only, indicated the waste effect. Comparative indices would take this aspect into account. An example of a commonly used index is that of Bray and Curtis (1957), which measures dissimilarity between two sites a and b:

$$D = \frac{\sum_{i=1}^{s} X_{ia} - X_{ib}}{\sum_{i=1}^{s} (X_{ia} + X_{ib})} \tag{12.8}$$

where X_{ia} and X_{ib} are the number of individuals in the i_{th} species at the respective stations. Other similarity/comparative indices are given by Hellawell (1977) and Washington (1984).

The Shannon approximate index (H'' or $\bar{d}$) has been used most often to evaluate the effect of wastewater on diversity. Although the index has been criticized, Wilhm and Dorris (1968) demonstrated that $\bar{d}$ can be reliable in determining the effects of wastes using benthic invertebrates (Table 12.10). Near waste outfalls, $\bar{d}$ was typically ≈ 1. Intermediate values were found downstream and values farther downstream suggested recovery.

Table 12.10 Comparison of $\bar{d}$ in response to various types of wastes

Waste	$\bar{d}$		
	Above outfall	Outfall	Downstream
Domestic and oil		0.84	1.59	3.44
Domestic and oil	3.75	0.94	2.43	3.80
Oil brines	3.36	1.58		3.84
Oil refinery		0.98	2.79	3.17
Total dissolved solids		0.55		3.01
Oil brines		1.49	2.50	
Oil brines		1.44	2.70	
Storm sewer		1.45	2.81	

Source: Wilhm and Dorris (1968).

Further study of a variety of environments and waste types showed $\bar{d}$ to provide consistent results separating polluted from unpolluted conditions as follows (Wilhm and Dorris, 1968): Clean water was represented by similar values for 22 widely varying environments, mean $\bar{d} = 3.30$ (range 2.63–4.00); waste affected areas in 21 environments and widely varying waste types showed consistently low values, mean $\bar{d} = 0.95$ (range 0.42–1.60).

As a result of these observations Wilhm and Dorris suggested the following general guidelines for distinguishing the state of a stream: heavy pollution, $\bar{d} < 1.0$; Moderate pollution, $\bar{d} = 1.0$–3.0; Clean water, $\bar{d} > 3.0$.

Having absolute values for judging the conditions of streams would be highly desirable especially in situations where control sites are unavailable or pre-pollution data are lacking. However, the index has been used more often for site-to-site comparison, rather than relying too heavily on the absolute values, and in combination with other indices such as a simple abundance of species or taxa.

Diversity can be used effectively whether identification is to species or some other hierarchical level. This has been demonstrated by several workers. The effect of organic waste was easily recognized in Plum Creek, Ohio, USA as $\bar{d}$ decreased downstream from the wastewater input regardless of whether the taxonomic level used was genus, order or class (Egloff and Brakel, 1973) (Figure 12.14). The waste loading was rather mild as BOD increased on the average from 3.5 to $9.0 \, \text{mg} \, \text{l}^{-1}$ and DO was usually $>6 \, \text{mg} \, \text{l}^{-1}$ downstream. The number of genera decreased from 16–22 upstream to 7–9 downstream. Hughes (1978) and Hellawell (1977) have also shown the close agreement among diversity values using either species, genera or family. However, sensitivity declined greatly if class or order were used similar to that shown in Figure 12.14. The family level was useful in describing conditions in two tributary streams to Lake Washington – one with a heavily developed watershed (7 invertebrate families) and the other

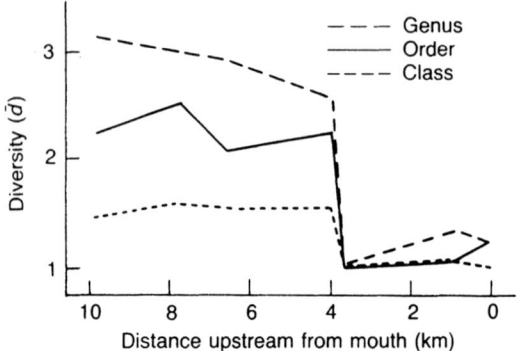

Figure 12.14 Diversity indexes for three taxonomic levels of macroinvertebrates in response to an organic waste (Egloff and Brakel, 1973).

relatively undeveloped (14 families). The range in $\bar{d}$ at three sites ($n = 9$) in the former stream was 1.75–0.29 and 2.97–1.60 ($n = 9$) in the latter (Knutzen, 1975).

The evenness factor in diversity indices may lead to difficulties in some situations. Species richness can decrease, while evenness actually increases due to other interacting factors in the environment (Hulbert, 1971). An interesting example of this is from Spring Creek, near State College, Pennsylvania, USA (Cole, 1973). The stream received a secondary treated effluent that contributed 5–20% of the stream's flow. BOD increased only slightly from 0.9–1.0 mg l^{-1} upstream to 1.0–2.2 mg l^{-1} downstream, although SRP increased from 50 to 300 µg l^{-1}. The treatment plant replaced individual septic systems and created a point source of enrichment, which apparently resulted in an increased abundance of macrophytes (*Elodea*, *Chara* and *Potamogeton*) downstream from the source producing large diurnal swings in DO. The number of species in riffles decreased from 46 to 27 (41%) downstream from the wastewater source; however, $\bar{d}$ decreased only 15% from 2.74 to 2.26 in spite of DO concentrations <3 mg l^{-1}. The author's explanation for failure of $\bar{d}$ to show an appropriate magnitude of effect was the increased habitat diversity created by the macrophytes allowing the evenness factor to compensate for decreased species richness. Such interference with evenness by other factors has been shown by others (Godfrey, 1978).

With regard to toxicants, Winner *et al.* (1975) found that the non-formulated indices of number of species and individuals each showed a predictable response to a gradient in Cu from 120 to 23 µg l^{-1} downstream from an experimental addition of CuSO$_4$. However, the Shannon index (Equation 12.4) and Margelef's index (Equation 12.3) showed a much poorer correlation with the Cu gradient. Moreover, four species were eliminated at sites with 120 and 66 µg l^{-1} and such observations must be considered in addition to the use of numerical indices.

The change in specific taxa is included in the Bray–Curtis dissimilarity index (Equation 12.8) and was found to represent the response to Cu in laboratory channels better than the Shannon index (Perkins, 1983). Other comparative indices also gave a favourable response. The number of species decreased linearly over a log series of Cu concentrations, while the Bray–Curtis index increased with near linearity after 14 and 28 days' exposure of an acclimated macroinvertebrate community. The Shannon index, on the other hand, first increased and then decreased as Cu increased. This behaviour was apparently related to interference by the evenness factor. The Bray–Curtis index also provided the most meaningful description of the response to toxicity in other laboratory microcosms (Pontasch et al., 1989).

Diversity and comparative indices were also evaluated following a gasoline spill in Wolf Lodge Creek, Idaho, USA (Pontasch and Brusven, 1988). The Bray–Curtis index (Equation 12.8), as well as another comparative index (x^2) gave the most meaningful evaluation of the extent of effects and recovery from the gas spill. However, as stressed by Pontasch and Brusven, all indices have biases and none should be used to the exclusion of sound ecological judgement.

With the myriad formulations in the literature, the investigator may understandably be confused as to which one(s) to use to evaluate macroinvertebrate response in a stream receiving a waste that may be either toxic, enriching, alter the habitat or some combination thereof. Much of the criticism levelled at the diversity indices, especially Shannon's, is from the standpoint of biological relevance. Yet, diversity indices continue to be used, and used effectively in many cases, to describe effects of pollution. The principal reason for their popularity is probably that they require less knowledge about the organisms present than do biotic indices. However, diversity (as well as similarity) indices are also more general than biotic indices, which tend to be specific to certain geographical areas and usually to only one type of waste (organic). The purpose here is to represent the types of indices available, emphasizing those that have been used more frequently with some success, while indicating their shortcomings. The investigator should not rely on any one index, but try several in an attempt to describe the population changes due to wastewater inputs to the stream in question. As Washington (1984) points out, some that have promise, such as Hulbert's PIE, should be used more frequently.

Furthermore, the definition of reference conditions is often critical to the interpretation of biological survey data (Barbour et al., 1999). Reference conditions are typically either site-specific or regional. Site-specific reference conditions are usually measured at a site upstream from a point source of pollution or from a 'paired' watershed, whereas regional reference conditions are derived from least disturbed sites within the same ecoregion. The ecoregion approach is based on the assumption that water bodies reflect the lands they drain and that similar lands should produce similar water bodies in terms of water quality and biota (Omernik, 1987). Ecoregion boundaries are based on regional patterns in soils, vegetation, geology, climate and physiography. Ecoregions have been found to be helpful in conducting environmental assessments, setting regional resource management goals, and developing biological criteria and water quality standards (Omernik, 1995).

12.10.3 Sampling problems

Although the abundance of benthic invertebrates is highly variable, diversity is usually less so and should require fewer samples. Therefore, to use diversity to study waste effects in streams, stations should be selected that are of similar substrate type, velocity, etc. and random samples should be collected from either a grid or a transect in either a transverse or longitudinal direction. The sample should also be biased to high variety, or species richness, by selecting riffle areas where habitat is more diverse, thereby providing more sensitivity.

Relatively few samples are usually necessary to define diversity adequately (Wilhm, 1972). The values for $\bar{d}$ were found to approach a maximum in the first four samples out of ten in 13 different habitats. Thus, the mean diversity from three samples provided a reasonable estimate of the maximum diversity in each section of stream, assuming that the number of individuals (N) in each species is sufficiently great. Using the Brillouin index (Brillouin, 1962) (Equation 12.5), an N of only 10 was found to be adequate (95% confidence) to detect the effect of an acid spill, acid mine waste and domestic sewage, when 25 random samples were collected from a larger sample (Kaesler *et al.*, 1978). However, H did not approach the maximum until N was >100. This was similarly shown by Hughes (1978) who successively pooled $50 \times 0.1\text{-m}^2$ samples. In this case, $\bar{d}$ approached the maximum in the whole $5\,\text{m}^2$ when $10 \times 0.3\text{-m}^2$ samples were analysed.

12.11 Invasive, non-native invertebrates

Invasive benthic invertebrates have also created problems in non-native habitats. One recent example is the zebra mussel (*Dreissena polymorpha*) invasion in North America. A small bivalve native to the Black Sea, the zebra mussel, invaded the Great Lakes in the mid-1980s, probably through the discharge of ballast water from an oceangoing ship (Mills *et al.*, 1994). The zebra mussel has caused problems in Europe for more than 100 years. Since its introduction to the Great Lakes, it has already obstructed water intake systems for power plants and municipal and industrial treatment plants, with the cost for mitigating the damages estimated at US$5 billion. The zebra mussel's rapid growth and lack of predators have led to peak densities of over $300\,000\,m^{-2}$, comprising up to 70% of the macroinvertebrate biomass in some places. The zebra mussel has altered the lakes' characteristics by filtering large amounts of plankton ($1\,l\,animal^{-1}\,day^{-1}$), thus transferring nutrients from the water column to the sediment. While their water filtering is believed to have increased water clarity in parts of the Great Lakes (a possible benefit), water column primary and secondary production is also diverted from pathways that support fish growth and the lake's economically important fishery. Other negative effects of zebra mussels in North America include biofouling of fishing gear and navigation buoys (Nalepa et al., 2000), accelerated extinction rates of native freshwater bivalves by about 10-fold (Ricciardi et al., 1998) and stimulation of the cyanobacterium *Microcystis aeruginosa* blooms in summer (Lavrentyev et al., 1995). While multiple negative consequences of the zebra mussel invasion are obvious, one possible benefit is to several species of diving ducks for which zebra mussels have become an important prey item (Hamilton and Ankney, 1994).

Since 1990, the zebra mussel has spread throughout the Mississippi River basin and is expected to invade much of North America over time. Control measures to mitigate the effects of zebra mussels are being developed and implemented (Nalepa and Schloesser, 1993; D'Itri, 1997). Whether efforts to prevent the spread of zebra mussels into the western part of the USA will be effective remains to be seen.

Another invertebrate invader to northern Wisconsin and Minnesota is the rusty crayfish, which were introduced to the USA from anglers' bait buckets (Lodge et al., 1985). Rusty crayfish displace native crayfish from their burrow, consume game fish eggs (thus incidentally reducing their main predators) and decimate macrophyte beds. Commercial harvesting of rusty crayfish, primarily for export to Scandinavia, has increased the spread of rusty crayfish throughout the region. Conversely, non-native crayfish from North America decimated a highly prized native crayfish (*Astacus astacus*) in mainland Europe by introducing a fungal disease (*Aphanomyces astaci*), to which the European species have no immunity (Mason, 2002), and by competing with native species (Vorburger and Ribi, 1999). The crayfish plague has spread throughout Europe, the Middle East and England. The continued introduction of more non-native crayfish to replace the declining natives has only compounded the problem.

The Asian tiger mosquito was introduced into the USA in used automobile tyres that were imported from Asia for re-treading and resale.

The mosquito larvae resided in the water that collected in the tyres and emerged as the adults, which feed on most mammals and birds in the USA. The mosquito is a vector for eastern equine encephalitis, an often-fatal viral infection of people as well as horses (Craig, 1993). First established in the 1980s, mosquitoes had spread throughout 25 states by 1992.

Chapter 13

Fish

Fish are most frequently used as the so-called target organisms in developing criteria and setting standards for water quality. That is, the criteria for acceptable water quality, except for eutrophication, are usually based on fish, especially if they have economic or recreational value. Specific standards exist for dissolved oxygen (DO) and temperature that have been experimentally determined with fish and will be discussed in detail. The standards for toxicants are usually undefined, however, and the limits are determined by employing bioassays, either *in situ* or *in vitro* with the relevant water and species. That is due to variability in the effect of a given toxicant from water-to-water and species-to-species. Nevertheless, published criteria for concentrations of particular toxicants are often used in the absence of on-site bioassay results to evaluate toxicity.

Fish populations are also used to assess water quality with indices similar to those used with macroinvertebrates (Section 12.10). For example, Karr *et al.* (1986) developed an index of biological integrity using fish that includes trophic as well as diversity and tolerance attributes.

13.1 Dissolved oxygen criteria

Recommended limits for DO were given by the National Committee on Water Quality in the USA in their 'green book' and 'blue book' (WQC, 1968, 1973) to protect communities and populations of fish and aquatic life against mortalities as well as to prevent adverse effects on eggs, larvae and population growth. A distinction has been made between cold- and warm-water environments. According to these documents, DO should be $>5\,\text{mg}\,l^{-1}$ nearly all the time in warm water, assuming periods of much higher concentrations. For short periods, concentrations of $4-5\,\text{mg}\,l^{-1}$ should be tolerated, but it should never become $<4\,\text{mg}\,l^{-1}$. In cold-water environments, concentrations should be $>7\,\text{mg}\,l^{-1}$ for successful spawning and egg and larval development. DO $>6\,\text{mg}\,l^{-1}$ should exist for growth and $>5-6\,\text{mg}\,l^{-1}$ for survival over short periods of time. These recommended limits have been changed recently and are published in the 'gold book'

(WQC, 1986): 7-day mean minimum DOs of $4\,\text{mg}\,\text{l}^{-1}$ for warm water and $5\,\text{mg}\,\text{l}^{-1}$ for cold water with 7-day means somewhat higher at $6\,\text{mg}\,\text{l}^{-1}$ and $9.5\,\text{mg}\,\text{l}^{-1}$, respectively. As will be shown, these values may not be sufficiently restrictive.

Minimum levels for adult survival may be surprisingly low even for fish of low tolerance to DO depletion. For example, $1-3\,\text{mg}\,\text{l}^{-1}$ is sufficient for survival for a short time even with cold-water species such as salmon. These same fish could feed and reproduce at $3\,\text{mg}\,\text{l}^{-1}$ (Doudoroff and Shumway, 1967). However, the level required to sustain long-term survival and normal production of highly desired species, such as salmon and trout, is much higher.

Early field evidence by Ellis (1937) showed that DO concentrations $>5\,\text{mg}\,\text{l}^{-1}$ were associated with populations of fish of good abundance and species diversity. This was largely the basis for the early setting of the $5\,\text{mg}\,\text{l}^{-1}$ standard, which has often been misinterpreted as a 'requirement' for fish. The limit of $5-6\,\text{mg}\,\text{l}^{-1}$ that has come to be accepted in standards is probably a compromise for either warm- or cold-water fish. Requirements for some stages of a fish's life history are much higher than $5\,\text{mg}\,\text{l}^{-1}$. To ensure maximum production of many desired species of fish, a DO content near air saturation (e.g. $9.2\,\text{mg}\,\text{l}^{-1}$ at $20°C$) should be maintained for the critical stages of life (Doudoroff and Shumway, 1967; WQC, 1973). However, many non-game fish species, such as goldfish (*Carassius*), have their normal activity limited at much lower concentrations of DO than do the salmonids (Macon, 1974).

Before proceeding to a discussion of effects of low DO on the various life stages and activities of fish, it would be well to make a point about the units with which to express DO. Although fish obtain oxygen from the water according to the partial pressure difference between the surrounding water and blood circulating through the gills, the desired unit is concentration rather than partial pressure or percent of saturation (Doudoroff and Shumway, 1967). This is because the metabolic demand for oxygen by fish increases as the temperature increases. At the same time the actual concentration at, e.g. 100% saturation, decreases. That is, the amount of DO necessary to maintain a constant partial pressure (say 100%) decreases as temperature increases.

That fish actually need more DO (concentration) as temperature increases to carry out normal activity, and as a result much more than a constant partial pressure, is shown in Figure 13.1, which was interpolated from data by Graham (1949). It can be seen that the standard metabolic rate of brook trout, in terms of O_2 consumed, increases dramatically as temperature increases to near the lethal limit. At the same time, the minimum DO concentration to sustain that metabolic rate increases from about $2-4\,\text{mg}\,\text{l}^{-1}$. The minimum saturation level to sustain metabolism, of course, increases to an even greater extent. Thus, the overriding effect of temperature on

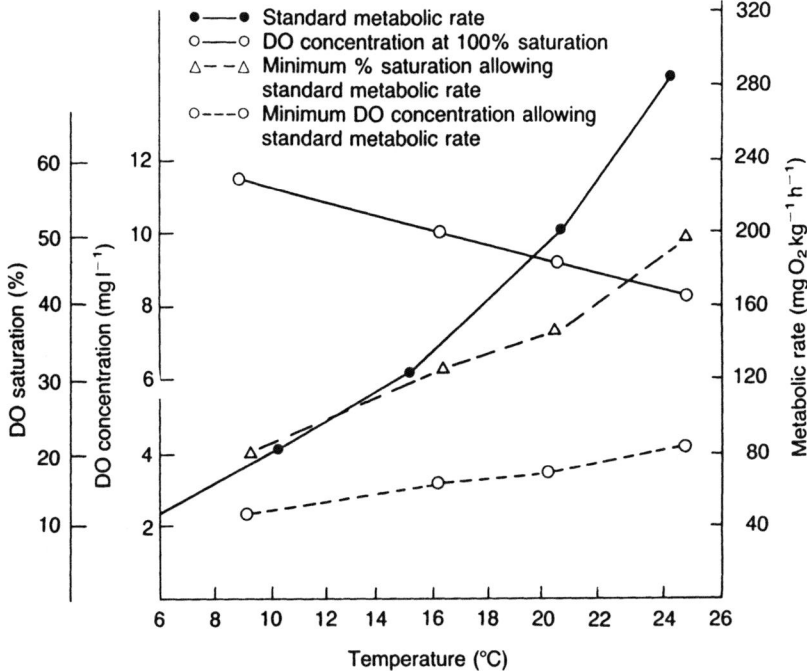

Figure 13.1 Standard metabolism of brook trout (*Salvelinus fontinalis*) in relation to minimum DO needed in concentration and percentage saturation (modified from Graham, 1949).

metabolic demand for DO is apparent and maintenance of a constant percentage saturation as temperature increases would be detrimental to sensitive fish species.

13.1.1 Embryonic and larval development

The growth of salmonid (coho) embryos and the size of fry that emerge from the gravels in streams are limited by the supply of DO as shown by Shumway *et al.* (1964) in Figure 13.2. The dashed lines indicate embryo growth at 80 and 67% of the maximum.

Note that there is an interaction between DO and water velocity on the size of the emerging fry. Growth to a given level can be sustained by either a velocity increase with declining DO or a DO increase with declining velocity. At a constant high or low velocity, growth shows a progressive, although gradual, decrease below air-saturation values to a level of $\approx 5\,\mathrm{mg\,l^{-1}}$ after which the decline in growth is more rapid.

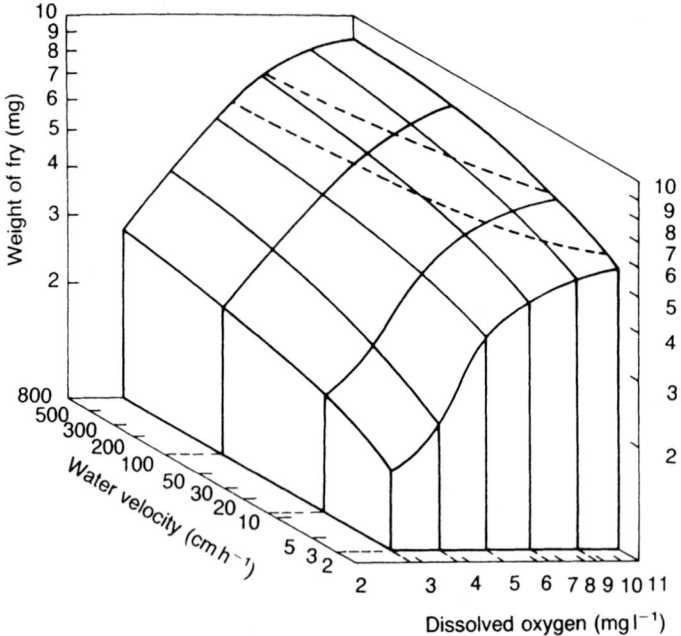

Figure 13.2 Development of coho salmon embryos at different concentrations of DO and at different velocity levels. Dashed lines indicate DO and velocity combinations producing 20 and 33% reduction in growth (Shumway et al., 1964).

Subsequent work has shown a similar effect of DO on embryonic development to the larval stage (Carlson and Siefert, 1974; Carlson and Herman, 1974; Siefert and Spoor, 1973). For most species studied, which included a range from cold-water to warm-water fishes, the time to hatching increased, growth decreased and survival decreased as DO was reduced. The greatest effect occurred below ≈50% saturation, which was usually 5–6 mgl^{-1} at the test temperatures used. With species such as the white sucker, detrimental effects were not observed unless the DO level was <50% saturation. A greater tolerance was shown by larval survival for all species than by other indicators (Figure 13.3). Although it is clear that survival dropped off most precipitously at concentrations below ≈3–5 mgl^{-1} for all species, three species (salmon, trout and catfish) showed poorer survival at high concentrations. Two of those three species also showed a reduction in survival as DO was reduced below the highest level. Furthermore, delayed hatching, delayed feeding and/or slower growth were also shown with these species at levels between 100 and 50% saturation. These results, therefore, agree with those of Shumway et al. (1964); that is, detrimental effects on larval development

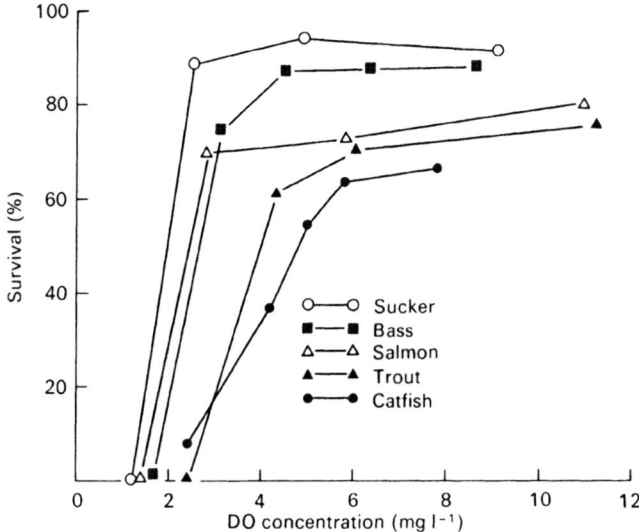

Figure 13.3 Effect of DO concentration on survival of fish larvae from the embryo stages of four species. Time periods and temperature were as follows: channel catfish, 19 days at 25°C; lake trout, 131 days at 7°C; coho salmon, 119 days at 7–10°C; largemouth bass, 20 days at 20°C; and white sucker, 22 days at 18°C. (Data from Siefert and Spoor, 1973; Carlson and Siefert, 1974; Carlson *et al.*, 1974)

will probably occur with both cold- and warm-water species when any reduction occurs in DO to levels <100% saturation even though the greatest damage occurs at concentrations <5 mg l^{-1}.

13.1.2 Swimming performance

The swimming performance of coho salmon also begins to decline progressively below 100% saturation (about 9–10 mg l^{-1}) (Figure 13.4). Again there is no clear threshold of effect, although the decrease in performance, as with larval development, is greatest at DO concentrations <5 mg l^{-1}. The points represent velocities at which the salmon failed to maintain themselves in a current (testing details are given in Davis *et al.*, 1963). Therefore, at DOs <9–10 mg l^{-1} the ability of migratory fish to traverse velocity obstacles or for them to avoid predators under any conditions would be reduced (Davis *et al.*, 1963).

13.1.3 Food consumption and growth

The amount of food consumed, as well as growth, is also progressively impaired below DO saturation (9–10 mg l^{-1} at test temperatures). This was

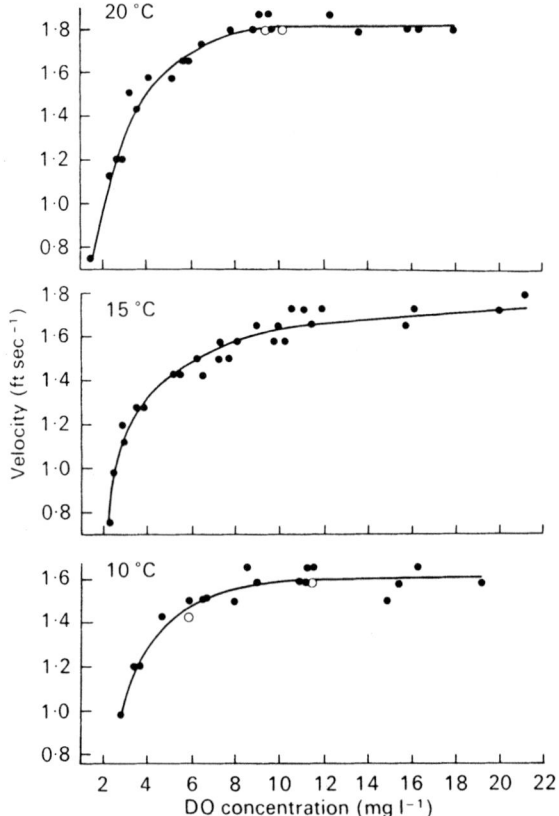

Figure 13.4 Relationships between DO concentration and swimming velocity in coho salmon at 20, 15, and 10°C. Each point represents velocity at which first underling failed to maintain orientation (Davis et al., 1963).

observed for juvenile largemouth bass (a warm-water fish) as well as salmon (Warren, 1971). Figure 13.5 shows the relationship of food-consumption rate with DO concentration for bass (Stewart et al., 1967). Again, the advantage of DO at saturation appears to hold for both warm-water and cold-water species.

Fish probably respond to the minimum daily DO rather than the daily average as shown in Figure 13.6. For juvenile salmon held at DO concentrations that fluctuated diurnally, the resulting growth was not very different from the growth of salmon held at a constant DO near the minimum for those exposed to a diurnally fluctuating DO. At unrestricted food ration, salmon clearly required a relatively high DO concentration for maximum growth. This is largely because the more food consumed, the more oxygen necessary for full utilization of that food.

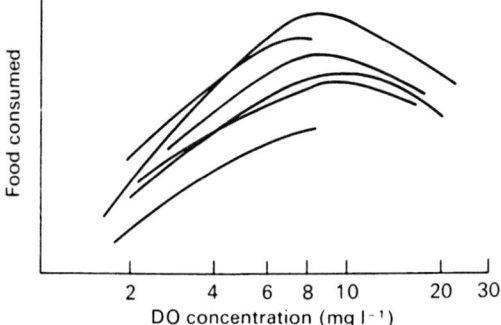

Figure 13.5 Food consumption in largemouth bass related to DO concentration. Each curve represents a separate experiment (modification from Stewart et al., 1967).

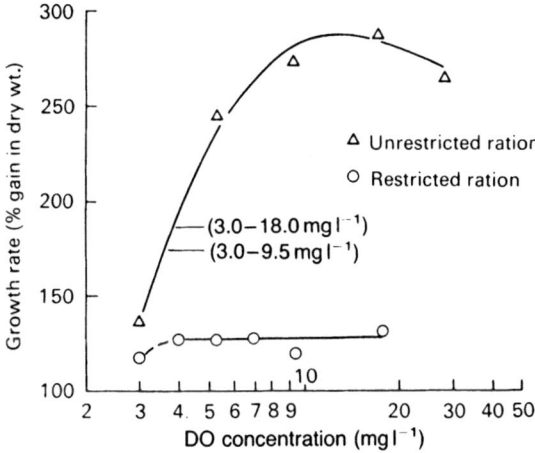

Figure 13.6 The growth rate of juvenile coho salmon fed restricted and unrestricted rations at constant and diurnally fluctuated oxygen concentrations (Doudoroff and Shumway, 1967).

The requirements for DO do vary among different species and seasons. However, the previously cited information indicates that near-saturated concentrations of DO are required during reproductive periods and periods of high food availability, consumption and growth if maximum production of cold-water as well as warm-water fish is to be realized. Levels much lower than saturation can be tolerated during periods of reduced feeding and growth is nearly independent of DO if not feeding. The minimum limits of DO at 5–6 mg l^{-1}, often cited in standards, carry an assumption that

Table 13.1 Example of recommended minimum concentrations of DO for selected levels of fish protection

Estimated natural seasonal minimum DO	Recommended minimum DO			
	Nearly maximum	High	Moderate	Low
5	5	4.7	4.2	4.0
6	6	5.6	4.8	4.0
7	7	6.4	5.3	4.0
8	8	7.1	5.8	4.3
9	9	7.7	6.2	4.5
10	10	8.2	6.5	4.6
12	12	8.9	6.8	4.8
14	14	9.3	6.8	4.9

Source: WQC (1973).

normal variations result in concentrations being much higher than $5\,\text{mg}\,\text{l}^{-1}$ most of the time. Doudoroff and Shumway (1967) have suggested that biologists should refrain from giving a value that is supposed to 'sustain maximum production of natural populations' unless it is air saturation ($9-10\,\text{mg}\,\text{l}^{-1}$ for cold-water species). Of course, it is recognized that such high values are not needed all the time nor does DO remain at or above those levels in all unpolluted water. In order to develop more specific standards, detailed seasonal requirements for various species would need to be determined.

Table 13.1 shows a working guideline based on the hypothesis of progressive detrimental effect below saturation (WQC, 1973). Minimum levels are suggested to provide high, moderate and low levels of protection based on the existing natural minimum. Note that $4\,\text{mg}\,\text{l}^{-1}$ is suggested as a floor.

These suggested levels are based on the concept that some damage will result from any reduction in the average daily minimum DO below air saturation. From the appearance of many of the relationships (Figures 13.2–13.6), damage may not be as linearly related to DO reduction as this model (Table 13.1) would suggest. Nevertheless, this is an approach to a more reasonable guideline than the often stated minimum requirement of $5\,\text{mg}\,\text{l}^{-1}$.

13.1.4 DO and sediment

Sediment may interact with low DO to adversely affect reproduction in salmonid fishes. Sediment enters streams from watershed erosion resulting from several types of disturbance. Of prime importance are irrigation systems, urban runoff and deforestation and the accompanying road-building activities.

With the filling in of the interstices within the bottom substrata with fine sediment, the intragravel water velocities are reduced. As those velocities decrease, the supply rate of DO to developing embryos is also reduced. The expected effect is as suggested in Figure 13.2. At the same time the DO of the intragravel water is apt to decrease because of the increased retention time of intragravel water (or slower replenishment of oxygen-saturated surface water) and the possible increased oxygen demand of organic matter within the sediments. The effects of reduced DO supply on developing trout eggs through sedimentation from stream-bank erosion caused by increased peak flows and sediment input from irrigation return flows were evaluated in Bluewater Creek, Montana, USA (Peters, 1967). More than 80% mortality of trout eggs buried within the stream gravels was observed at stations where water carried a mean suspended sediment content of only 200 mg l^{-1}. The physical effect was similar to that caused by runoff patterns from urban watersheds with a high proportion of impervious surface (paved), although the substratum instability caused by increased flows may also result in physical damage to eggs. May *et al.* (1997) found lower DO and higher sediment levels in spawning gravel in urbanized streams relative to rural streams in the Puget Sound region.

Clear-cutting watersheds of small streams has resulted in DO decreases of 40% in the intragravel water with the effect persisting for at least three years after logging (Hall and Lanz, 1969; Ringler and Hall, 1975). Pre- and post-logging (three years after) intragravel DO averaged 10.5 and 6.2 mg l^{-1} respectively, during the winter-spring period. That resulted in a 73% reduction in the resident cut-throat trout population. Immediately after logging, intragravel DO dropped to 1.3 mg l^{-1} and stream temperature reached 24°C in summer. In a partially clear-cut watershed (25%), where buffer strips were left along the stream in affected areas and trees were not felled in the stream, DO decreased only ≈10% and no damage to fish was observed. Emergence of young coho salmon was apparently unaffected by the physical changes.

In Bluewater Creek the banks were stabilized with vegetation and irrigation return-flow ditches were lined to cut down on sediment input. After 2 years of stream recovery following the improvements, the ratio of trout to rough fish increased sixfold (Table 13.2).

Vegetated riparian zones would also help maintain stream water temperature and water quality in urban streams. As urbanization increases, riparian buffers decrease. In 22 Puget Sound lowland streams, May *et al.* (1997) found an inverse relationship between riparian buffer area (width >30 m) and urbanization, as measured by per cent total impervious area (% TIA).

This illustrates the significance of stream-bank stabilization, vegetation protection and flow control, all of which should be employed to preserve the quality of urban streams. Even if inadvertently damaged, most streams can show some degree of recovery from sedimentation if proper corrections are installed. Probably the most pronounced recovery would come in those

Table 13.2 Sediment load reduction and fish population improvement in Bluewater Creek, Montana

Station	Sediment reduction (ton day^{-1})	Reduction (%)	Pretreatment trout/rough fish	Post-treatment trout/rough fish
2	1.9	32		
3	14.0	52	39/61	78/22
4	10.5	44	12/88	51/49

Source: After Marcuson (1970).

Note
Stations numbered in a downstream direction.

streams affected by a high discharge velocity and with only a moderate sediment load naturally. Those with a natural depositing substrata would probably show less improvement.

13.2 Temperature criteria

The thermal tolerance of fish can be illustrated by Figure 13.7. A very large zone of tolerance can be defined within which all life-history stages of a species can proceed normally. The lower and upper tolerance limits, often referred to

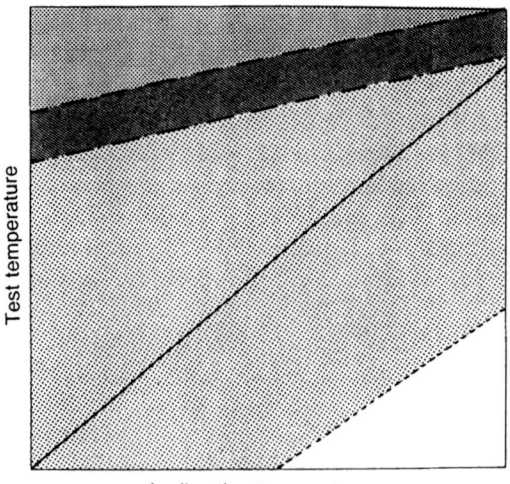

Figure 13.7 Zones of tolerance and resistance of fish as affected by changing acclimation temperature. — —, Upper lethal limit; — · —, upper incipient lethal level; ---, lower incipient lethal level (modified from Brett, 1960).

as the upper and lower incipient lethal levels, increase with increases in the acclimation temperature within the range of tolerance. At high temperatures there is a zone of resistance that is rather narrow between the incipient lethal level (long term) and the short-term lethal limit. The increase in the short-term lethal level for roach was found to be 1°C for each 3°C increase in acclimation temperature over the range of tolerance (Cocking, 1959).

13.2.1 Lethal temperature

Although the lethal limit is not very useful as a maximum limit to protect all life stages of a fish from the effects of heated water, the values are determined with a high level of precision and do indicate the variation in species tolerance over a wide temperature range. Figure 13.8 shows the distribution of fish families to lethal temperatures over the range of tolerance. Several points are apparent. Smelt (Osmeridae), salmonids and white-fish (Corigonidae) are the most intolerant of high temperature, whereas catfish (Ichtaluridae) and some minnows (Cyprinidae) are the most tolerant. The

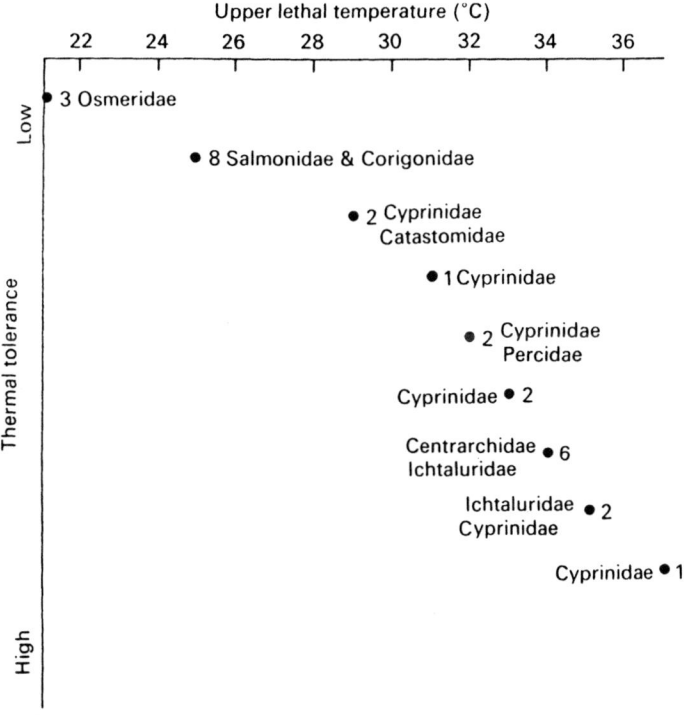

Figure 13.8 Families of fishes including number of species and lethal temperatures (compiled by Welch and Wojtalik, 1968).

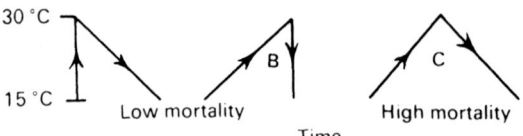

Figure 13.9 Effect of exposure to three kinds of temperature cycles on mortality in juvenile salmon (Redrawn from Templeton et al., 1969).

minnows are, however, quite variable in their tolerance. That the members of that family are found in (and tolerate) a wide variety of thermal regimes is not surprising because there are so many species of minnows.

The lethal limit is usually expressed as an LT_{50} and is determined experimentally. Test fish are exposed at a constant temperature over time and the interpolated temperature at which 50% of the test fish survive for 96 h is the LT_{50}^{96}. If the temperature is fluctuating daily, slightly higher maximums can be tolerated for short periods if temperature is lowered below that level for longer periods. This has been demonstrated with juvenile salmon exposed to raised and lowered temperature (Figure 13.9). In this case, example C produced the greatest mortality, probably because the mean temperature over the exposure period was greatest even though in each case the fish were exposed to greater than the lethal temperatures for a short period.

Although fish can resist temperatures above the 96-h lethal limit for short periods, such thermal shocks have been shown to be detrimental to salmon even if not thermally lethal (Coutant, 1969). At temperatures from 28 to 29°C, salmon showed measurably shorter times to 'equilibrium loss' than to death and this equilibrium loss resulted in increased predation on the juvenile salmon.

13.2.2 Limits for normal activity and growth

Obviously, thermal standards or limits should not be set solely on lethality or 'equilibrium loss' in environments where fish are expected to grow and reproduce. The 'scope for activity' is the temperature range at which fish can optimally feed, swim and avoid predators. The scope for activity (as determined by respiration rate) occurs over a temperature range that is much less than the lethal limit (25°C) for brook trout, whereas for bullhead catfish the maximum scope occurs near the lethal limit, which is 37°C. This example illustrates a difference between cold- and warm-water fishes (Figure 13.10). The position of the maximum scope for activity for brown trout is similar to that for the bullhead, but that trout's lethal limit is near 25°C, similar to brook trout. Brown trout are generally considered more tolerant of higher sublethal temperatures than brook trout (Brett, 1956). This may partly explain why brown trout often tend to dominate in apparently marginal waters.

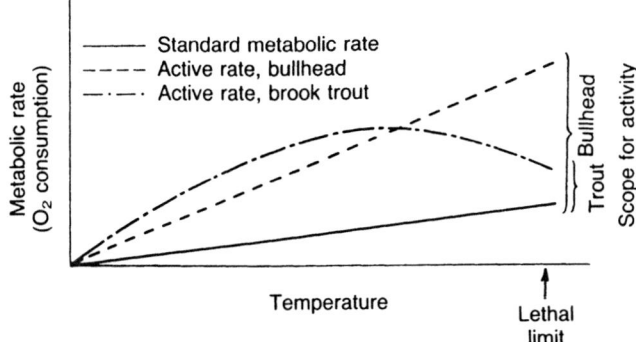

Figure 13.10 Relation of temperature to metabolic rate and scope for activity in brook trout and bullhead catfish. For brook trout, lethal temperature is 25°C and peak scope for activity is at 19°C. For bullhead, lethal and peak scope temperatures are at 37°C. For these species, scope for activity is difference between active and standard metabolic rate. Note that scope for activity is much greater for bullhead than trout near the lethal limit (from Brett, 1956).

Warren (1971) has discussed how the 'scope for growth' for salmon should be maximum at temperatures much less than the lethal limit, but the scope for growth is also a function of food availability. Growth will decrease irrespective of temperature when food is restricted; however, the temperature range for the optimum scope will become narrower when food is restricted than if unrestricted. Averett (1969) has supported this hypothesis with experiments using juvenile coho salmon, showing that the optimum temperature or maximum scope for growth is in the range 14–17°C (Figure 13.11). At unrestricted ration, the amount for growth, after the necessary metabolic demands were met, decreased above 17°C, which is 7°C below the lethal limit for coho. The normal metabolic demands or losses were cost of food handling, standard metabolism and excretion.

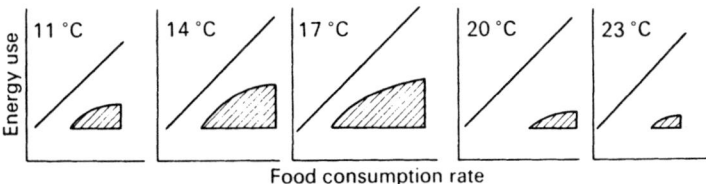

Figure 13.11 Partitioning of food energy consumed in coho salmon at varying temperatures. Cross-hatched area shows energy as growth, diagonal line indicates total energy consumed (modified from Averett, 1969).

Although the temperatures required for maximum scope for growth and activity would be desirable as criteria for all important and sensitive species, such extensive data may be difficult to obtain. A more easily obtained index experimentally is 'preferred temperature', or the temperature willingly selected by a species exposed to an experimental temperature gradient. That temperature (or range) tends to approximate the temperature range for the maximum scope for growth and activity. Table 13.3 shows lethal and preferred temperatures for cold- and warm-water species. Note that the preferred temperature for brook trout is 14–16°C, very close to the temperature for maximum scope for growth of coho salmon (14–17°C). These data are plotted in Figure 13.12 and show an interesting phenomenon; the preferred temperature becomes closer to the lethal limit as thermal tolerance increases from cold- to warm-water species.

That fish select a preferred temperature in nature was demonstrated in a study of fish distribution in the Wabash River, Indiana, USA in response to a heated-water discharge (Gammon, 1969). The effluent was on average 8°C higher than the river temperature and the river-effluent mixed temperature within 1.2 km downstream was 2–3°C above the ambient (upstream) temperature. Good agreement was observed between temperature preference determined in the laboratory and the distribution of species in thermal zones in the river (Table 13.4). Although a clear segregation of species was observable during the warm season in relation to mean temperature, during winter all

Table 13.3 Final temperature preferred and some upper lethal temperatures for a range in species tolerance

Species	Final preference (°C)	Upper lethal (°C)
Lake trout	12	
Lake whitefish	12.7	
Rainbow trout	13.6	
Brook trout	14–16	25
Brown trout	12.4–17.6	
Yellow perch	21	
Burbot	21.2	
Muskellunge	24	
Yellow perch	24.2	32
Grass pickerel	26.6	
Small-mouth bass	29	
Goldfish	28.1	34
Pumpkinseed	31.5	
Carp	32	
Large-mouth bass	30–32	34
Blue-gill	32.3	34

Source: Welch and Wojtalik (1968).

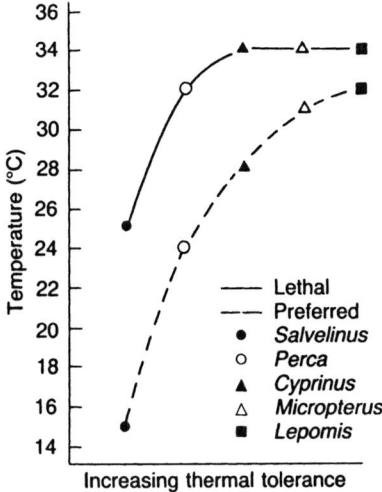

Figure 13.12 Preferred and lethal temperatures of five genera of fishes (data compiled by Welch and Wojtalik, 1968).

species were attracted to the effluent area because the resulting temperature there was closer to their preferred temperature than that of the normal river.

The abundance of all species in the river in relation to the mean daily maximum temperature is shown in Figure 13.13. The number of species was substantially reduced in the zone of maximum temperature increase. Although the temperature considerably downstream of the maximum heated

Table 13.4 Distribution of representative species of fish in the Wabash River among several definable thermal zones

Avoided all increases	Thermal gradient in the river		
	Normal to 27°C	32°C	35°C
Golden redhorse	Sauger	Gizzard shad	All fish left the area, no mortality
Shorthead redhorse	White bass	Carp	
Smallmouth bass	Northern river carpsucker	Buffalo fish	
Spotted bass		Longnose gar	
		Shortnose gar	
		Channel cat	
		Flathead cat	

Source: Data from Gammon (1969).

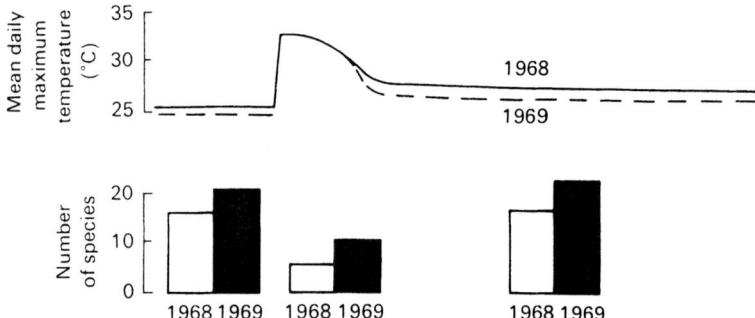

Figure 13.13 Relationship of fish species number to temperature increase in the Wabash River, Indiana, caused by heated water from a power plant (after Gammon, 1969).

zone was higher than the normal river, more species were actually found there. Fish do not necessarily exist only in areas of their preferred temperature nor are they excluded from environments where temperatures are less than preferred. Although all species sampled were present in the river before heat addition, most apparently preferred a slightly elevated temperature than the normal. An important point for management in this case, however, is that the most desirable species was one of the most sensitive. Smallmouth bass, a highly prized game species, avoided all temperature increases.

Based largely on preferred and lethal temperatures, the prediction of fish community changes in the Columbia River (cold water) and the Tennessee River (warm water) were hypothesized with increasing (mean daily) maximum temperature (Bush et al., 1974). If the percentage of species lost is plotted versus temperature for both rivers and compared with the normal maximum temperatures (Figure 13.14), it is again indicated that fish in warm-water environments are living closer to their lethal limit than most species in cold water (Figure 13.12). That is, a 2°C increase in the Tennessee River would result in more species lost than the same increase in the Columbia River. In reality then, a greater 'assimilative capacity' for heat with respect to fish tolerance does not occur in warm water just because those species are more tolerant. On the other hand, there is not that much assimilative capacity available in cold water either because the maximum scope for growth and activity tends to be much less than the lethal limit in contrast to warm-water species. Moreover, salmon would be the first to disappear with a 2°C increase in the Columbia River. Nevertheless, more species would be lost by an equivalent increase in warm water, although the margin available for temperature increase in cold water may be only a few degrees, because of a few important species like salmon and trout.

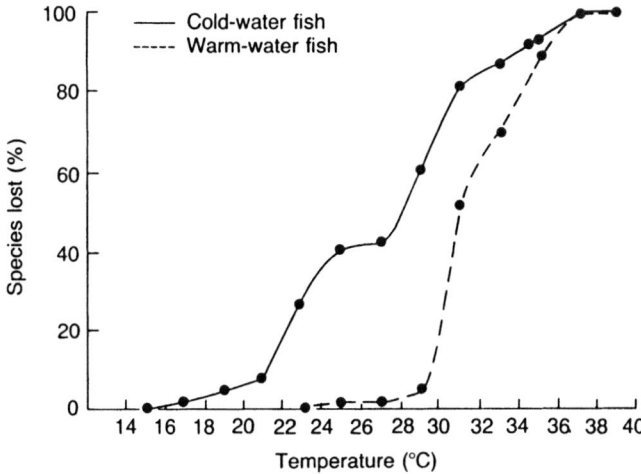

Figure 13.14 Predicted species loss rate from a cold-water system (Columbia River, maximum mean daily temperature 21°C) and a warm-water system (Tennessee River, maximum mean daily temperature 30°C) with increasing temperature (Bush et al., 1974).

This type of analysis may have value in assessing the potential effects of global warming on aquatic organisms. As previously indicated, fish tend to represent a narrower range of thermal tolerance than invertebrates. Increases in the average temperature of 3–4°C may cause extinction of several fish species in the Great Plains, for example, because migration north could be blocked and insufficient time would likely exist for genetic adaptation (Matthews and Zimmerman, 1990).

Thermal limits for reproduction have been determined rather precisely for salmon but less data are available for warm-water species. The results given in Table 13.5 for salmon were summarized by Brett (1956). The following are upper limits (maximum weekly means) for spawning suggested for the representative warm-water species (WQC, 1986): largemouth bass, 21°C; smallmouth bass, 17°C; yellow perch, 12°C; northern pike, 11°C.

Table 13.5 Thermal limits for salmon reproduction

Species	Limits for successful hatching and survival of young (°C)
Sock-eye	4.4–5.8 to 12.8–14.2
Chinook	5.6–9.4 to 14.4
All salmon	5.8 to 12.8

Source: Brett (1956).

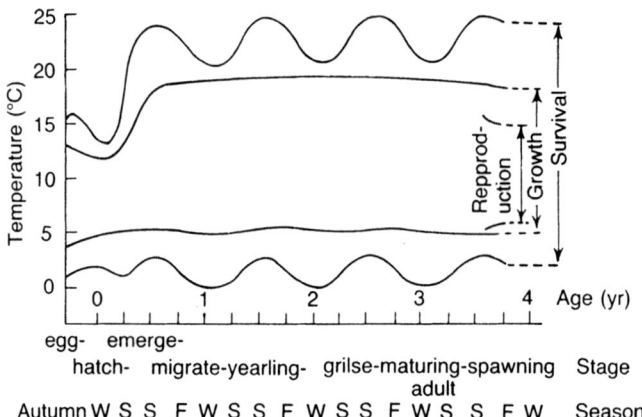

Figure 13.15 Schematic representation of thermal requirements for different life processes that characterize Pacific salmon (Brett, 1960).

The critical nature of the reproductive life-stage is evident in the seasonal thermal requirements for pacific salmon (Figure 13.15) summarized by Brett (1960). Note that the most restricted range is during egg development and that a gradual increase is recommended in spring for normal growth to result. The temperature for optimum growth is suggested to remain rather constant after the larval stage, but the lethal temperature for post-larval fish increases in spring and summer, probably as a result of acclimation to rising temperature and a longer photoperiod.

13.3 Temperature standards

The standards for temperature vary from state to state in the USA, but in general are not greatly different from those recommended in WQC (1968) and will be discussed here. Standards for cold-waters are usually stated separately from those for warm-water environments. Warm water was defined by a >25°C maximum daily mean temperature, which is actually the upper lethal limit for most salmonids. The recommendations were that the rise above ambient (normal, before heat addition) should not exceed 2.8°C in streams and 1.7°C in lakes. The maximum temperature at any time should not exceed 32–34°C, depending on area. The rate of rise should not exceed $3°C\,h^{-1}$. Cold water was conversely defined by a <25°C maximum daily mean temperature. The rise above ambient should not exceed 2.8°C in streams and 3°C in lakes. The maximum temperature at any time should not exceed 20–21°C.

More recently, species-specific criteria were recommended for survival, growth and reproduction (WQC, 1986). For growth the maximum weekly mean temperatures were calculated as the optimum temperature plus one-third of the ultimate incipient lethal level minus the optimum. These values would fall roughly one-third of the distance between the lower and upper curves in Figure 13.12.

13.4 Comments on standards

The recommended maximum temperatures of 34°C and 21°C actually fall between the lethal and preferred temperatures for warm- and cold-water species as indicated above. If the limit for warm water (34°C) persisted constantly for as long as a day or two, mortality could result or, more likely, many species would leave the area. However, fish would not be exposed to 34°C constantly. If the maximum daily temperature were kept below that level, temperature would be lower at night.

The maximum rise above ambient, or ΔT, is the second most critical factor and must be set from consideration of the following.

1. If the maximum temperature at any time is not allowed to exceed the recommended limit (e.g. 34°C), the length of exposure to that summer maximum will increase and may become critical as ΔT increases. Reduced growth may result if they are exposed too long to above the maximum scope for growth.
2. Reproduction may be offset from food availability if the ΔT is too high so that the optimum temperature for reproduction would occur earlier than the onset of food production which, if dependent on phytoplankton production, may be controlled by light intensity.
3. The amount of day-to-day fluctuation that would occur as a result of power-plant problems, flow change or climate change could result in day-to-day fluctuations that progressively increase.

Considering point 3, a hypothetical effect of flow and climate change is illustrated in Table 13.6. Such rapid temperature changes (drops) are possible in heated portions of rivers and would create stress in developing fry. Such changes have been shown to cause warm-water fish, such as smallmouth bass, to desert their nests (Rawson, 1945).

A logical approach to developing standards for temperature, as well as for DO, would be to make them less general and more specific for a purpose. The specific purpose would probably be protection of the most recreationally/economically important fish species, which is effectively what is usually done. The aquatic community in general would probably be protected by protecting those most sensitive fish, which has been suggested (Mount, 1969; Becker, 1972; Bush et al., 1974). This seems reasonable if one considers the position of invertebrates compared to fish in the species frequency distribution with lethal temperatures. The lethal temperature mode for freshwater

Table 13.6 Effect of flow and climate change on the flow-weighted temperature in a heated portion of a river

	Flow (m³ s⁻¹)	°C	Flow-weighted	Mixed °C
Normal conditions during spring reproduction period				
Effluent	2	38	$\dfrac{76 + 84}{6}$	$= 26.7$
River	4	21		
Sudden increase in flow only				
Effluent	2	38	$\dfrac{76 + 252}{14}$	$= 23.4$
River	12	21		
Sudden flow increase and normal river temperature drop				
Effluent	2	38	$\dfrac{76 + 216}{14}$	$= 21.9$
River	12	18		

invertebrates is to the right of that for fish in Figure 13.16. Assuming that standards to protect fish would also protect the aquatic community in general, specific guidelines, such as those suggested in Table 13.7, could be utilized in setting limits for maximum temperature and ΔT in different environments.

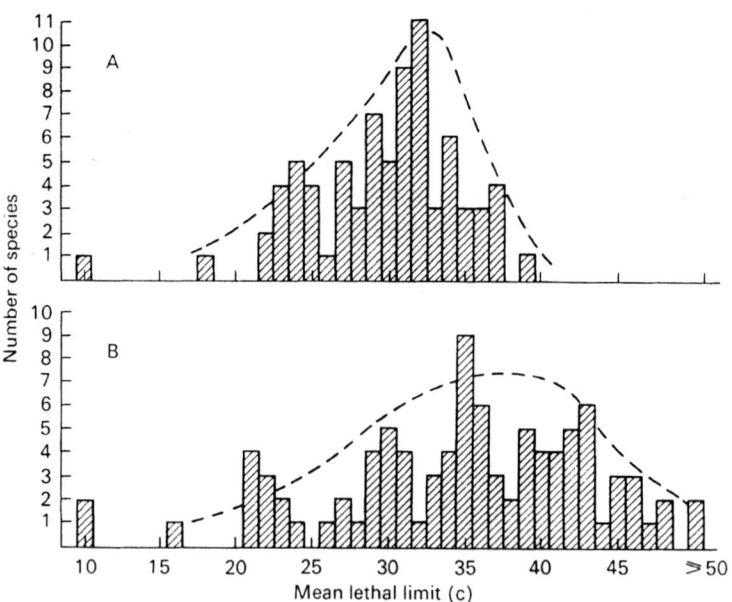

Figure 13.16 Distribution of freshwater fish (A) and invertebrates (B) according to tolerance to temperature. Dashed line shows approximate mode (Bush et al., 1974 with permission of the American Chemical Society).

Table 13.7 Maximum temperatures for various life stages of important fish species

Maximum temperature (°C)	Life stages
34	Growth: catfish, gar, white and yellow bass, buffalo, carpsucker, shad
32	Growth: largemouth bass, drum, bluegill, and crappie
29	Growth: pike, perch, walleye, smallmouth bass, and sauger
26.7	Spawning and egg development of largemouth bass, white and yellow bass, and spotted bass
20	Growth or migration routes of salmonids and for egg development of perch and smallmouth bass
12.8	Spawning and egg development of salmon and trout
9	Spawning and egg development of lake trout, walleye, northern pike, sauger, and Atlantic salmon

Source: WQC (1968).

With respect to points one and two earlier, one could then use these maxima with an ambient temperature cycle, which normally increases gradually from the winter low to the summer high, to judge a safe increment rise above ambient (ΔT) for individual species environments. Using this approach for the Tennessee River, it can be shown that 2.8°C may be a safe limit for ΔT while 5.6°C is not (Figure 13.17).

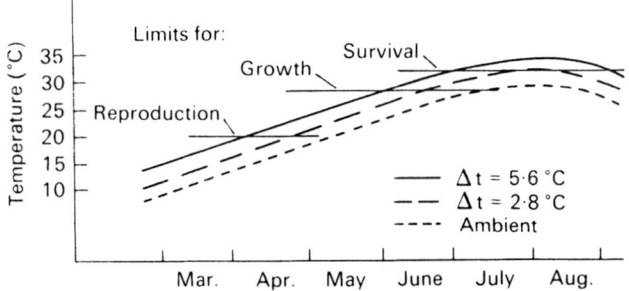

Figure 13.17 Effect of 2.8°C and 5.6°C temperature increments on the normal temperature cycle of the Tennessee River (Welch, 1969b). Periods for life-cycle activities that could be affected are shown according to the standards for smallmouth bass and yellow perch (WQC, 1968).

A ΔT rise above ambient of 5.6°C appears excessive for smallmouth bass, a highly prized sports species, in the Tennessee River during high-temperature, low-flow years. The periods of reproduction and growth, according to temperature maxima for those activities in Table 13.7 (20°C and 29°C maxima, respectively), would occur from one to two months earlier and either mortality, substantial weight loss or a complete evacuation of the species from the area could be expected in summer even if the maximum were restricted to 34°C. In fact, Horning and Pearson (1973) have noted that whereas the best growth for smallmouth bass occurred at 26–29°C, a temperature even lower was preferred, so accordingly 29 or 30°C would be the maximum tolerable temperature for the species. This technique of an optimized thermal regime based on a consideration of the various life-stage requirements has also been illustrated similarly for largemouth bass in WQC (1973).

Interestingly enough, a ΔT of 5.6°C (10°F) was accepted by most states in the Tennessee Valley in the late 1960s, but under pressure from the then federal Water Pollution Control Administration, the standard was reduced to 2.8°C (5°F) which seems appropriate from the above analysis. Subsequently, a series of whole-ecosystem, 112-m-long artificial streams were constructed at the Browns Ferry Power Station to develop temperature criteria for populations of smallmouth bass (Wrenn, 1980). Survival and growth of bass through a full year (1977–1978) was not different among four treatment temperatures: ambient (maximum = 30°C), ambient +3°C, ambient +6°C and ambient +9°C. As a result, a maximum weekly mean temperature of 32–33°C for growth and a short-term maximum of 35°C were recommended for smallmouth bass (Wrenn, 1980). These limits contrast significantly with the previously reported range for optimum (25–29°C) and the reported upper lethal temperature (35°C). Apparently, a greater tolerance to raised temperature was shown for this species when exposed throughout a full year in a whole-stream environment than had been the case when exposed under laboratory conditions.

13.5 Global warming

Over the past 20 years, the concept that human activities can change the climate on a global scale has become widely accepted by scientists and the public alike. Global warming refers to the rising temperature of the earth due to the increase in greenhouse gases (GHGs) from human activities. GHGs include water vapour, carbon dioxide, ozone, methane, nitrous oxide and chlorofluorocarbons. The increase in carbon dioxide, which accounts for approximately 55% of the global warming effect, is primarily due to combustion of fossil fuels, as well as deforestation. Carbon dioxide concentrations have increased 31% since 1750 (IPCC, 2001). Methane (from digestive processes of ruminants such as sheep, cows and goats, as well as anaerobic decay of organic matter) and chlorofluorocarbons (synthetic chemicals used in refrigeration systems) are the other primary contributors to global warming contributing 15 and 24%, respectively. The globally averaged surface temperature is projected to increase by 1.4–5.8°C over the period 1990–2100 (IPCC, 2001). This temperature increase will have significant effects on aquatic ecosystems resulting in serious environmental, societal and economic consequences. These effects will be linked to the response of the hydrologic cycle to climate change. In turn, hydrologic processes will influence climate response.

Although site-specific hydrologic responses may be difficult to predict, it is likely that global warming will cause decreased river flows and lake levels, contraction of wetlands, decreased soil moisture, reduced atmospheric humidity, increased evaporation, increased frequency of drought, variable precipitation, increased incidence of floods and erosion, increased sea level and subsequent coastal flooding, and decreased groundwater recharge and storage (IPCC, 2001). These hydrologic changes will result in altered water quality and aquatic communities.

A warming over recent decades has been documented in a number of lakes (Gerdeaux, 1998). Ice breakup dates from 1968–1988 were 0.82 day earlier per year for lakes in southern Wisconsin and 0.45 day per year for northern Wisconsin lakes (Anderson *et al.*, 1996). Water temperatures have increased by 2°C over 20 years and ice-free season has increased by 3 weeks in the Experimental Lakes Area of Canada (Schindler, 1997).

As discussed in the previous sections, fish are extremely responsive to temperature. Therefore, fish communities are especially vulnerable to the effects of global warming. Some species of cold-water fish may become locally extinct from waters that warm to temperatures above the tolerance limit. Vulnerable populations include southern populations of Atlantic salmon (*Salmo salar*) in northern Spain and south-west France (McCarthy and Houlihan, 1997). The range of other stenothermic fish species may shift to northern or higher-altitude colder regions (Magnuson *et al.*, 1997). Fish that inhabit very restricted ranges such as desert pools and streams will become extinct following dry-out of their habitat (Carpenter *et al.*, 1992). On the other hand, warm-water species may expand their range moving into higher latitudes. For example, 27 species of Cyprinidae and Centrarchidae are predicted to move into the Great Lakes as waters warm (Mandrak, 1989).

Global warming will also affect fish by lowering water quality and quantity. During droughts, decreased DO combined with increased temperature can severely stress and kill fish (Everard, 1996). Salmon populations will be particularly affected by prolonged drought (Armstrong *et al.*, 1998). In the Pacific Northwest region of North America, the amount of water held in springtime snow packs of the mountains has declined steadily (as much as 60% in some places) between 1950 and 1992. Smaller snow packs mean lower stream flows and higher temperature for fish in the rivers (Melack *et al.*, 1997; Mote *et al.*, 1999; Miles *et al.*, 2000). Although global warming will likely result in warmer, wetter winters in the Pacific Northwest, the precipitation will fall as rain instead of snow, reducing the mountain snow pack, which provides the region's river flows and hydropower. Ironically, the effects of global warming in this region mean higher river flows in the late winter and spring and more flooding, but lower river flows in the summer and fall, when fish are especially dependent on sufficient, cool, well-oxygenated water. Furthermore, lower river flows could hinder spawning of adult salmon and desiccate spawning gravels. Adverse effects on nutrient cycles and plankton productivity were also predicted to decrease food supply to juvenile sockeye salmon in the Fraser River drainage basin, thus decreasing smolt to adult fish survival (Henderson *et al.*, 1992). The salmon populations of the Pacific Northwest have precipitously declined due to over-harvesting, habitat degradation, competition with hatchery fish and hydropower dams that block fish passage. Climate variability is bringing additional stress to systems, such as the rivers of the Pacific Northwest, which are already stressed by human activities.

13.6 Toxicants and toxicity

The problems caused by toxicants in wastewater and their effects on fish and aquatic invertebrates have plagued fisheries managers and pollution control agencies in the USA for over 50 years. Recently, the threat of toxicants and other hazardous substances to humans and the clean-up of abandoned waste dumps have received much attention. The pathway to the human population from these toxic waste dumps is usually through the groundwater, so aquatic organisms in surface waters are often not an issue at hazardous waste sites. Nevertheless, the >1000 abandoned sites deemed sufficiently hazardous to require clean-up (Miller, 1988) represent a more recent addition to society's legacy of less than effective control of toxicants.

While many problems of toxicity in surface waters have been controlled (e.g. treatment of point-source industrial effluents and banning the use of some pesticides), there are continually new ones caused by the increased production of synthetic organic chemicals. Although detailed analysis of the effects of each of these compounds is required before marketing is approved, aspects of their long-term effects and fate/recycling in the environment are often not revealed for many years. Food-chain bioaccumulation effects of chlorinated hydrocarbon insecticides (DDT, dieldrin, etc.), polychlorinated biphenyls (PCBs) and methylated mercury (CH_3Hg^+) are examples of unforeseen problems.

Because of the variation in tolerance among species to be protected and variation in effects caused by interacting environmental factors, toxic effects are usually not predictable in any quantitative sense. That is, the response of fish populations cannot be predicted, especially in the long term, from given inputs of toxicants to a specific stream. Moreover, acceptable levels for toxicants, as is the case for temperature, DO and pH, are usually not given in enforcement standards for water quality. What is usually offered in place of specific standards is a statement to the effect that 'no contaminant (pollutant) can be added to waters that is detrimental to fish and aquatic life or other water uses'. The onus is left on the wastewater discharger and/or enforcement agency to assess each problem on its own merits. Such an assessment most appropriately employs an experimental approach, but may only entail a simple comparison of existing levels of toxicants in the water in question with published criteria (e.g. WQC, 1986; USEPA, 2002a). Application of test results from one water to another, with differing chemical characteristics, may not be appropriate without correction or qualification as will be discussed later. Criteria for some toxicants (e.g. Cu, Zn, Cd, ammonia) are interpreted as standards in some states.

13.6.1 Determination of toxicity

Controlling waste loading to waters to prevent detrimental effects without specific standards usually necessitates some type of investigation into the toxicity existing in the water or the potential toxicity from wastes either added already or planned for addition. The investigation usually entails a standardized toxicity bioassay, employing the waste, the receiving water and an experimental animal, usually a fish, but may include a crustacean or alga. The results from such a bioassay (or test) provide information about the acute or short-term toxicity from which a lethal concentration can be determined. Based on the lethal concentration and some knowledge of chronic effects, a 'safe level' can be estimated.

Procedures for acute toxicity bioassays (or tests) are given in Standard Methods (APHA, 1985). Typically, the exposure period is 2–4 days to a series of concentrations with the number of surviving animals recorded daily or for shorter intervals. Response is commonly plotted as either percentage survival (arithmetic scale or transformed, e.g. as probits) with time being held constant (4 days), or as time to 50% survival, on the ordinant and concentration of toxicant on the abscissa (Figure 13.18). With time constant, the concentration producing 50% mortality (survival) is interpolated and is termed the lethal concentration producing 50% survival after 96 h or the 96-hour LC_{50}. If log concentration is plotted, the relationship is more linear.

With time variable, the 'lethal threshold concentration' can be estimated by approximating the toxicant concentration at which the resistance time becomes long (asymptotic) at an additional, relatively small decrease in toxicant concentration (Figure 13.18).

LC values at other percentage survivals (e.g. LC_1, LC_{10}) can also be estimated from the data by plotting results on probit paper on which survival versus concentration is a straight line (APHA, 1985). However,

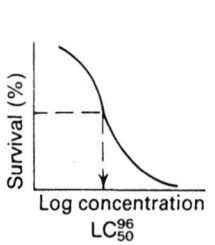 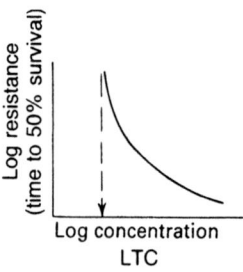

Figure 13.18 Two graphical methods of determining the lethal concentration of a toxicant. LC_{50}^{96} (medium lethal concentration after 96-h exposure) and LTC (lethal threshold concentration) indicated by broken-line arrows.

the 50%, or median, survival level is most commonly used because that estimate is least variable.

Bioassays can be either static, with no exchange of test water during the 96-h period, or continuous flow, entailing either several solution changes per day or simple renewal of the bulk solution daily. The continuous-flow procedure is considered more appropriate, because in nature organisms are continuously exposed to renewed solutions. In static laboratory solutions, animals can effectively alter the toxicant concentrations and either detoxify the solutions or cause self-contamination with waste products. The lethal concentrations from continuous-flow experiments are apt to be lower than from static ones. For example, Brungs (1969) reported a 96-hour LC_{50} for Zn and fathead minnows (*Pimaphales promelas*) in static tests of 12–13 mg l^{-1}, but it was only 8.4 mg l^{-1} in continuous-flow tests.

Knowing the median lethal concentration, or the concentration that kills 50% of the fish, is not an acceptable goal to control toxicity in a receiving water. Conventionally, an application factor (AF) is used to proportionately reduce the lethal concentration to one that would allow complete survival of all stages of the animal's life history. The 'safe concentration' is the product of the 96-hour LC_{50} and the AF.

AFs vary from 0.5 to 0.01, depending on the toxicant. Although in many instances the AF value is arbitrary, AFs have been developed for many toxicants that are based on long-term (1–2 months) chronic bioassays using such life-history functions as growth, fecundity (egg production), hatching success, food conversion and behaviour as the response. With the chronic effect level determined for the most sensitive life history stage, the AF can be estimated as:

$$AF = \text{chronic } LC_{50}/LC_{50}^{96}$$

Results of AF determinations for such toxicants as pentachlorophenol, cyanide, dieldrin, 2,4-D, malathion, copper and pulp-mill effluent, using growth and reproductive success most often, have ranged from 0.02 to 0.5 (Warren, 1971). Additional values are given in WQC (1973) for LAS (detergent), chromium, chlorine, sulphide, nickel, lead and zinc. As an example of the procedure, Brungs (1969) showed that fecundity in fathead minnows was more sensitive to chronic levels of Zn^{2+} than growth or egg hatching success (Figure 13.19). Using the calculated 50% reduction in fecundity and the associated concentration of Zn^{2+}, the estimated AF could be calculated:

$$AF = 0.088/9.2 \, mg \, l^{-1} \; LC_{50}^{96} = 0.009$$

In the case of Zn, the more conventional AF of 0.1 would be tenfold too high.

This procedure of estimating a safe level is very effective in accounting for the variability in water quality characteristics within a receiving water, as well as the variability among waters. Those variations are accounted for in the short-term acute bioassay using the receiving water in question. The more difficult to obtain AF is determined from long-term chronic bioassays conducted earlier.

Faster procedures for determining chronic toxicity have been developed for testing complex effluents. These are a 7-day subchronic test using larval and embryo-larval stages of fathead minnows (Norberg and Mount, 1985), and a three-brood, life-cycle test using the cladoceran *Ceriodaphnia* (Mount and Norberg, 1984). Chronic concentrations for the larval fathead minnow tests are estimated as either the LC_1 or the mean of the concentrations giving no observable effect and the lowest observable effect using growth and

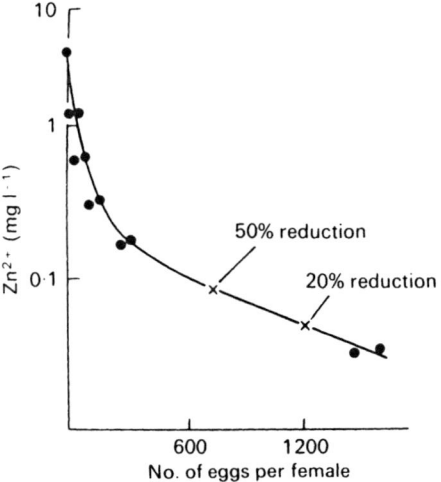

Figure 13.19 Effect of Zn^{2+} on fecundity in fathead minnows (Brungs, 1969).

survival over 7 days. Precision was greater and threshold concentrations lower for LC_1 estimates than for means of no effect and lowest effect using two toxicants (Pickering, 1988). Short-term chronic tests such as these were developed for use in the permission system for wastewater discharge.

There is still the problem of extrapolation from one test species to the myriad species in the receiving water. Because it is not possible to conduct long-term (or even short-term) chronic tests with every species in a receiving water, there are three alternatives (Woelke, 1968): select the most sensitive species with which to determine the AF; assume proportionality in AFs between an easily studied test species (e.g. fathead minnow) and the most sensitive species in the receiving water; or select the most economically important species even if it may not be the most sensitive. Woelke (1968) used the reproductive stage of the economically important Pacific oyster and subsequently showed that it is as sensitive as many other shellfish species to pulp-mill waste as well as oil-spill dispersants and is much more sensitive than adult fish. In that case, the oyster is an ideal test species being both sensitive and economically important.

Another serious problem if the $LC_{50} \times AF$ procedure is used is the complexity of most wastes which often contain numerous toxicants that can interact. Mixtures of toxicants can display either additive, antagonistic or synergistic effects (Figure 13.20). If strictly additive, there is no difference in the LC_{50} whether the mixture is mostly compound A or B. If the effect is antagonistic, the more equal the mixture of A and B, the lower the toxicity (higher the LC_{50}). Antagonism exists between Zn^{2+} and S^-, where toxicity of the mixture is least when the stochiometric ratio is one (Hendricks, 1978). Synergism results in greater toxicity as the mixture of A and B is equalized. Cu^{2+} and Zn^{2+} together is a classic example of synergism. There is a no interaction area where the effect is due to either A or B, so toxicity declines with less concentration.

Assuming an additive or slightly synergistic effect is one approach to estimating the effect of a mixture of toxicants. An additive model has been

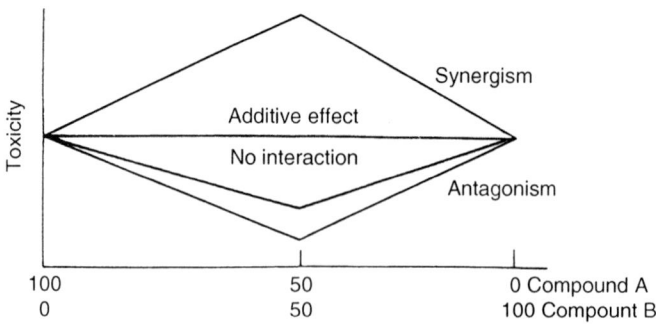

Figure 13.20 Relative effect of toxicant mixtures on toxicity.

used to predict lethal effects of mixtures of toxicants (Brown, 1968; Brown et al., 1970):

$$C_1/LC_{50(1)} + C_2/LC_{50(2)} + C_n/LC_{50(n)} = 1.0$$

where $C_1, C_2 \ldots C_n$ are the concentrations of various toxicants in the mixture. Thus, in a mixture where the fractions of all toxicant concentrations, relative to their respective LC_{50} values, sum to 1.0, 50% mortality should result. In actual applications, Brown et al. (1970) found that the additive model underestimated the acute toxicity in two polluted rivers and an estuary. That is, the respective fractions produced 50% mortality when the sum was <1.0 of the order of two-thirds. Therefore, one might wish to stipulate that the sum of the fractions of the respective toxicants should not exceed some value <1.0, say 0.5–0.8. A model to approximate long-term safe concentrations of a wastewater with a mixture of toxicants in a receiving water would be (WQC, 1973):

$$C_1/LC_{50(1)}AF_1 + C_2/LC_{50(2)}AF_2 \ldots C_n/LC_{50(n)}AF_n < 1.0 \text{ (e.g. 0.8)}$$

A simple hypothetical example of estimating an acceptable volume addition of a wastewater using the above model is as follows. River X flows at $5 \, m^3 \, s^{-1}$ during summer low flow and effluent Y flows at $0.1 \, m^3 \, s^{-1}$ with a Cu content of $1.0 \, mg \, l^{-1}$ and a Zn content of $1.5 \, mg \, l^{-1}$. Actual short-term bioassays are usually conducted with a sensitive species and the receiving water in question.

The mixed, effluent-river concentrations ($0.1/5 \times 1.0$ and 1.5) are $0.02 \, mg \, l^{-1}$ Cu and $0.03 \, mg \, l^{-1}$ Zn. Assuming slight synergism and determined LC_{50} values of 0.08 and $0.5 \, mg \, l^{-1}$ for Cu and Zn, respectively, suggests that short-term mortality should not result:

$$0.02/0.08 + 0.03/0.5 = 0.31$$

But there could be a chronic effect assuming AFs for Cu and Zn of 0.1 and 0.01, respectively:

$$0.02/(0.08 \times 0.1) + 0.03/(0.5 \times 0.01) = 8.5$$

Therefore, a recommendation would be to reduce the effluent flow during low river flow by 90% ($1 - 0.8/8.5$). Or the discharger could opt to remove the toxic metals through treatment or to pond the wastewater and wait for higher river flows such that the mixed river concentrations do not exceed $1.9 \, \mu g \, l^{-1}$ Cu and $2.8 \, \mu g \, l^{-1}$ Zn ($0.8/8.5 \times 20 \, \mu g \, l^{-1}$ and $30 \, \mu g \, l^{-1}$).

13.6.2 Heavy metals

Elements such as Zn, Cd, Cu, Pb, Cr, Hg, Ag and Ni are well-known heavy metals that can occur in a variety of wastes and cause either acute or chronic effects on organisms in receiving waters. Although one or more of these metals can originate from a variety of industrial activities, they can also occur in significantly high concentrations in municipal effluents, urban run-off and with Pb and Hg through atmospheric deposition. Recommended criteria for maximum and continuous concentrations for each of these metals (as dissolved concentrations) are given in Table 13.8. These values are based on rather extensive chronic bioassays and adherence to these criteria should provide minimal risk to even the most sensitive species.

The toxic effect of a given concentration of a metal can vary greatly depending on the form of the metal, which depends on pH, temperature and the presence of dissolved organic substances. The mode of action of metals that produces the lethal effect is an irritation to the fish's gills, causing a secretion of mucus and internal deterioration of the gill lamellae. Death is due to suffocation as the capacity of the gill system to effectively exchange gases between the ambient environment and the fish's blood is impaired. Anionic imbalance can also result as excretion and ion exchange across the gill membrane are adversely affected.

Table 13.8 Recommended water quality criteria for representative toxic pollutants in freshwater (USEPA, 2002a)

Toxicant	CMC^a ($\mu g\, l^{-1}$)	CCC^b ($\mu g\, l^{-1}$)
Metals (dissolved)		
Cd^c	2.0	0.25
Cr (VI)	16	11
Cu^c	13	9.0
Pb^c	65	2.5
Hg	1.4	0.77
Ni^c	470	52
Zn^c	120	120
Other inorganics		
NH_3	15^d	
Cl_2	19	11^e
H_2S	–	2.0
HCN	22	5.2
Organics		
Total PCBs	–	0.014
DDT	1.1	0.001
Chlordane	2.4	0.0043
Toxaphene	0.73	0.0002

Notes
a CMC = criterion maximum concentration.
b CCC = criterion continuous concentration.
c At 100 mg l^{-1} hardness.
d Criterion assumes the presence of salmonid fish and pH 7 (see Appendix C, USEPA, 2002a).
e Criterion depends on presence/absence of fish early life stages, pH and temperature (see Appendix C, USEPA, 2002a).

Lloyd (1960) showed with rainbow trout killed by exposure to $ZnSO_4$ that most of the Zn was contained in gills and mucus. The mucus, however, was produced from the body surface and not the gills. Cytological breakdown of the gill tissue was observed within a few hours at high concentrations. Tissue hypoxia resulting from cytological breakdown of the gills has been further demonstrated as the cause for acute toxicity from Zn (Burton et al., 1972).

Usually the effect of heavy metals is considered to be due more to the dissolved ionic form of a metal (e.g. Cu^{2+} or Zn^{2+}) than to other hydroxide complexes (e.g. $Cu(OH)^+$, $Cu(OH)_2$, $Zn(OH)^+$ or carbonate complexes. The relative amount in the ionic form increases with a decrease in pH. For example, Cairns (1971) showed for a range of Zn concentrations from 10–32 mg l^{-1} that mortality of only 0–10% occurred at pH values from 7.3 to 8.8, but mortality was complete at pH values from 5.7 to 7.0. However, Mount (1966) found that toxicity of Zn to fathead minnows increased with increasing pH. The LC_{50} values increased by a factor of nearly 3 when pH was decreased from 8.6 to 5.0. This was thought to be due to an effect of continuous flow bioassay systems maintaining precipitated Zn in suspension. A direct effect of pH on the toxicity to fish from Cd, Zn and Cu was also shown by Cusimano et al. (1986). Toxicity was reduced at pH 4.7 compared to pH 7 and was explained as H^+ acting as a competitor for binding sites on the gill surface (see discussion below).

Thus, there is considerable uncertainty in estimating the toxic effect of a metal (or metals) based on determinations of total acid-digestible content of either filtered or unfiltered water. The latter (i.e. unfiltered) would tend to overestimate availability of the ionic form while the former may underestimate availability from equilibrium considerations or over-estimate availability if an organic complex is involved.

McKnight (1981) used a specific ion electrode to determine the effect of the copper ion, following a $CuSO_4$ treatment of a water supply reservoir in which there was a complexing interference to toxicity from humic substances. Laegrid et al. (1983) also used the specific ion electrode to show that the free metal ion accounted for the toxicity of Cd in waters with high humic substances, but also found that complexation with low-molecular weight organic compounds, apparently secreted by phytoplankton, increased toxicity beyond that due to Cd^{2+} alone. A similar approach with Cd was used by Sherman et al. (1987), who found that Cd additions to a pond environment did not produce the expected toxicity based on Cd^{2+} concentrations in laboratory experiments, because high pond pH caused precipitation as $CdCO_3$.

Because the toxicity of metals is more directly related to the dissolved ionic form, the freshwater and saltwater criteria for metals are now expressed as dissolved metal concentration in the water column (USEPA, 2002a). These water quality criteria were calculated by using the previous aquatic life criteria expressed as total recoverable metal (WQC, 1986) and multiplying it by a conversion factor (CF). The CFs for dissolved metals can be found in Appendix A of USEPA (2002a). Furthermore, the acute and chronic criteria are now expressed as criteria maximum concentrations (CMCs) and criteria continuous concentrations (CCCs), respectively. CMC is the highest concentration of a material in surface water to which an aquatic community can be exposed briefly without resulting in an unacceptable effect. CCC is the highest concentration of a material in surface water to which an aquatic community can be exposed indefinitely without resulting in an unacceptable effect.

Another important factor that affects the toxicity of metals is hardness (Ca + Mg), which reduces the toxic effect of a metal through competitive inhibition at the gill surface. The non-toxic Ca^{2+} and Mg^{2+} ions compete with the toxic metals for binding sites. If Ca or Mg occupy the sites, the gill lamellae are protected from deterioration. Pagenkopf (1983) has shown that the competitive interaction (inhibition) factor (CIF) can be estimated according to:

$$\text{CIF} = S^{n-}/S_T = 1/(1 + K_M[M^{2+}])$$

which represents the available fraction (S^{n-}) of the total (S_T) gill sites. K_M and $[M^{2+}]$ are, respectively, the equilibrium solubility coefficient and molar metal concentration. The effective toxicant concentration (for Cu in this case, ETC_{Cu}) includes the CIF and the available toxic metal concentration (the ionic form):

$$\text{ETC}_{Cu} = [\text{Cu}_T]\alpha \text{Cu}_i/(1 + K_M[M^{2+}])$$

where Cu_T and αCu_i are total and fractional coefficient for available $[\text{Cu}^{2+} + \text{Cu(OH)}^{+} + \text{Cu(OH)}_2]$ Cu concentrations. Pagenkopf considered competition to be effective if hardness was >1 mM (100 mg l^{-1} CaCO$_3$). By correcting LC$_{50}$ values for hardness by computing ETC values from several data sets using Cu, Zn and Cd, the variability among experiments was reduced by an average of 35%.

CMCs for metals (Table 13.8) can be corrected for hardness and converted to dissolved concentrations using the appropriate CFs in the following empirical equations (WQC, 1986; USEPA, 2002a):

$$\text{Cd} = \exp\{1.0166[\ln \text{ hard}] - 3.924\}(\text{CF})$$

where $\text{CF}_{Cd} = 1.136672 - [(\ln \text{ hard})(0.041838)]$

$$\text{Cu} = \exp\{0.9422[\ln \text{ hard}] - 1.700\}(\text{CF})$$

where $\text{CF}_{Cu} = 0.960$

$$\text{Zn} = \exp\{0.8473[\ln \text{ hard}] + 0.884\}(\text{CF})$$

where $\text{CF}_{Zn} = 0.978$

Using the above equations, recommended CMCs for Cd, Cu and Zn for hardness levels of 50, 100 and 200 mg l^{-1} are, respectively: 0.9, 2.0, 3.9; 7.0,

13, 26; and 65, 120, 211, all in $\mu g\,l^{-1}$ (USEPA, 2002a). The values listed in Table 13.8 are calculated for $100\,mg\,l^{-1}$ hardness.

Temperature can affect the toxicity of metals and other toxicants as well (DeSylva, 1969). LC_{50} values can be expected to decrease with increased temperature even within the animals optimum range. This may be due the increase in the fish's metabolic rate, enhancing the rate of toxicant uptake or the stress from increased temperature may lower the fish's resistance.

Toxicity can also be increased by low DO. Pickering (1968), for example, showed a lowering in the 96-hour LC_{50} for bluegill sunfish exposed to Zn; an LC_{50} of $10-12\,mg\,l^{-1}$ at a DO content of $5.6\,mg\,l^{-1}$ compared to an LC_{50} near $7\,mg\,l^{-1}$ at a DO of $1.8\,mg\,l^{-1}$.

At this point it may be interesting to compare the response of different species/tests to representative metals (Table 13.9). The acute criteria (WQC, 1986; USEPA 2002a) are recommended maximum, 1-h mean concentrations and are not much lower than the LC_{50} values for rainbow trout. LC_{50} values for *Daphnia* are quite similar to those for trout. Several authors have shown the similar response of fish and *Daphnia* (Atwater *et al.*, 1983; Doherty, 1983). The sensitivity of *Selenastrum* to metals, comparable to that for fish and *Daphnia*, is not surprising considering that Cu (as $CuSO_4$ and organic complexes) and Zn (organic complexes) are used as algicides. Some of the relatively fast tests used for toxicity screening for potential toxicity, such as the Microtox and Techan procedures, are relatively insensitive compared to the responses by other organisms.

Over the long term, however, metals can be taken up at sublethal concentrations and accumulated within body tissues. In general, uptake and accumulation of a metal at low concentration can occur more readily if it is complexed in organic form and relatively non-ionic. An example of the relative effectiveness of an ionic and non-ionic form was shown with Hg. Ethyl mercury

Table 13.9 Comparison of acute toxic levels (designated) for different organisms and tests

Species/test	Acute toxic level ($\mu g\,l^{-1}$)				Index
	Cr	Cd	Cu	Zn	
WQC (1986)	16	2	13	120	Acute max. for $100\,mg\,l^{-1}$ hardness
Rainbow trout	17	10	80	500	LC_{50}
Daphnia	50	5	60	1000	LC_{50} (approx.)
Selenastrum	238	57	54	51	EC_{50}
Microtox[a]	651	236	8	107	$EC_{50} \times 10^{-3}$
Techan[b]	124	140	54	30	$EC_{50} \times 10^{-3}$

Notes
a Bacterial luminescent activity after 5 min.
b Marine alga O_2 uptake and luminescent activity after 1 h. From WQC (1973, 1986), Turbak *et al.* (1986), McFeters *et al.* (1983) and USEPA (2002a).

phosphate (EMP) showed a marked difference in acute toxicity to fish whether added with or without Cl^- as described by Amend et al. (1969):

Test chemicals	Mercurial produced	Per cent mortality
EMP + H_2O	$CH_3CH_2Hg^+$	0–5
EMP + NaCl	CH_3CH_2HgCl	25–60
EMP + $CaCl_2$	CH_3CH_2HgCl	15–65
EMP + $Ca(NO_3)_2$	–	0
NaCl, $CaCl_2$ + $Ca(NO_3)_2$	–	0

With Cl^- added, an un-ionized organic mercury complex was produced, which can be taken up more readily and therefore produce a greater mortality. As the above indicates, and as suggested by Laegrid et al. (1983), such an effect is most pronounced by low-molecular weight compounds.

Compounds of intermediate water solubility seem to be most toxic, i.e. produce an effect at a lower concentration. If a compound is very water-soluble (polar), it is readily available because of its solubility in water and, if in high enough concentration, it can produce toxicity through an irritation effect on gills. However, the ionized form does not move across lipid membrane surfaces as readily as non-polar low-water-soluble compounds and they would not be readily absorbed. The non-polar low-water-soluble substance, on the other hand, will not be as available for uptake. Because of the ease of uptake of organic complexes, high body concentrations have been shown to result in fish exposed to very low concentrations in water.

The primary anthropogenic sources of Hg to the environment are fossil fuel combustion (e.g. coal-burning power plants), waste incineration and chlor-alkali industry, which uses Hg as a cathode in the production of chlorine (USEPA, 2001). Mercury has been used in dental applications, paints, pharmaceuticals, precision instruments and fungicides. Natural sources of Hg include volcanic emissions and rock weathering.

One of the most toxic compounds is methyl mercury (CH_3Hg^+) which is an organic Hg complex. This compound can be produced in nature through microorganism-mediated reactions in the sediment regardless of the form in which Hg is added. Once an environment is contaminated with Hg, the sediments, through the continued microbial production of CH_3Hg^+, can contaminate fish for many years. It would not be unusual for a receiving water to require from 20 to 30 years to recover from the detrimental effects of continued recycling of CH_3Hg^+ from sediments to organisms even after the Hg input has ceased (Håkanson, 1975; Shin and Krenkel, 1976). Although organisms will lose Hg from tissues once the absorption rate has decreased, with a half-life body decay of about 2 years, CH_3Hg^+ nevertheless can be concentrated in tissue by a factor of thousands of times from water concentrations as low or lower than $1 \mu g l^{-1}$ (Peakall and Lovett, 1972). The sediment effect of CH_3Hg^+ production could last from 10 to

100 years, but normally the sedimentation process should cover the Hg reserve sufficiently to result in recovery to acceptable levels in organisms in from 15 to 30 years (Håkanson, 1975).

Biomagnification of Hg along the food chain has been documented in various aquatic ecosystems. In Onondaga Lake, New York, there was a marked increase in Hg from phytoplankton and invertebrates to fish, with the highest concentrations in piscivorous fish such as walleye (Becker and Bigham, 1995). Bioconcentration factors ranged from 8.3×10^4 for benthic macroinvertebrates to 3.7×10^6 for piscivorous fish. There has also been a problem with Hg biomagnification in the Florida Everglades, where CH_3Hg^+ concentrations in fish tissue exceed $30\,ng\,g^{-1}$ (Cleckner et al., 1998). There is some evidence that high body burdens of CH_3Hg^+ cause reduced hatching success and larval heart rate in fish (e.g. Latif et al., 2001). Adverse effects of Hg have been documented in piscivorous birds such as common loons (Barr, 1986) and great white herons (Spalding et al., 1994). A review of the effects of Hg in aquatic ecosystems is provided by Boening (2000).

Mercury also causes human health problems that include neurological and developmental disorders. Thus, fish consumption advisories are issued throughout the world to limit uptake of Hg through ingestion of contaminated fish and shellfish. The European Union and USEPA have recommended a maximum average mercury content of $300\,\mu g\,kg^{-1}$ in fish consumed by humans. Fish advisories have increased steadily in the USA from 899 advisories by 27 states in 1993 to 2242 advisories by 41 states in 2000 (USEPA, 2001a).

Pb is another heavy metal of concern due to its toxicity to aquatic life, wildlife and humans. Pb concentrates in organisms but does appear to biomagnify along food chains (Spry and Wiener, 1991). Although some sources of Pb (e.g. leaded gasoline) have been controlled, Pb is still used for many purposes and enters the environment through a myriad of pathways. The historical use of Pb shot pellets for hunting is of particular concern in aquatic systems. Huge numbers of pellets are deposited in the sediment of marshes in hunting areas. Waterfowl such as ducks, geese, coots and swans inadvertently ingest the pellets as grit for their crop and millions die annually due to lead poisoning. Less than 10 lead pellets can be fatal to waterfowl. In the USA, there are at least 2.4 million waterfowl fatalities due to lead poisoning, out of a total North American population of 100 million, each year (Mason, 2002). The USA has phased out the use of lead shot; replacing it with steel shot; however, lead fishing weights are still used (Dodds, 2002). In Britain, lead for angling has been banned since 1987, resulting in a much lower incidence of lead poisoning in waterfowl (Mason, 2002).

13.6.3 Other inorganic toxicants

Examples of common toxic non-metal, inorganic compounds include ammonia (NH_3), hydrogen cyanide (HCN), hydrogen sulphide (H_2S) and chlorine

(Cl_2). All of these toxicants are most toxic in their undissociated, molecular state.

The quantity of the toxic undissociated (un-ionized) ammonia (NH_3) form is a function of pH, as shown in Figure 13.21. Ammonium (NH_4^+) is the most abundant form and accordingly analyses should be reported as NH_4^+-N. The pH controlled equilibrium is:

$$K = [NH_4^+][OH^-]/[NH_3]$$

Therefore, as pH increases so does the un-ionized ammonia fraction.

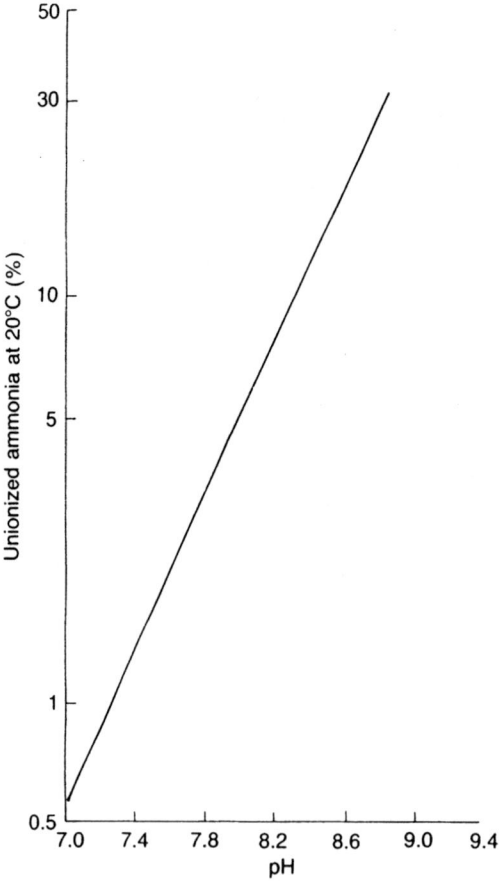

Figure 13.21 Percentage of un-ionized ammonia (NH_3) as a function of pH (WQC, 1973).

The lethal concentration for un-ionized NH_3 is $\approx 1\,mg\,l^{-1}$ for most species of fish. Concentrations of even total NH_4^+ seldom reach that level naturally. However, ammonium may reach or exceed that quantity in anoxic eutrophic environments where nitrification is impeded. Also, several milligrams per litre are common in domestic sewage and some industrial wastes. In fact, ammonium-N in effluents from secondary sewage treatment plants can approach $20\,mg\,l^{-1}$. However, one can readily see the overriding importance of pH from Figure 13.21. About $200\,mg\,l^{-1}$ ammonium would be required for lethality ($1\,mg\,l^{-1}\,NH_3$) at pH 7, while $<3\,mg\,l^{-1}$ would be sufficient at pH 9. Hence, the importance of algal blooms (eutrophication) in waters receiving high ammonium wastewater, due to the potential for photosynthetically caused high pH (Section 4.4).

The recommended maximum limit for NH_3 as a 4-day mean was $0.02\,mg\,l^{-1}$ (WQC, 1986). This value may be unnecessarily low, if calculated from the ambient water pH, because fish tend to lower the pH of water surrounding their gills by respiring CO_2 (Szumski et al., 1982). However, Solbe and Shurben (1989) have shown that rainbow trout eggs, if exposed from shortly after fertilization to 2.5 months, sustained high mortality at NH_3 concentrations as low as $0.027\,mg\,l^{-1}$.

USEPA recently revised the freshwater aquatic life criteria for ammonia (USEPA, 1999). The revised acute criterion for ammonia is now dependent on pH and fish species, whereas the chronic criterion is dependent on pH and temperature. At lower temperatures, the chronic criterion is also dependent on the presence or absence of early life stages of fish (USEPA, 1999).

HCN and H_2S are also undissociated forms that are controlled by pH and are also the toxic fraction (Doudoroff et al., 1966). In contrast to NH_3, the undissociated HCN and H_2S fractions increase with decreasing pH. For example, at pH 7 one half is the toxic H_2S and one half is HS^-, but at pH 5, 99% is H_2S (WQC, 1986). H_2S is unstable in the presence of O_2 and readily converts to SO_4^{2-} (Section 4.3). Thus, H_2S would usually not present a problem where DO is otherwise adequate for fish. However, a critical condition could occur for developing eggs and larvae near sediments where H_2S would originate. Recommended maximum CCCs for HCN and H_2S are 5.2 and $2\,\mu g\,l^{-1}$, respectively (WQC, 1986; USEPA, 2002a).

Chlorine became increasingly important during the early and mid-1970s as investigators discovered that effluents from sewage treatment plants were toxic to fish due to maintained residuals of chlorine for disinfection. The toxicity of free and/or combined (chloramines) chlorine, referred to as total residual chlorine ($TRCl_2$), is also controlled by pH. The most toxic form, hypochlorous acid (HOCl), increases as pH decreases.

The problem was reviewed by Brungs (1973), who cited work in Michigan, USA with caged fish held downstream from treatment-plant effluents. When the effluent was not chlorinated, no mortality resulted, but when chlorinated the LC_{50} values for 96 h ranged from 14 to $29\,\mu g\,l^{-1}$ as $TRCl_2$. In some areas, such as the lower James River, massive fish kills were reported from effluent chlorine residuals as high as $2.2\,\text{mg}\,l^{-1}$ (Bellanca and Bailey, 1977). Similarly, Tsai (1973) investigated streams receiving treated sewage effluent from 149 plants and found fishless waters where total residual chlorine was $0.37\,\text{mg}\,l^{-1}$ or greater and observed levels as high as $1.5\,\text{mg}\,l^{-1}$.

The recommended maximum CCC of $TRCl_2$ is $11\,\mu g\,l^{-1}$ (WQC, 1986; USEPA, 2002a). Hematological work has shown that $3\,\mu g\,l^{-1}$ is the maximum $TRCl_2$ concentration that would be safe for coho salmon (Buckley *et al.*, 1976; Buckley, 1977).

13.6.4 Organic toxicants

The myriad synthetic organic chemicals produced annually may pose the major threat, as well as challenge, to society's ability to prevent the degradation of environmental quality resulting from toxicity. The threat of organic toxicants may be greatest because of their more neutrally charged state and therefore greater mobility to transfer across cellular membranes (i.e. more soluble in lipids), as well as a high resistance of many compounds to degradation. Such characteristics create the problem of long-lasting residues and food-chain magnification that has rendered populations of fish unsafe for human consumption.

The most famous compound in this regard is the organochlorine insecticide DDT and its metabolites DDE and DDD. Because DDT is fat-soluble, but rather insoluble in water ($\approx 1\,\mu g\,l^{-1}$), it can be readily absorbed by organisms from water or in food resulting in body residues that are 10^6-times the water concentration (Reinart, 1970). While DDT can be absorbed from water or ingested with food, the two pathways are not additive (Chadwick and Brocksen, 1969). While other organochlorine insecticides are famous for their toxicity and are even more toxic to fish than DDT (e.g. endrin and dieldrin), DDT has received relatively more attention and has had a greater toxic impact due to its greater use. Over one-half million metric tons of DDT have been used in the USA and it continues to be used in other countries (Hileman, 1988).

The recommended maximum CCCs for organochlorine insecticides range from 0.001 to $0.004\,\mu g\,l^{-1}$ (WQC, 1986; USEPA, 2002a). Although such concentrations are one or two orders of magnitude below the lethal concentrations, they should not be considered over restrictive considering the biomagnification potential of these substances. For example, fish in Lake Michigan apparently concentrate DDT and dieldrin from the part per trillion level in water to the part per million level (Reinart, 1970).

The most detrimental subacute effect of organochlorine compounds to fish is on the reproduction stage. That was first demonstrated by Johnson (1967) for the Japanese medaka (*Oryzias latipes*). He found that whereas adults exposed to endrin concentrations of $0.3\,\mu g\,l^{-1}$ or less survived in apparent good health, the hatching fry from those fish showed severe behavioural changes and poor survival which were directly related to the concentration and time of adult exposure. That was true even though the eggs were incubated in endrin-free water, indicating that residues in the adult fish were transferred to the eggs.

About 2.3 million metric tons of pesticides are used annually worldwide (Nowell *et al.*, 1999). In the USA alone approximately 630 different chemicals are used. The majority of the insecticides and herbicides (78–80%) are used on crops, such as corn, cotton, wheat and soybeans, with the remainder being used on lawns and golf courses (Miller, 1998). While pesticides have provided some benefits, their use has resulted in high environmental costs including $1.8 billion for groundwater remediation, $24 million in fishery losses and $2.1 billion in losses of birds (Pimental *et al.*, 1992).

Pesticides have created major problems in the environment because most of what is applied never reaches the target organism (Younos and Weigmann, 1988), but are available to non-target species. Many years of education and persistence on the part of fisheries and pollution control agencies (and dead fish) were required before the documented damage caused by application of organochlorine insecticides had an impact on their use. As a result, relatively shorter-lived and less toxic insecticides have since replaced many of the organochlorine and organophosphate compounds, whose use has been greatly reduced or curtailed altogether since the 1970s. Nevertheless, much of their residues still remain in the environment due to their relative resistance to degradation. Residue levels of organochlorine pesticides (e.g. DDT and dieldrin) in fish and birds have decreased significantly in the USA following curtailment of use – demonstrating that, even for highly resistant and biomagnifying substances, availability in the environment and body residues will decrease if the input to the environment decreases. Nevertheless, highly toxic organochlorine insecticides such as chlordane and mirex (CCCs of 0.0043 and 0.001 $\mu g\,l^{-1}$, WQC, 1986) can still be used and a thousand metric tons of pesticides in general are used in agriculture daily in the USA (Younos and Weigmann, 1988). Atrazine is a herbicide that is widely used in the midwestern USA (32 million kg applied annually) for weed control. Although it does not bioconcentrate, concerns about its toxicity, carcinogenicity and ability to disrupt the endocrine system (see Section 13.6.5) have recently emerged (Carder and Hoagland, 1998; Nowell *et al.*, 1999). Toxaphene is another pesticide of concern due to its biomagnification and movement through the atmosphere. Kidd *et al.* (1995) found unusually high concentrations of toxaphene in fish from a remote subarctic lake.

PCBs, which were used extensively in industry and appear to have similar effects and persistence as the organochlorine insecticides (WQC, 1973), have become increasingly important in the 1970s and 1980s. Although their use was banned in the late 1970s, their residues in land-fills, river and lake sediments and transformers continue to present problems of dispersal in the environment and biomagnification in the food chain, e.g. the Hudson River (Brown *et al.*, 1985), the Great Lakes (Hileman, 1988), and the Arctic (Pearce, 1997). The global movement and biomagnification of PCBs is described in Colborn *et al.* (1996).

In contrast to the organochlorine insecticides, most herbicides, such as 2,4-D, 2,4,5-T and endothal, are less acutely toxic and degrade more rapidly. The herbicide 2,4-D persists in some trophic levels and sediment for one to three months but apparently has produced little adverse effect on aquatic animals (Wojtalik *et al.*, 1971). On the other hand, 2,4,5-T may contain dioxin and presents a residue problem (Galston, 1979).

Wastewater from the bleaching of pulp and paper contains organochlorine compounds that are highly toxic, including dioxin. Leach and Thakore (1975) found that chlorinated compounds derived from resin acids in Kraft pulp-mill wastewater were the cause of most of the effluent toxicity. Five identified compounds had 96-hour LC_{50} values ranging from 0.32 to 1.5 mg l^{-1}. Organochlorine compounds from Kraft mill bleaching effluents have been distributed over broad areas in the northern Baltic Sea (Sodergren *et al.*, 1988) (Figure 3.13). Detrimental effects on perch (*Perca fluviatilis*) reproduction and physiology were demonstrated in laboratory experiments to collaborate the increased tissue levels of organochlorine compounds and reduced biomass levels of perch observed up to 10 km from effluent sources.

13.6.5 Endocrine-disrupting chemicals

The human endocrine system includes several glands (e.g. thyroid, pituitary, pineal, ovaries, testes) and the hormones that they produce (e.g. adrenalin, oestrogen, testosterone). There is growing concern that some synthetic organic chemicals that humans release into the environment can disrupt this system. Endocrine-disrupting chemicals can be defined as *substances that cause adverse health effects in an organism, or its progeny, consequent to endocrine function*. Many of these chemicals influence the normal functioning of hormones. The focus of most of the research to date has been on the disruption of sex hormones (such as oestrogen), but other hormone systems are likely to be affected. Thus, these chemicals may disrupt the reproductive, endocrine, immune and nervous systems of animals, as well as adversely effect embryonic and early postnatal development (Colborn and Clement, 1992). In the simplest terms, it is thought that these chemicals mimic the body's natural hormones, binding with receptor molecules and thus interfering with the regulation of the endocrine system (Trussel, 2001). Because there is some uncertainty as to whether the observed effects of endocrine disrupters are due to their hormonal properties or to some other toxicological mechanism, the National Research Council (1999) recommends the term 'hormonally active agent' instead of endocrine-disrupting chemical.

Many substances have been implicated in endocrine disruption. Some are naturally occurring such as phyto-oestrogens, which are found in plants, and sex hormones that enter the environment via sewage effluent discharges. Synthetic chemicals that can potentially disrupt the endocrine system include organochlorine compounds such as PCBs and dioxins, phenolic compounds, phthalates (used to make plastics flexible) and some polycyclic aromatic hydrocarbons (PAHs). Some pesticides are also endocrine disrupters including DDT and its metabolites, triazines, pentachlorophenol, malathion, diledrin and methoxychlor. Some heavy metals (mercury, lead) can also disrupt the endocrine system (Colborn and Clement, 1992; National Research Council, 1999). Tributyl tin, which was widely used as an anti-fouling agent on boat hulls, is extremely toxic and also disrupts endocrine function (Champ and Seligman, 1996). Another group of chemicals of emerging concern is the pharmaceutically active chemicals (e.g. pain killers, antibiotics, synthetic steroids used in contraceptives, caffeine, growth hormones), which also enter the environment through wastewater discharges and runoff. Some pharmaceuticals are also endocrine disrupters, e.g. the growth hormones used in livestock production (National Research Council, 1999).

Endocrine disrupters have been reported to have a variety of effects on humans and wildlife that include developmental abnormalities, declining sperm counts, feminized wildlife populations and increased hormone-related cancers (e.g. breast and testicular cancers). In particular, the chemicals that mimic oestrogen, commonly referred to as 'eco-oestrogens' or environmental oestrogens, have been implicated in causing these effects (Colborn et al., 1993). While the eco-oestrogens are the most obvious and notorious of the endocrine disrupters, they are likely to represent only one of the many ways in which chemicals can mimic hormones, neurotransmitters, growth agents and other important biological functions (McLachlan and Arnold, 1996).

Feminization and other reproductive and developmental abnormalities in wild fish due to exposure to oestrogen mimickers in sewage effluent have been reported in a number of studies (e.g. Purdom et al., 1994; Jobling et al., 1998; Rodgers-Gray et al., 2000). These effects are not limited to fish. Developmental abnormalities in embryos and hatchlings of snapping turtles from the Great Lakes Region were found to be related to dibenzodioxins and dibenzofurans (Bishop et al., 1998). Developmental and reproductive abnormalities in hatchling and juvenile alligators (*Alligator mississippiensis*) in Lake Apopka, Florida, were linked to exposure to endocrine-disrupting chemicals during embryo development (Guillette et al., 1999). Endocrine disruption in birds and mammals has also been reported (Colborn et al., 1993; Tyler et al., 1998).

There is much to be learned about the effects of endocrine disrupters on humans and wildlife. Many of these chemicals have not been routinely monitored in the environment and little is known about the endocrine-disrupting mechanisms of specific chemicals (Matthiessen, 2000). Organisms are also exposed to a milieu of chemicals in the environment, making isolation of causative agents and their effects extremely difficult. The sources of these chemicals are diverse and widespread, including common items such as plastic bottles and wrappers, contraceptives, coffee and lawn care products. The effects of endocrine-disrupting chemicals in the environment have been critically reviewed by Sumpter (1998) and Tyler *et al.* (1998).

13.6.6 Acidification

Acidification of lakes and streams due to acid precipitation is a global pollution problem affecting large areas of the eastern USA, eastern Canada, Scandinavia and northern Europe. Acidification is the loss of acid neutralizing capacity (ANC) or alkalinity (Section 4.4). The loss is caused by the replacement of HCO_3^-, which is essentially equivalent to ANC (Alk) in waters with pH <8.3, with either SO_4^{2-} or NO_3^-. The latter are the strong acid anions in precipitation resulting from the combustion of fossil fuels and the resulting atmospheric emission of SO_2 and NO_x, which are ultimately oxidized to the strong acids (H_2SO_4 and HNO_3) and can be deposited far downwind from sources. When acid in precipitation is deposited on to a watershed, it becomes neutralized (H^+ is consumed by bases) by weathering processes to a greater or lesser degree. In watersheds dominated by sedimentary bedrock rich in $CaCO_3$, acidity is readily neutralized. However, in those dominated by igneous bedrock such as granite and basalt, which is overlaid by a thin soil mantle, the supply of neutralizing bases through weathering is low (Wright and Snekvik, 1976; Henriksen, 1980). The effectiveness of neutralization through weathering decreases as the ratio of lake area to watershed area increases, because the relative supply of bases decreases. Due to its more conservative nature, SO_4^{2-} represents the major cause for lake and stream acidification (Henriksen, 1980; Brakke *et al.*, 1988).

The chemical process of acidification with strong acid can be described by considering the cation–anion charge balance in freshwater:

$$2[Ca^{2+}] + 2[Mg^{2+}] + [Na^+] + [K^+] + [H^+]$$
$$= 2[CO_3^{2-}] + [HCO_3^-] + [Cl^-] + 2[SO_4^{2-}] + [OH^-] + [NO_3^-] \quad (13.1)$$

To obey neutrality, the cation equivalents must equal the anion equivalents. If H_2SO_4 is added to a water body, H^+ and SO_4^{2-} will increase at the expense of HCO_3^- according to:

$$H_2SO_4 + Ca(HCO_3^-) \rightarrow CaSO_4 + 2CO_2 + 2H_2O \quad (13.2)$$

Equation 13.1 can be simplified to the following in most dilute freshwater with near neutral pH:

$$[HCO_3^-] = 2[Ca^{2+} + Mg^{2+}] - 2[SO_4^{2-}] \quad (13.3)$$

Na^+ is relatively scarce without marine salt contamination and K^+ and NO_3^- are relatively low. Thus in unacidified waters, Ca^{2+} and Mg^{2+} are the principal cations and HCO_3^- is the principal anion so that significant departures from a near 1:1 relation between the non-marine equivalents of $Ca^{2+} + Mg^{2+}$ and HCO_3^- has indicated the extent of acidification (Henriksen, 1980). As SO_4^{2-} increases, HCO_3^- decreases (Equation 13.3). This led to the following equation (in equivalents) to evaluate how much acidification has occurred (Wright, 1983):

$$\Delta ALK = 0.91(Ca^* + Mg^*) - Alk + H^+ + Al \qquad (13.4)$$

where Alk on the right-hand side is the existing level and above pH 5, H^+ and Al are insignificant and an asterisk denotes non-marine.

The following was used to estimate the amount of non-marine SO_4 added:

$$\text{net } SO_4^* = SO_4^* - [0.09(Ca^* + Mg^*) + Na^* + K^*] \qquad (13.5)$$

where SO_4^* on the right-hand side is existing SO_4^*, and the bracketed term is background SO_4^* before acidification. Thus, net SO_4^* should approximate ΔALK and Wright (1983) showed that in 206 Adirondack lakes net SO_4^* averaged 119 μequiv. l^{-1} and ΔALK averaged 103 μequiv. l^{-1}. That is, for the average lake, ≈ 1 μequiv. l^{-1} alkalinity (HCO_3^-) was removed by the addition of ≈ 1 μequiv. $l^{-1}SO_4^*$ through acid deposition.

While this is an easily explained equilibrium model that has been effective at understanding and predicting the general pattern of acidification, there are problems when individual lakes are considered. For example, $Ca^* + Mg^*$ may increase with acidification through weathering and thereby compensate for some of the SO_4^* added. However, the maximum effect of such compensation is considered to be $\approx 40\%$ (Wright and Henriksen, 1983).

The effects of acidification are due to decreased pH and increased Al. As ANC is depleted, pH decreases; very slowly at first because of the buffering effect of HCO_3^- (H^+ is consumed), but rapidly as ANC nears depletion. Usually, at a pH of 5.2, all the ANC is gone and a further decrease in pH occurs with a continued increase in strong acid. Also as pH approaches 5, Al is dissolved from watershed soils and rocks and the concentration begins to increase in runoff water while providing some buffering against further decreases in pH. Al may occur in various forms, but the free Al^{3+} and hydroxides, $Al(OH)^{2+}$ and $Al(OH)_2^+$, are principal forms of the labile monomeric Al fraction that is considered most toxic (Driscoll et al., 1980; Henriksen et al., 1984). Organic-Al complexes are not considered toxic.

Damage to fish populations may occur if lake pH drops below ≈ 6 although effects are more pronounced at pH 5.5 and below. While a lake or stream with pH 5 would be considered completely acidified (all ANC gone), damage to fish populations occurs at pH values >5. Much of the adverse effect at pH values between 6 and 5 observed in natural waters is probably due to Al toxicity as well as to H^+ (Baker and Schofield, 1982). For example, Muniz and Leivestad (1980) showed that Cl^- loss from the blood of brown trout did not increase until pH was decreased below 4.6, while Cl^- loss was much greater at pH 5.1 and 5.5 than at 4.0 in the presence of 900 $\mu g\, l^{-1}$ Al. Others have found toxic effects due to Al to be greatest around pH 5.2–5.4, possibly due to Al hydroxides precipitating on gill surfaces (Schofield and Trojnar, 1980; Baker and Schofield, 1982). Henriksen et al. (1984) showed mortalities to juvenile salmon when labile monomeric Al concentrations increased to between 50 and 70 $\mu g\, l^{-1}$ following stream-flow-increase events even though pH did not drop below 5. Jensen and Snekvik (1972) considered the limit for damage due to pH effects alone to be between 4.5 and 5.0, although Schindler et al. (1986) reported long-term effects on lake trout production resulting from food web modifications at pH values <6 without Al. Also, Rahel and Magnuson (1983) found some fish species to be missing from Wisconsin lakes with pHs below 6.2.

The problems caused by acid precipitation are considered global in proportion; effects on fish populations have been most pronounced in southern Norway (Figure 13.22) and south-western Sweden, eastern Canada and north-eastern USA, especially New York state. Although acid precipitation affects a much larger area, acidification is most damaging in waters of low buffering (ANC) capacity (see above). Western Sweden and Norway have many lakes in watersheds composed of resistant rock and thin soils and are thus highly sensitive to acid precipitation (i.e. have low ANC). For example, Wright and Snekvik (1978) found that 80% of 700 lakes surveyed

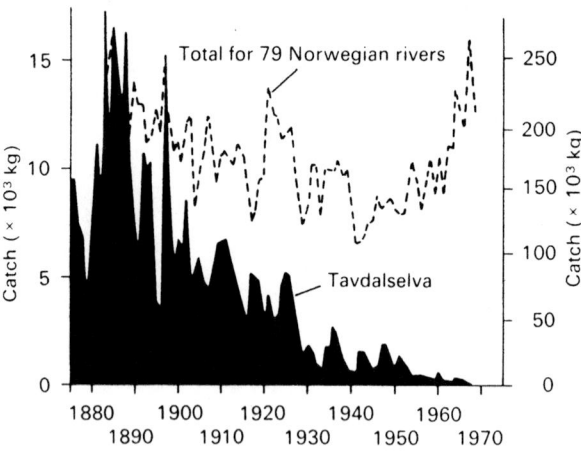

Figure 13.22 Salmon-catch records for 79 Norwegian rivers unaffected by acid rain and for a southern Norway acidic river during 1880–1970 (Wright et al., 1976).

in southern Norway during 1974–75 were acidified to the point that they were fishless or had sparse fish populations. Figure 13.22 shows the dramatic decline and elimination of salmon from the Tovdal River. Leivestad and Muniz (1976) found that brown trout mortality in the Tovdal River was inversely related to Cl^- content of the blood. A resurvey of the southern Norwegian lakes in 1986 showed that pH had changed little in the intervening years, but that sulphate had decreased and nitrate had increased significantly (Henriksen et al., 1988).

An extensive, stratified lake survey in acidification-sensitive regions of the USA was reported by Landers et al. (1988). Results showed that 4.6% (326) of the 7096 lakes of 4–200 ha in the north-east had ANC<0, while 138 (11%) of Adirondack lakes had <0 ANC (results based on statistical projections). Fewer (148, 1.7%) of the 8501 lakes in the upper midwest USA had ANC<0. The Adirondack lakes had lower median ANC and pH and higher Al than lakes in other areas. Although SO_4^{2-} in the Adirondacks was not higher than other areas, the median level was still of the order of ANC (see results above by Wright, 1983). There is general agreement that acidification of north-eastern and upper midwestern lakes in the USA has been caused largely by SO_4^{2-} deposition and replacement of ANC. That is indicated by trends of decreasing ANC/(Ca + Mg) associated with increasing H^+ deposition (Brakke et al., 1988; Eilers et al., 1988).

Based on the above discussion of effects of pH and Al, the 474 north-eastern and midwestern USA lakes estimated to have ANC<0 would also be expected to have damaged fish populations. Damage could be expected in even more lakes because, as indicated above, adverse effects of pH and Al have been observed at pH levels between 5 and 6 before ANC is exhausted, especially if Al is present. For example, while only 3.4% (240) of the north-eastern lakes had pH<5, 12.9% (916) had pH<6. In the upper midwest, only 1.5% (130) had pH<5, while 9.6% (818) of the lakes had pH<6. Also, small lakes, which are known to be more susceptible to acidification, were under-represented in the eastern survey.

Sulfate deposition has declined in the 1980s and 1990s in Europe and North America as a result of reduced emissions and there have been signs of lake recovery (Wright and Schindler, 1995). However, emission of NO_x has continued to increase and may cause acidification problems.

13.7 Invasive, non-native fish

The introduction of many fish species to non-native habitats has increased throughout the world due to both intentional introduction for aquaculture and angling and accidental introduction from ballast water transfers and aquarium releases. Introduced fish species may extirpate native species, stunt native fish through competition, alter food webs and nutrient cycling, and cause disease. Characteristics of successful fish invaders include production of large numbers of eggs, production of live young or mouth-brooded eggs or young (which increases survival rate), wide environmental tolerances, flexible habitat requirements, opportunistic feeding, and aggressive or flexible behaviour (Arthington and Mitchell, 1986).

In many cases, native species have been eliminated by introduced fish, resulting in reduced regional biodiversity and homogenization of the world's fish faunas (Moyle et al., 1986; Vander Zanden et al., 1999; Rahel, 2000). Rahel (2000) documented the homogenization of fish faunas across the USA by comparing fish faunal lists for the 48 coterminous states and found that, on average, pairs of states have 15.4 more species in common now than were present before Europeans settled North America. Most of these introductions were intentional and occurred from east to west, as western waters lacked the game fish (e.g. northern pike, walleye, bass, sunfish, catfish) that were desired by people moving west from eastern USA. The stocking of common carp (*Cyprinus carpio*), which is native to Asia, in the Great Lakes began in the 1870s, but by the 1890s they were already considered a problem due to their effects on nearshore benthic areas used by other fish and waterfowl (Mills et al., 1994). Carp are established throughout the USA and portions of Canada and much money has been spent to remove carp from waters where they compete with game fish. Carp also cause adverse water quality problems including sediment re-suspension, which increases turbidity and nutrient release. Approximately 61 non-native fish species have become established in the USA (Benson, 2000).

The introduction of the Nile perch (*Lates niloticus*) to Lake Victoria in East Africa, beginning in 1954, illustrates the dramatic effects of a non-native fish on native species. More than 300 species of endemic cichlid fish species are believed to have evolved in Lake Victoria, the world's second largest freshwater lake, during the past 12 500 years (Johnson et al., 1996). The cichlids exhibited remarkably different feeding habits and ecological niches. Nile perch persisted at low densities in Lake Victoria until the early 1980s when the population exploded throughout the lake for unknown reasons (Kaufman, 1992). Approximately 200 species of the cichlids have vanished within the last two decades primarily due to predation by the introduced Nile perch as well as overfishing and eutrophication (Barel, 1985; Goldschmidt et al., 1993; Seehausen et al., 1997). While biodiversity in the lake has dramatically decreased with the loss of the cichlids, the Nile perch fishery is four times larger than the cichlid fishery, providing benefits to the local communities. However, as eutrophication progresses, the long-term viability of the Nile perch fishery is questionable (Kaufman, 1992). This is an example of an introduction that had positive economic benefits, but at the expense of lost species and diversity.

Even salmonids, which are typically highly valued components of their ecosystem, can become a 'pest' when introduced elsewhere (Dextrase and Coscarelli, 2000). An example of this is the lake trout (*Salvelinus namaycush*), which has had negative effects on native fish species when introduced outside its native range (Crossman, 1995). The recent illegal introduction of lake trout to Yellowstone Lake (Montana) has caused a significant reduction in the native Yellowstone cutthroat (*Oncorhynchus clarki bouvieri*) population (Kaeding et al., 1996). Brook trout (*Salvelinus fontinalis*), which are native to eastern USA, are another example of a highly valued species that caused reductions in native salmonids when introduced into waters of the western USA. Brook trout in turn have been negatively affected through competition and predation by the introduction of the German brown trout (*Salmo trutta*) and the western rainbow trout (*Oncorhynchus mykiss*) in eastern waters (Larson and Moore, 1985; Krueger and May, 1991). While brown trout introductions to New Zealand before the 1900s have created excellent fisheries, these aggressive predators had a negative effect on the native *Galaxias* spp., replacing them as the dominant vertebrate predator in many streams (Crowl et al., 1992). This switch, due to predation effects on grazing invertebrates, was observed to result in higher biomass of periphyton in streams with trout than those with *Galaxias* (Biggs et al., 2000).

Largemouth bass (*Micropterus salmoides*) is a very popular game fish and is probably the most widely stocked warm-water species in the USA (Smith and Reeves, 1986). Although its introduction has created some outstanding sport fisheries, negative ecological impacts have been associated with many introductions. Lassuy (1995) identified largemouth bass as a contributing factor in the decline of 21 native fish species listed under the US Endangered Species Act. This fish is also believed to have been a factor in the extinction of several native North American fish taxa (Miller et al., 1989).

Introduction of fish species outside their native range continues to occur, but the primary source of these introductions has shifted from government actions to illegal and inadvertent releases (Rahel, 2000). For example, illegal introduction of northern pike and walleye has occurred throughout the Pacific North-west (McMahon and Bennett, 1996). Bait minnows have also been introduced by anglers to waters far outside their native range (Litvak and Mandrak, 2000).

Accidental introductions continue to occur and result in far-reaching ecological impacts. The release of ship ballast water was the cause of the recent establishment of the round goby (*Neogobius melanostomous*) and the ruffe (*Gymnocephalus cernuus*) in the Great Lakes (Ricciardi and MacIsaac, 2000). The sea lamprey (*Petromyzon marinus*) is another example of an inadvertent invasion with significant adverse ecological effects. Following the opening of a canal in 1829 that linked Lake Erie to Lake Ontario, and the rest of the Great Lakes to the Atlantic Ocean, the sea lampreys invaded the Great Lakes. The sea lamprey is an ectoparasite that spawns in freshwater and can kill its fish hosts within a week. The invasion of the sea lampreys caused a precipitous decline in lake trout and the collapse of commercial fisheries in many areas of Lake Huron and Michigan (Mills et al., 1994). The decline of the lake trout may have been partially due to pollutants.

Appendices

Appendix A
Description of benthic macroinvertebrates and comments on their biology

The following is a brief account of the taxonomy and biology of the life stages of macroinvertebrate fauna that inhabit fresh and, to some extent, brackish water. For more details see Usinger (1956).

A. Body with three pairs of legs (Insecta); abdomen with two or three long tail-like cerci; thorax and abdomen usually with plate-, feather-, tassel-, or finger-like tracheal gills.

1   Mayflies (Ephemeroptera);

    (a) Taxonomy: tarsi with one claw; gills plate-, feather- or tassel-like and present on one or more of first abdominal segments;
    (b) Food habits: herbivorous grazers and detrital feeders;
    (c) Habitat: depositing substrata, Ephemeridae (burrowers); eroding substrata, Heptageniidae (clingers); eroding and depositing substrata, Baetidae (sprawlers on bottom, agile, free-ranging and trash- and moss-inhabiting);
    (d) Life history: three-stage cycle (egg → naiad → adult); usually <1–1 year.

2   Stone flies (Plecoptera);

    (a) Taxonomy: tarsi with two claws; gills usually present and finger-like and may occur on abdomen, thorax or labium
    (b) Food habits: herbivorous mostly, some carnivorous (Perlidae, Perlodidae);
    (c) Habitat: mostly swift currents ($\geq 1\,\text{ft}\,\text{s}^{-1}$), eroding substrata;
    (d) Life history: three-stage cycle; 1–3 years.

B. Body worm-like, elongate and cylindrical to short and obese, or flattened and oval with appendages and sclerotized head capsule.

1. Caddis flies (Trichoptera);

    (a) Taxonomy: thorax with three pairs of jointed legs and segment with a pair of hook-bearing appendages; antennae inconspicuous, one-segmented;
    (b) Food habits: herbivorous (including detrital) and carnivorous (Rhyacophilidae);
    (c) Habitat: usually eroding substrata (free living, net spinners and case builders); depositing substrata (some case builders);
    (d) Life history: four-stage cycle (egg → larvae → pupae → adult); 1–2 years.

2. True flies (Diptera);

    (a) Taxonomy: thorax without jointed legs but often with fleshy prolegs; head capsule distinct, at least anteriorly; abdomen with gills or a breathing tube at posterior end;
    (b) Food habits: detrital herbivorous feeders;
    (c) Habitat: eroding substrata (cling by permanent attachment); vegetation (cling with hooks); depositing substrata (tube builders);
    (d) Life history: four-stage cycle; <1–1 year.

C. Body not worm-like, wing rudiments as external flap-like appendages, mouthparts consisting of long and scoop-like labium, which covers other mouthparts when folded.

1. Dragonflies and damselflies (Odonata);

    (a) Food habits: all predaceous;
    (b) Habitat: depositing substrata and vegetation (sprawlers, climbers and burrowers);
    (c) Life history: three-stage cycle; 1–5 years.

D. Animals living within hard carbonate shell: Mollusca (snails and clams).

1. Snails (Gastropoda): shells entire, usually spiral;

    (a) Food habits: mostly detrital feeders and grazers;
    (b) Habitat: mainly in depositing substrata, but also in eroding substrata;
    (c) Life history: three-stage cycle, <1–1 year.

2. Clams, bivalves (Pelecypoda): shells consisting of two hinged halves;

    (a) Food habits: detrital and herbivorous filter feeders;
    (b) Habitat: mainly depositing substrata;
    (c) Life history: three-stage cycle; <1–1 year.

E. Body worm-like and divided into many small segments much wider than long; bristles frequently present.

1   Worms and leeches (Annelida);

    (a) Food habits: mostly detrital feeders, some predators and parasites;
    (b) Habitat: usually depositing substrata;
    (c) Life history: two- or three-stage cycle; $<1-1$ year.

F. Body not worm-like, with at least five pairs of legs (Crustacea).

1   Body flattened laterally (Amphipoda, scuds);

    (a) Food habits: detrital;
    (b) Habitat: eroding and depositing substrata with vegetation;
    (c) Life history: three-stage cycle; <1 year.

2   Body flattened horizontally: Isopoda (aquatic sow bugs), eyes not on stalks; Decapoda (crayfish), eyes on stalks; biology similar to Amphipoda.

Appendix B

Study questions and answers

Short-Answer Questions

1. Explain how *two phenomena* that occur as a result of increasing primary productivity of diatoms and/or green algae due to eutrophication would tend to favour blue-greens.
2. Explain verbally, with a steady-state equation, why the inflow concentration of limiting nutrient to a continuous culture of algae is so closely related to the concentration of biomass in the outflow. Also, with a steady-state equation, explain why growth rate increases with dilution rate.
3. Eutrophic lakes A and B have similar areas and wind speed and fetch, but have mean depths of 10 m and 3 m, respectively. Internal loading of phosphorus was shown by mass balance to be of a similar magnitude in both lakes.

    (a) Describe two sediment release mechanisms for each lake that may explain their internal loading.
    (b) Why might you expect concentrations of algae to be greater in lake B than A?

4. Calculate the expected net internal phosphorus loading in a lake with the following characteristics:
    flushing rate = 2.25/year; mean depth = 15 m; external loading = 2.0 g/m^2 yr and mean annual TP = 60 μg/l (mg/m^3).
    Would you expect this lake's summer epilimnetic TP to be greater or less than 60 μg/l? Why?
5. Give two explanations why eutrophic lakes tend to have more small than large bodied zooplankton. What is the consequence of that condition in terms of algal abundance and transparency?
6. Two streams have similar incident light, temperature and rubble substratum, but have different average velocities; A = 35 cm/s and B = 10 cm/s. In which stream would you expect the concentration of SRP (soluble reactive phosphorus) limiting periphytic algal growth to be lowest? Why?

7 Is Lake Tahoe, as a whole, more or less sensitive to a given inflow concentration of phosphorus (assuming P is limiting) than Lake Washington? Explain with an equation. Water detention times: L. Tahoe = 700 years; L. Washington = 3 years.
8 A eutrophic lake has a mean summer chl a concentration of 25 µg/l and a transparency of 1.0 m. The lake has a maximum depth of 10 m, a mean depth of 6 m, and it stratifies thermally with the thermocline at about 3 m. Is it possible to reduce algal biomass so that the average is less than 10 µg/l by completely mixing the lake with compressed air during the summer? Assume nutrients will not limit.
9 The following data are available from a stream receiving wastewater from a point source.

	upstream	waste ↓	1 km	5 km
AFDW/Chl a	200		400	1000
No. species (macroinvertebrates)	25		5	10
No./m² (macroinvertebrates)	500		200	5000

Describe the type(s) of wastewater that could produce such an effect and give your reasons with regard to each variable.
10 Thick mats of *Cladophora*, a filamentous alga, were observed immediately downstream from a wastewater effluent input. Given the alternatives of (a) secondary treated sewage, (b) untreated sewage, (c) pulp mill waste liquor, and (d) an organic insecticide, which type(s) of waste would most likely cause this problem and why? Which would not likely be the cause and why?
11 (a) Determine, using the steady state version of Vollenweider's model (or a modified version) how effective (in %) an in-lake treatment should be at reducing internal loading to a loading level that would recover the lake to a mesotrophic state (i.e. $\leq 25\,\mu g/l$ TP). The lake has the following characteristics:

$$L_{ext} = 300\,\text{mg/m}^2 \cdot y,\ \rho = 0.4/y,\ \overline{TP} = 35\,\mu g/l\ (\text{mg/m}^3),\ \bar{z} = 15\,\text{m}.$$

(b) What is the major assumption you must make (other than completely mixed, constant volume, accuracy of external load, etc.) in order to use the model in this way?

12 Give two physical processes that reduce the biomass and diversity (no. of species) of invertebrate animals in streams with urbanized watersheds and briefly describe their effect.
13 Calculate the steady state biomass in a nutrient limited continuous culture system, which is characterized by the following parameters; maximum growth rate = 1.3/day; half saturation constant = 0.3; inflow nutrient concentration = 4.0 mg/l; dilution rate = 1.0/day; and the biomass to

nutrient yield ratio = 64. Would a higher or lower maximum biomass result if these parameters were applied to a batch culture and why?

14 Based on the turbidity of inflow water, a proposed reservoir is projected to have a Secchi disc transparency averaging 2 m during the spring and summer. Assuming complete mixing, how deep should the reservoir be on average to prevent the maximum algal biomass (chl a) from exceeding 10 µg/l? Also assume that the Secchi disc disappears at 15% surface intensity (I_0), the respiration: P_{max} ratio is 0.1 and the light extinction coefficient for chl a is 0.025 m²/mg.

15 Explain why buoyancy, possessed by some blue greens such as *Aphanizomenon*, should provide a competitive advantage in a relatively shallow, polymictic eutrophic lake.

16 Compare the expected effects of wind on the internal loading of phosphorus (release of P from sediments) in two 300 ha (750 acre) eutrophic lakes; A is stratified and B is polymictic.

17 Is the expected relationship between the LC_{50}s (with fish) determined for heavy metals and water hardness inverse or direct? Give an explanation for the phenomenon.

18 For a soft water stream receiving wastewater containing the heavy metals Cu and Zn that raised stream levels to double the maximum acceptable concentrations (MACs = 7 µg Cu/l; 59 µg Zn for 50 mg/l hardness), speculate on the direction of effect (increase, decrease, no change) and a reason for your choice for each of the following variables.

(a) AFDW/chl a;
(b) no. sp. macroinvertebrates;
(c) $\bar{d}$ for macroinvertebrates;
(d) D for macroinvertebrates.

19 The lower Deschutes River in Oregon has a summer flow that is maintained, partly due to releases from a dam, at around 100 m³s (3500 cfs), and the mid-summer water temperature reaches a maximum of about 17°C. Explain the probable effects of increases of 3°C and 7°C above ambient on growth and survival of the highly prized rainbow trout population.

20 Explain 'top-down' control of plankton algal biomass in lakes and give two methods by which it may be enhanced to improve the quality of eutrophic lakes.

Multiple-Choice Questions

1 Net productivity at steady state is

(a) dx/dt
(b) μx

(c) $\mu x - Dx$
(d) $\dfrac{dx/dt}{x}$

2 In the nitrogen budget of a eutrophic lake, the important losses are

(a) denitrification and sedimentation
(b) denitrification and fixation
(c) nitrification and fixation
(d) nitrification and sedimentation

3 In the P budget of a eutrophic lake, in which external loading = 30 kg, gain in the lake = 50 kg, and outflow = 20 kg, internal loading is

(a) −40 kg
(b) −60 kg
(c) +60 kg
(d) +40 kg

4 Lake acidification can be most simply defined as

(a) a decrease in pH
(b) an increase in H^+
(c) a replacement of HCO_3^- with strong acid
(d) a replacement of $Ca^{2+} + Mg^{2+}$ with strong acid

5 Iron redox reactions at the sediment surface in lakes result in

(a) ferrous iron and soluble P under oxic conditions
(b) ferrous iron and soluble P under anoxic conditions
(c) ferric iron and particulate P under anoxic conditions
(d) ferric iron and soluble P under oxic conditions

6 The Q_{10} rule predicts, for a 10°C rise in temperature, that biological activity (e.g. growth rate) will increase by

(a) 200%
(b) 100%
(c) 50%
(d) 25%

7 Photosynthesis in the water column of lakes tends to be directly related to light intensity at

(a) intensities $< I_K$
(b) intensities $> I_K$
(c) rates $< P_{max}$
(d) rates $> P_{max}$

8 Compensation depth is the depth at which

   (a) light intensity is enough for $P > R$
   (b) light intensity $= 1\% I_0$
   (c) $P = R$ with mixing
   (d) light intensity is sufficient for $P = R$

9 Lake ageing (or filling) may accelerate as rooted macrophtes establish because they

   (a) outcompete plankton algae for nutrients
   (b) increase recycling of nutrients from sediments
   (c) prevent resuspension of particulate matter
   (d) encourage bottom covering projects by lake-shore residents

10 $R = \dfrac{1}{1 + \sqrt{\rho}}$ represents an estimate of

   (a) P sedimentation rate
   (b) relative thermal resistance to mixing
   (c) P sedimentation rate coefficient
   (d) P retention coefficient

11 Match the process names with the appropriate reactions by placing the correct letters in the blanks.

   _____ nitrification
   _____ sulfate reduction
   _____ denitrification
   _____ nitrogen fixation
   _____ bacterial photosynthesis

   (a) $CO_2 + 2H_2S \rightarrow CH_2O + H_2O + 2S$
   (b) $NH_4^+ + 2O_2 \rightarrow NO_3^- + 2H^+ + H_2O$
   (c) $SO_4^{2-} + 2CH_2O + 2H^+ \rightarrow H_2S + 2CO_2 + 2H_2O$
   (d) $5S + 6NO_3^- + 2H_2O \rightarrow 5SO_4^{2-} + 3N_2 + 4H^+$
   (e) $2N + 3H_2 \rightarrow 2NH_3$

12 Acidification of lakes, defined as alkalinity reduction, frequently occurs in nearly direct proportion to

   (a) pH in precipitation minus uptake in the watershed
   (b) pH in precipitation
   (c) SO_4^{2-} in precipitation
   (d) $NO_3^- + SO_4^{2-}$ in precipitation

13 Algae may self limit their productivity in lakes because

   (a) CO_2 accumulates faster than it can escape to the atmosphere
   (b) O_2 becomes supersaturated

(c) CO_2 decreases and pH increases
(d) CO_2 increases and pH increases

14 Phosphorus is usually limiting in lakes because

(a) P occurs in smaller concentration than N or C
(b) N/P ratios are usually greater than 10/1 in oligotrophic-mesotrophic lakes
(c) N/P ratios are usually less than 10/1 except in hypereutrophic lakes
(d) N/P ratios are usually greater than 20/1 in hypereutrophic lakes

15 Algal growth (yield) of non-N fixers will likely be limited by nitrogen if the N:P ratio in water is

(a) more than that in algal cells
(b) less than that in algal cells
(c) more than 10:1
(d) more than 20:1

16 A mass balance of phosphorus in lakes will indicate a net internal loading if

(a) output exceeds input
(b) output is less than input
(c) output plus lake P increase exceeds input
(d) output plus lake P increase is less than input

17 dx/dt represents net productivity of plankton algae when there are

(a) no losses due to dilution or grazing
(b) losses due to dilution or grazing
(c) no respiration losses
(d) nutrient limitations

18 Plankton algal growth is said to be light limited if the critical depth

(a) exceeds the compensation depth
(b) exceeds the mixing depth
(c) is less than the compensation depth
(d) is less than the mixing depth

19 Photosynthesis in the water column is considered to be light limited at intensities less than

(a) 50% I_0
(b) 100% I_0

(c) I_k
(d) μ_{max}

20 P_{max} (mg C/mg chl-hr) in the water column is subject to limitation by

   (a) temperature and nutrients
   (b) temperature but not nutrients
   (c) nutrients but not temperature
   (d) grazing

21 The growth of individual species of plankton algae in response to temperature is best described by

   (a) the RGT rule
   (b) the Q_{10} rule
   (c) a nearly linear model
   (d) an exponential model

22 Carbon

   (a) never limits plankton algal growth
   (b) may limit when photosynthesis is high and alkalinity is low
   (c) may limit when nighttime respiration is low
   (d) may limit when photosynthesis is high and pH is low

23 According to the Vollenweider model (completely mixed, constant volume, etc.), lakes with

   (a) short detention times are tolerant of high inflow phosphorus concentrations
   (b) short detention times are tolerant to any inflow phosphorus concentration
   (c) long detention times are intolerant to high inflow phosphorus concentrations
   (d) long detention times are tolerant to high inflow phosphorus concentrations

24 Lakes may theoretically age (fill in) and 'die' due to cultural eutrophication at rates of

   (a) 1–5 m per 1000 years
   (b) 1–5 m per 100 years
   (c) 1–5 m per 10 years
   (d) 1–5 m per 10 000 years

25 The capacity of zooplankton to control phytoplankton biomass in lakes is indicated by a ratio of

   (a) zooplankton production: phytoplankton production $= 1.0$
   (b) zooplankton production: phytoplankton production > 0.1

(c) zooplankton consumption: phytoplankton production = 1.0
(d) zooplankton consumption: phytoplankton production > 0.1

26 *Sphaerotilus* growth in streams is largely dependent on

(a) the limiting nutrient content, whether N or P
(b) DOC content above 15 mg/l
(c) light availability
(d) DOC content from 1–30 mg/l

27 Barring large, non-algal sources of particulate matter, transparency can usually be improved more per unit chl *a* reduction in

(a) oligotrophic than eutrophic lakes
(b) eutrophic than oligotrophic lakes
(c) mesotrophic than oligotrophic
(d) eutrophic than mesotrophic

28 Compared to the P concentration producing a nuisance biomass level in a eutrophic lake (~ 10 mg chl a/m^3), a nuisance biomass level in streams (~ 150 mg chl a/m^2) could probably occur at a P concentration

(a) greater than that in the eutrophic lake
(b) less than that in the eutrophic lake
(c) about the same as that in the eutrophic lake
(d) that is uncomparable to that in the eutrophic lake

29 If the hypolimnetic DO is 10 mg/l on 31 May and 2 mg/l on 31 August, and the hypolimnetic area and volume are, respectively, 1.25×10^6 m^2 and 5×10^6 m^3, the ODR in mg/m$^2 \cdot$ day is about

(a) 250
(b) 200
(c) 350
(d) 400

30 N/P ratios can be expected to decrease as lakes become eutrophic because

(a) wastewater has an N/P $\simeq 3$ and nitrification becomes more important
(b) wastewater has an N/P $\simeq 13$ and denitrification becomes more important
(c) wastewater has an N/P $\simeq 13$ and nitrification becomes more important
(d) wastewater has an N/P $\simeq 3$ and denitrification becomes more important

31 The expected flux rate of TP to the sediment in a lake with a TP loading of 512 mg/m$^2 \cdot$ y, a detention time of two years, a $\overline{TP}$ concentration of 42 mg/m^3, and a mean depth of 10 m, is about

(a) 300 mg/m$^2 \cdot$ y
(b) 0.71/y
(c) 212 mg/m$^2 \cdot$ y
(d) 604 mg/m^2

32 Organisms most characteristic of a septic zone in a stream receiving organic waste are

(a) *Sphaerotilus* and protozoans
(b) *Sphaerotilus* and facultative bacteria
(c) facultative bacteria and protozoans
(d) facultative bacteria and fecal bacteria

33 Organisms most characteristic of the recovery zone of a stream receiving organic waste are

(a) filamentous algae
(b) *Sphaerotilus*
(c) facultative bacteria
(d) protozoans

34 Rooted, submersed macrophytes have been shown to contribute internal loading of P in lakes primarily by

(a) root uptake and leaf excretion
(b) root uptake and plant senescence
(c) leaf uptake and leaf excretion
(d) leaf uptake and plant senescence

35 Macroinvertebrate communities normally found in streams of low order tend to be composed of

(a) burrowers and small particulate selectors
(b) clingers and large particulate shredders
(c) burrowers and large particulate shredders
(d) clingers and small particulate collectors

36 Phosphorus internal loading in the shallow unstratified lakes may occur as a result of

(a) iron *reduction* under *oxic* conditions at the sediment–water interface
(b) high pH in the *photic zone*
(c) excretion of SRP from *zooplankton*
(d) microbial decomposition of organic matter at the sediment–water interface

37 The concept of cascading trophic interactions suggests that algal biomass may be controlled by

   (a) increased piscivory resulting in decreased planktivory
   (b) increased planktivory resulting in decreased piscivory
   (c) decreased piscivory resulting in increased planktivory
   (d) decreased planktivory resulting in increased piscivory

38 Blue green buoyancy occurs with

   (a) increased CO_2, decreased pH and decreased light
   (b) decreased CO_2, increased pH and decreased light
   (c) decreased CO_2, decreased pH and decreased light
   (d) decreased CO_2, increased pH and increased light

39 N/P ratios in lakes decrease with increasing eutrophication because

   (a) nitrification and sediment phosphorus release increase under anoxic conditions
   (b) denitrification and sediment phosphorus release increase under oxic conditions
   (c) denitrification and sediment phosphorus release increase under anoxic conditions
   (d) nitrification and sediment phosphorus release increase under oxic conditions

40 Phosphorus is considered the macronutrient that usually limits primary production in most lakes because

   (a) there is little recycling of phosphorus
   (b) lakes are a net sink for phosphorus
   (c) lakes have more N and C than phosphorus
   (d) N and C have atmospheric sources

41 Oxygen in stratified lakes decreases as hypolimnetic depth

   (a) increases and as eutrophication increases
   (b) decreases and as eutrophication increases
   (c) decreases and as eutrophication decreases
   (d) increases and as eutrophication decreases

42 Blue greens that fix nitrogen are species of

   (a) *Oscillatoria* and *Anabaena*
   (b) *Anabaena* and *Aphanizomenon*
   (c) *Oscillatoria* and *Microcystis*
   (d) *Microcystis* and *Aphanizomenon*

43 The greatest increase in periphytic algae in streams would most likely occur from a 10 µg/l increase in SRP if current velocity were

(a) ≤ 15 cm/s and initial SRP were 20 µg/l
(b) >15 cm/s and initial SRP were 2 µg/l
(c) >15 cm/s and initial SRP were 20 µg/l
(d) ≤ 15 cm/s and initial SRP were 2 µg/l

44 The threshold concentration of BOD_5 that produces nuisance biomass levels of filamentous bacteria in streams is around

(a) 1 mg/l
(b) 5 mg/l
(c) 10 mg/l
(d) 50 mg/l

45 By suggesting that the number of invertebrate species will increase as the maximum daily mean temperature increases from 15°C to 30°C assumes that

(a) there is ample recruitment of new species that are tolerant to the new temperature
(b) the more tolerant species are already present
(c) the present species acclimated to a 15°C maximum will readapt to a 30°C maximum
(d) invertebrates cannot acclimate

46 Submersed, rooted macrophytes colonize lakes to a maximum depth where I_z (radiation at depth) is

(a) $\geq 1\% I_O$
(b) $\geq 50\% I_O$
(c) $\geq 10\% I_O$
(d) $\geq 75\% I_O$

47 Chlorophyll a in periphytic algal mats in streams may not correlate well with TP in the water, as algal chl a does in lakes, because

(a) N usually limits in streams
(b) the chl a/area is correlated with supply rate (TP × flow)
(c) only SRP is utilized
(d) the chl a/area is not restricted to the P in a given volume

48 Filamentous bacterial biomass will dominate the periphyton in streams if there is a

(a) DOC (dissolved organic carbon) source even without N and P
(b) N and P source even without DOC

(c) DOC source with N and P
(d) DOC source with N

49 ODR (oxygen deficit rate) is related to phosphorus loading in stratified lakes because

    (a) P is related to algal biomass, which demands oxygen when it sinks into the hypolimnion
    (b) P is related to algal biomass, which demands oxygen in the epilimnion
    (c) phosphate can be oxidized
    (d) P is related to algal and macrophyte biomass, which demands oxygen when it sinks into the hypolimnion

50 Some aquatic worms, especially *Tubifex*, are thought to tolerate, and even thrive in, organically enriched streams primarily because they

    (a) can live anaerobically
    (b) resist low *DO*
    (c) resist low *DO* and reproduce rapidly
    (d) are not preyed upon

51 The depositing substrata in a stream selects for

    (a) stoneflies, caddisflies and mayflies
    (b) worms, midges, and burrowing mayflies
    (c) midges and burrowing stoneflies
    (d) worms, snails and caddisflies

52 There is evidence that organic enrichment of streams, resulting in mats of *Sphaerotilus*, has

    (a) always benefited fisheries
    (b) sometimes been a benefit and sometimes a detriment to fisheries
    (c) never benefited fisheries
    (d) benefited some fisheries but not trout

53 Regarding Vollenweider's steady state model for phosphorus, sedimentation flux rate

    (a) and the sedimentation rate coefficient (σ) decrease with an increase in flushing rate
    (b) increases, but the sedimentation rate coefficient (σ) decreases, with an increase in flushing rate
    (c) and the sedimentation rate coefficient (σ) increase with an increase in influshing rate
    (d) decreases, but the sedimentation rate coefficient (σ) increases, with an increase in flushing rate

54 Submersed macrophyte populations often decrease as highly eutrophic or hypereutrophic states are reached in lakes as a result of

(a) shading by planktonic and periphytic algae
(b) shading by planktonic algae only
(c) increasing organic content of the sediments
(d) increasing toxicants in sediments

55 The autotrophic index (AFDW/chl a)

(a) decreases with inorganic nutrient enrichment because heterotrophs increase
(b) increases with organic enrichment because heterotrophs increase
(c) increases with organic enrichment because autotrophs increase
(d) decreases with organic enrichment because autotrophs increase

56 The ΔT experienced by invertebrate organisms entrained in cooling waters is

(a) more important to their survival than the final temperature
(b) of equal importance with final temperature
(c) less important than the final temperature
(d) of little concern in the receiving stream

57 The net increase in biomass of periphytic algae in streams is equal to

(a) net productivity
(b) net productivity minus losses to grazing and scouring
(c) gross productivity
(d) gross productivity minus losses to grazing and scouring

58 Dissolved oxygen concentration (mg/l) should be used as the unit in water quality standards rather than per cent saturation because

(a) transport into the fish's blood is a function of concentration rather than partial pressure
(b) analytical methods are more appropriate and criteria research results are largely in concentration
(c) % saturation gives waste dischargers too much of an advantage
(d) fish need more DO as temperature increases

59 The toxicity of heavy metals to fish is affected physiologically by H^+ in a manner

(a) proportional to that of hardness ions
(b) opposite to that of hardness ions
(c) consistent with that of hardness ions
(d) opposite to that of dissolved oxygen

60 In stratified lakes, trout are limited by eutrophication more than bass, crappie and pike because of (directly or indirectly)

   (a) oxygen
   (b) temperature
   (c) pH
   (d) NH_3

61 Excess *Cladophora* biomass in streams and nearshore areas in lakes and even estuaries is most likely caused by increases in

   (a) DOC
   (b) N
   (c) P
   (d) BOD

62 Excess *Sphaerotilus* biomass in streams is most likely caused by increases in

   (a) treated sewage
   (b) treated pulp mill waste
   (c) untreated pulp mill waste
   (d) agricultural runoff

63 Low level enrichment of an oligotrophic stream

   (a) may benefit fish growth, but will always decrease macroinvertebrate (M) diversity
   (b) may benefit fish growth and increase M diversity
   (c) will probably not affect either fish growth or M diversity
   (d) will probably not affect fish growth, but will always decrease M diversity

64 To use diversity indices (e.g. d, $\bar{d}$) requires

   (a) an experimental control and taxonomic knowledge to species
   (b) a taxonomic knowledge to sp., but no control
   (c) a control, but no taxonomic knowledge to sp.
   (d) neither a control or taxonomic knowledge to sp.

65 Macrophyte distribution and abundance in lakes have been associated with

   (a) sediment type and light availability more than nutrients
   (b) sediment type and nutrients more than light
   (c) light and nutrients more than sediment type
   (d) nutrients more than sediment type and light

66 A DO concentration decrease from 9 mg/l to 6 mg/l may result in

  (a) mortality in juvenile bluegill sunfish
  (b) mortality in juvenile coho salmon
  (c) growth impairment in juvenile coho salmon
  (d) growth impairment in common carp

67 The cause for mortality of fish exposed to lethal concentrations of heavy metals is asphyxiation due to

  (a) gill tissue deterioration
  (b) gill tissue deterioration and mucous accumulation
  (c) cellular respiration failure internally
  (d) a reduction in red blood cell count

68 Natural factors that may favour high diversity in macroinvertebrates are

  (a) eroding substratum and moderate organic food supply
  (b) eroding substratum and high organic food supply
  (c) depositing substratum and moderate organic food supply
  (d) depositing substratum and high organic food supply

Answers

Short-answer questions

1 The two phenomena of choice are:

  (a) increased algal abundance increases turbidity and light attenuation, thus decreasing the photic zone depth;
  (b) increased productivity (photosynthesis) raises the pH and lowers the CO_2 concentration.

The lower CO_2 level favours BGs, because of their lower half saturation constant and buoyancy is also increased as a result of reducing photosynthesis at depth where CO_2 and light are low and N and P high – vacuoles expand due to reduced turgor pressure or consumed photosynthate (polysaccharide) ballast. The resulting buoyancy provides more light to BGs in an otherwise low-light environment.

Alternatively, there are the hypotheses of Si depletion with increased production of diatoms eventually limiting their capacity to use the increased N and P. For this to work one still needs the light/CO_2 aspect to restrict green algae. Declining N/P ratios also occur with eutrophication, due to increased P internal loading and denitrification, with low

NO$_3$ favouring N fixers, but not all BGs that dominate in eutrophic lakes are N fixers (e.g. *Microcystis*). Best correlations of BGs with nutrient changes have involved transparency as well. (Nonetheless credit was given for these answers.)

2 $X = Y\left[N_i - \dfrac{K_n D}{\mu_m - D}\right] = Y(N_i - N)$

This shows that biomass X is about equal to the difference between inflow and outflow nutrient concentration with the biomass having converted the available soluble nutrient into biomass; most of N_i is removed and converted to biomass so long as D is low enough to permit enough time for uptake.

$N = \dfrac{K_n D}{\mu_m - D}$

This shows that as D increases, N increases, which results in higher μ according to

$\mu = \dfrac{\mu_m N}{K_n + N}$

or at steady state $\mu X = DX$, so as D increases, μ increases.

3 (a) A stratifies resulting in an anoxic hypolimnion, thus P can release from sediment due to iron reduction and PO$_4$ solubilization and diffusion into anoxic overlying water and/or release from bacterial cells in surficial sediment under anoxia.

  B does not stratify although periods of calm weather can result in anoxic water layer overlying sediment accumulating P due to iron reduction with subsequent mixing entraining the high-P layer. Higher temperature at sediment surface due to complete mixing enhances microbial decomposition of organic matter. BG migration from oxic sediments, macrophyte decomposition and P released that was translocated from roots, and high pH caused by intense photosynthesis if such high pH water reaches the sediment surface are all possibilities. High pH can act together with wind mixing solubilizing P from entrained particulate P.

  (b) The thermocline may act as a barrier to P entrainment to the photic zone, which is probably restricted to the epilimnion in eutrophic lakes (due to high light attenuation), while in unstratified lakes, internally loaded P is readily available for uptake.

4   $60 = \left[\dfrac{L_{ext}}{15 \times 2.25} \dfrac{1}{1 + 1/\sqrt{2.25}}\right] + \dfrac{L_{int}}{15 \times 2.25}$

$60 = 35.6 + \dfrac{L_{int}}{15 \times 2.25}$

$L_{int} = 825\,\text{mg/m}^2 \cdot \text{y}$

Less, because external input generally declines with low flow in summer and having 15 m $\bar{z}$, internally loaded P in the hypolimnion would have only a moderate impact to the epilimnion until turnover, unless storms eroded the metalimnion.

5   Eutrophication promotes dominance by blue greens which are inefficiently grazed resulting in increased amounts of detritus and bacteria that favour small bodied zooplanktons.

Increased planktivory can also remove the large bodied zooplankton because they are selected over the small bodied ones. Low DO in hypolimnion may deny the large zooplankton a refuge during the daylight hours.

Removal of the large-bodied, efficient-grazing zooplankton results in higher algal biomass and poorer transparency, which in turn favours more blue greens.

6   In A, because with higher velocity, the boundary layer surrounding the periphyton mat would be smaller affording a large gradient in SRP for uptake and, thus, uptake efficiency per unit ambient SRP would be greater.

7   $TP = \dfrac{TP_i}{1 + \sqrt{\tau}}$ shows that L. Tahoe is less sensitive because long τ allows for much greater loss of inflow TP due to sedimentation.

$TP = TP_i \dfrac{1}{1 + \sqrt{700}} = TP_i\, 0.036$

$\quad\; = TP_i \dfrac{1}{1 + \sqrt{3}} = TP_i\, 0.366$

Tahoe can tolerate 10X the TP_i as Lake Washington.

8   $\ln 100/15 = [(0.025 \times 25) + K_w]1$
$1.9 = 0.625 + K_w \qquad K_w = 1.27$

$C_{max} = \dfrac{1}{K_c}\left[\dfrac{F}{r24Z_m} - K_w\right] = \dfrac{1}{0.025}\left[\dfrac{2.7 \times 12}{0.1 \times 24 \times Z_m} - 1.27\right]$

$\quad\quad = 3.2\ \mu\text{g/l for 10 m and 39 for 6 m}$

alternatively $Z_{cr} = \dfrac{2.7 \times 12}{0.1 \times 24[(10 \times 0.025) + 1.27]} = 8.9\,\text{m}$

$\phantom{\text{alternatively } Z_{cr}}= 7.4$ for 10 hr light

Thus, Z_{mix} needed to have $<10\,\mu\text{g/l}$ chl a is about 9 m. Although $39\,\mu\text{g/l}$ chl a represents the maximum that can occur with the mean being somewhat lower, it is unlikely that the mean can be maintained at $10\,\mu\text{g/l}$.

9 The waste is toxic initially, because both sp. no. and biomass decline (biomass would not decline if strictly organic) – the effect is mild because populations still exist – typical effect of inhibition exceeding beneficial effects of increased food. Organic is secondary effect, because although both sp. no. and biomass increase, biomass increases much more than sp. no., which is still less than control – now increased food effect on biomass, while adverse physical–chemical environment maintains low sp. no. (or still could be lingering effect of toxicants on intolerant forms). Increased AI could be due to toxicity to autotrophs in first stage, then the large increase above the 400 level due to organic response of heterotrophs.

10 *Cladophora* mats could be caused by 'a', due to the inorganic nutrients (produces autotrophic growth) or by insecticide that would have killed grazers allowing algae to proliferate. Being an autotroph, *Cladophora* would not respond to organic substances for energy like heterotrophs. And it would not be adversely affected by an insecticide – the effect of periphytic algae proliferating if grazers are killed with insecticide was shown in artificial channels and has been observed in streams so affected.

Cladophora mats would not result from untreated sewage immediately downstream because of competition with faster growing filamentous bacteria, nor from a pulp mill waste for the same reason, plus there is little N or P in p.m.w. Recall the course of events in the Manawatu R., NZ.

11 (a) $L_{ext} + L_{int} = TP\overline{Z}(\rho + \rho^{0.5}) = 35 \times 15(0.4 + 0.4^{0.5})$

$\phantom{(a) L_{ext} + L_{int}} = 542\,\text{mg/m}^{2-}\text{y}$

$L_{int} = L_{int+ext} - L_{ext} = 542 - 300 = 242\,\text{mg/m}^2 \cdot \text{y}$

$L_{int} + L_{ext} = 25 \times 15(0.4 + 0.4^{0.5}) = 387\,\text{mg/m}^2 \cdot \text{y}$ for mesotrophy

$100 - [(387 - 300)/242] = 64\%$

Answer is the same if you use $TP = L_{ext}/[\overline{Z}(\rho + \rho^{0.5})] + L_{int}/\overline{Z}\rho$

(b) That the empirical sedimentation rate coefficient $\rho^{0.5}$ applies to the lake in question.

12 Scouring of bottom substrata by high peak flows – unstable gravels/rubbles is undesirable habitat (even depositing substrata adversely affected).

Deposition of high sediment loads transported from impervious surfaces reduces the habitat in substrata for diverse communities.

Toxicants, BOD, DO, etc are chemical effects

13 $$x = y\left[N_i - K_N \frac{D}{\mu_{max} - D}\right] = 64\left[4 - 0.3\frac{1.0}{1.3 - 1.0}\right]$$
$$= 192\,\text{mg/l}$$

Higher, because 1 mg/l nutrients in the continuous culture remains unused due to D being so near μ_{max}. In a batch culture, more time would be available for biomass to convert nearly all soluble nutrient into biomass.

14 $I_Z = I_0 e^{-K_w Z}$ $\quad C = \frac{1}{K_c}\left[\frac{Fi\lambda}{r24Z_{mix}} - K_w\right]$

$\ln\dfrac{I_Z}{I_O} = -K_w Z$ $\quad CK_c + K_w = \dfrac{Fi\lambda}{r24Z_{mix}}$

$\ln\dfrac{15}{100} = -K_w Z$ $\quad Z_{mix} = \dfrac{Fi\lambda}{r24(CK_c + K_w)}$

$K_w = 0.95/\text{m}$ $\quad\quad\quad\quad = \dfrac{2.7 \cdot 12}{0.1 \cdot 24(10 \cdot 0.025 + 0.95)}$

$\quad\quad\quad\quad\quad\quad\quad\quad\quad = 11.3\,\text{m}$

$\quad\quad\quad\quad\quad\quad\quad\quad\quad 9.4\,\text{m for}\,\lambda = 10\,\text{hr}$

15 Buoyant algae would obtain more light than non-buoyant species in eutrophic lakes where transparency is low and optimum light would occur only near the surface. Soluble nutrients would be depleted near the surface, during weakly stratified periods, but exit at higher concentrations near the bottom and/or below the photic zone. Blue greens can easily traverse the water column in a shallow lake in less than a day optimizing their growth conditions, while non-buoyant species would sink out of the photic zone.

16 Wind will increase internal loading in B, but not in A, by entraining water overlying sediments that have become anoxic and enriched with P during stable conditions preceding the wind. By replacing high-P

water overlying sediment with low-P water in that wind-off, wind-on process, sediment P release rates are increased by increasing the sediment-water concentration gradient. This same sediment–water interface renewal process occurs if cause for internal loading is high pH or temperature controlled microbial decomposition. Thermocline erosion in a stratified lake entrains P into the epilimnion and increases the availability of internally loading P, but that does not increase internal loading *per se*.

17 The relationship is direct; that is LC_{50s} increase (solution becomes less toxic) as hardness increases. An explanation is that the hardness ions (Ca^{2+} and Mg^{2+}) compete for binding sites on the gill surface, blocking access of heavy metals to the sites, resulting in less gill deterioration and irritation (= less mucus buildup).

18 Double the MACs should result in stressful conditions producing low survival in some sp. – also the effect is compounded because of the $\geq$ additive effect of Zn and Cu.

(a) Increase, probably, because algae affected more than heterotrophs; could see no change if both heterotrophs and algae affected equally.
(b) Decrease, because metals are inhibitors – sensitive sp. will be eliminated – only most tolerant will remain.
(c) Decrease – richness (no. sp.) declines, although the magnitude of change may be less than # sp. due to evenness.
(d) Increase – the decreased no. of sp. and abundance changes of individual sp. increases the dissimilarity of the two stations.

19 Growth will be reduced w/+ 3°, because the max. temp. is already at the upper range of the growth scope for salmonids. Growth will be reduced much more at +7° because the fish will be stressed and lose weight for most of the period of insect emergence (although insect growth/emergence may adjust to earlier in the year) and also because 24° is right at their 4-day lethal temp. In both cases they will seek lower temp.

20 Refers to control of planktivorous fish by either reducing their populations or providing refuges for large efficiently grazing zooplankton, thereby causing a reduction in algal content. Aerate hypolimnion to provide a refuge for large zoops or introduce high density of piscivorous fish. Poisoning all fish of course will work, but is usually not compatible with recreational use.

Multiple-choice questions

1. b
2. a
3. d
4. c
5. b
6. b
7. a
8. d
9. b
10. d
11. (a) b
    (b) c

   (c) d
   (d) e
   (e) a
12. d
13. c
14. b
15. b
16. c
17. a
18. d
19. c
20. a
21. c
22. b
23. d
24. a
25. c
26. d
27. a
28. b

29. c
30. d
31. a
32. d
33. a
34. b
35. b
36. d
37. a
38. b
39. c
40. d
41. b
42. b
43. b
44. b
45. a
46. c
47. d
48. c

49. a
50. c
51. b
52. b
53. d
54. a
55. b
56. c
57. b
58. d
59. a
60. b
61. c
62. c
63. b
64. c
65. a
66. c
67. b
68. a

Appendix C

Glossary*

Absorption penetration of one substance into the body of another.
Acclimation the process of adjusting to change, e.g. temperatures, in an environment.
Acute involving a stimulus severe enough to rapidly induce a response; in bioassay tests, a response observed within 96 hours typically is considered an acute one.
Adaptation a change in the structure, form, or habit of an organism resulting from a change in its environment.
Adsorption the taking up of one substance at the surface of another.
Aerobe an organism that can live and grow only in the presence of free oxygen.
Aerobic the condition associated with the presence of free oxygen in an environment.
Algae (alga) simple plants, many microscopic, containing chlorophyll. Most algae are aquatic and may produce a nuisance when conditions are suitable for prolific growth.
Algicide a specific chemical highly toxic to algae. Algicides are often applied to water to control nuisance algal blooms.
Alkalinity acid-neutralizing capacity of water, primarily composed of bicarbonate, carbonate and hydroxides.
Allelopathy production of chemicals by one species that inhibit the growth or behaviour of another species.
Allocthonous said of food material reaching an aquatic community from the outside in the form of organic detritus.
Alluvial transported and deposited by running water.
Amictic a lake that almost never mixes.
Amphipods, *see* Scuds.
Anadromous fish fish that typically inhabit seas or lakes but ascend streams at more or less regular intervals to spawn; e.g. salmon, steel-head, or American shad.

*Modified from WQC, 1968, 1973.

Anaerobe an organism for whose life processes a complete or nearly complete absence of oxygen is essential.

Anaerobic the condition associated with the lack of free oxygen in an environment.

Annelids segmented worms, as distinguished from the non-segmented roundworms and flatworms. Most are marine; however, many live in soil or freshwater. Aquatic forms may establish dense populations in the presence of rich organic deposits. Common examples of segmented worms are earthworms, sludgeworms, and leeches.

Anoxic depleted of free oxygen; anaerobic.

Antagonism the power of one toxic substance to diminish or eliminate the toxic effect of another; interactions of organisms growing in close association, to the detriment of at least one of them.

Application factor a factor applied to a short-term or acute toxicity test to estimate a concentration of waste that would be safe in a receiving water.

Assimilation the transformation and incorporation of substances (e.g. nutrients) by an organism or ecosystem.

Autochonous said of food material entering an environment from inside in the form of photosynthetic (autotrophic) production.

Autotrophic organism an organism capable of constructing organic matter from inorganic substances.

Benthic region the bottom of a body of water. This region supports the benthos.

Benthos aquatic bottom-dwelling organisms including: (1) sessile animals, such as the sponges, barnacles, mussels, oysters, some worms, and many attached algae; (2) creeping forms, such as insects, snails, and certain clams; and (3) burrowing forms which include most clams and worms.

Best management practices (BMPs) methods or practices that reduce non-point sources of pollution. BMPs include, but are not limited to, structural and non-structural controls and operations and maintenance procedures.

Bioaccumulation uptake and retention of environmental substances by an organism from its environment, as opposed to uptake from its food.

Bioassay a determination of the concentration or dose of a given material necessary to affect a test organism under stated conditions.

Biochemical oxygen demand (BOD) the amount of O_2 required by bacteria while oxidizing decomposable organic matter under aerobic conditions.

Biodiversity the number of different species or organisms in a region.

Biomagnification *see* Trophic accumulation.

Biomass the living weight of a plant or animal population, usually expressed on a unit area basis.

Biopollution an introduced, non-native species that causes harm to an ecosystem.

Biota all living organisms of a region.

Biotic index   a numerical index using various aquatic organisms to determine their degree of tolerance to differing water conditions.
Biotoxin   toxin produced by a living organism; the biotoxin which causes paralytic shellfish poisoning is produced by certain species of dinoflagellate algae.
Bioturbation   stirring of sediments by movement and activity of sediment-dwelling organisms.
Bivalve   an animal with a hinged two-valve shell; examples are the clam and oyster.
Black liquor   waste liquid remaining after digestion of rags, straw, and pulp.
Bloom   an unusually large number of organisms per unit of water, usually algae, made up of one or a few species.
Blue-green bacteria   a group of bacteria (formerly called algae) with a blue pigment, in addition to the green chlorophyll. This group is more appropriately referred to as 'cyanobacteria'. A stench is often associated with the decomposition of dense blooms of blue-greens in fertile lakes.
Body burden   the total amount of a substance present in the body tissues and fluids of an organism.
Buffer capacity   the ability of a solution to maintain its pH when stressed chemically.

Carrying capacity   the maximum biomass that a system is capable of supporting continuously.
Catadromous fishes   fishes that feed and grow in freshwater, but return to the sea to spawn. The best known example is the American eel.
Chelate   to combine with a metal ion and hold it in solution preventing it from forming an insoluble salt.
Chemolithotrophy   autotrophy with the energy sources of inorganic chemical bonds and inorganic substances as electron donors (chemosynthesis); also called chemoautotrophy.
Chronic   involving a stimulus that lingers or continues for a long period of time, often one-tenth of the life span or more.
Clean Water Association   an association of organisms, usually characterized by many different kinds (species). These associations occur in natural unpolluted environments.
Coagulation   a water treatment process in which chemicals are added to combine with or trap suspended and colloidal particles to form rapidly settling aggregates.
Coarse or rough fish   those species of fish considered to be of poor fighting quality when taken on tackle and of poor food quality. These fish may be undesirable in a given situation, but at times may be classified differently, depending upon their usefulness. Examples include carp, goldfish, gar, sucker, bowfin, gizzard shad, goldeneye and mooneye.

Coelenterate a group of aquatic animals that have gelatinous bodies, tentacles, and stinging cells. These animals occur in great variety and abundance in the sea and are represented in freshwater by a few types. Examples are hydra, corrals, sea anemones, and jellyfish.

Cold-blooded animals (poikilothermic animals) animals that lack a temperature regulating mechanism that offsets external temperature changes. Their temperature fluctuates to a large degree with that of their environment. Examples are fish, shellfish, and aquatic insects.

Coliform bacteria a group of bacteria inhabiting the intestines of animals including man, but also found elsewhere. It includes all the aerobic, non-spore forming, rod-shaped bacteria that produce from lactose fermentation within 48 hours at 37°C.

Conservative pollutant a pollutant that is relatively persistent and resistant to degradation, such as PCB and most chlorinated hydrocarbon insecticides.

Consumers organisms that consume solid particles of organic food material. Protozoa are consumers.

Crustacea mostly aquatic animals with rigid outer coverings, jointed appendages, and gills. Examples are crayfish, crabs, barnacles, water fleas, and sow bugs.

Cumulative brought about or increased in strength by successive additions.

Cyanotoxins diverse group of chemicals produced by cyanobacteria that are toxic; includes neurotoxins, hepatotoxins and dermatoxins.

Daphnia mostly microscopic swimming crustaceans, often forming a major portion of the zooplankton population. The second antennae are very large and are used for swimming.

Demersal living or hatching on the bottom, as fish eggs that sink to the bottom.

Denitrification conversion of nitrate to N_2 gas by microorganisms; a form of respiration that uses nitrate rather than O_2 to oxidize organic carbon under anaerobic conditions.

Dermatitis any inflammation of the skin. One type may be caused by the penetration beneath the skin of a cercaria found in water; this form of dermatitis is commonly called 'swimmers' itch'.

Detention time the time necessary for lake water replacement.

Detritivores organisms that eat detritus; also called saprophytes.

Detritus unconsolidated sediments comprised of both inorganic and dead and decaying organic material.

Dimictic a lake that mixes twice each year.

Disinfection-by-products compounds such as trihalomethanes produced by the interaction of organic matter and disinfectants such as chlorine and ozone.

Diurnal occurring once a day, i.e. with a variation period of 1 day; occurring in the daytime or during a day.

Diversity the abundance in numbers of species in a specified location.
Drift the material that washes downstream, particularly invertebrates.
Dystrophic said of brownwater lakes and streams usually with a low lime content and a high organic content; often lacking in nutrients.

Ecology the science of the interrelations between living organisms and their environment.
Ecoregion an area with similar soils, vegetation, geology and physiography as well as distinct assemblage of communities.
Effluent the wastewater released from a sewage plant, factory or other point source.
Emergent aquatic plants plants that are rooted at the bottom but project above the water surface. Examples are cattails and bulrushes.
Endocrine-disrupting chemicals substances that cause adverse health effects in an organism, or its progeny, consequent to endocrine function.
Environment the sum of all external influences and conditions affecting the life and the development of an organism.
Epilimnion the surface waters in a thermally stratified body of water; these waters are characteristically well mixed.
Epiphytic living on the surface of other plants.
Estuary commonly an arm of the sea at the lower end of a river. Estuaries are often enclosed by land except at channel entrance points.
Euphotic zone the lighted region that extends vertically from the water surface to the level at which photosynthesis fails to occur because of ineffective light penetration.
Eurytopic organisms organisms with a wide range of tolerance to a particular environmental factor. Examples are sludgeworms and bloodworms.
Eutrophic abundant in nutrients and having high rates of productivity frequently resulting in oxygen depletion below the surface layer.
Eutrophication the intentional or unintentional enrichment of water.
Eutrophic waters waters with a good supply of nutrients. These waters may support rich organic productions, such as algal blooms.

Facultative able to live under different conditions, as in facultative aerobes and facultative anaerobes.
Facultative aerobe an organism that although fundamentally an anaerobe can grow in the presence of free oxygen.
Facultative anaerobe an organism that although fundamentally an aerobe can grow in the absence of free oxygen.
Fall overturn a physical phenomenon that may take place in a body of water during the early autumn. The sequence of events leading to fall overturn include: (1) Cooling of surface waters, (2) density change in surface waters producing convection currents from top to bottom,

(3) circulation of the total water volume by wind action, and (4) vertical temperature equality, 4 °C. The overturn results in a uniformity of the physical and chemical properties of the water.

Fauna the entire animal life of a region.

Fecal coliform bacteria bacteria of the coliform group of fecal origin (from intestines of warm-blooded animals) as opposed to coliforms from non-fecal sources.

Finfish that portion of the aquatic community make up of the true fishes as opposed to invertebrate shellfish.

Flatworms (Platyhelminthes) non-segmented worms, flattened from top to bottom. In all but a few of the flatworms, complete male and female reproductive systems are present in each individual. Most flatworms are found in water, moist earth, or as parasites in plants and animals.

Floating aquatic plants plants that wholly or in part float on the surface of the water. Examples are water lilies, water shields, and duckweeds.

Flocculation the process by which suspended colloidal or very fine particles are assembled into larger masses or floccules which eventually settle out of suspension.

Flora the entire plant life of a region.

Flushing rate the fraction of lake water replaced per time.

Food chain the transfer of food energy from plants or organic detritus through a series of organisms, usually four or five, consuming and being consumed.

Food web the interlocking pattern formed by a series of interconnecting food chains.

Free residual chlorination chlorination that maintains the presence of hypochlorous acid (HOCl) or hypochlorite ion (OCl^-) in water.

Fry (sac fry) the stage in the life of a fish between the hatching of the egg and the absorption of the yolk sac. From this stage until they attain a length of 1 inch, the young fish are considered advanced fry.

Fungi (fungus) simple or complex organisms without chlorophyll. The simpler forms are one celled; the higher forms have branched filaments and complicated life cycles. Examples of fungi are moulds, yeasts, and mushrooms.

Fungicide substances or a mixture of substances intended to prevent, destroy, or mitigate any fungi.

Game fish those species of fish considered to possess sporting qualities on fishing tackle. These fish may be classified as undesirable, depending upon their usefulness. Examples of freshwater game fish are salmon, trout, grayling, black bass, muskellunge, walleye, northern pike, and lake trout.

Geosmin an organic chemical produced by some bacteria and algae that causes taste and odour problems in water supplies.

Green algae algae that have pigments similar in colour to those of higher green plants. Common forms produce floating algal mats in lakes.
Gross production total energy fixed including that used for respiration.

Half-life the period of time in which a substance loses half of its active characteristics (used especially in radiological work); the time required to reduce the concentration of a material by half.
Hemostasis the cessation of the flow of blood in the circulatory system.
Herbicide substances or a mixture of substances intended to control or destroy any vegetation.
Herbivore an organism that feeds on vegetation.
Heterocyst specialized cyanobacterial cell in which nitrogen fixation occurs at high rates.
Heterotrophic organism organisms that are dependent on organic matter for food.
Higher aquatic plants flowering aquatic plants. (These are separately categorized herein as Emergent, Floating, and Submerged Aquatic Plants.)
Histopathologic occurring in tissue due to a diseased condition.
Holomictic lakes lakes that are completely circulated to the deepest parts at time of winter cooling.
Hydrophobic unable to combine with or dissolve in water.
Hydrophytic growing in or in close proximity to water; e.g. aquatic algae and emergent aquatic vascular plants.
Hypereutrophic extremely productive, with very high primary producer biomass.
Hypertrophy non-tumorous increase in the size of an organ as a result of enlargment of constituent cells without an increase in their number.
Hypolimnion the region of a body of water that extends from below the thermocline to the bottom of the lake; it is thus removed from much of the surface influence.
Hypoxic oxygen-depleted, less than $2\,\text{mg}\,\text{l}^{-1}$ DO.

Insecticide substances or a mixture of substances intended to prevent, destroy, or repel insects.
Interstitial between particles.
Invertebrates animals without backbones.
Isothermal with the same temperature.

Labile unstable and likely to change under certain influences.
Laminar flow flow all in one direction, with little lateral mixing (as opposed to turbulent flow).
Lentic or lenitic environment standing water and its various intergrades. Examples of lenitic environments are lakes, ponds, and swamps.

Lethal involving a stimulus or effect causing death directly.
Life cycle the series of stages in the form and mode of life of an organism: i.e. the stages between successive recurrences of a certain primary stage such as the spore, fertilized egg, seed, or resting cell.
Limnetic zone the open-water region of a lake. This region supports plankton and fish as the principal plants and animals.
Limnology the study of inland (continental) waters (i.e. lakes, rivers, wetlands).
Lipophilic having an affinity for fats or other lipids.
Littoral zone the shallow shoreward region of a body of water having light penetration to the bottom; frequently occupied by rooted plants.
Lotic environment running waters, such as streams or rivers.

Macronutrient a chemical element necessary in large amounts for the growth and development of plants.
Macro-organisms plants, animals, or fungal organisms visible to the unaided eye.
Macrophyte the larger aquatic plants, as distinct from the microscopic plants, including aquatic mosses, liverworts and larger algae as well as vascular plants; no precise toxonomic meaning; generally used synonymously with aquatic vascular plants.
Marl an earthy, unconsolidated deposit formed in freshwater lakes, consisting chiefly of calcium carbonate mixed with clay or other impurities in varying proportions.
Median lethal concentration (LC_{50}) the concentration of a test material that causes death to 50 per cent of a population within a given time period.
Median lethal dose (LD_{50}) the dose of a test material, ingested or injected, that kills 50 per cent of a group of test organisms.
Median tolerance limit (TL_{50}) the concentration of a test material in a suitable diluent (experimental water) at which just 50 per cent of the test animals are able to survive for a specified period of exposure.
Meromictic lakes lakes in which dissolved substances create a gradient of density differences in depth, preventing complete mixing or circulation of the water.
Mesotrophic having a nutrient load resulting in moderate productivity.
Metabolites products of metabolic processes.
Metalimnion the intermediate zone in a stratified lake in which the temperature change with depth is rapid (also referred to as thermocline, which is the preferred term).
Methylation combination with the methyl radical (CH_2).
Microcystins cyclic peptides produced by cyanobacteria that cause liver damage (hepatotoxins).
Micronutrient chemical element necessary in only small amounts for growth and development; also known as trace elements.

Microorganism   any minute organism invisible or barely visible to the unaided eye.
Molluscicide   substances or a mixture of substances intended to destroy or control snails.
Mollusc   (mollusca) a large animal group including those forms popularly called shellfish (but not including crustaceans). All have a soft unsegmented body protected in most instances by a calcareous shell. Examples are snails, mussels, clams, and oysters.
Monomictic   a lake that mixes once a year.
Moss   any bryophytic plant characterized by small, leafy, often tufted stems bearing sex organs at the tips.
Motile   exhibiting or capable of spontaneous movement.

Naiad   another term for the larva of an aquatic insect.
Nanoplankton   plankton cells between approximately 3 and 50 µm.
Nekton   swimming organisms able to navigate at will.
Nematoda   unsegmented roundworms or threadworms. Some are free living in soil, freshwater, and salt water; some are found living in plant tissue; others live in animal tissue as parasites.
Net productivity   energy fixed into organic matter by green plants and available for transfer in food web. Respiration of plants excluded.
Neuston   organisms resting or swimming on the surface film of the water.
Nitrification   bacterial processes that convert nitrate to ammonium, yielding energy.
Non-conservative pollutant   a pollutant that is quickly degraded and lacks persistence, such as most organophosphate insecticides.
Non-point source   a diffuse source of pollution from the landscape (e.g. urban and agricultural runoff, atmospheric deposition).
Nutrients   organic and inorganic chemicals necessary for the growth and reproduction of organisms.
Nymph   another term for larva of an aquatic insect.

Oligotrophic   having a small supply of nutrients and thus supporting little organic production, and seldom if ever becoming depleted of oxygen.
Organic detritus   the particulate remains of disintegrated plants and animals.
Oxic   with O_2 (aerobic).
Oxygen-debt   a phenomenon that occurs in an organism when available oxygen is inadequate to supply the respiratory demand. During such a period the metabolic processes result in the accumulation of breakdown products that are not oxidized until sufficient oxygen becomes available.

Parasite   an organism that lives on or in a host organism from which it obtains nourishment at the expense of the latter during all or part of its existence.

Parthenogenesis a form of reproduction in which populations comprised of females produce diploid copies of themselves.

Pelagic zone the free-water region of a sea. (Pelagic refers to the sea, limnetic refers to bodies of freshwater.)

Periphyton associated aquatic organisms attached or clinging to stems and leaves of rooted plants or other surfaces projecting above the bottom of a water body.

Pesticide any substance used to kill plants, insects, algae, fungi, and other organisms; includes herbicides, insecticides, algalcides, fungicides, and other substances.

Photosynthesis the process by which simple sugars and starches are produced from carbon dioxide and water by living plant cells, with the aid of chlorophyll and in the presence of light.

Phytoplankton plant plankton that live unattached in water.

Piscicide substances or a mixture of substances intended to destroy or control fish populations.

Plankton (plankter) organisms of relatively small size, mostly microscopic, that have either relatively small powers of locomotion or that drift in that water with waves, currents, and other water motion.

Point source a clearly defined source of pollution that typically discharges via a pipe to a water body (e.g. sewage outfall).

Polymictic a lake that almost continuously circulates or that has multiple mixing periods annually.

Pool zone the deep-water area of a stream, where the velocity of current is reduced. The reduced velocity provides a favourable habitat for plankton. Silt and other loose materials that settle to the bottom of this zone are favourable for burrowing forms of benthos.

Producers organisms, for example, plants, that synthesize their own organic substance from inorganic substances.

Productivity the rate of storage of organic matter in tissue by organisms including that used by the organisms in maintaining themselves.

Profundal zone the deep and bottom-water area beyond the depth of effective light penetration. All of the lake floor beneath the hypolimnion.

Protozoa organism consisting either of a single cell or of aggregates of cells, each of which performs all the essential functions in life. They are mostly microscopic in size and largely aquatic.

Pycnocline a layer of water that exhibits rapid change in density, analogous to thermocline.

Rapids zone the shallow-water area of a stream, where velocity of current is great enough to keep the bottom clear of silt and other loose materials, thus providing a firm bottom. This zone is occupied largely by specialized benthic or periphytic organisms that are firmly attached to or cling to a firm substrate.

Red tide a visible red-to-orange coloration of an area of the sea caused by the presence of a bloom of certain 'armoured' flagellates.

Reducers organisms that digest food outside the cell wall by means of enzymes secreted for this purpose. Soluble food is then absorbed into the cell and reduced to a mineral condition. Examples are fungi, bacteria, protozoa, and non-pigmented algae.

Refractory resisting ordinary treatment and difficult to degrade.

Riffle a section of a stream in which the water is usually shallower and the current of greater velocity than in the connecting pools; a riffle is smaller than a rapid and shallower than a chute.

Riparian related to or located on the bank of a stream or river.

Rotifers (rotatoria) microscopic aquatic animals, primarily free-living, freshwater forms that occur in a variety of habitats. Approximately 75 per cent of the known species occur in the littoral zone of lakes and ponds. The more dense populations are associated with a substance of submerged aquatic vegetation. Most forms ingest fine organic detritus for food, whereas others are predaceous.

Safety factor a numerical value applied to short-term data from other organisms in order to approximate the concentration of a substance that will not harm or impair the organism being considered.

Saprophytes heterotrophs that decompose organic carbon.

Scuds (amphipods) macroscopic aquatic crustaceans that are laterally compressed. Most are marine and estuarine. Dense populations are associated with aquatic vegetation. Great numbers are consumed by fish.

Secchi disc a device used to measure visibility depths in water. The upper surface of a circular metal plate, 20 centimetres in diameter, is divided into four quadrants and so painted that two quadrants directly opposite each other are black and the intervening ones white. When suspended to various depths of water by means of a graduated line, its point of disappearance indicates the limit of visibility.

Sediment the particulate matter that settles to the bottom of a liquid.

Seiche a form of perodic current system, described as a standing wave, in which some stratum of the water in a basin oscillates about one or more nodes.

Sessile organisms organisms that sit directly on a base without support, attached or merely resting unattached on a substrate.

Seston suspended particles and organisms between 0.0002 and 1 mm in diameter.

Shellfish a group of molluscs usually enclosed in a self-secreted shell; includes oysters and clams.

Sludge a solid waste fraction precipitated by a water treatment process.

Smolt a young fish, usually a salmonid, as it begins and during the time it makes its seaward migration.

Sorption a general term for the processes of absorption and adsorption.

Species (both singular and plural) a natural population or group of populations that transmit specific characteristics from parent to off-spring. They are reproductively isolated from other populations with which they might breed. Populations usually exhibit a loss of fertility when hybridizing.

Sphaerotilus a slime-producing, non-motile, sheathed, filamentous, attached bacterium. Great masses are often broken from their 'holdfasts' by currents and are carried floating downstream in gelatinous flocks.

Sponges (porifera) one of the sessile animals that fasten to piers, pilings, shells, rocks, etc. Most live in the sea.

Spore the reproductive cell of a protozoan, fungus, alga, or bryophyte. In bacteria, spores are specialized resting cells.

Spring overturn a physical phenomenon that may take place in a body of water during the early spring. The sequence of events leading to spring overturn include: (1) melting of ice cover, (2) warming of surface waters, (3) density change in surface waters producing convection currents from top to bottom, (4) circulation of the total water volume by wind action, and (5) vertical temperature equality, 4°C. The overturn results in a uniformity of the physical and chemical properties of the water.

Standing crop quantitative abundance of organisms, either mass, volume or numbers, present at any one time.

Stenotopic organisms organisms with a narrow range of tolerance for a particular environmental factor. Examples are trout, stonefly nymphs, etc.

Stoichiometric the mass relationship in a chemical reaction.

Stratification the phenomenon occurring when a body of water becomes divided into distinguishable layers.

Subacute involving a stimulus not severe enough to bring about a response speedily.

Sublethal involving a stimulus below the level that causes death.

Sublittoral zone the part of the shore from the lowest water level to the lower boundary of plant growth.

Submerged aquatic plant a plant that is continuously submerged beneath the surface of the water. Examples are the pondweed and coontail.

Succession the orderly process of community change in which a sequence of communities replaces one another in a given area until a climax community is reached.

Swimmers' itch a rash produced on bathers by a parasitic flatworm in the cercarial stage of its life cycle. The organism is killed by the human body as soon as it penetrates the skin; however, the rash may persist for a period of about 2 weeks.

Symbiosis two organisms of different species living together, one or both of which may benefit and neither is harmed.

Synergistic interactions of two or more substances or organisms producing a result that any was incapable of independently.

Thermocline a layer in a thermally stratified body of water in which the temperature changes rapidly relative to the remainder of the body.

Tolerant association an association of organisms capable of withstanding adverse conditions within the habitat. It is usually characterized by a reduction in species (from a clean water association) and an increase in individuals representing a particular species.

Trophic accumulation passing of a substance through a food chain such that each organism retains all or a portion of the amount in its food and eventually acquires a higher concentration in its flesh than in the food.

Trophic level a scheme of categorizing organisms by the way they obtain food from primary producers or organic detritus involving the same number of intermediate steps.

Trophogenic region the superficial layer of a lake in which organic production from mineral substances takes place on the basis of light energy.

Tropholytic region the deep layer of a lake, where organic dissimilation predominates because of light deficiency.

Turnover rate fractional replacement of standing crop per time.

Turnover time time necessary to replace the standing crop.

Ultraoligotrophic extremely unproductive system.

Univoltine producing one generation annually.

Warm and cold-water fish warm-water fish include black bass, sunfish, catfish, gar, and others; whereas cold-water fish include salmon and trout, whitefish and others. The temperature factor determining distribution is set by adaptation of the eggs and larvae to warm or cold water.

Watershed area above a point in a stream that catches the water that flows down to that point; also called drainage basin or a catchment in Europe.

Wetland areas inundated or saturated by water at a frequency and duration sufficient to support a prevalence of vegetation adapted for life in saturated soil conditions.

Zooglea bacteria embedded in a jellylike matrix formed as the result of metabolic activities.

Zooplankton protozoa and other animal microorganisms living unattached in water. These include small crustacea, such as *Daphnia* and *Cyclops*.

References

Adams, J. R. (1969) Ecological investigations related to thermal discharges. *Pacific Gas and Electric Company Report*, Emeryville, CA.

Adrian, R. and Deneke, R. (1996) Possible impact of mild winters on zooplankton succession in eutrophic lakes of the Atlantic European area. *Freshwater Biol.*, **36**, 757–70.

Ahlgren, G. (1977) Growth of *Ocillatoria agardii* Gom. in chemostat culture. I. Investigation of nitrogen and phosphorus requirements. *Oikos*, **29**, 209–24.

Ahlgren, G. (1978) Response of phytoplankton and primary production to reduced nutrient loading in Lake Norrviken. *Verh. Int. Verein. Limnol.*, **20**, 840–5.

Ahlgren, G. (1987) Temperature functions in biology and their application to algal growth constants. *Oikos*, **49**, 177–90.

Ahlgren, G., Lundstedt, L., Brett, M. T. and Forsberg, C. (1990) Lipid composition and food quality of some freshwater phytoplankton for cladoceran zooplankters. *J. Plankton Res.*, **12**, 809–18.

Ahlgren, I. (1977) Role of sediments in the process of recovery of a eutrophicated lake, in *Interactions Between Sediments and Freshwater* (ed. H. L. Golterman), Dr W. Junk, The Hague, pp. 372–7.

Ahlgren, I. (1978) Response of Lake Norrviken to reduced nutrient loading. *Verh. Int. Verein. Limnol.*, **20**, 846–50.

Ahlgren, I. (1979) Lake metabolism studies and results at the Institute of Limnology in Uppsala. *Arch. Hydrobiol. Beih.*, **13**, 10–30.

Ahlgren, I. (1980) A dilution model applied, to a system of shallow eutrophic lakes after diversion of sewage effluents. *Arch. Hydrobiol.*, **89**, 17–32.

Ahlgren, I. (1988) Nutrient dynamics and trophic state response of two eutrophicated lakes after reduced nutrient loading, in *Eutrophication and Lake Restoration: Water Quality and Biological Impacts* (ed. G. Balvay), Thononles-Bains, pp. 79–97.

Ahlgren, I., Bostrom, B. and Petersson, A.-K. (1988a) Seasonal variation of the microbial community in the sediments of a hypereutrophic lake. *Verh. Int. Verein. Limnol.*, **23**, 460–1.

Ahlgren, I., Frisk, T. and Kamp-Nielsen, L. (1988b) Empirical and theoretical models of phosphorus loading, retention and concentration vs. lake trophic state. *Hydrobiologia*, **170**, 285–303.

Allison, E. M. and Walsby, A. E. (1981) The role of potassium in the control of turgor pressure in a gas vacuolate blue-green alga. *J. Exp. Bot.*, **32**, 241–9.

Ambühl, H. (1959) Die bedeutung der strömung als ökologischer factor. *Schweiz. Z. Hydrol.*, **21**, 133–264.

Amend, D., Yasutake, W. and Morgan, R. (1969) Some factors influencing susceptibility of rainbow trout to the acute toxicity of an ethyl mercury phosphate formulation (Timsan). *Trans. Am. Fish. Soc.*, **98**, 419–25.

Amy, G. L., Thompson, J. M., Tan, L., Davis, M. K. and Krasner, S. W. (1990) Evaluation of THM precursor contributions from agricultural drains. *J. Am. Water Works Assoc.*, **82**, 57–64.

Andersen, J. M. (1975) Influence of pH on release of phosphorus from lake sediments. *Arch. Hydrobiol.*, **76**, 411–19.

Anderson, C. W., Tanner, D. Q. and Lee, D. B. (1994) Water-quality data for the South Umpqua River basin, Oregon, 1990–1992. *U.S. Geological Survey Open-File Report 94–40*, Portland, OR.

Anderson, E. L., Welch, E. B., Jacoby, J. M., Schimek, G. M. and Horner, R. R. (1999) Periphyton removal related to phosphorus and grazer biomass level. *Freshwater Biol.*, **41**, 633–51.

Anderson, J. B. and Mason, W. T. (1968) A comparison of benthic macroinvertebrates collected by dredge and basket sampler. *J. Water Pollut. Control Fed.*, **40**, 252–9.

Anderson, W. L., Robertson, D. M. and Magnuson, J. J. (1996) Evidence of recent warming and El Nino-related variations in ice breakup of Wisconsin lakes. *Limnol. Oceanogr.*, **41**, 815–21.

Andersson, G., Berggen, H., Cronberg, G. and Gelin, C. (1978) Effects of planktivorous and benthivorous fish on organisms and water chemistry in eutrophic lakes. *Hydrobiologia*, **59**, 9–16.

Antia, N. J., McAllister, C. D., Parsons, T. R., Stephens, K. and Strickland, J. D. H. (1963) Further measurements of primary production using a large volume plastic sphere. *Limnol. Oceanogr.*, **8**, 166–83.

APHA (1985) *Standard Methods for the Examination of Water and Wastewater*, 16th edn, APHA, Washington, DC, 1136pp.

Armstrong, J. D., Braithwaite, V. A. and Fox, M. (1998) The response of wild Atlantic salmon parr to acute reductions in water flow. *J. Anim. Ecol.*, **67**, 292–7.

Arnold, D. E. (1971) Ingestion, assimilation, survival, and reproduction by *Daphnia pulex* fed seven species of blue-green algae. *Limnol. Oceanogr.*, **16**, 906–20.

Arrhenius, S. (1889) Uber die Reakionsgeschwindigkeit bei der Inversion von Rohrzucker durch Sauren. *Z. Phys. Chem.*, **4**, 226–34.

Arruda, J. A. and Fromm, C. H. (1989) The relationship between taste and odor problems and lake enrichment from Kansas lakes in agricultural watersheds. *Lake Reserv. Manage.*, **5**, 45–52.

Arthington, A. H. and Mitchell, D. S. (1986) Aquatic invading species, in *Ecology of Biological Invasions* (eds R. H. Groves and J. J. Burdon), Cambridge University Press, Cambridge, pp. 34–53.

Atwater, J. W., Jasper, S., Mavinic, D. S. and Koch, F. A. (1983) Experiments using *Daphnia* to measure landfill leachate toxicity. *Water Res.*, **17**, 1855–61.

Auer, M. T. and Canale, R. P. (1982) Ecological studies and mathematical modeling of *Cladophora* in Lake Huron. 2. Phosphorus uptake kinetics. *J. Great Lakes Res.*, **8**, 84–92.

Auer, M. T., Doerr, S. M., Effler, S. W. and Owens, E. M. (1997) A zero degree of freedom total phosphorus model. 1. Development for Onondaga Lake, New York. *Lake Reserv. Manage.*, **13**, 118–30.

Averett, R. C. (1961) Macroinvertebrates of the Clark Fork River, Montana – a pollution survey. *Montana Board of Health and Fish and Game Department Report.*, No. 61–1.

Averett, R. C. (1969) Influence of temperature on energy and material utilization by juvenile coho salmon. PhD thesis, Oregon State University, Corvallis.

AWWA (1987a) *Current Methodology for the Control of Algae in Surface Reservoirs* American Water Works Association (AWWA) Research Foundation, Denver, CO.

AWWA (1987b) *Identification and Treatment of Tastes and Odors in Drinking Water* (eds J. Mallevialle and I. H. Suffet), American Water Works Association (AWWA) Research Foundation, Denver, CO.

Ayles, B. G., Lark, J. G. I., Barica, J. and Kling, H. (1976) Seasonal mortality of rainbow trout (*Salmo gairdnerii*) planted in small eutrophic lakes of Central Canada. *J. Fish Res. Board Canada*, **33**, 647–55.

Baalsrud, K. (1967) Influence of nutrient concentration on primary production, in *Pollution and Marine Ecology* (eds T. A. Olson and F. J. Burgess), John Wiley and Sons, New York, pp. 159–69.

Babin, J., Prepas, E. E., Murphy, T. P., Serediak, M., Curtis, P. J., Zhang, Y. and Chambers, P. A. (1994) Impact of lime on sediment phosphorus release in hardwater lakes: the case of hypereutrophic Halfmoon Lake, Alberta. *Lake Reserv. Manage.*, **8**, 131–42.

Bachmann, R. W., Hoyer, M. V. and Canfield, D. E., Jr. (1999) The restoration of Lake Apopka in relation to alternate stable states. *Hydrobiologia*, **394**, 219–32.

Bachmann, R. W., Hoyer, M. V. and Canfield, D. E., Jr. (2000) Internal heterotrophy following the switch from macrophytes to algae in Lake Apopka, Florida. *Hydrobiologia*, **418**, 217–27.

Baden, S. P., Loo, L.-O., Pihl, L. and Rosenberg, R. (1990) Effects of eutrophication on benthic communities including fish: Swedish west coast. *Ambio*, **19**, 113–22.

Bahr, T. G., Cole, R. A. and Stevens, H. K. (1972) *Recycling and Ecosystem Response to Water Manipulation*, Technical Report No. 37, Inst. Water Res., Michigan State University, East Lansing.

Baker, J. P. and Schofield, C. L. (1982) Aluminium toxicity to fish in acid waters. *Water, Air and Soil Poll.*, **18**, 289–309.

Barbiero, R. P. (1991) Sediment-water transport of phosphorus by the blue-green alga *Gloeotrichia echinulata*. PhD dissertation, Dept Civil Eng., University of Washington, Seattle.

Barbiero, R. P. and Kann, J. (1994) The importance of benthic recruitment to the population development of *Aphanizomenon flos-aquae* and internal loading in a shallow lake. *J. Plankton Res.*, **16**, 1581–8.

Barbiero, R. P. and Welch, E. B. (1992) Contribution of benthic blue-green algal recruitment to lake populations and phosphorus translocation. *Freshwater Biol.*, **27**, 249–60.

Barbour, M. T., Gerritsen, J., Snyder, B. D. and Stribling, J. B. (1999) *Rapid Bioassessment Protocols for Use in Streams and Wadeable Rivers: Periphyton, Benthic Macroinvertebrates and Fish*, 2nd edn. EPA 841-B-99-002. US Environmental Protection Agency, Office of Water, Washington, DC.

Barel, C. D. N., Dorit, R., Greenwood, P. H., Fryer, G., Hughes, N., Jackson, P. B. N., Kawanabe, H., Lowe-McDonnell, R. H., Nagoshi, M., Ribbink, A. J., Trewavas, E., Witte, F. and Yamaoka, K. (1985) Destruction of fisheries in Africa's lakes. *Nature*, 315, 19–20.

Barica, J. (1984) Empirical models for prediction of algal blooms and collapses, winter oxygen depletion and a freeze-out effect in lakes: summary and verification. *Verh. Int. Verein. Limnol.*, 22, 309–19.

Barko, J. W. (1983) The growth of *Myriophyllum spicatum* L. in relation to selected characteristics of sediment and solution. *Aquat. Bot.*, 15, 91–103.

Barko, J. W. and Smart, R. M. (1986) Sediment-related mechanisms of growth limitation in submersed macrophytes. *Ecology*, 67, 1328–40.

Barr, J. F. (1986) Population dynamics of the common loon (*Gavia immer*) associated with mercury-contaminated waters in northwestern Ontario. Canadian Wildlife Service, Occasional Paper No. 56.

Barrett, S. C. H. (1989) Waterweed invasions. *Sci. Am.*, Oct., 90–7.

Bartsch, A. F. and Ingram, W. M. (1959) Stream life and the pollution environment. *Public Works*, 90, 104–10.

Beak, T. W. (1965) A biotic index of polluted streams and the relationship of pollution to fisheries. *Adv. Water Pollut. Res., Proc. 2nd Int. Conf.*, 1, 191–210.

Beck, W. M. (1955) Suggested method for reporting biotic data. *Sewage Ind. Wastes*, 27, 1193–7.

Becker, D. (1972) Columbia River thermal effects study: reactor effluent problems. *J. Water Pollut. Cont. Fed.*, 45, 850–69.

Becker, D. S. and Bigham, G. N. (1995) Distribution of mercury in the aquatic food web of Onondaga Lake, New York. *Water Air Soil Pollut.*, 80, 563–71.

Beeton, A. M. (1965) Eutrophication of the St. Lawrence and Great Lakes. *Limnol. Oceanogr.*, 10, 240–54.

Beeton, A. M. and Edmondson, W. T. (1972) The eutrophication problem. *J. Fish Res. Board Canada*, 29, 673–82.

Bělehrádek, J. (1926) Influence of temperature on biological processes. *Nature (Lond.)*, 118, 117–18.

Bělehrádek, J. (1957) Physiological aspects of heat and cold. *Ann. Rev. Physiol.*, 198, 59–82.

Bellanca, M. A. and Bailey, D. S. (1977) Effects of chlorinated effluents on aquatic ecosystem in the lower James River. *J. Water Pollut. Control Fed.*, 49, 639–45.

Bengtsson, L., Fleischer, S., Lindmark, G. and Ripl, W. (1975) The Lake Trummen restoration project. I. Water and sediment chemistry. *Verh. Int. Verein. Limnol.*, 19, 1080–7.

Benndorf, J. (1990) Conditions for effective biomanipulation; conclusions derived from whole-lake experiments in Europe. *Hydrobiologia*, 200/201, 187–203.

Benndorf, J., Kneschke, H., Kossatz, K. and Penz, E. (1984) Manipulation of the pelagic food web by stocking with predacious fishes. *Int. Rev. Ges. Hydrobiol.*, 69, 407–28.

Benndorf, J. and Putz, K. (1987) Control of eutrophication of lakes and reservoirs by means of pre-dams. II. Validation of the phosphate removal model and size optimization. *Water Res.*, 21, 839–42.

Benson, A. J. (2000) Documenting over a century of aquatic introductions in the United States, in *Nonindigenous Freshwater Organisms* (eds R. Claudi and J. H. Leach), Lewis Publishers, Boca Raton, FL, pp. 1–31.

Bernhardt, H. (1981) Recent developments in the field of eutrophication prevention. *Z Wasser Abwasser Forsch.*, **17**, 14–26.

Best, M. D. and Mantai K. E. (1978) Growth of *Myriophyllum spicatum*: sediment or lake water as a source of nitrogen and phosphorus. *Ecology*, **59**, 1075–80.

Bierman, V., Verhoff, V., Poulson, T. and Tenney, M. (1973) Multinutrient dynamic model of algal growth and species competition in eutrophic lakes, in *Modeling the Eutrophication Process* (ed. J. M. Middlebrooks), Water Research Laboratory, Utah State University, pp. 89–109.

Bierman, V. J. (1976) Mathematical model of selective enrichment of blue-green by nutrient enrichment, in *Modeling of Biochemical Processes in Aquatic Ecosystems* (ed. R. P. Canale), Ann Arbor Science Publ., Inc., Ann Arbor, MI.

Bierman, V. J., Jr., Dolan, D. M., Kasprzk, R. and Clark, J. L. (1984) Retrospective analysis of the response of Saginaw Bay, Lake Huron, to reductions in phosphorus loadings. *Environ. Sci. Technol.*, **18**, 23–31.

Biggs, B. J. (1985). Algae – a blooming nuisance in rivers. *Soil Water*, **21**, 27–31.

Biggs, B. J. F. (1988) Algal proliferations in New Zealand's shallow, stony foothills-fed rivers; toward a predictive model. *Verh. Int. Verein. Limnol.*, **23**, 1405–11.

Biggs, B. J. F. (1989) Biomonitoring of organic pollution using periphyton, South Branch, Canterbury, New Zealand. *N. Z. J. Marine Freshwater Res.*, **23**, 263–74.

Biggs, B. J. F. (1995) The contribution of flood disturbance, catchment geology and land use to the habitat template of periphyton in streams. *Freshwater Biol.*, **33**, 419–38.

Biggs, B. J. F. (2000) Eutrophication of streams and rivers: dissolved nutrient–chlorophyll relationships for benthic algae. *J. N. Am. Benthol. Soc.*, **19**, 17–31.

Biggs, B. J. F. and Close, M. E. (1989) Periphyton biomass dynamics in gravel bed rivers: the relative effects of flows and nutrients. *Freshwater Biol.*, **22**, 209–31.

Biggs, B. J. F. and Hickey, C. W. (1994) Periphyton responses to a hydraulic gradient in a regulated river in New Zealand. *Freshwater Biol.*, **32**, 49–59.

Biggs, B. J. F. and Price, G. M. (1987) A survey of filamentous algal proliferations in New Zealand rivers. *N. Z. J. Marine Freshwater Res.*, **21**, 175–91.

Biggs, B. J. F., Goring, D. G. and Nikora, V. I. (1998) Subsidy and stress responses of stream periphyton to gradients in water velocity as a function of community growth form. *J. Phycol.*, **34**, 598–607.

Biggs, B. J. F., Smith, R. A. and Duncan, M. J. (1999) Velocity and sediment disturbance of periphyton in head water streams: biomass and metabolism. *J. N. Am. Benthol. Soc.*, **18**, 222–41.

Biggs, B. J. F., Francoeur, S. N., Huryr, A. D., Young, R., Arbuckle, C. J. and Townsend, C. R. (2000) Trophic cascades in streams: effects of nutrient enrichment on autotrophic and consumer benthic communities under two different fish predation regimes. *Can. J. Fish. Aquat. Sci.*, **57**, 1380–94.

Birch, P. B. (1976) The relationship of sedimentation and nutrient cycling to the trophic status of four lakes in the Lake Washington drainage basin. Ph.D. Dissertation, University of Washington, Seattle.

Bishop, C. A., Ng, P., Pettit, K. E., Kennedy, S. W., Stegeman, J. J., Norstrom, R. J. and Brooks, R. J. (1998) Environmental contamination and developmental abnormalities in eggs and hatchlings of the common snapping turtle (*Chelydra serpentina*) from the Great Lakes–St Lawrence River basin (1989–91). *Environ. Pollut.*, **101**, 143–56.

Björk, S. (1974) *European lake Rehabilitation Activities* Institute of Limnology Report, University of Lund, 23 pp.

Björk, S. (1985) Lake restoration techniques, in *Lake Pollution and Recovery*, International Congress European Water Pollution Control Association, Rome, pp. 293–301.

Björk, S. *et al.* (1972) Ecosystem studies in connection with the restoration of lakes. *Verh. Int. Verein. Limnol.*, **18**, 379–87.

Blindow, I., Andersson, G., Hargeby, A. and Johansson, S. (1993) Long-term pattern of alternative stable states in two shallow eutrophic lakes. *Freshwater Biol.*, **30**, 159–67.

Bloomfield, J. A., Park, R. A., Scavia, D. and Zahorcak, C. S. (1973) Aquatic modeling in the EDFB, US-IBP, in *Modeling the Eutrophication Process* (eds E. J. Middlebrooks, D. H. Falkenborg and T. W. Maloney), Ann Arbor Science Publ., Inc., Ann Arbor, MI, pp. 139–58.

Boening, D. W. (2000) Ecological effects, transport, and fate of mercury: a general review. *Chemosphere*, **40**, 1335–52.

Boers, P. (1991) *The Release of Dissolved Phosphorus from Lake Sediments*. Ph.D. Dissertation, University of Wageningen, The Netherlands.

Bole, J. B. and Allan, J. R. (1978) Uptake of phosphorus from sediment by aquatic plants, *Myriophyllum spicatum* and *Hydrilla verticillata*. *Water Res.*, **12**, 353–8.

Booker, M. J. and Walsby, A. E. (1981). Bloom formation and stratification by a planktonic blue-green alga in an experimental water column. *Br. Phycol. J.*, **16**, 411–21.

Boorman, G. A., Dellarco, V., Dunnick, J. K., Chapin, R. E., Hunter, S., Hauchman, F., Gardner, H., Cox, M. and Sills, R. C. (1999) Drinking water disinfection byproducts: review and approach to toxicity evaluation. *Environ. Health Perspect.*, **107**, 207–17.

Borchardt, M. A., Hoffman, J. P. and Cook, P. W. (1994) Phosphorus uptake kinetics of *Spirogyra fluviatilis* (Charophyceae) in flowing water. *J. Phycol.*, **30**, 403–17.

Borman, F. H. and Likens, G. E. (1967) Nutrient cycling. *Science*, **155**, 474–29.

Boström, B. (1984) Potential mobility of phosphorus in different types of lake sediments. *Int. Rev. Ges. Hydrobiol.*, **69**, 454–74.

Boström, B., Ahlgren, I. and Bell, R. T. (1985) Internal nutrient loading in a eutrophic lake, reflected in seasonal variations of some sediment parameters. *Verh. Int. Verein. Limnol.*, **22**, 3335–9.

Boström, B., Janson, M. and Forsberg, C. (1982) Phosphorus release from lake sediments. *Arch. Hydrobiol. Beih. Ergebn. Limnol.*, **18**, 5–59.

Bothwell, M. L. (1985) Phosphorus limitation of lotic periphyton growth rates: an intersite comparison using continuous-flow toughs (Thompson River System, British Columbia). *Limnol. Oceanogr.*, **30**, 527–42.

Bothwell, M. L. (1988) Growth rate responses of lotic periphytic diatoms of experimental phosphorus enrichment: the influence of temperature and light. *Can. J. Fish. Aquat. Sci.*, **45**, 261–70.

Bothwell, M. L. (1989) Phosphorus-limited growth dynamics of lotic periphytic diatom communities: areal biomass and cellular growth rate responses. *Can. J. Fish. Aquat. Sci.*, **46**, 1293–301.

Bourassa, N. and Cattaneo, A. (1998) Control of periphyton biomass in Laurentian streams, Québec. *J. N. Am. Benthol. Soc.*, **17**, 420–9.

Bray, J. R. and Curtis, J. T. (1957) An ordination of the upland forest communities of southern Wisconsin. *Ecol. Monogr.*, **27**, 325–49.

Brett, J. R. (1956) Some principles in the thermal requirements of fishes. *Q. Rev. Biol.*, **31**, 75–87.

Brett, J. R. (1960) Thermal requirements of fish – three decades of study, 1940–70, in *Biological Problems in Water Pollution*. Trans. 2nd Seminar 1959, USPHS, Cincinnati, Ohio, pp. 111–17.

Brett, M. T. and Goldman, C. R. (1996) A meta-analysis of the freshwater trophic cascade. *Proc. Nat. Acad. Sci. USA*, **93**, 7723–6.

Brett, M. T. and Goldman, C. R. (1997) Consumer versus resource control in freshwater pelagic food webs. *Science*, **275**, 384–6.

Brett, M. T. and Müller-Navarra, D. C. (1997) The role of highly unsaturated fatty acids in aquatic foodweb processes. *Freshwater Biol.*, **38**, 483–99.

Brett, M. T., Müller-Navarra, D. C. and Park, S.-K. (2000) Empirical analysis of the effect of phosphorus limitation on algal food quality for freshwater zooplankton. *Limnol. Oceanogr.*, **45**, 1564–75.

Breukelaar, A. W., Lammens, E. H. R. R., Breteler, J. G. P. K. and Tátrai, I. (1994) Effects of benthivorous bream (*Abramis brama*) and carp (*Cyprinus carpio*) on sediment resuspension and concentrations of nutrients and chlorophyll-a. *Freshwater Biol.*, **32**, 113–21.

Brezonik, P. L. and Lee, G. F. (1968) Denitrification as a nitrogen sink in Lake Mendota, Wisc. *Environ. Sci. Technol.*, **2**, 120–5.

Brillouin, L. (1962) *Science and Information Theory*, 2nd edn, Academic Press, New York.

Brinkhurst, R. O. (1965) Observations on the recovery of a British River from gross organic pollution. *Hydrobiolia*, **25**, 9–51.

Bristow, J. W. and Whitcombe, M. (1971) The role of roots in the nutrition of aquatic vascular plants. *Am. J. Bot.*, **58**, 8–13.

Brock, D. B. (1970) *Biology of Microorganisms*. Prentice-Hall, Englewood Cliffs, NJ.

Brooks, J. L. and Dodson, S. (1965) Predation, body size and composition of plankton. *Science*, **150**, 28–35.

Brown, M. P., Werner, M. B., Sloan, R. J. and Simpson, K. W. (1985) Polychlorinated biphenyls in the Hudson River. *Environ. Sci. Technol.*, **19**, 656–61.

Brown, V. M. (1968) The calculation of the acute toxicity of mixtures of poisons to rainbow trout. *Water Res.*, **2**, 723–33.

Brown, V. M., Shurben, D. G. and Shaw, D. (1970) Studies on water quality and the absence of fish from some polluted English rivers. *Water Res.*, **4**, 363–82.

Brungs, W. (1969) Chronic toxicity of zinc to the Fathead Minnow, *Pimephales promelas* Rafinesque. *Trans. Am. Fish. Soc.*, **98**, 272–9.

Brungs, W. (1973) Effects of residual chlorine on aquatic life. *J. Water Pollut. Control Fed.*, **45**, 2180–93.

Bryan, A., Marsden, K. and Hanna, S. (1975) A summary of observations on aquatic weed control methods. British Columbia Department of Environment, unpublished report, 73pp.

Buckley, J. A. (1971) Effects of low nutrient dilution water and mixing on the growth of nuisance algae. MS thesis, University of Washington, Seattle, 116pp.

Buckley, J. A. (1977) Heinz body hemolytic anemia in coho salmon (*Oncorhynchus kisutch*) exposed to chlorinated waste water. *J. Fish Res. Board Canada*, **34**, 215–24.

Buckley, J. A., Whitmore, C. M. and Matsuda, R. I. (1976) Changes in blood chemistry and blood-cell morphology in coho salmon (*Oncorhynchus kisutch*) following exposure to sublethal levels of total residual chlorine in municipal waste water. *J. Fish Res. Board Canada*, **33**, 776–82.

Burns, C. W. (1968) The relationship between body size of filter-feeding cladocera and the maximum size of particle ingested. *Limnol. Oceanogr.*, **13**, 675–8.

Burns, C. W. (1969) Relation between filtering rate, temperature, and body size in four species of *Daphnia*. *Limnol. Oceanogr.*, **14**, 693–700.

Burns, C. W. (1987) Insights into zooplankton-cyanobacteria interactions derived from enclosure studies. *N. Z. J. Marine Freshwater Res.*, **21**, 477–82.

Burton, D. J., Jones, A. H. and Cairns, J., Jr. (1972) Acute zinc toxicity to rainbow trout (*Salmo gairdnari*): confirmation of the hypothesis that death is related to tissue hypoxia. *J. Fish Res. Board Canada*, **29**, 1463–6.

Bush, M. B. (2003) *Ecology of a Changing Planet*, 3rd edn, Upper Saddle River, NJ.

Bush, R. M., Welch, E. B. and Buchanan, R. J. (1972) Plankton associations and related factors in a hypereutrophic lake. *Water, Air Soil Pollut.*, **1**, 257–74.

Bush, R. M., Welch, E. B. and Mar, B. W. (1974) Potential effects of thermal discharges on aquatic systems. *Environ. Sci. Technol.*, **8**, 561–8.

Butcher, R. W. (1933) Studies on the ecology of rivers. I. On the distribution of macrophytic vegetation in the rivers of Britain. *J. Ecol.*, **21**, 58–91.

Butkus, S. R., Welch, E. B., Horner, R. R. and Spyridakis, D. E. (1988) Lake response modeling using biologically available phosphorus. *J. Water Pollut. Control Fed.*, **60**, 1663–9.

Cairns, J., Jr. (1956) Effects of increased temperatures on aquatic organisms. *Ind. Wastes.*, **1**, 150–2.

Cairns, J., Jr. (1971) The effects of pH, solubility, and temperature upon the acute toxicity of zinc to the bluegill sunfish (*Lepomis macrochirus* Raf.). *Trans. Kans. Acad. Sci.*, **74**, 81–92.

Cairns, J., Jr., Crossman, J. S., Dickson, K. L. and Herricks, E. E. (1971) The recovery of damaged streams. *Assoc. Southeastern Biol. Bull.*, **18**, 79–106.

Cairns, J., Jr. and Dickson, K. L. (1971) A simple method for the biological assessment of the effects of waste discharges on aquatic bottom-dwelling organisms. *J. Water Pollut. Control Fed.*, **43**, 755–72.

Canfield, D. E., Jr. and Bachmann, R. W. (1981) Prediction of total phosphorus concentrations, chlorophyll *a*, and Secchi depths in natural and artificial lakes. *Can. J. Fish. Aquat. Sci.*, **38**, 414–23.

Canfield, D. E. Jr., Langeland, K. A., Linda S. B. and Haller, W. T. (1985) Relations between water transparency and maximum depth of macrophyte colonization in lakes. *J. Aquat. Plant Management*, **23**, 25–8.

Canfield, D. E., Shireman, J. V., Colle, D. E., Haller, W. T., Watkins, C. E. II and Maceina, M. J. (1984) Prediction of chlorophyll *a* concentrations in Florida lakes: importance of aquatic macrophytes. *Can. J. Fish. Aquat. Sci.*, **41**, 497–501.

Carder, J. P. and Hoagland, K. D. (1998) Combined effects of alachlor and atrazine on benthic algal communities in artificial streams. *Environ. Toxicol. Chem.*, **17**, 1415–20.

Carignan, R. and Kalff, J. (1980) Phosphorus sources for aquatic weeds, water or sediments. *Science*, **207**, 987–9.

Carlson, A. R. and Herman, L. J. (1974) Effects of lowered dissolved oxygen concentrations on channel catfish (*Ictalurus punctatus*) embryos and larvae. *Trans. Am. Fish. Soc.*, **103**, 623–6.

Carlson, A. R. and Siefert, R. E. (1974) Effects of reduced oxygen on the embryos and larvae of lake trout (*Salvelinus namaycush*) and largemouth bass (*Micropterus salmoides*). *J. Fish Res. Board Canada*, **31**, 1393–6.

Carlson, K. L. (1983) The effects of induced flushing on water quality in Pelican Horn, Moses Lake, Wa. MS thesis, University of Washington, Seattle.

Carlson, R. E. (1977) A trophic state index for lakes. *Limmol. Oceanogr.*, **22**, 361–8.

Carmichael, W. W. (1986) Algal toxins. *Adv. Bot. Res.*, **12**, 47–101.

Carmichael, W. W. (1994) The toxins of cyanobacteria. *Sci. Am.*, **270**, 78–86.

Carpenter, S. R. (1980a) Enrichment of Lake Wingra, Wisconsin by submersed macrophyte decay. *Ecology*, **61**, 1145–55.

Carpenter, S. R. (1980b) The decline of *Myriphyllum spicatum* in a eutrophic Wisconsin USA lake. *Can. J. Bot.*, **58**, 527–35.

Carpenter, S. R. (1981) Submersed vegetation: an internal factor in lake ecosystem succession. *Am. Naturalist*, **118**, 372–83.

Carpenter, S. R. and Adams, M. S. (1977) The macrophyte nutrient pool of a hardwater eutrophic lake: implications for macrophyte harvesting. *Aquat. Bot.*, **3**, 239–55.

Carpenter, S. R., Fisher, S. G., Grimm, N. B. and Kitchell, J. F. (1992) Global change and freshwater ecosystems. *Annu. Rev. Ecol. Syst.*, **23**, 119–39.

Carpenter, S. R. and Kitchell, J. F. (1992) Trophic cascade and biomanipulation: interface of research and management. *Limnol. Oceanogr.*, **37**, 208–13.

Carpenter, S. R., Kitchell, J. F. and Hodgson, J. R. (1985) Cascading trophic interactions and lake productivity. *Bioscience*, **35**, 634–9.

Carpenter, S. R., Kitchell, J. F., Hodgson, J. R., Cochran, P. A., Elser, J. J., Elser, M. M., Lodge, D. M., Kretchmer, D., He, X. and von Ende, C. N. (1987) Regulation of lake primary productivity by food web structure. *Ecology*, **68**, 1863–76.

Carr, J. F. and Hiltunen, J. K. (1965) Changes in the bottom fauna of western Lake Erie from 1930 to 1961. *Limnol. Oceanogr.*, **10**, 551–69.

Carroll, J. (2003) *Moses Lake Total Maximum Daily Load Study*, Washington Department of Ecology, Olympia, WA.

Chadwick, G. G. and Brocksen, R. W. (1969) Accumulation of dieldrin by fish and selected fish-food organisms. *J. Wildlife Management*, **33**, 693–700.

Chambers, P. A. and Kalff, J. (1985) Depth distribution and biomass of submersed aquatic macrophyte communities in relation to Secchi depth. *Can. J. Fish. Aquat. Sci.*, **42**, 701–9.

Chambers, P. A., Prepas, E. E., Bothwell, M. L. and Hamilton, H. R. (1989) Roots versus shoots in nutrient uptake by aquatic macrophytes in flowing waters. *Can. J. Fish. Aquat. Sci.*, **46**, 435–9.

Champ, M. A. and Seligman, P. F. (1996) *Organotin: Environmental Fate and Effects*, Chapman & Hall, London.

Chandler, J. R. (1970) A biological approach to water quality management. *Water Pollut. Control*, **69**, 415–21.

Chapra, S. C. (1975) Comment on 'An empirical method of estimating retention of phosphorus in lakes' by W. B. Kirchner and P. J. Dillon. *Water Resources Res.*, **11**, 1033–4.

Chapra, S. C. and Canale, R. P. (1991) Long-term phenomenological model of phosphorus and oxygen for stratified lakes. *Water Res.*, **25**, 707–15.

Chapra, S. C. and Reckhow, K. H. (1979) Expressing the phosphorus loading concept in probabilistic terms. *J. Fish Res. Board Canada*, **36**, 225–9.

Chapra, S. C. and Tarapchak, S. J. (1976) A chlorophyll *a* model and its relationship to phosphorus loading plots for lakes. *Water Resources Res.*, **12**, 1260–4.

Charlson, R. J. and Rodhe, H. (1982) Factors controlling the acidity of natural rainwater. *Nature*, **295**, 683–5.

Chen, C. W. (1970) Concepts and utilities of an ecological model. *J. Sanit. Eng. Div. Proc. Amer. Soc. Civil Eng.*, **96**, 1085–97.

Chessman, B. C., Hutton, P. E. and Burch, J. M. (1992) Limiting nutrients for periphyton growth in sub-alpine, forest, agricultural and urban streams. *Freshwater Biol.*, **28**, 349–61.

Chetelat, J., Pick, F. R., Morin, A. and Hamilton, P. B. (1999) Periphyton biomass and community composition in rivers of different nutrient status. *Can. J. Fish. Aquat. Sci.*, **56**, 560–9.

Chorus, I. (ed.) (2001) *Cyanotoxins*, Springer-Verlag, Berlin, Germany, 357 pp.

Chorus, I. and Bartram, J. (eds) (1999) *Toxic Cyanobacteria in Drinking Water: A Guide to their Public Health Consequences, Monitoring and Management*, Published on behalf of WHO by E & FN Spon, London, 416pp.

Chorus, I., Falconer, I. R., Salas, H. J. and Bartram, J. (2000) Health risks caused by freshwater cyanobacteria in recreational waters. *J. Toxicol. Environ. Health. B. Crit. Rev.*, **4**, 323–47.

Christensen, M. H. and Harremoes, P. (1972) *Biological Denitrification in Water Treatment*, Rep. 72–2, Department of Sanitary Engineers, Technical University of Denmark.

Chrystal, G. (1904) Some results in the mathematical theory of seiches. *Proc. R. Soc. Edinb.*, **25**, 328–37.

Chu, F. S., Huang, X. and Wei, R. D. (1990) Enzyme-linked immunosorbent assay for microcystins in blue-green algal blooms. *J. Assoc. Off. Anal. Chem.*, **73**, 451–6.

Chu, F. S. and Wedepohl, R. (1994) Algal toxins in drinking water? Research in Wisconsin. *LakeLine*, April, 41–2.

Chutter, F. M. (1969) The effects of silt and sand on the invertebrate fauna of streams and rivers. *Hydrobiolia*, **34**, 57–75.

Chutter, F. M. (1972) An empirical biotic index of the quality of water in South African streams and rivers. *Water Res.*, **6**, 19–30.

Cleckner, L. B., Garrison, P. J., Hurley, J. P., Olson, M. L. and Krabbenhoft, D. P. (1998) Trophic transfer of methyl mercury in the northern Florida Everglades. *Biogeochemistry*, **40**, 347–61.

Clubb, R. W., Gaufin, A. R. and Lords, J. L. (1975) Acute cadmium toxicity studies upon nine species of aquatic insects. *Environ. Research*, **9**, 332–41.

Cocking, A. W. (1959) The effects of high temperature on roach (*Rutilus rutilus*). II. The effects of temperature increasing at a known constant rate. *J. Exp. Biol.*, **36**, 217–36.

Coffey, B. T. and McNabb, C. D. (1974) Eurasian water milfoil in Michigan. *Mich. Bot.*, **13**, 159–65.

Colborn, T. and Clement, C. (eds) (1992) *Chemically-Induced Alterations in Sexual and Functional Development: The Wildlife/Human Connection*, Princeton Scientific Publishing Company, Princeton, New Jersey.

Colborn, T., Dumanoski, D. and Myers, J. P. (1996) *Our Stolen Future*, Dutton, New York.

Colborn, T., vom Saal, F. S. and Soto, A. M. (1993) Developmental effects of endocrine-disrupting chemicals in wildlife and humans. *Environ. Health Perspect.*, **101**, 378–84.

Cole, R. A. (1973) Stream community response to nutrient enrichment. *J. Water Pollut. Control Fed.*, **45**, 1875–88.

Collins, G. B. and Weber, C. I. (1978) Phycoperiphyton (algae) as indicators of water quality. *Trans. Am. Microscop. Soc.*, **97**, 36–43.

Cooke, G. D. and Carlson, R. E. (1986) Water quality management in a drinking water reservoir. *Lake Reserv. Manage.*, **2**, 363–71.

Cooke, G. D. and Carlson, R. E. (1989) *Reservoir Management for Water Quality and THM Precursor Control*. AWWA Research Foundation and American Water Works Association, 387 pp.

Cooke, G. D., Heath, R. T., Kennedy, R. H. and McComas, M. R. (1978) The effect of sewage diversion and aluminum sulfate application on two eutrophic lakes. *Ecol. Res. Ser.* EPA 600/3-78-003, Cinn., OH, pp. 101.

Cooke, G. D., Heath, R. T., Kennedy, R. H. and McComas, M. R. (1982) Change in lake trophic state and internal phosphorus release after aluminum sulfate application. *Water Res. Bull.*, **18**, 699–705.

Cooke, G. D. and Kennedy, R. H. (2001) Managing drinking water supplies. *Lake Reserv. Manage.*, **17**, 157–74.

Cooke, G. D., Welch, E. B., Peterson, S. A. and Newroth, P. R. (1986) *Lake and Reservoir Restoration*, Butterworths, Boston, USA.

Cooke, G. D., Welch, E. B., Peterson, S. A. and Newroth, P. R. (1993) *Restoration and Management of Lakes and Reservoirs*, 2nd edn, CRC Press, Inc., Boca Raton, FL.

Cornett, R. J. and Rigler, F. H. (1979) Hypolimnetic oxygen deficits: their predictions and interpretation. *Science*, **205**, 580–1.

Coutant, C. C. (1966) Alteration of the community structure of periphyton by heated effluents. Report to AEC for contract AT (45–1)–1830, Battelle Memorial Inst.

Coutant, C. C. (1969) Temperature, reproduction, and behavior. *Chesapeake Sci.*, **10**, 261–74.

Craig, G. B., Jr. (1993) The diaspora of the Asian tiger mosquito, in *Biological Pollution: The Control and Impact of Invasive Exotic Species* (ed. B. N. McKnight), Indiana Academy of Sciences, Indianapolis, pp. 101–20.
Cronberg, G., Gelin, C. and Larsson, K. (1975) The Lake Trummen restoration project. II. Bacteria, phytoplankton, and phytoplankton productivity. *Verh. Int. Verein. Limnol.*, **19**, 1088–96.
Crossman, E. J. (1995) Introduction of the lake trout (*Salvelinus namaycush*) in areas outside its native distribution: a review. *J. Great Lakes Res.*, **21**(Suppl. 1), 17–29.
Crowl, T. A., Townsend, C. R. and McIntosh, A. R. (1992) The impact of introduced brown and rainbow trout on native fish: the case of Australasia. *Rev. Fish Biol. Fish*, **2**, 217–41.
Csanady, G. T. (1970) Dispersal of effluents in the Great Lakes, in *Water Research*, Vol. 4, Pergamon Press, Elmsford, NY, pp. 79–114.
Cuker, B. E. (1987) Field experiment on the influences of suspended clay and P on the plankton of a small lake. *Limnol. Oceanogr.*, **32**, 840–847.
Cuker, B. E., Gama, P. and Burkholder, J. M. (1990) Type of suspended clay influences lake productivity and phytoplankton community response to phosphorus loading. *Limnol. Oceanogr.*, **35**, 830–839.
Cullen, P. and Forsberg, C. (1988) Experiences with reducing point sources of phosphorus to lakes. *Hydrobiologia*, **170**, 321–36.
Cummins, K. W. (1974) Structure and function of stream ecosystems. *Bioscience*, **24**, 631–41.
Cummins, K. W., Klug, J. J., Wetzel, R. G., Petersen, R. C., Superkropp, K. F., Manny, B. A., Wuycheck, J. C. and Howard, F. O. (1972) Organic enrichment with leaf leachate in experimental lotic ecosystems. *Bioscience*, **22**, 719–22.
Curtis, E. J. and Harrington, D. W. (1971) The occurrence of sewage fungus in rivers of the United Kingdom. *Water Res.*, **5**, 281–90.
Cusimano, R. F., Brakke, D. F. and Chapman, G. A. (1986) Effects of pH on the toxicities of cadmium, copper and zinc to steelhead trout (*Salmo gairdneri*). *Can. J. Fish. Aquat. Sci.*, **43**, 1497–503.
Danielsdottir, M. and Brett, M. T. The impact of algal food quality, nutrients and zooplanktivory on planktonic food web interactions. unpublished data.
Danielson, T. J. (1998) *Wetland Bioassessment Fact Sheets*. EPA843-F-98-001. US Environmental Protection Agency, Office of Wetlands, Oceans, and Watersheds, Wetlands Division, Washington, DC.
Davis, G. E., Foster, J. and Warren, C. E. (1963) The influence of oxygen concentration on the swimming performance of juvenile Pacific salmon at various temperatures. *Trans. Am. Fish. Soc.*, **92**, 111–24.
Davison, W., Reynolds, C. S. and Finlay, B. J. (1985) Algal control of lake geochemistry; redox cycles in Rostherne Mere, U.K. *Water Res.*, **19**, 265–67.
DeGasperi, C. L., Spyridakis, D. E. and Welch, E. B. (1993) Alum and nitrate as controls of short-term anaerobic sediment phosphorus release: an *in vitro* comparison. *Lake Reserv. Manage.*, **8**, 49–59.
Delwiche, C. C. (1970) The nitrogen cycle. *Sci. Amer.*, **223**, 136–47.
DeMelo, R., France, R. and McQueen, D. J. (1992) Biomanipulation: hit or myth? *Limnol. Oceanogr.*, **37**, 192–207.

DeSylva, D. P. (1969) Theoretical considerations of the effects of heated effluents on marine fishes, in *Biological Aspects of Thermal Pollution* (eds P. A. Krenkel and F. C. Parker), Vanderbilt Univ. Press, Nashville, TN, pp. 229–93.

Dextrase, A. J. and Coscarelli, M. A. (2000) Intentional introductions of nonindigenous freshwater organisms in North America, in *Nonindigenous Freshwater Organisms* (eds R. Claudi and J. H. Leach), Lewis Publishers, Boca Raton, FL, pp. 61–98.

Dillon, P. J. (1975) The phosphorus budget of Cameron Lake, Ontario: The importance of flushing rate relative to the degree of eutrophy of a lake. *Limnol. Oceanogr.*, **29**, 28–39.

Dillon, P. J. and Rigler, F. H. (1974a) A test of a simple nutrient budget model predicting the phosphorus concentration in lake water. *J. Fish Res. Board Canada*, **31**, 1771–8.

Dillon, P. J. and Rigler, F. H. (1974b) The phosphorus–chlorophyll relationship in lakes. *Limnol. Oceanogr*, **19**, 767–73.

Dillon, P. J. and Rigler, F. H. (1975) A simple method for predicting the capacity of a lake for development based on lake trophic state. *J. Fish Res. Board Canada*, **32**, 1519–31.

Dimond, J. B. (1967) *Pesticides and Stream Insects*. Bulletin 23, Maine Forest Service and the Conservation Foundation, Washington, DC.

D'Itri, F. M. (ed.) (1997) *Zebra Mussels and Aquatic Nuisance Species*, Lewis Publishers, Boca Raton, FL, 638 pp.

Dodds, W. K. (1989) Photosynthesis of two morphologies of *Nostoc parmelioides* (Cyanobacteria) as related to current velocities and diffusion patterns. *J. Phycol.*, **25**, 258–62.

Dodds, W. K. (2002) *Freshwater Ecology: Concepts and Environmental Applications*, Academic Press, San Diego, CA.

Dodds, W. K., Jones, J. R. and Welch, E. B. (1998) Suggested classification of stream trophic state: distributions of temperate stream types by chlorophyll, total nitrogen, and phosphorus. *Water Res.*, **32**, 1455–62.

Dodds, W. K., Smith, V. H. and Zander, B. (1997) Developing nutrient targets to control benthic chlorophyll levels in streams: a case study of the Clark Fork River. *Water Res.*, **31**, 1738–50.

Doherty, F. G. (1983) Interspecies correlations of acute aquatic median lethal concentrations for four standard testing species. *Environ. Sci. Technol.*, **17**, 661–5.

Dominie, D. R. (1980) Hypolimnetic aluminum treatment of soft water Annabessacook Lake, in *Restoration of Lakes and Inland Waters* EPA-440/5-81-010, Cinn, OH, pp. 417–23.

Doudoroff, P. and Shumway, D. L. (1967) Dissolved oxygen criteria for the protection of fish, in *A Symposium on Water Quality Criteria to Protect Aquatic Life* American Fish Society, Special Publication No. 4, pp. 13–19.

Doudoroff, P., Leduc, G. L. and Schneider, C. R. (1966) Acute toxicity to fish of solutions containing complex metal cyanides in relation to concentrations of molecular hydrocyanic acid. *Trans. Am. Fish. Soc.*, **95**, 6–22.

Drenner, R. W., Strickler, J. R. and Obrien, W. J. (1978) Capture probability – role of zooplankter escape in selective feeding of planktivorous fish. *J. Fish Res. Board Can.*, **35**, 1370–3.

Driscoll, C. T., Jr., Baker, J. P., Bisogni, J. J., Jr. and Schofield, C. L. (1980) Effect of aluminium speciation on fish in dilute acidified waters. *Nature*, **284**, 161–84.
Droop, M. R. (1973) Some thought on nutrient limitation in algae. *J. Phycol.*, **9**, 264–72.
Duarte, C. M. and Kalff, J. (1986) Littoral slope as a predictor of the maximum biomass of submerged macrophyte communities. *Limnol. Oceanogr.*, **31**, 1072–80.
Dugdale, R. C. (1967) Nutrient limitation in the sea: dynamics, identification, and significance. *Limnol. Oceanogr.*, **12**, 658–95.
Dunst, R. C. (1981) Dredging activities in Wisconsin's lake renewal program, in *Restoration of Lakes and Inland Waters* EPA-440/5-81-010, Washington, DC, pp. 86–8.
Dunst, R. C., Born, S. M., Uttormark, P. D., Smith, S. A., Nichols, S. A., Peterson, J. O., Knauer, D. R., Serns, S. L., Winter, D. R. and Wirth, T. L. (1974) *Survey of Lake Rehabilitation Techniques and Experiences* Technical Bulletin 75, Department of Natural Resources, Madison, WI, USA.
Edmondson, W. T. (1966) Changes in the oxygen deficit of Lake Washington. *Verh. Int. Verein. Limnol.*, **16**, 153–8.
Edmondson, W. T. (1969) Eutrophication in North America, in *Eutrophication: Causes, Consequences, and Correctives*. National Academy of Science, Washington, DC.
Edmondson, W. T. (1970) Phosphorus, nitrogen, and algae in Lake Washington after diversion of sewage. *Science*, **169**, 690–1.
Edmondson, W. T. (1972) Nutrients and phytoplankton in Lake Washington, in *Nutrients and Eutrophication: The Limiting Nutrient Controversy* (ed G. E. Likens), Special Symposium, *Limnol. and Oceanogr.*, **1**, 172–93.
Edmondson, W. T. (1978) Trophic equilibrium of Lake Washington. *Ecol. Res. Ser.*, EPA-600/3-77-087, Cinn., OH, 36pp.
Edmondson, W. T. (1994) Sixty years of Lake Washington: a curriculum vitae. *Lake Reserv. Manage.*, **10**, 75–84.
Edmondson, W. T. and Lehman, J. R. (1981) The effect of changes in the nutrient income on the condition of Lake Washington. *Limnol. Oceanogr.*, **26**, 1–28.
Edmondson, W. T. and Litt, A. H. (1982) *Daphnia* in Lake Washington. *Limnol. Oceanogr.*, **27**, 272–93.
Edmondson, W. T., Anderson, G. C. and Peterson, D. R. (1956). Artificial eutrophication of Lake Washington. *Limnol. Oceanogr.*, **1**, 47–53.
Egloff, D. A. and Brakel, W. H. (1973) Stream pollution and a simplified diversity index. *J. Water Pollut. Control Fed.*, **45**, 2269–75.
Ehrlich, G. G. and Slack, K. V. (1969) *Uptake and Assimilation of Nitrogen in Microbiological Systems*, ASTM STP 488, Am. Soc. Test. Mat., Philadelphia, pp. 11–23.
Eichenberger, E. and Schlatter, A. (1978) Effect of herbivorous insects on the production of benthic algal vegetation in outdoor channels. *Verh. Int. Verein Limnol.*, **20**, 1806–10.
Eilers, J. M., Brakke, D. F. and Landers, D. N. (1988) Chemical and physical characteristics of lakes in the upper mid-west United States. *Environ. Sci. Technol.*, **22**, 164–72.
Einsele, W. (1936) Uber die Beziehungen des Eisenkreislaufs zumphosphatkreislauf im entrophen see. *Arch. Hydrobiol.*, **29**, 664–86.

Ellis, M. M. (1937) Detection and measurement of stream pollution. Bulletin 22, US, Bureau of Fish, **48**, 365–537.

Elser, H. J. (1967) Observations on the decline of water milfoil and other aquatic plants: Maryland, 1962–7. Unpublished report., Department Chesapeake Bay Affairs, 14 pp.

Elton, C. S. (1958) *The Ecology of Invasions by Animals and Plants*, Methuen and Company, London.

Elwood, J. W., Newbold, J. D., Trimble, A. F. and Stork, R. W. (1981) The limiting role of phosphorous in a woodland stream ecosystem: Effects of P enrichment on leaf decomposition and primary producers. *Ecology*, **62**, 146–58.

Enell, M. and Löfgren, S. (1988) Phosphorus in interstitial water: Methods and dynamics. *Hydrobiologia*, **170**, 103–32.

Entranco Engineers (1993) *Lake Youngs Water Quality Study, Phase I*, Prepared by Entranco Engineers, Bellevue, WA. Prepared for Seattle Water Department Water Quality Divison, Seattle, WA.

Eppley, R. W. (1972) Temperature and phytoplankton growth in the sea. *Fisheries Bull.*, **70**, 1063–85.

Everard, M. (1996) The importance of periodic droughts for maintaining diversity in freshwater environments, *Freshwater Forum*, **7**, 33–50.

Falconer, I. R. (ed.) (1993) *Algal Toxins in Seafood and Drinking Water*, Academic Press, London.

Falconer, I. R. (1996) Potential impact on human health of toxic cyanobacteria. *Phycologia*, **35**(Suppl. 6); 6–11.

Falconer, I. R. and Humpage, A. R. (1996) Tumour promotion by cyanobacterial toxins. *Phycologia*, **35**(Suppl. 6); 74–9.

Falconer, I. R., Beresford, A. M. and Runnegar, M. T. C. (1983) Evidence of liver damage by toxin from a bloom of the blue-green alga, *Microcystis aeruginosa*. *Med. J. Aust.*, **1**, 511–14.

Falconer, I., Bartram, J., Chorus, I., Kuiper-Goodman, T., Utkilen, H. and Codd, G. (1999) Safe levels and practices, in *Toxic Cyanobacteria in Water: A Guide to their Public Health Consequences, Monitoring and Management* (eds I. Chorus and J. Bartram), Published on behalf of WHO by E & FN Spon, London, pp. 155–78.

Fallon, R. D. and Brock, T. D. (1984) Overwintering of *Microcystis* in Lake Mendota. *Freshwater Biol.*, **11**, 217–26.

Fastner, J., Neumann, U., Wirsing, B., Weckesser, J., Wiedner, C., Nixdorf, B. and Chorus, I. (1999) Microcystins (hepatotoxic heptapeptides) in German fresh water bodies. *Environ. Toxicol.*, **14**, 13–22.

Fisher, S. G. and Likens, G. E. (1972) Stream ecosystem: organic energy budget. *Bioscience*, **22**, 33–5.

Fitzgerald, G. P. (1964) The biotic relationships within water blooms, in *Algae and Man* (ed. D. F. Jackson), Plenum Press, New York, pp. 300–6.

Fitzgerald, G. P. and Nelson, T. C. (1966) Extractive and enzymatic analyses for limiting or surplus phosphorus in algae. *J. Phycol.*, **2**, 32–7.

Fjerdingstad, E. (1964) Pollution of streams estimated by benthal phytomicroorganisms. I. A saprobic system based on communities of organisms and ecological factors. *Int. Rev. Geol. Hydrobiol.*, **49**, 63–131.

Fogg, G. E. (1965) *Algal cultures and phytoplankton ecology.* Univ. Wisconsin Press, Madison, pp. 126.

Fore, L. S., Karr, J. R. and Wisseman, R. W. (1996) Assessing invertebrate responses to human activities: evaluating alternative approaches. *J. N. Am. Benthol. Soc.*, **15**, 212–31.

Forel, F. A. (1895) *Le Leman: monographie limnologique. Tome 2, Mecanique, Chimie, Thermique, Optique, Acoustique.* F. Rouge, Lausanne, 651 pp.

Fox, H. M. (1950) Hemoglobin production in *Daphnia. Nature (Lond)*, **166**, 609–10.

Fox, H. M., Simmonds, B. G. and Washbourn, R. (1935) Metabolic rates of emphemerid nymphs from swiftly flowing and still waters. *J. Exp. Biol.*, **12**, 179–84.

Francis, G. (1878) Poisonous Australian lake. *Nature*, May 2nd, 11–12.

Freeman, M. C. (1986) The role of nitrogen and phosphorus in the development of *Cladophora glomerata* (L.) Kutzing in the Manuwatu River, New Zealand. *Hydrobiologia*, **31**, 23–30.

Frodge, J. D. Thomas, G. L. and Pauley, G. B. (1987) Impact of triploid grass carp (*Ctenopharyngodon idella*) on water quality – evaluation of aquatic plants on water quality, in *An Evaluation of Triploid Grass Carp on Lakes in the Pacific Northwest* (eds Pauley, G. B. and G. L. Thomas), WA Coop. Fish. Unit, Univ. of Wash., Seattle, WA.

Frodge, J. D., Thomas, G. L. and Pauley, G. B. (1990) Effects of canopy formation by floating and submergent aquatic macrophytes on the water quality of two shallow Pacific Northwest lakes. *Aquat. Bot.*, **38**, 231–48.

Fuhs, G. W., Demmerle, S. D., Canelli, E. and Chew, M. (1972) Characterization of phosphorus limited plankton, in *Nutrients and Eutrophication: The Limiting Nutrient Controversy*, (ed. G. E. Likens), Special Symposium, *Limnol. Oceanogr.*, **1**, 113–33.

Gabrielson, J. O., Perkins, M. A. and Welch, E. B. (1984) The uptake, translocation and release of phosphorus by *Elodea densa. Hydrobiologia*, **111**, 43–48.

Gächter, V. R. (1976) Die tiefenwasserableitung, ein weg zur sanierung von seen. *Zchweiz. Z. Hydro.*, **38**, 1–28.

Gächter, R. and Mares, A. (1985) Does settling seston release soluble reactive phosphorus in the hypolimnion of lakes? *Limnol. Oceanogr.*, **30**, 364–71.

Gächter, R., Meyer, J. S. and Mares, A. (1988) Contribution of bacteria to release and fixation of phosphorus in lake sediments. *Limnol. Oceanogr.*, **33**, 1542–58.

Gächter, R. and Müller, B. (2003) Why the phosphorus retention of lakes does not necessarily depend on the oxygen supply to their sediment surface. *Limnol. Oceanogr.*, **48**, 929–33.

Galston, A. W. (1979) Herbicides: a mixed blessing. *Biol. Sci.*, **29**, 85–90.

Gameson, A. L. H., Barrett, M. J. and Shewbridge, J. S. (1973) The aerobic Thames estuary. *Adv. Water Pollut. Res.* Proc. Sixth Internat. Conf. (ed. S. H. Jenkins), Pergamon Press, Oxford and NY. 843–50.

Gammon, J. R. (1969) Aquatic life survey of the Wabash River, with special reference to the effects of thermal effluents on populations of macroinvertebrates and fish. Unpublished report, DePauw University, Greencastle, Ind., 65pp.

Ganf, G. G. (1983) An ecological relationship between *Aphanizomenon* and *Daphnia pulex. Aust. J. Mar. Freshwater Res.*, **34**, 755–73.

Gannon, J. E. and Beeton, A. M. (1971) The decline of the large zooplankton, *Limnocalanus macrurus* Sars (Copepoda: Calanoida), in Lake Erie. *Proc. 14th Conf. Great Lakes Res.* Internat. Assoc. Gt. Lakes Res., pp. 27–38.

Gaufin, A. R. (1958) The effects of pollution on a midwestern stream. *Ohio J. Sci.*, **58**, 197–208.

Gaufin, A. R. and Tarzwell, C. M. (1956) Aquatic macroinvertebrate communities as indicators of organic pollution in Lytle Creek. *Sewage Ind. Wastes*, **28**, 906–24.

Gerdeaux, D. (1998) Fluctuations in lake fisheries and global warming, in *Management of Lakes and Reservoirs During Global Climatic Change* (eds J. G. Jones, P. Puncochar, C. S. Reynolds and D. W. Sutcliffe), Kluwer Academic, Dordrecht, pp. 263–72.

Gerloff, G. C. (1975) Nutritional ecology of nuisance aquatic plants. *Ecol. Res. Ser.*, EPA-660/3-75-027, 78 pp.

Gerloff, G. C. and Krombholz, P. H. (1966) Tissue analysis to determine nutrient availability. *Limnol. Oceanogr.*, **11**, 529–37.

Gerloff, G. C. and Skoog, F. (1954) Cell contents of nitrogen and phosphorus as a measure of their availability for growth of *Microcystis aeruginosa*. *Ecology*, **35**, 348–53.

Gessner, F. (1959) *Hydrobotanik. Die Physiologischen Grundlagen der Pflanzenverbreitung im Wasser. II. Stoffhausholt*. VEB Deutscher Verlag der Wissenschaften, Berlin, 701 pp.

Gilmartin, M. (1964) The primary production of a British Columbia fjord. *J. Fish Res. Board Canada*, **21**, 505–38.

Gliwicz, Z. M. (1975) Effect of zooplankton grazing on photosynthetic activity and composition of phytoplankton. *Verh. Int. Verein. Limnol.*, **19**, 1490–7.

Gliwicz, Z. M. and Hillbricht-Ilkowska, A. (1973) Efficiency of the utilization of nanoplankton primary production by communities of filter-feeding animals measured *in situ*. *Verh. Int. Verein. Limnol.*, **18**, 197–212.

Godfrey, P. J. (1978) Diversity as a measure of benthic macroinvertebrate community response to water pollution. *Hydrobiologia*, **57**, 111–22.

Goldman, C. R. (1960a) Molybdenum as a factor limiting primary productivity in Castle Lake, Calif. *Science*, **132**, 1016–17.

Goldman, C. R. (1960b) Primary productivity and limiting factors in three lakes of the Alaskan Peninsula. *Ecol. Monogr.*, **30**, 207–30.

Goldman, C. R. (1962) Primary productivity and micronutrient limiting factors in some North American and New Zealand lakes. *Verh. Int. Verein. Limnol.*, **15**, 365–74.

Goldman, C. R. (1981) Lake Tahoe: two decades of change in a nitrogen deficient oligotrophic lake. *Verh. Int. Verein. Limnol.*, **21**, 45–70.

Goldman, C. R. and Carter, R. (1965) An investigation by rapid carbon-14 bioassay of factors affecting the cultural eutrophication of Lake Tahoe, California-Nevada. *J. Water Pollut. Control Fed.*, **37**, 1044–59.

Goldman, C. R. and Horne, A. J. (1983) *Limnology* McGraw-Hill, Inc., New York.

Goldman, C. R. and Wetzel, R. G. (1963) A study of the primary productivity of Clear Lake, Lake County, Calif. *Ecology*, **44**, 283–94.

Goldman, J. C. (1973) Carbon dioxide and pH: effect on species succession of algae. *Science*, **182**, 306–7.

Goldman, J. C. and Carpenter, E. J. (1974) A kinetic approach to the effect of temperature on algal growth. *Limnol. Oceanogr.*, **19**, 756–66.

Goldman, J. C., Porcella, D. B., Middlebrooks, E. J. and Toerien, D. F. (1971) *The Effect of Carbon on Algal Growth – Its Relationship to Eutrophication*. Occasional Paper 6, Utah State University, Water Research Lab. and College of Engineering, April, 56 pp.

Goldschmidt, T., Witte, F. and Wanink, J. (1993) Cascading effects of the introduced Nile Perch on the detritivorous/phytoplanktivorous species in the sublittoral areas of Lake Victoria. *Conserv. Biol.*, **7**, 686–700.

Golterman, H. L. (1972) Vertical movement of phosphate in fresh water, in *Handbook of Environmental Phosphorus* (eds E. J. Griffith, A. M. Beeton, J. Spencer and D. Mitchell), John Wiley & Sons, New York, pp. 509–37.

Golterman, H. L. (2001) Phosphate release from anoxic sediments or 'What did Mortimer really write?' *Hydrobiologia*, **450**, 99–106.

Goodman, D. (1975) The theory of diversity-stability relationships in ecology. *Q. Rev. Biol.*, **50**, 237–66.

Gorham, E. and Boyce, F. M. (1989) Influence of lake surface area and depth upon thermal stratification and the depth of the summer thermocline. *J. Great Lakes Res.*, **15**, 233–45.

Grafius, E. and Anderson, N. H. (1972) *Literature Review of Foods of Aquatic*

Graham, J. M. (1949) Some effects of temperature and oxygen pressure on the metabolism and activity of the speckled trout, *Salvelinus fontinalis*. *Can. J. Res. D*, **27**, 270–88.

Grandberg, K. (1973) The eutrophication and pollution of Lake Päijäne, Central Finland. *Ann. Bot. Finn*, **10**, 267–308.

Grant, N. G. and Walsby, A. E. (1977) The contribution of photosynthate to turgor pressure rise in the planktonic blue-green alga *Anabaena flos-aquae*. *J. Exp. Bot.*, **28**, 409–15.

Grassle, J. F. and Grassle, J. P. (1974) Opportunistic life histories and genetic systems in marine benthic polychaetes. *J. Marine Res.*, **32**, 253–84.

Gray, N. F. and Hunter, C. A. (1985) Heterotrophic slimes in Irish rivers – evaluation of the problem. *Water Res.*, **6**, 685–91.

Graynoth, E. (1979) Effects of logging on stream environments and faunas in Nelson, N. Z. *N. Z. J. Marine Freshwater Res.*, **13**, 79–109.

Great Phosphorus Controversy, The (1970) *Environ. Sci. Technol.*, **4**, 725–6.

Green, G. H. and Hargrave, B. T. (1966) Primary and secondary production in Bas d'Or Lake, Nova Scotia, Canada. *Verh. Int. Verein. Limnol.*, **16**, 333–40.

Gross, M. L. (1983) Response of largemouth bass and black crappie to summer drawdown of Long Lake, Kitsap County, Washington, MS thesis, University of Washington.

Guillette, L. J., Jr., Brock, J. W., Rooney, A. A. and Woodward, A. R. (1999) Serum concentrations of various environmental contaminants and their relationship to sex steroid concentrations and phallus size in juvenile American alligators. *Arch. Environ. Contam. Toxicol.*, **36**, 447–55.

Gulati, R. D., Lammens, E. H. R. R., Meijer, M.-L. and van Donk, E. (eds) (1990) *Biomanipulation Tool for Water Management*. Proceedings of an International

Conference held in Amsterdam, The Netherlands, 8–11 August 1989, Kluwer Academic Publishers, Dordrecht, Boston, London.

Haines, T. A. (1973) Effects of nutrient enrichment and a rough-fish population (carp) on a game-fish population (smallmouth bass). *Trans. Am. Fish. Soc.*, **102**, 346–54.

Håkanson, L. (1975) Mercury in Lake Vänern – present status and prognosis. Swedish Environ. Prot. Bd., NLU, Report No. 80, 121 pp.

Hall, D. J., Cooper W. and Werner, E. (1970) An experimental approach to the production dynamics and structure of freshwater animal communities. *Limnol. Oceanogr.*, **15**, 839–928.

Hall, J. D. and Lanz, R. L. (1969) The effects of logging on the habitat of coho salmon and cutthroat trout in coastal streams. *Symposium on Salmon and Trout in Streams* (ed. T. G. Northcote), H. R. MacMillan Lectures in Fisheries, University of British Columbia, Vancouver, pp. 355–75.

Halsey, T. G. (1968) Autumnal and overwinter limnology of three small eutrophic lakes with particular reference to experimental circulation and trout mortality. *J. Fish Res. Board Canada*, **25**, 81–99.

Hamilton, D. J. and Ankney, C. D. (1994) Consumption of zebra mussel (*Dreissena polymorpha*) by diving ducks in Lakes Erie and St. Clair. *Wildfowl*, **45**, 159–66.

Hamilton, R. D. and Preslan, J. (1970) Observations on heterotrophic activity in the eastern tropical Pacific. *Limnol. Oceanogr.*, **15**, 395–401.

Hansson, L.-A. (1995) Diurnal recruitment patterns in algae: effects of light cycles and stratified conditions. *J. Phycol.*, **31**, 540–6.

Hansson, L.-A., Rudstam, L. G., Johnson, T. B., Soranno, P. and Allen, Y. (1994) Patterns in algal recruitment from sediment to water in a dimictic, eutrophic lake. *Can. J. Fish. Aquat. Sci.*, **51**, 2825–33.

Harris, G. P. (1994) Pattern, process and prediction on aquatic ecology. A limnological view of some general ecological problems. *Freshwater Biol.*, **32**, 143–60.

Hartman, R. T. and Brown, D. L. (1967) Changes in internal atmosphere of submerged vascular hydrophytes in relation to photosynthesis. *Ecology*, **48**, 252–8.

Havel, J. E. and Hebert, P. D. N. (1993) *Daphnia lumholtzi* in North America: another exotic zooplankter. *Limnol. Oceanogr.*, **38**, 1823–7.

Havens, K. E. (1991) Fish-induced sediment resuspension: effects on phytoplankton biomass and community structure in a shallow hypereutrophic lake. *J. Plankton Res.*, **13**, 1163–76.

Havens, K. E. (1993) Responses to experimental fish manipulations in a shallow, hypereutrophic lake: the relative importance of benthic nutrient recycling and trophic cascade. *Hydrobiologia*, **254**, 73–80.

Hawkes, H. A. (1962) Biological aspects of river pollution, in *River Pollution. Two: Causes and Effects* (ed. L. Klein), Butterworths, London, pp. 311–432.

Hawkes, H. A. (1969) Ecological changes of applied significance induced by the discharge of heated waters, in *Engineering Aspects of Thermal Pollution* (eds F. L. Parker and P. A. Krenkel), Vanderbilt University Press, Nashville, Tennessee, pp. 15–57.

Hays, F. R. and Phillips, J. E. (1958) Lake water and sediment. III: Radiophosphorus equilibrium with mud, plants, and bacteria under oxidized and reduced conditions. *Limnol. Oceanogr.*, **3**, 459–75.

Healy, F. P. (1978) Physiological indicators of nutrient deficiency in algae. *Mitt. Verh. Int. Verein. Limnol.*, **21**, 34–41.

Heinle, D. R. (1969a) Effects of elevated temperature on zooplankton, *Chesapeake Sci.*, **10**, 186–209.
Heinle, D. R. (1969b) *Thermal Loading and the Zooplankton Community*. Patuxent Thermal Studies, Supplementary reports, Nat. Res. Inst., Ref. No. 69–8, University of Maryland.
Hellawell, J. M. (1977) Change in natural and managed ecosystems: detection, measurement and assessment. *Proc. R. Soc. Lond. B*, **197**, 31–57.
Hellström, B. (1941) *Wind Effect on Lakes and Rivers*, Kungliga Tekniska Högskolan, Avhandling 26, Stockholm.
Henderson, M. A., Levy, D. A. and Stockner, J. S. (1992) Probable consequences of climate change on freshwater production of Adams River sockeye salmon (*Oncorhynchus nerka*). *GeoJournal*, **28**, 51–9.
Hendrey, G. R. (1973) Productivity and growth kinetics of natural phytoplankton communities in four lakes of contrasting trophic state. PhD dissertation, University of Washington, Seattle.
Hendrey, G. R. and Welch, E. B. (1974) Phytoplankton productivity in Findley Lake. *Hydrobiologia*, **45**, 45–63.
Hendricks, A. C. (1978) Response of *Selenastrum capricornutum* to zinc sulfides. *J. Water Pollut. Control Fed.*, **50**, 163–8.
Henrikson, A. (1980) Acidification of freshwaters – a large scale titration, in *Ecological Impact of Acid Precipitation* (eds D. Drabløs and A. Tollen), SNF-project, NISK, 1432–Ås, pp. 68–74.
Henriksen, A. Skogheim, O. K. and Rosseland, B. O. (1984) Episodic changes in pH and aluminium – speciation kill fish in a Norwegian salmon river. *Vatten*, **40**, 255–60.
Henriksen, A., Lein, L., Traaen, T. S. *et al.* (1988) Lake acidification in Norway – Present and predicted chemical status. *Ambio*, **17**, 259–66.
Herbert, D., Elsworth, R. and Telling, R. C. (1956) The continuous culture of bacteria, a theoretical and experimental study. *J. Gen. Microbiol.*, **14**, 601–22.
Herrera Environmental Consultants (1996) *Lake Youngs Reservoir Taste and Odor Control Study*. Technical Memorandum. Prepared by Herrera Environmental Consultants, Seattle, WA. Prepared for the Seattle Water Department, Seattle, WA.
Hester, F. E. and Dendy, J. B. (1962) A multiple-plate sampler for aquatic macroinvertebrates. *Trans. Am. Fish. Soc.*, **91**, 420.
Hieltjes, A. H. M. and Lijklema, L. (1980) Fractionation of inorganic phosphates in calcareous sediments. *J. Environ. Qual.*, **9**, 405–7.
Hileman, B. (1988) The Great Lakes cleanup effort. *Chem Eng. News*, Am. Chem. Soc., Washington, DC, pp. 22–39.
Hillbricht-Ilkowska, A. I. (1972) Interlevel energy transfer efficiency in planktonic food chains. *International Biological Programme* – Section PH, 13 December 1972, Reading, England.
Hilsenhoff, W. L. (1977) *Use of Arthropods to Evaluate Water Quality of Streams*. Technical Bulletin No. 100, US Department of Nature Research, 16 pp.
Hitzfeld, B. C., Höger, S. J. and Dietrich, D. R. (2000) Cyanobacterial toxins: removal during drinking water treatment, and human risk assessment. *Environ. Health Perspect.*, **108**, 113–22.
Hodgson, R. H. and Otto, N. E. (1963) Pondweed growth and response to herbicides under controlled light and temperature. *Weed Sci.*, **11**, 232–7.

Hoehn, R. C., Barnes, D. B., Thompson, B. C., Randall, C. W., Grizzard, T. J. and Shaffer, P. T. B. (1980) Algae as sources of trihalomethane precursors. *J. Am. Water Works Assoc.*, **72**, 344–50.

Hoehn, R. C., Stauffer, J. R., Masnik, M. T. and Hocutt, C. H. (1974) Relationships between sediment oil concentrations and the macroinvertebrates present in a small stream following an oil spill. *Environ. Lett.*, **7**, 345–52.

Hogg, I. D. and Williams, D. D. (1996) Response of stream invertebrates to a global warming thermal regime: an ecosystem-level manipulation. *Ecology*, **77**, 395–407.

Holton, H., Brettum, P., Holton, G. and Kjellberg, G. (1981) *Kolbotnvatn med Fillop: Sammerstilling av Undersokelsesresultates 1978–1979*. Norsh Inst. for Vannfors., Norway, Report No. 0-78007.

Horne, A. J. and Goldman, C. R. (1972) Nitrogen fixation in Clear Lake, Calif. I. Seasonal variation and the role of heterocysts. *Limnol. Oceanogr.*, **17**, 678–92.

Horner, R. R. and Welch, E. B. (1981) Stream periphyton development in relation to current velocity and nutrients. *Can. J. Fish. Aquat. Sci.*, **38**, 449–57.

Horner, R. R., Welch, E. B., Seeley, M. R. and Jacoby, J. M. (1990) Responses of periphyton to changes in current velocity, suspended sediment and phosphorus concentration. *Freshwater Biol.*, **24**, 215–32.

Horner, R. R., Welch, E. B. and Veenstra, R. B. (1983) Development of nuisance periphytic algae in laboratory streams in relation to enrichment and velocity, in *Periphyton of Freshwater Ecosystems* (ed. R. G. Wetzel), *Dev. Hydrobiol.*, **17**, 121–34.

Horning, W. B., II and Pearson, R. E. (1973) Growth temperature requirements and lower lethal temperatures for juvenile smallmouth bass (*Micropterus dolomieu*). *J. Fish Res. Board Canada*, **30**, 1226–30.

Howard, D. L., Frea, J. I., Pfister, R. M. and Dugan, P. R. (1970) Biological nitrogen fixation in Lake Erie. *Science*, **169**, 61–2.

Hrbacek, J., Dvorakova, M., Korinek, V. and Prochazkova, L. (1961) Demonstration of the effect of the fish stock on the species composition of zooplankton and the intensity of metabolism of the whole plankton assemblage. *Verh. Int. Verein. Limnol.*, **14**, 192–5.

Hughes, B. D. (1978) The influence of factors other than pollution on the value of Shannon's Diversity Index, for macroinvertebrates in streams. *Water Res.*, **12**, 359–64.

Hughs, J. C. and Lund, J. W. G. (1962) The rate of growth of *Asterionella formosa* Hass. in relation to its ecology. *Arch. Microbiol.*, **42**, 117–29.

Hulbert, S. H. (1971) The nonconcept of species diversity: a critique and alternative parameters. *Ecology*, **52**, 577–86.

Hutchinson, G. E. (1957) *A Treatise on Limnology*, Vol. 1. John Wiley & Sons, New York.

Hutchinson, G. E. (1967) *A Treatise on Limnology*, Vol. 2, John Wiley & Sons, New York.

Hutchinson, G. E. (1970a) The biosphere. *Sci. Am.*, **223**, 44–53.

Hutchinson, G. E. (1970b) The chemical ecology of three species of *Myriophyllum* (Angiospermae, Haloragaceae). *Limnol. Oceanogr.*, **15**, 1–5.

Hutchinson, G. E. and Bowen, V. T. (1950) Limnological studies in Connecticut. 9. A quantitative radiochemical study of the phosphorus cycle in Lindsley Pond. *Ecology*, **31**, 194–203.

Hyenstrand, P., Blomqvist, P. and Pettersson, A. (1998) Factors determining cyanobacterial success in aquatic systems – a literature review. *Arch. Hydrobiol. Spec. Issues Adv. Limnol.*, **51**, 41–62.
Hynes, H. B. N. (1960) *The Biology of Polluted Water*. Liverpool University Press, Liverpool, England.
Infante, A. and Abella, S. E. B. (1985) Inhibition of *Daphnia* by *Oscillatoria* in Lake Washington. *Limnol. Oceanogr.*, **30**, 1046–52.
Infante, A. and Litt, A. H. (1985) Differences between two species of *Daphnia* in the use of 10 species of algae in Lake Washington. *Limnol. Oceanogr.*, **30**, 1053–9.
IPCC (2001) *Climate Change 2001: The Scientific Basis*. Contribution of Working Group I to the Third Assessment Report of the Intergovernmental Panel on Climate Change (eds J. T. Houghton, Y. Ding, D. J. Griggs, M. Noguer, P. J. van der Linden, X. Dai, K. Maskell and C. A. Johnson), Cambridge University Press, Cambridge, United Kingdom and New York, NY, USA, 881pp.
Isaac, G. W., Matsuda, R. I. and Welker, J. R. (1966) A limnological investigation of water quality conditions in Lake Sammamish. *Water Quality Series No. 2*, Municipality of Metropolitan Seattle, pp. 47.
Jacoby, J. M. (1985) Grazing effects on periphyton by *Theodoxis fluviatillis* (Gastropoda) in a lowland stream. *J. Freshwater Ecol.*, **3**, 265–74.
Jacoby, J. M. (1987) Alterations in periphyton characteristics due to grazing in a Cascade foothills stream. *Freshwater Biol.*, **18**, 495–508.
Jacoby, J. M., Bouchard, D. D. and Patmont, C. R. (1990) Response of periphyton to nutrient enrichment in Lake Chelan, WA. *Lake Reservoir Management*, **7**, 33–43.
Jacoby, J. M., Collier, D. C., Welch, E. B., Hardy, F. J. and Crayton, M. (2000) Environmental factors associated with a toxic bloom of *Microcystis aeruginosa*. *Can. J. Fish. Aquat. Sci.*, **57**, 231–40.
Jacoby, J. M., Gibbons, H. L., Hanowell, R. and Bouchard, D. D. (1994) Wintertime blue-green algal toxicity in a mesotrophic lake. *J. Freshwater Ecol.*, **9**, 241–51.
Jacoby, J. M., Lynch, D. D., Welch, E. B. and Perkins, M. A. (1982) Internal phosphorus loading in a shallow, eutrophic lake. *Water Res.*, **16**, 911–19.
Jacoby, J. M., Welch, E. B. and Wertz, I. (2001) Alternate stable states in a shallow lake dominated by *Egeria densa*. *Verh. Int. Verein. Limnol.*, **27**, 3805–10.
Jasper, S. and Bothwell, M. L. (1986) Photosynthetic characteristics of lotic periphyton. *Can. J. Fish. Aquat. Sci.*, **43**, 1960–9.
Jensen, H. S., Kristensen, P., Jeppesen, E. and Skytthe, A. (1992) Iron–phosphorus ratio in surface sediments as an indicator of phosphate release from aerobic sediments in shallow lakes. *Hydrobiologia*, **235/236**, 731–43.
Jensen, K. W. and Snekvik, E. (1972) Low pH levels wipe out salmon and trout populations in southernmost Norway. *Ambio*, **1**, 223–5.
Jensen, L. D. and Gaufin, A. R. (1964) Effects of ten organic insecticides on two species of stonefly naiads. *Trans. Am. Fish. Soc.*, **93**, 27–34.
Jeppesen, E., Kristensen, P., Jensen, J. P., Søndergaard, M., Mortensen, E. and Lauridsen, T. (1991) Recovery resilience following a reduction in external phosphorus loading of shallow, eutrophic Danish lakes: duration, regulating factors and methods for overcoming resilience. *Mem. Ist. Ital. Idrobiol.*, **48**, 127–48.

Jeppesen, E., Søndergaard, M., Søndergaard, M. and Christoffersen, K. (eds) (1998) *The Structuring Role of Submerged Macrophytes in Lakes*, Springer-Verlag, New York, NY.

Jobling, S., Nolan, M., Tyler, C. R., Brighty, G. and Sumpter, J. P. (1998) Widespread sexual disruption in wild fish. *Environ. Sci. Technol.*, **32**, 2498–506.

Jochimsen, E. M., Carmichael, W. W., An, J. S., Cardo, D. M., Cookson, S. T., Holmes, C. E. M., de C. Antunes, M. B., de Melo Filho, D. A., Lyra, T. M., Barreto, V. S., Azevedo, S. and Jarvis, W. R. (1998) Liver failure and death after exposure to microcystins at a hemodialysis center in Brazil. *N. Engl. J. Med.*, **338**, 873–8.

Johnson, H. E. (1967) The effects of endrin on the reproduction of a freshwater fish (*Oryzias latipes*). PhD dissertation, University of Washington, Seattle, pp. 136.

Johnson, T. C., Scholz, C. A., Talbot, M. R., Kelts, K., Ricketts, R. D., Ngobi, G., Beuning, K., Ssemmanda, I. and McGill, J. W. (1996) Late Pleistocene desiccation of Lake Victoria and rapid evolution of cichlid fishes. *Science*, **273**, 1091–2.

Johnston, B. R. and Jacoby, J. M. (2003) Cyanobacterial toxicity and migration in a mesotrophic lake in western Washington (USA). *Hydrobiologia*, **495**, 79–91.

Jones, C. A. and Welch, E. B. (1990) Internal phosphorus loading related to mixing and dilution in a dendritic, shallow prairie lake. *J. Water Pollut. Control Fed.*, **62**, 847–52.

Jones, G. J. and Orr, P. T. (1994) Release and degradation of microcystin following algicide treatment of a *Microcystis aeruginosa* bloom in a recreational lake, as determined by HPLC and protein phosphatase inhibition assay. *Water Res.*, **28**, 871–6.

Jones, J. G., Gardener, S. and Simon, B. M. (1983) Bacterial reduction of ferric iron in a stratified eutrophic lake. *J. Gen. Microbiol.*, **129**, 131–9.

Jones, J. R. and Bachmann, R. W. (1976) Prediction of phosphorus and chlorophyll levels in lakes. *J. Water Pollut. Control Fed.*, **48**, 2176–82.

Jones, J. R., Smart, M. M. and Burroughs, J. N. (1984) Factors related to algal biomass in Missouri Ozark streams. *Verh. Int. Verein. Limnol.*, **22**, 1867–75.

Jones, R. I. (1979) Notes on the growth and sporulation of a natural population of *Aphanizomenon flos-aquae*. *Hydrobiologia*, **62**, 55–8.

Jorgensen, S. E., Friis, M. B., Henrikson, J. *et al.* (1979) *Handbook of Environmental Data and Ecological Parameters*. Pergamon Press, Oxford.

Joubert, M. (17 June 2003) Personal communication (e-mail to J. Jacoby), Seattle Public Utilities, Seattle, WA.

Kadlec, R. H. and Knight, R. L. (1996) *Treatment Wetlands*, Lewis Publishers, Boca Raton, FL.

Kaeding, L. R., Boltz, G. D. and Carty, D. G. (1996) Lake trout discovered in Yellowstone Lake threatens native cutthroat trout. *Fisheries*, **21**, 16–20.

Kaesler, R. L., Herricks, E. E. and Crossman, J. S. (1978) Use of indices of diversity and hierarchical diversity in stream surveys, in *Biological Data in Pollution Assessment: Quantitative and Statistical Analyses* (eds K. L. Dickson, J. Cairns, Jr. and R. J. Livingston), ASTM STP 652, Amer. Soc. Test. Mat., Philadelphia, pp. 92–112.

Kamp-Nielsen, C. (1974) Mud-water exchange of phosphate and other ions in undisturbed sediment cores and factors affecting the exchange rates. *Arch. Hydrobiol.*, **73**, 218–37.

Kann, J. and Smith, V. H. (1999) Estimating the probability of exceeding elevated pH values critical to fish populations in a hypereutrophic lake. *Can. J. Fish. Aquat. Sci.*, **56**, 1–9.

Kappers, F. I. (1976) Blue-green algae in the sediment of the lake Brielse Meer. *Hydrobiol. Bull.*, **10**, 164–71.

Karr, J. R. (1981) Assessment of biotic integrity using fish communities. *Fisheries*, **6**, 21–7.

Karr, J. R. (1991) Biological integrity: a long-neglected aspect of water resource management. *Ecol. Appl.*, **1**, 66–84.

Karr, J. R. (1998) Rivers as sentinels: using the biology of rivers to guide landscape management, in *River Ecology and Management: Lessons from the Pacific Coastal Ecoregion* (eds R. J. Naiman and R. E. Bilby), Springer-Verlag, New York, pp. 502–28.

Karr, J. R. and Chu, E. W. (1999) *Restoring Life in Running Waters: Better Biological Monitoring*, Island Press, Washington, DC.

Karr, J. R., Fausch, K. D., Angermeier, P. L. *et al.* (1986) *Assessing biological integrity in running waters: a method and its rationale.* Illinois Natural History Survey, Champaigne, Ill., Special publ. 6.

Kaufman, L. (1992) Catastrophic change in species-rich freshwater ecosystems. *Bioscience*, **42**, 846–58.

Keating, K. I. (1977) Blue-green algal inhibition of diatom growth: transition from mesotrophic to eutrophic community structure. *Science*, **199**, 971–3.

Kennefick, S. L., Hrudey, S. E., Peterson, H. G. and Prepas, E. E. (1993) Toxin release from *Microcystis aeruginosa* after chemical treatment. *Water Sci. Technol.*, **27**, 433–40.

Kerr, P. C., Paris, D. F. and Brockway, D. L. (1970) The interrelation of carbon and phosphorus in regulating heterotrophic and autotrophic populations in aquatic ecosystems. *EPA Report* 16060 FGS 07/70, Raleigh, NC.

Keup, L. E. (1966) Stream biology for assessing sewage treatment plant efficiency. *Water and Sewage Works*, **113**, 411–17.

Kidd, K., Schindler, A. D. W., Muir, D. C. G., Lockhart, W. L. and Hesslein, R. H. (1995) High concentrations of toxaphene in fishes from a subarctic lake. *Science*, **269**, 240–2.

Kilham, S. S. (1975) Kinetics of silicon-limited growth in the freshwater diatom *Asterionella formosa. J. Phycol.*, **11**, 396–9.

Kimmel, B. L. and Groeger, A. W. (1984) Factors controlling primary production in lakes and reservoirs: a perspective, in *Proc. Int. Conf. Lake and Reservoir Management*, North American Lake Management Society, EPA 440/5/84-001, pp. 277–81.

King, D. L. (1970) Role of carbon in eutrophication. *J. Water Pollut. Control Fed.*, **42**, 2035–51.

King, D. L. (1972) Carbon limitation in sewage lagoons, in *Nutrients and Eutrophication*, Special Symposium, Vol. 1, Amer. Soc. Limnol. Oceanogr., pp. 98–110.

King, D. L. and Novak, J. T. (1974) The kinetics of inorganic carbon-limited algal growth. *J. Water Pollut. Control Fed.*, **46**, 1812–16.

King, W. D., Dodds, L. and Allen, A. C. (2000) Relation between stillbirth and specific chlorination by-products in public water supplies. *Environ. Health Perspect.*, **108**, 883–7.

King County (2003) *Lake Washington Existing Conditions Report.* Prepared by H. Gibbons, S. Nobel and E. B. Welch, Tetra Tech ISG, Seattle; J. Good, Parametrix, Inc., Seattle; D. Bouchard, S. Coughlin and J. Frodge, King County Department of Natural Resources and Parks, Seattle, WA.

Kirchner, W. B. and Dillon, P. J. (1975) An empirical method of estimating the retention of phosphorus in lakes. *Water Resources Res.*, **11**, 182–3.

Klein, L. (1962) *River Pollution, Two: Causes and Effects*, Butterworths, London.

Kleindl, W. J. (1995) *A Benthic Index of Biotic Integrity for Puget Sound Lowland Streams, Washington, USA.* M.S. Thesis, University of Washington, College of Forest Resources, Seattle, WA.

Klemer, A. R. (1973) Factors affecting the vertical distribution of a blue-green alga. PhD dissertation, University of Minnesota.

Klemer, A. R. and Kanopka, A. E. (1989) Causes and consequences of blue green algal (Cyanobacteria) blooms. *Lake Reservoir Management*, **5**, 9–20.

Klemer, A. R., Detenbeck, N., Grover, J. and Fung-Brasino, T. (1988) Macronutrient interactions involved in cyanobacterial bloom formation. *Verh. Int. Verein. Limnol.*, **23**, 1881–5.

Klemer, A. R., Feuillade, J. and Feuillade, M. (1982) Cyanobacterial blooms: carbon and nitrogen limitation have opposite effects on the buoyancy of *Oscillatoria*. *Science*, **215**, 1629–31.

Klemer, A. R., Pierson, D. C. and Whiteside, M. C. (1985) Blue-green algal (cyanobacterial) nutrition, buoyancy and bloom formation. *Verh. Internat. Verein. Limnol.*, **22**, 2791–8.

Knapp, R. T. (1943) Density currents: their mixing characteristics and their effect on the turbulence structure of the associated flow. *Eng. Bull.*, **27**, University Iowa Studies, Iowa City.

Knoechel, R. and Kalff, J. (1975) Algal sedimentation: the cause of a diatom-blue-green succession. *Verh. Int. Verein. Limnol.*, **19**, 745–54.

Knutzen, J. (1975) Course project in ecological effects of wastewater, 20pp.

Kolar, C. S. and Lodge, D. M. (2001) Progress in invasion biology: Predicting invaders. *Trends Ecol. Evol.*, **16**, 199–204.

Kolar, C. S. and Wahl, D. H. (1998) Daphnid morphology deters fish predators. *Oecologia*, **116**, 556–64.

Kolkwitz, R. and Marsson, M. (1908) Okologie der pflanzlichen Saprobien. *Berichte Deutsch. Bot. Gesellschaft*, **26a**, 505–19.

Koncsos, L. and Somlyódy, L. (1994) Analysis on parameters of suspended sediment models for a shallow lake, in *Water Quality International*, 94 IAWQ 17th Biennial International Conference, Budapest, Hungary.

Kormondy, E. J. (1969) *Concepts of Ecology* Prentice-Hall, Englewood Cliffs, NJ.

Koski-Vähälä, J. and Hartikainen, H. (2001) Assessment of the risk of phosphorus loading due to resuspended sediment. *J. Environ. Qual.*, **30**, 960–6.

Kotak, B. G., Lam, A. K.-Y., Prepas, E. E., Kenefick, S. L. and Hrudey, S. E. (1995) Variability of the hepatotoxin microcystin-LR in hypereutrophic drinking water lakes. *J. Phycol.*, **31**, 248–63.

Kotak, B. G., Zurawell, R. W., Prepas, E. E. and Holmes, C. F. B. (1996) Microcystin-LR concentration in aquatic food web compartments from lakes of varying trophic status. *Can. J. Fish. Aquat. Sci.*, **53**, 1974–85.

Krasner, S. W., Sclimenti, M. J. and Means, E. G. (1994) Quality degradation: implications for DBP formation. *J. Am. Water Works Assoc.*, **86**, 34–47.

Kristensen, P., Søndergaard, M. and Jeppesen, E. (1992) Resuspension in a shallow eutrophic lake. *Hydrobiologia*, **228**, 101–9.

Krueger, C. C. and May, B. (1991) Ecological and genetic effects of salmonid introductions in North America. *Can. J. Fish. Aquat. Sci.*, **48**, 66–77.
Kuentzel, L. E. (1969) Bacteria, carbon dioxide, and algal blooms, *J. Water Pollut. Control Fed.*, **41**, 1737–47.
Kvarnäs, H. and Lindell, T. (1970) *Hydrologiska studier i Ekoln* Rapport over hydrologisk verksamhet inom Naturvårdsverkets Limnologiska Undersökning januari-augusti 1969. UNGI Rapport 3, Uppsala.
Laegrid, M., Alstad, J., Klaveness, D. and Seip, H. M. (1983) Seasonal variation of cadmium toxicity toward the alga *Selenastrum capricornutum* Printz in two lakes with different humus content. *Environ. Sci. Technol.*, **17**, 357–61.
Lager, J. A. and Smith, W. G. (1975) *Urban Stormwater Management and Technology: An Assessment* EPA-670/2-74-040, Washington, DC.
Lake Erie Report (1968) *A Plan for Water-Pollution Control*, Department of the Interior, Fed. Water Poll. Admin., 107pp.
Lam, A. K.-Y., Prepas, E. E., Spink, D. and Hrudey, S. E. (1995) Chemical control of hepatotoxic phytoplankton blooms: Implications for human health. *Water Res.*, **29**, 1845–54.
Lamberti, G. A. and Resh, V. H. (1983) Stream periphyton and insect herbivores: an experimental study of grazing by a caddisfly population. *Ecology*, **64**, 1124–35.
Lamberti, G. A., Ashkenas, L. R., Gregory, S. V. and Steinman, A. D. (1987) Effects of three herbivores on periphyton communities in laboratory streams. *J. N. Am. Benthol. Soc.*, **6**, 92–104.
Lampert, W. (1981a) Toxicity of the blue-green *Microcystis aeruginosa*: Effective defense mechanism against grazing pressure by *Daphnia*. *Verh. Int. Verein. Limnol.*, **21**, 1436–40.
Lampert, W. (1981b) Inhibitory and toxic effects of blue-green algae on *Daphnia*. *Int. Rev. Ges. Hydrobiol.*, **66**, 285–8.
Lampert, W. (1987) Laboratory studies on zooplankton cyanobacteria interactions. *N. Z. J. Marine Freshwater Res.*, **21**, 483–90.
Lampert, W. (1989) The adaptive significance of diel vertical migration of zooplankton. *Funct. Ecol.*, **3**, 21–7.
Lampert, W. and Sommer, U. (1997) *Limnoecology: The Ecology of Lakes and Streams*, Oxford University Press, New York, USA.
Landers, D. H. (1982) Effects of naturally senescing aquatic macrophytes on nutrient chemistry and chlorophyll *a* of surrounding water. *Limnol. Oceanogr.*, **27**, 428–39.
Landers, D. H., Overton, W. S., Linhurst, R. A. and Brakke, D. F. (1988) Eastern lake survey – regional estimates at lake chemistry. *Environ. Sci. Technol.*, **22**, 128–35.
La Point, T. W., Melancon, S. M. and Morris, M. K. (1984) Relationships among observed metal concentrations, criteria, and benthic community structural responses in 15 streams. *J. Water Pollut. Control Fed.*, **56**, 1030–8.
Larsen, D. P., Malueg, K. W., Schultz, D. W. and Brice, R. M. (1975) Response of eutrophic Shagawa Lake, Minnesota, USA, to point-source phosphorus reduction. *Verh. Int. Verein. Limnol.*, **19**, 884–92.
Larsen, D. P. and Mercier, H. T. (1976) Phosphorus retention capacity of lakes. *J. Fish Res. Board Canada*, **33**, 1742–50.
Larsen, D. P., Schultz, D. W. and Malueg, K. W. (1981) Summer internal phosphorus supplies in Shagawa Lake, Minnesota. *Limnol. Oceanogr.*, **26**, 740–53.

Larsen, D. P., Van Sickle, J., Malueg, K. W. and Smith, P. D. (1979) The effect of wastewater phosphorus removal on Shagawa Lake, Minnesota: phosphorus supplies, lake phosphorus and chlorophyll a. *Water Res.*, **13**, 1259–72.

Larson, D. W., Dahm, C. N. and Geiger, N. S. (1987) Vertical partitioning of the phytoplankton assemblage in ultraoligotrophic Crater Lake, Oregon, USA. *Freshwater Biol.*, **18**, 429–42.

Larson, G. L. and Moore, S. E. (1985) Encroachment of exotic rainbow trout into stream populations of native brook trout in the southern Appalachian Mountains. *Trans. Am. Fish. Soc.*, **114**, 195–203.

Lassuy, D. (1995) Introduced species as a factor in extinction and endangerment of native fish species, in *American Fisheries Society Symposium 15: Proceedings of the International Symposium and Workshop on the Uses and Effects of Cultured Fishes in Aquatic Ecosystems* (eds H. L. Schramm, Jr. and R. G. Piper), American Fisheries Society, Bethesda, MD, pp. 391–6.

Lathrop, R. C. (1988) Phosphorus trends in the Yahara lakes since the mid-1960s. Research Mngt. Findings, Wisconsin, Department of Natural Resources, Madison, WI.

Lathrop, R. C. (1990) Response of Lake Mendota (Wisconsin, USA) to decreased phosphorus loadings and the effect on downstream lakes. *Ver. Internat. Verein. Limnol.*, **24**, 457–63.

Lathrop, R. C., Carpenter, S. R., Stow, C. A., Soranno, P. A. and Panuska, J. C. (1998) Phosphorus loading reductions needed to control blue-green algal blooms in Lake Mendota. *Can. J. Fish. Aquat. Sci.*, **55**, 1169–78.

Latif, M. A., Bodaly, R. A., Johnston, T. A. and Fudge, R. J. P. (2001) Effects of environmental and maternally derived methylmercury on the embryonic and larval stages of walleye (*Stizostedion vitreum*). *Environ. Pollut.*, **111**, 139–48.

Lavrentyev, P. J., Gardner, W. S., Cavaletto, J. F. and Beaver, J. R. (1995) Effects of the zebra mussel (*Dreissena polymorpha* Pallas) on protozoa and phytoplankton from Saginaw Bay, Lake Huron. *J. Great Lakes Res.*, **21**, 545–57.

Lawton, G. W. (1961) Limitation of nutrients as a step in ecological control. *Algae and Metropolitan Wastes*, R. A. Taft San. Eng. Center, Tech. Rept. W61-3.

Leach, J. and Dawson, H. (1999) *Crassula helmsii* in the British Isles – an unwelcome invader. *Br. Wildlife*, **10**, 234–9.

Leach, J. M. and Thakore, A. N. (1975) Isolation and identification of constituents toxic to juvenile rainbow trout (*Salmo gairdneri*) in caustic extraction effluents from kraft pulp mill bleach plants. *J. Fish Res. Board Canada*, **32**, 1249–57.

Leivstad, H. and Muniz, I. P. (1976) Fish kills at low pH in a Norwegian river. *Nature.*, **259**, 391–2.

Lennox, L. J. (1984) Lough Ennell: laboratory studies on sediment phosphorus release under varying mixing, aerobic, anaerobic conditions. *Freshwater Biol.*, **14**, 183–7.

Lijklema, L., Gelencsér, P., Szilágyi, F. and Somlyódy, L. (1986) Sediment and its interaction with water, in *Modeling and Managing Shallow Lake Eutrophication With Application to Lake Balaton* (eds L. Somlyódy and G. van Straten), Springer-Verlag, Berlin, pp. 156–82.

Likens, G. E. and Borman, F. H. (1974) Linkages between terrestrial and aquatic ecosystems. *Bioscience*, **24**, 447–56.

Lin, C. (1971) Availability of phosphorus for *Cladophora* growth in Lake Michigan. *Proc. 14th Conf. Great Lakes Res.*, Intern. Assoc. Great Lakes Des., pp. 39–43.

Lind, O. T. and Dávalos-Lind, L. O. (2002) Interaction of water quantity with water quality: the Lake Chapala example. *Hydrobiologia*, 467, 159–67.

Lindell, T. (1975) Vänern, in *Vänern, Vättern, Mälaren, Hjälmaren – en översikt* Statens Naturvårdsverk, Publikationer 1976: 1, Stockholm.

Lindeman, R. L. (1942) Trophic dynamic aspect of ecology. *Ecology*, 23, 399–418.

Litvak, M. K. and Mandrak, N. E. (2000) Baitfish trade as a vector of aquatic introductions, in *Nonindigenous Freshwater Organisms* (eds R. Claudi and J. H. Leach), Lewis Publishers, Boca Raton, FL, pp. 163–80.

Livingstone, D. M. and Imboden, D. M. (1996) The prediction of hypolimnetic oxygen profiles: a plea for a deductive approach. *Can. J. Fish. Aquat. Sci.*, 53, 924–32.

Lloyd, R. (1960) The toxicity of zinc sulfate to rainbow trout. *Ann. Appl. Biol.*, 48, 84–94.

Lock, M. A. and John, P. H. (1979) The effect of flow patterns on uptake of phosphorus by river periphyton. *Limnol. Oceanogr.*, 24, 376–83.

Lodge, D. M., Magnuson, J. J. and Beckel, A. M. (1985) Lake-bottom tyrant. *Nat. Hist.*, 94, 32–7.

Loeb, S. L. (1986) Algal biofouling of oligotrophic Lake Tahoe: causal factors affecting production, in *Algal Biofouling* (eds L. V. Evans and K. D. Hoagland), Elsevier Science Publishers B. V., Amsterdam, The Netherlands, pp. 159–73.

Leob, S. L., Reuter, J. E. and Goldman, C. R. (1983) Littoral zone production of oligotrophic lakes, in *Periphyton of Freshwater Ecosystems* (ed. R. G. Wetzel) (*Developments in Hydrobiologia*, No. 17), Dr W. Junk Publishers, The Hague, pp. 161–7.

Löfgren, S. (1987) Phosphorus retention in sediments – implications for aerobic phosphorus release in shallow lakes. PhD dissertation, Uppsala University, Sweden.

Lohman, K., Jones, J. R. and Baysinger-Daniel, C. (1991) Experimental evidence for nitrogen limitation in northern Ozark streams. *J. N. Am. Benthol. Soc.*, 10, 14–23.

Lohman, K., Jones, J. R. and Perkins, B. D. (1992) Effects of nutrient enrichment and flood frequency on periphyton biomass in northern Ozark streams. *Can. J. Fish. Aquat. Sci.*, 49, 1198–205.

Long, E. B. (1976) The interaction of phytoplankton and the bicarbonate system. PhD dissertation, Kent State University, Ohio.

Lorenzen, M. W. and Fast, A. W. (1977) *A Guide to Aeration/Circulation Techniques for Lake Management*, EPA-600 13-77-004, USEPA, Washington, DC.

Luettich, R. A., Jr., Harleman, D. R. F. and Somlyódy, L. (1990) Dynamic behaviour of suspended sediment concentrations in a shallow lake perturbed by episodic wind events. *Limnol. Oceanogr.*, 35, 1050–67.

Lund, J. W. G. (1950) Studies on *Asterionella formosa* Hass. II. Nutrient depletion and the spring maximum. *J. Ecol.*, 38, 1–35.

Lynch, M. and Shapiro, J. (1981) Predation, enrichment, and phytoplankton community structure. *Limnol. Oceanogr.*, 26, 86–102.

Maciolek, J. A. and Maciolek, M. G. (1968) Microseston dynamics in a simple Sierra Nevada lake-stream system. *Ecology*, 49, 60–75.

Mack, R. N., Simberloff, D., Lonsdale, W. M., Evans, H., Clout, M. and Bazzaz, F. A. (2000) Biotic invasions: causes, epidemiology, global consequences, and control. *Ecol. Appl.*, **10**, 689–710.

Macon, T. T. (1974) *Freshwater Ecology* John Wiley & Sons, New York, 343pp.

Magnuson, J. J. (1976) Managing with exotics – A game of chance. *Trans. Am. Fish. Soc.*, **105**, 1–9.

Magnuson, J. J., Webster, K. E., Assel, R. A., Bowser, C. J., Dillon, P. J., Eaton, J. G., Evans, H. E., Fee, E. J., Han, R. I., Mortsch, L. R., Schindler, D. W. and Quinn, F. H. (1997) Potential effects of climate changes on aquatic systems: Laurentian Great Lakes and Precambrian Shield region. *Hydrol. Process.*, **11**, 825–71.

Mandrak, N. E. (1989) Potential invasion of the Great Lakes by fish species associated with climatic warming. *J. Great Lakes Res.*, **15**, 306–16.

Mann, K. H. (1965) Heated effluents and their effects on the invertebrate fauna of rivers. *Proc. Soc. Water Treat. Exam.*, **14**, 45–53.

Marcuson, P. E. (1970) *Stream Sediment Investigation Progress Report* Mont. Fish and Game Dept., Helena.

Margalef, R. (1958) Temporal succession and spatial heterogeneity in phytoplankton. *Perspectives in Marine Ecology*. University of California Press, p. 323.

Margalef, R. (1969) Diversity and stability: a practical proposal and a model of interdependence. *Symposium on Diversity and Stability in Ecological Systems* Brookhaven National Lab., Upton, NY.

Markowski, S. (1959) The cooling water of power stations: new factor in the environment of marine and freshwater invertebrates. *J. Animal Ecol.*, **28**, 243–58.

Marshall, P. T. (1958) Primary production in the arctic. *J. Cons. Int. Explor. Mer.*, **23**, 173–7.

Martin, J. B., Jr., Bradford, B. N. and Kennedy, H. G. (1969) *Factors Affecting the Growth of Najas in Pickwick Reservoir*. TVA, National Fertilizer Development Center Report.

Mason, C. (2002) *Biology of Freshwater Pollution*, 4th edn, Prentice Hall, London.

Mason, W. T., Jr., Anderson, J. B., Kreis, R. D. and Johnson, W. C. (1970) Artificial substrate sampling, macroinvertebrates in a polluted reach of the Klamath River, Oregon, *J. Water Pollut. Control Fed.*, **42**, R315–27.

Matthews, W. J. and Zimmerman, E. G. (1990) Potential effects of global warming on native fishes of the southern Great Plains and the southwest. *Fisheries*, **15**, 26–32.

Matthiessen, P. (2000) Is endocrine disruption a significant ecological issue? *Ecotoxicology*, **9**, 21–4.

May, C. W., Welch, E. B., Horner, R. R., Karr, J. R. and Mar, B. W. (1997) *Quality Indices for Urbanization Effects in Puget Sound Lowland Streams*, Department of Civil Engineering, University of Washington Water Resources Series Technical Report, No. 154. Seattle, WA.

McBride, G. B. and Pridmore, R. D. (1988) Prediction of [chlorophylla-*a*] in impoundments of short hydraulic retention time: mixing effects. *Verh. Int. Verein. Limnol.*, **23**, 832–6.

McCarthy, I. D. and Houlihan, D. F. (1997) The effects of temperature on protein metabolism in fish: the possible consequences for wild Atlantic salmon (*Salmo salar* L.) stocks in Europe as a result of global warming, in *Global Warming:*

Implications for Freshwater and Marine Fish (eds C. M. Wood and D. G. McDonald), Cambridge University Press, Cambridge, pp. 51–77.

McConnell, J. W. and Sigler, W. F. (1959) Chlorophyll and productivity in a mountain river. *Limnol. Oceanogr.*, **4**, 335–51.

McCormick, P. V. and Stevenson, R. V. (1989) Effects of snail grazing on benthic algal community structure in different nutrient environments. *J.N. Am. Benthol. Soc.*, **8**, 162–72.

McFeters, G. A., Bond, P. J., Olson, S. B. and Tchan, Y. T. (1983) A comparison of microbial bioassays for the detection of aquatic toxicants. *Water Res.*, **17**, 1757–62.

McGauhey, P. H., Rohlich, G. A. and Pearson, E. P. (1968) *Eutrophication of Surface Waters – Lake Tahoe: Bioassay of Nutrient Sources*. First Progress Report, Lake Tahoe Area Council, 178 pp.

McIntire, C. D. (1966) Some effects of current velocity on periphyton communities in laboratory streams. *Hydrobiologia*, **37**, 559–70.

McIntire, C. D., Colby, J. A. and Hall, J. D. (1975) The dynamics of small lotic ecosystems: a modelling approach. *Verh. Int. Verein. Limnol.*, **19**, 1599–609.

McIntire, C. A. and Phinney, H. K. (1965) Laboratory studies of periphyton production and community metabolism in lotic environments. *Ecol. Monogr.*, **35**, 237–58.

McKinnon, S. L. and Mitchell, S. F. (1994) Eutrophication and black swan (*Cygnus atratus* Latham) populations: tests of two simple relationships. *Hydrobiologia*, **279/280**, 163–70.

McKnight, D. (1981) Chemical and biological processes controlling the response of a freshwater ecosystem to copper stress: A field study of the $CuSO_4$ treatment of Mill Pond Reservoir, Burlington, Massachusetts. *Limnol. Oceanogr.*, **26**, 518–31.

McLachlan, J. A. and Arnold, S. F. (1996) Environmental estrogens. *Am. Sci.*, **84**, 452–61.

McMahon, J. W. and Rigler, F. H. (1965) Feeding rate of *Daphnia magna* Straus on different foods labeled with radioactive phosphorus. *Limnol. Oceanogr.*, **10**, 105–13.

McMahon, T. E. and Bennett, D. H. (1996) Walleye and northern pike: boost or bane to northwest fisheries? *Fisheries*, **21**, 6–13.

McNaught, D. C. (1975) A hypothesis to explain the succession from calanoids to cladocerans during eutrophication. *Verh. Int. Verein. Limnol.*, **19**, 724–31.

McQueen, D. J. and Post, J. R. (1988) Limnocorral studies of cascading trophic interactions. *Verh. Int. Verein. Limnol.*, **23**, 739–47.

McQueen, D. J., Post, J. R. and Mills, E. L. (1986) Trophic relationships in freshwater pelagic ecosystems. *Can. J. Fish. Aquat. Sci.*, **43**, 1571–81.

Meijer, M.-L, de Boois, I., Scheffer, M., Portielje, R. and Hosper, H. (1999) Biomanipulation in shallow lakes in The Netherlands: an evaluation of 18 case studies. *Hydrobiologia*, **408/409**, 13–30.

Meijer, M.-L., de Haan, M. W., Breukelaar, A. W. and Buiteveld, H. (1990) Is reduction of the benthivorous fish an important cause of high transparency following biomanipulation in shallow lakes? *Hydrobiologia*, **200/201**, 303–15.

Meijer, M.-L., Jeppesen, E., van Donk, E., Moss, B., Scheffer, M., Lammens, E., van Nes, E., van Berkum, J. A., de Jong, G. J., Faafeng, B. A. and Jensen, J. P. (1994) Long-term responses to fish-stock reduction in small shallow lakes: interpretation of five-year results of four biomanipulation cases in The Netherlands and Denmark. *Hydrobiologia*, **275/276**, 457–66.

Melack, J. M., Dozier, J., Goldman, C. R., Greenland, D., Milner, A. M. and Naiman, R. J. (1997) Effects of climate change on inland waters of the Pacific Coastal Mountains and Western Great Basin of North America. *Hydrol. Process.*, **11**, 971–92.

Menzel, D. W. and Ryther, J. H. (1964) The composition of particulate organic matter in the western North Atlantic. *Limnol. Oceanogr.*, **9**, 179–86.

Messer, J. and Brezonik, P. L. (1983) Comparison of denitrification rate estimation techniques in a large, shallow lake. *Water Res.*, **17**, 631–40.

METRO (1987) Municipality of Metropolitan Seattle, Seattle, WA.

Micheli, F. (1999) Eutrophication, fisheries, and consumer–resource dynamics in marine pelagic ecosystems. *Science*, **285**, 1396–8.

Mihursky, J. (1969) *Patuxent Thermal Studies. Summary and Recommendations*. Nat. Res. Inst. Spec. Rept. No. 1, University of Maryland, College Park, MD.

Mihursky, J. and Kennedy, V. S. (1967) Water temperature criteria to protect aquatic life, in *Symposium on Water Quality Criteria* Am. Fish Soc. Spec. Pub. No. 4, pp. 20–32.

Milbrink, G. (1980) Oligochaete communities in pollution biology: the European situation with special reference to lakes in Scandinavia, in *Aquatic Oligochaete Biology* (eds R. O. Brinkhurst and D. G. Cook), Plenum Press, NY.

Miles, E. L., Snover, A. K., Hamlet, A. F., Callahan, B. and Fluharty, D. (2000) Pacific Northwest regional assessment: The impacts of climate variability and climate change on the resources of the Columbia River basin. *J. Am. Water Res. Assoc.*, **36**, 399–420.

Miller, G. T., Jr. (1988) *Living in the Environment; an Introduction to Environmental Science*, 5th edn, Wadsworth, Belmont, CA.

Miller, G. T., Jr. (1998) *Living in the Environment*, 10th edn, Wadsworth, Belmont, CA.

Miller, R. R., Williams, J. D. and Williams, J. E. (1989) Extinctions of North American fishes during the past century. *Fisheries*, **14**, 22–38.

Miller, T. G., Melancon, S. M. and Janika J. J. (1983) Site specific water quality assessment, Prickley Pear Creek, Montana, US EPA 600/X-82-013, Las Vegas, NV.

Miller, W. E., Greene, J. C. and Shirogama, T. (1978) The *Salenastrum capricornutum* Printz algal assay bottle test. *US EPA Report* EPA-600-/9-78-018, Cincinnati, OH.

Miller, W. E., Maloney, T. E. and Greene, J. C. (1974) Algal productivity in 49 lake waters as determined by algal assays. *Water Res.*, **8**, 667–79.

Mills, E. L. and Forney, J. L. (1983) Impact on *Daphnia pulex* of predation by young yellow perch in Oneida Lake, New York. *Trans. Amer. Fish. Soc.*, **112**, 154–61.

Mills, E. L., Leach, J. H., Carlton, J. T. and Secor, C. L. (1994) Exotic species and the integrity of the Great Lakes. *Bioscience*, **44**, 666–76.

Minshall, G. W. (1978) Autotrophy in stream ecosystems. *Bioscience*, **28**, 767–71.

Miranda, L. E., Hargreaves, J. A. and Raborn, S. W. (2001) Predicting and managing risk of unsuitable dissolved oxygen in a eutrophic lake. *Hydrobiologia*, **457**, 177–85.

Misra, R. D. (1938) Edaphic factors in the distribution of aquatic plants in the English Lakes. *J. Ecol.*, **26**, 411–51.

Mitchell, S. F. (1989) Primary production in a shallow eutrophic lake dominated alternatively by phytoplankton and by macrophytes. *Aquat. Bot.*, **33**, 101–10.

Mooney, H. A. and Drake, J. A. (eds) (1986) *Ecology of Biological Invasions of North America and Hawaii*, Springer-Verlag, New York, NY.

Moore, L. and Thornton, K., eds (1988) *Lake and Reservoir Restoration Guidance Manual* EPA 440/5-88-002.

Mortimer, C. H. (1941) The exchange of dissolved substances between mud and water in lakes (parts I and II). *J. Ecol.*, **29**, 280–329.
Mortimer, C. H. (1942) The exchange of dissolved substances between mud and water in lakes (parts III, IV, summary, and references). *J. Ecol.*, **30**, 147–301.
Mortimer, C. H. (1952) Water movements in lakes during summer stratification; evidence from the distribution of temperature in Lake Windermere, with an appendix by M. S. Longuet-Higgins. *Phil. Trans. Ser. B*, **236**, 355–404.
Mortimer, C. H. (1971) Chemical exchanges between sediments and water in the Great Lakes – speculations on probable regulatory mechanisms. *Limnol. Oceanogr.*, **16**, 387–404.
Moss, B., Madgwick, J. and Phillips, G. (1996) *A Guide to the Restoration of Nutrient-enriched Shallow Lakes*, W.W. Hawes, London, UK.
Moss, B., Stansfield, J. and Irvine, K. (1990) Problems in the restoration of a hypereutrophic lake by diversion of a nutrient-rich inflow. *Verh. Int. Verein. Limnol.*, **24**, 568–72.
Mote, P., Holmberg, M. and Mantua, N. (1999) *Impacts of Climate Change: Pacific Northwest*. A summary of the Pacific Northwest Regional Assessment Group for the US Global Change Research Program (eds A. K. Snover, E. Miles and the JISAO/SMA Climate Impacts Group), JISAO/SMA Climate Impacts Group, University of Washington, Seattle, WA.
Mount, D. I. (1966) The effect of total hardness and pH on acute toxicity of zinc to fish. *Air Water Poll. Int. J.*, **10**, 49–56.
Mount, D. I. (1969) Developing thermal requirements for freshwater fishes, in *Biological Aspects of Thermal Pollution* (eds P. A. Krenkel and F. L. Parker), Vanderbilt University Press, Nashville, Tennessee, pp. 140–7.
Mount, D. I. and Norberg, T. J. (1984) A seven-day life-cycle cladoceran toxicity test. *Environ. Toxicol. Chem.*, **3**, 425–34.
Moyle, P. B., Li, H. W. and Barton, B. A. (1986) The Frankenstein effect: Impact of introduced fish on native fishes in North America, in *Fish Culture in Fisheries Management* (ed. R. H. Stroud), American Fisheries Society, Bethesda, MD, pp. 415–26.
Moyle, P. B. and Light, T. (1996) Biological invasions of fresh water: empirical rules and assembly theory. *Biol. Conserv.*, **78**, 149–61.
Müller-Navarra, D. C. (1995) Evidence that a highly unsaturated fatty acid limits *Daphnia* growth in nature. *Arch. Hydrobiol.*, **132**, 297–307.
Müller-Navarra, D. C., Brett, M. T., Liston, A. and Goldman C. R. (2000) A highly unsaturated fatty acid predicts carbon transfer between primary producers and consumers. *Nature*, **403**, 74–77.
Mulligan, H. F. and Barnowski, A. (1969) Growth of phytoplankton and vascular aquatic plants at different nutrient levels. *Verh. Int. Verein. Limnol.*, **17**, 302–10.
Muniz, I. P. and Leivestad, H. (1980) Acidification – effects on freshwater fish, in *Proc. Int. Conf. Ecol. Impact of Acid Precipitation* (eds D. Drobløs and A. Tollan), Oslo, Norway, pp. 84–92.
Murphy, G. I. (1962) Effect of mixing depth and turbidity on the productivity of fresh-water impoundments. *Am. Fish. Soc.*, **91**, 69–76.
Nalepa, T. F., Fahnenstiel, G. L. and Johengen, T. H. (2000) Impacts of the zebra mussel (*Dreissena polymorpha*) on water quality: A case study in Saginaw Bay,

Lake Huron, in *Nonindigenous Freshwater Organisms* (eds R. Claudi and J. H. Leach), Lewis Publishers, Boca Raton, FL, pp. 255–72.

Nalepa, T. F. and Schloesser, D. W. (1993) *Zebra Mussels: Biology, Impacts, and Control*, Lewis Publishers, Boca Raton, FL, 832pp.

NALMS (2001) *Managing Lakes and Reservoirs*, North American Lake Management Society (NALMS), Terrene Institute, Madison, WI and Alexandria, VA.

National Research Council (1996) *Stemming the Tide: Controlling Introductions of Nonindigenous Species by Ships' Ballast Water*, National Academy Press, Washington, DC.

National Research Council (1999) *Hormonally Active Agents in the Environment*, National Academy Press, Washington, DC.

Nebeker, A. V. (1971a) Effect of water temperature on nymphal feeding rate, emergence, and adult longevity of the stone fly *Pteronarcys dorsata*. *J. Kans. Entomol. Soc.*, **44**, 21–6.

Nebeker, A. V. (1971b) Effect of high winter water temperatures on adult emergence of aquatic insects. *Water Res.*, **5**, 777–83.

Neil, J. H. and Owen, G. E. (1964) Distribution, environmental requirements, and significance of *Cladophora* in the Great Lakes. *Proc. 7th Conf. Great Lakes Res.*, **11**, 113–21.

Newbold, J. D., Elwood, J. W., O'Neill, R. V. and Van Winkle, W. (1981) Measuring nutrient spiralling in streams. *Can. J. Fish. Aquat. Sci.*, **38**, 860–3.

Newbold, J. D., Erman, D. C. and Roby, K. B. (1980) Effects of logging on macroinvertebrates in streams with and without buffer strips. *Can. J. Fish. Aquat. Sci.*, **37**, 1076–85.

Newroth, P. R. (1975) Management of nuisance aquatic plants. Unpublished report, B. C. Dept. of the Environ., 13pp.

Nichols, D. S. and Keeney, D. R. (1976a) Nitrogen nutrition of *Myriophyllum spicatum*: variation of plant tissue nitrogen concentration with season and site in Lake Wingra. *Freshwater Biol.*, **6**, 137–44.

Nichols, D. S. and Keeney, D. R. (1976b) Nitrogen nutrition of *Myriophyllum spicatum*: uptake and translocation of ^{15}N by shoots and roots. *Freshwater Biol.*, **6**, 145–54.

Nichols, S. A. and Shaw, B. H. (1986) Ecological life histories of the three aquatic nuisance plants. *Myriophyllum spicatum, Potamogeton crispus* and *Elodea canadensis*. *Hydrobiologia*, **131**, 3–21.

Nõges, P. and Järvet, A. (1995) Water level control over light conditions in shallow lakes. *Report Series in Geophysics*, Vol. 32, University of Helsinki, pp. 81–92.

Nõges, P., Nõges, T., Haberman, J., Laugaste, R. and Kisdand, V. (1997) Tendencies and relations in plankton community and pelagic environment of Lake Võrtsjärv during three decades. *Proc. Acad. Sci. Estonia Ser. Ecology*, **46**(1/2), 40–58.

Noonan, T. A. (1986) Water quality in Long Lake, Minnesota following Riplox sediment treatment. *Lake Reservoir Management*, **2**, 131–7.

Norberg, T. J. and Mount, D. I. (1985) A new fathead minnow (*Pimephales promelas*) subacute toxicity test. *Environ. Toxicol. Chem.*, **4**, 711–18.

Nordin, R. N. (1985) *Water Quality Criteria for Nutrients and Algae* (Technical Appendix), British Columbia Ministry of the Environment, Victoria, BC, 104pp.

Novotny, V. and Olem, H. (1994) *Water Quality: Prevention, Identification, and Management of Diffuse Pollution*, Van Nostrand Rheinhold, New York.
Nowell, L. H., Capel, P. D. and Dileanis, P. D. (1999) *Pesticides in Stream Sediment and Aquatic Biota: Distribution, Trends, and Governing Factors*, Vol. 4, Pesticides in the Hydrologic System, Lewis Publishers, Boca Raton, FL.
Nūman, W. (1972) Predictions on the development of the salmonid communities in the European oligotrophic subalpine lakes during the next century. Unpublished manuscript. Staatliche Institut für Seenforschung und Seenbewirtschaftung, Langenargen Bodensee, Germany, 12 pp.
Nürnberg, G. K. (1984) The prediction of internal phosphorus load in lakes with anoxic hypolimnia. *Limnol. Oceanogr.*, **29**, 111–24.
Nürnberg, G. K. (1987a) A comparison of internal phosphorus loads in lakes with anoxic hypolimnia: Laboratory incubation versus in situ hypolimnetic phosphorus accumulation. *Limnol. Oceanogr.*, **32**, 1160–64.
Nürnberg, G. K. (1987b) Hypolimnetic withdrawal as lake restoration technique. *J. Environ Eng.*, **113**, 1006–17.
Nürnberg, G. K. (1988) Prediction of phosphorus release rates from total and reductant- soluble phosphorus in anoxic sediments. *Can. J. Fish. Aquat. Sci.*, **45**, 453–62.
Nürnberg, G. K. (1995a) The anoxic factor, a quantitative measure of anoxia and fish species richness in central Ontario lakes. *Trans. Am. Fish. Soc.*, **124**, 677–86.
Nürnberg, G. K. (1995b) Quantifying anoxia in lakes. *Limnol. Oceanogr.*, **40**, 1100–11.
Nürnberg, G. K. (1996) Trophic state of clear and colored, soft- and hardwater lakes with special consideration of nutrients, anoxia, phytoplankton and fish. *Lake Reserv. Manage.*, **12**, 432–47.
O'Brien, W. J. (1979) The predator-prey interaction of planktivorous fish and zooplankton. *Am Sci.*, **67**, 572–81.
Odum, E. P. (1959) *Fundamentals of Ecology* W. B. Saunders, Philadelphia.
Odum, E. P. (1969) The strategy of ecosystem development. *Science*, **164**, 264–70.
Odum, H. T. (1956) Primary production in flowing waters. *Limnol. Oceanogr.*, **1**, 102–17.
OECD (1982) *Eutrophication of Waters. Monitoring, Assessment and Control* OECD, Paris, 154 pp.
Ohle, W. (1953) Phosphor als Initialfaktor der Gewässereutrophierung. *Vom Wasser*, **20**, 11–23.
Ohle, W. (1975) Typical steps in the change of a limnetic ecosystem by treatment with therapeutica. *Verh. Int. Verein. Limnol.*, **19**, 1250.
Olson, T. A. and Rueger, M. E. (1968) Relationship of oxygen requirements to index-organism classification of immature aquatic insects. *J. Water Pollut. Control Fed.*, **40**, 188–202.
Oliver, B. G. and Shindler, D. B. (1980) Trihalomethanes from the chlorination of aquatic algae. *Environ. Sci. Technol.*, **14**, 1502–5.
Omernik, J. M. (1987). Ecoregions of the Conterminous United States. *Ann. Assoc. Am. Geographers*, **77**(1), 118–25.
Omernik, J. M. (1995). Ecoregions: A spatial framework for environmental management, in *Biological Assessment and Criteria: Tools for Water Resource Planning and Decision Making* (eds W. S. Davis and T. P Simon), Lewis Publishers, Boca Raton, FL, pp. 49–62.

Ormerod, J. G., Grynne, B. and Ormerod, K. S. (1966) Chemical and physical factors involved in heterotrophic growth response to organic pollution. *Verh. Int. Verein. Limnol.*, **16**, 906–10.

Osgood, R. A. (1988a) Lake mixis and internal phosphorus dynamics. *Arch. Hydrobiol.*, **113**, 629–38.

Osgood, R. A. (1988b) A hypothesis on the role of *Aphanizomenon* in translocating phosphorus. *Hydrobiologia*, **169**, 69–76.

Oskam, G. (1978) Light and zooplankton as algae regulating factors in eutrophic Biesbosch reservoirs. *Verh. Int. Verein. Limnol.*, **20**, 1612–18.

Paerl, H. W. (1988) Nuisance phytoplankton blooms in coastal, estuarine, and inland waters. *Limnol. Oceanogr.*, **33**, 823–47.

Paerl, H. W. (1996) A comparison of cyanobacterial bloom dynamics in freshwater, estuarine and marine environments. *Phycologia*, **35**(Suppl. 6), 25–35.

Paerl, H. W. and Ustach, J. F. (1982) Blue-green algal scums: An explanation for their occurrences during freshwater blooms. *Limnol. Oceanogr.*, **27**, 212–17.

Pagenkopf, G. K. (1983) Gill surface interaction model for trace-metal toxicity to fishes: role of complexation, pH and water hardness. *Environ. Sci. Technol.*, **17**, 342–7.

Palmstrom, N. S., Carlson, R. E. and Cooke, G. D. (1988) Potential links between eutrophication and the formation of carcinogens in drinking water. *Lake Reserv. Manage.*, **4**, 1–15.

Paloheimo, J. E. (1974) Calculation of instantaneous birth rate. *Limnol. Oceanogr.*, **19**, 692–4.

Parsons, T. R., Stephens, K. and Le Brasseuer, R. J. (1969) Production studies in the Strait of Georgia. Part I. Primary production under the Fraser River plume, February to May, 1967. *J. Exp. Mar. Biol. Ecol.*, **3**, 27–38.

Pastorak, R. A., Lorenzen, M. W. and Ginn, T. C. (1982) *Environmental Aspects of Artificial Aeration and Oxygenation of Reservoirs: A Review of Theory, Techniques and Experiences*. Tech. Rept. No. E-82-3, U.S. Army Corps of Engineers.

Patalas, K. (1970) Primary and secondary production in a lake heated by a thermal power plant. *Proc. Inst. Environ. Sci. 16th Annual Tech. Meet.*, pp. 267–71.

Patalas, K. (1972) Crustacean plankton and the eutrophication of St Lawrence Great Lakes. *J. Fish Res. Board Canada*, **29**, 1451–62.

Patten, B. C., Egloff, D. A., Richardson, T. H. *et al.* (1975) Total Ecosystem model for a cove in Lake Taxoma, in *Systems Analysis and Simulation in Ecology*, Vol. 3 (ed. B. C. Patten), Academic Press, NY.

Peakall, D. B. and Lovett, R. J. (1972) Mercury: its occurrence and effects in the ecosystem. *Bioscience*, **22**, 20–5.

Pearce, F. (1997) Why is the apparently pristine Arctic full of toxic chemicals that started off thousands of kilometres away? *New Scientist*, **31**, 24–7.

Pearsall, W. H. (1920) The aquatic vegetation of the English Lakes. *J. Ecol.*, **8**, 163–99.

Pearsall, W. H. (1929) Dynamic factors affecting aquatic vegetation. *Proc. Int. Cong. Plant Sci.*, **1**, 667–72.

Pechlaner, R. (1978) Erfahrungen mit Restaurierungsmassnahmen an eutrophierten Badeseen Tirols. *Dester Wasserwertsch*, **30**, 112–19.

Pederson, G. L., Welch, E. B. and Litt, A. H. (1976) Plankton secondary productivity and biomass: their relation to lake trophic state. *Hydrobiologia*, **50**, 129–44.

Peltier, W. H. and Welch, E. B. (1969) Factors affecting growth of rooted aquatic plants in a river. *Weed Sci.*, **17**, 412–16.
Peltier, W. H. and Welch, E. B. (1970) Factors affecting growth of rooted aquatic plants in a reservoir. *Weed Sci.*, **18**, 7–9.
Penn, M. R., Auer, M. T., Doerr, S. M., Driscoll, C. T., Brooks, C. M. and Effler, S. W. (2000) Seasonality in phosphorus release rates from the sediments of a hypereutrophic lake under a matrix of pH and redox conditions. *Can. J. Fish. Aquat. Sci.*, **57**, 1033–41.
Perakis, S. S., Welch, E. B. and Jacoby, J. M. (1996) Sediment-to-water blue-green algal recruitment in response to alum and environmental factors. *Hydrobiologia*, **318**, 165–77.
Percival, E. and Whitehead, H. (1929) A quantitative study of the fauna of some types of stream bed. *J. Ecol.*, **17**, 282–314.
Perkins, D., Kann, J. and Scoppettone, G. G. (2000) *The Role of Poor Water Quality and Fish Kills in the Decline of Endangered Lost River and Shortnose Suckers in Upper Klamath Lake.* U.S. Geological Survey, Biological Resources Division Report submitted to U.S. Bureau of Reclamation, Klamath Falls Project Office, Klamath Falls, Oregon. Contract 4-AA-29-12160.
Perkins, J. L. (1983) Bioassay evaluation of diversity and community composition indexes. *J. Water Pollut. Control Fed.*, **55**, 522–30.
Perkins, W. W., Welch, E. B., Frodge, J. and Hubbard, T. (1997) A zero degree of freedom total phosphorus model. 2. Application to Lake Sammamish, Washington. *Lake Reserv. Manage.*, **13**, 131–41.
Perrin, C. J., Bothwell, M. L. and Slaney, P. A. (1987) Experimental enrichment of a coastal stream in British Columbia: Effects of organic and inorganic additions on autotrophic periphyton production. *Can. J. Fish. Aquat. Sci.*, **44**, 1247–56.
Peters, J. C. (1960) *Stream Sedimentation Project Progress Report* Mont. Fish and Game Dept., Helena.
Peters, J. C. (1967) Effects on a trout stream of sediment from agricultural practices. *J. Wildlife Management*, **31**, 805–12.
Peterson, H. G., Hrudey, S. E., Cantin, I. A., Perley, T. R. and Kenefick, S. L. (1995) Physiological toxicity, cell membrane damage and the release of dissolved organic carbon and geosmin by *Aphanizomenon flos-aquae* after exposure to water treatment chemicals. *Water Res.*, **29**, 1515–23.
Peterson, S. A., Miller, W. E., Greene, J. C. and Callahan, C. A. (1985) Use of bioassays to determine potential toxicity effects of environmental pollutants, in *Perspectives on Nonpoint Source Pollution* EPA 440/5-85-001.
Phaup, J. D. and Gannon, J. (1967) Ecology of *Sphaerotilus* in an experimental outdoor channel. *Water Res.*, **1**, 523–41.
Phillips, G. L., Einson, D. and Moss, E. (1978) A mechanism to account for macrophyte decline in progressively eutrophicated fresh waters. *Aquatic Bot.*, **3**, 239–55.
Phillips, J. E. (1964) The ecological role of phosphorus in waters with special reference to microorganisms, in *Principles and Applications in Aquatic Microbiology* (eds H. Heukelekian and N. C. Doudero), John Wiley & Sons, New York, pp. 61–81.
Phinney, H. K. and McIntire, C. D. (1965) Effect of temperature on metabolism of periphyton communities developed in laboratory streams. *Limnol. Oceanogr.*, **10**, 341–4.

Pickering, Q. H. (1968) Some effects of dissolved oxygen concentrations upon the toxicity of zinc to the bluegill *Lepomis macrochirus* Raf. *Water Res.*, **2**, 187–94.

Pickering, Q. H. (1988) Evaluation and comparison of two short-term fathead minnow tests for estimating chronic toxicity. *Water Res.*, **22**, 883–93.

Pilotto, L. S., Douglas, R. M., Burch, M. D., Cameron, S., Beers, M., Rouch, G. R., Robinson, P., Kirk, M., Cowie, C. T., Hardiman, S., Moore, C. and Attwell, R. G. (1997) Health effects of exposure to cyanobacteria (blue-green algae) due to recreational water-related activities. *Aust. N. Z. J. Public Health*, **21**, 562–6.

Pimentel, D., Acquay, H., Biltonen, M., Rice, P., Silva, M., Nelson, J., Lipner, V., Giordana, S., Horowitz, A. and D'Amore, M. (1992) Environmental and economic costs of pesticide use. *Bioscience*, **42**, 750–60.

Plafkin, J. L., Barbour, M. T., Porter, K. D., Gross, S. K. and Hughes, R. M. (1989) *Rapid Bioassessment Protocols for Use in Streams and Rivers: Benthic Macroinvertebrates and Fish*. EPA 440-4-89-001. US Environmental Protection Agency, Assessment and Water Protection Division, Washington, DC.

Pomeroy, L. W. (1974) The ocean's food web, a changing paradigm. *Bioscience*, **24**, 499–503.

Pontasch, K. W. and Brusven, M. A. (1988) Diversity and community comparison indices: Assessing macroinvertebrate recovery following a gasoline spill. *Water Res.*, **22**, 619–26.

Pontasch, K. W., Smith, E. P. and Cairns, J. Jr. (1989) Diversity indices, community comparison indices and canonical discriminant analysis: interpreting the results of multispecies toxicity tests. *Water Res.*, **23**, 1229–38.

Porath, H. A. (1976) The effect of urban runoff on Lake Sammamish periphyton. MS thesis, University of Washington, Seattle, 74 pp.

Porcella, D. B., Peterson S. A. and Larsen, D. P. (1980) Index to evaluate lake restoration. *J. Environ. Eng.*, **106**, 1151–69.

Porter, G. W. (1977) The plant-animal interface in freshwater ecosystems. *Am. Sci.*, **65**, 159–69.

Porter, K. G. (1972) A method for the in situ study of zooplankton grazing effects on algal species composition and standing crop. *Limnol. Oceanogr.*, **17**, 913–17.

Porter, K. G. and McDonough, R. (1984) The energetic cost of response to blue-green algal filaments by cladocerans. *Limnol. Oceanogr.*, **29**, 365–9.

Post, J. R. and McQueen, D. J. (1987) The impact of planktivorous fish on the structure of a plankton community. *Freshwater Biol.*, **17**, 79–89.

Prairie, Y. T., Duarte, C. M. and Kalff, J. (1989) Unifying nutrient-chlorophyll relations in lakes. *Can. J. Fish. Aquat. Sci.*, **46**, 1176–82.

Prandtl, L. (1952) *Essentials of Fluid Dynamics*. Blackie & Son, Glasgow.

Prentki, R. T., Adams, M. S., Carpenter, S. R., Gasith, A., Smith, C. S. and Weiler, P. R. (1979) The role of submersed weedbeds in internal loading and interception of allochthonous materials in Lake Wingra, Wisconsin, USA. *Arch. Hydrobiol.*, **57**, 221–50.

Prescott, G. W. (1954) *How to Know the Freshwater Algae* Wm. C. Brown Co., Dubugue, Ia.

Pridmore, R. D. and McBride, G. B. (1984) Prediction of chlorophyll *a* concentrations in impoundments of shord hydraulic retention time. *J. Environ. Management*, **19**, 343–50.

Psenner, R., Boström, B., Dinka, M., Pettersson, K., Puckso, R. and Sager, M. (1988) Fractionation of phosphorus in suspended matter and sediment. *Arch. Hydrobiol. Suppl.*, **30**, 98–103.
Purdom, C. E., Hardiman, P. A., Bye, V. J., Eno, N. C., Tyler, C. R. and Sumpter, J. P. (1994) Estrogenic effects of effluents from sewage treatment works. *Chem. Ecol.*, **8**, 275–85.
Quinn, J. M. (1991) Guidelines for the control of undesired biological growths in water, New Zealand National Institute of Water and Atmospheric Research, Consultancy Report No. 6213/2.
Quinn, J. M. and McFarlane, P. N. (1987) Effects of slaughterhouse and dairy factory wastewaters on epilithon: a comparison in laboratory streams. *Water Res.*, **23**, 1267–73.
Rabalais, N. N., Turner, R. E. and Scavia, D. (2002) Beyond science into policy: Gulf of Mexico hypoxia and the Mississippi River. *Bioscience*, **52**, 129–42.
Raess, F. and Maly, E. J. (1986) The short-term effects of perch predation on a zooplankton prey community. *Hydrobiologia*, **140**, 155–60.
Rahel, F. J. (2000) Homogenization of fish faunas across the United States. *Science*, **288**, 854–6.
Rahel, F. J. and Magnuson, J. J. (1983) Low pH and the absence of fish species in naturally acidic Wisconsin lakes: inferences for cultural acidification. *Can. J. Fish. Aquat. Sci.*, **40**, 3–9.
Rast, W. and Lee, G. F. (1978) Summary analysis of the North American (US portion) OECD eutrophication project: Nutrient Loading – lake response relationships and trophic state indices. *Ecol. Res. Ser.* EPA-600 13-78-008, 455 pp.
Ravet, J. L., Brett, M. T. and Müller-Navarra, D. C. (2003) A test of the role of polyunsaturated fatty acids in phytoplankton food quality for *Daphnia* using liposome supplementation. *Limnol. Oceanogr.*, **48**, 1938–47.
Rawson, D. S. (1945) The experimental introduction of smallmouth bass into lakes of the Prince Albert National Park, Saskatchewan. *Trans. Am. Fish. Soc.*, **73**, 19–31.
Rawson, D. S. (1959) *Limnology and Fisheries of Cree and Wollaston Lakes in Northern Saskatchewan* Saskatchewan Dept. of Nat. Res., Fisheries Rept. No. 4.
Raymont, J. E. (1963) *Plankton and Productivity in the Oceans* Pergamon Press, Elmsford, NY.
Reckhow, K. H. and Chapra, S. C. (1983) *Data Analysis and Empirical Modeling Engineering Approaches for Lake Management* Vol. 1, Butterworths, Boston.
Redfield, A. C., Ketchum, B. H. and Richards, F. A. (1963) The influence of organisms on the composition of sea water, in *The Sea*, Vol. 2 (ed. N. Hill), Interscience, pp. 26–77.
Reinart, R. E. (1970) Pesticide concentrations in Great Lakes fish. *Pesticides Monitoring J.*, **3**, 233–40.
Repavich, W. M., Sonzogni, W. C., Standridge, J. H., Wedepohl, R. E. and Meisner, L. F. (1990) Cyanobacteria (blue-green algae) in Wisconsin waters: Acute and chronic toxicity. *Water Res.*, **24**, 225–31.
Resh, V. H. and Unzicker, J. D. (1975) Water quality monitoring and aquatic organisms: the importance of species identification. *J. Water Pollut. Control Fed.*, **47**, 9–19.
Revelle, P. and Revelle, C. (1988) *The Environment: Issues and Choices for Society*. Jones and Bartlett, Boston.

Reynolds, C. S. (1975) Interrelations of photosynthetic behavior and buoyancy regulation in a natural population of blue-green alga. *Freshwater Biol.*, **5**, 323–38.

Reynolds, C. S. (1984) *The Ecology of Freshwater Phytoplankton* Cambridge University Press, Cambridge.

Reynolds, C. S. (1987) Cyanobacterial water blooms, in *Advances in Botanical Research*, Vol. 3 (ed. P. Callow), Academic Press, London, pp. 67–143.

Reynolds, C. S., Jaworski, G. H. M., Cmiech, H. A. and Leedale, G. F. (1981) On the annual cycle of the blue-green alga *Microcystis aeruginosa* Kutz emend. Elenkin. *Phil. Trans. Soc. Lond. B*, **293**, 419–77.

Reynolds, C. S., Oliver, R. L. and Walsby, A. E. (1987) Cyanobacterial dominance: The role of buoyancy regulation in dynamic lake environments. *N. Z. J. Marine Freshwater Res.*, **21**, 379–90.

Rhee, G. Y. (1978) Effects of N:P atomic ratios and nitrate limitations on algal growth, cell composition and nitrate uptake. *Limnol. Oceanogr.*, **23**, 10–25.

Ricciardi, A. and MacIsaac, H. J. (2000) Recent mass invasion of the North American Great Lakes by Ponto-Caspian species. *Trends Ecol. Evol.*, **15**, 62–5.

Ricciardi, A., Neves, R. J. and Rasmussen, J. B. (1998) Impending extinctions of North American freshwater mussels (Unionidae) following the zebra mussel (*Dreissena polymorpha*) invasion. *J. Anim. Ecol.*, **67**, 613–9.

Rich, P. H. and Wetzel, R. G. (1978) Detritus in lake ecosystems. *Am. Mid. Nat.*, **112**, 57–71.

Richardson, C. J. and Qian, S. S. (1999) Long-term phosphorus assimilative capacity by freshwater wetlands: A new paradigm for sustaining ecosystem structure and function. *Environ. Sci. Technol.*, **33**, 1545–51.

Rickert, D. A., Petersen, R. R., McKenzie, S. W., Hines, W. G. and Wille, S. A. (1977) *Algal Conditions and the Potential for Future Algal Problems in the Willamette River*, Oregon US Geological Circular 715-G, 39 pp.

Riley, E. T. and Prepas, E. E. (1984) Role of internal phosphorus loading in two shallow lakes in Alberta, Canada. *J. Fish. Aquat. Sci.*, **41**, 845–55.

Ringler, N. H. and Hall, J. D. (1975) Effects of logging on water temperature and dissolved oxygen in spawning beds. *Trans. Am. Fish. Soc.*, **104**, 111–21.

Ripl, W. (1976) Biochemical oxidation of polluted lake sediments with nitrate – a new lake restoration method. *Ambio*, **5**, 132–5.

Ripl, W. (1986) Internal phosphorus recycling mechanisms in shallow lakes. *Lake Reservoir Management*, **2**, 138–42.

Robel, R. J. (1961) Water depth and turbidity in relation to growth of sago pondweed. *J. Wildlife Management*, **25**, 436–8.

Rock, C. A. (1974) The trophic status of Lake Sammamish and its relationship to nutrient income. PhD dissertation, University of Washington, Seattle.

Rodgers, G. K. (1966) *The Thermal Bar in Lake Ontario, Spring 1965 and Winter 1965–66* Great Lakes Res. Div., University of Michigan, Pub. No. 15, pp. 369–74.

Rodgers-Gray, T. P., Jobling, S., Morris, S., Kelly, C., Kirby, S., Janbakhsh, A., Harries, J. E., Waldock, M. J., Sumpter, J. P. and Tyler, C. R. (2000) Long-term temporal changes in the estrogenic composition of treated sewage effluent and its biological effects on fish. *Environ. Sci. Technol.*, **34**, 1521–8.

Rodhe, W. (1948) Environmental requirements of freshwater plankton algae. *Experimental Studies in the Ecology of Phytoplankton* Symbol. Bot. Upsalien 10, 149 pp.
Rodhe, W. (1966) Standard correlations between pelagic photosynthesis and light, in *Primary Productivity in Aquatic Environments* (ed. C. R. Goldman), University of California Press, pp. 365–82.
Rodhe, W. (1969) Crystallization of eutrophication concepts in northern Europe, in *Eutrophication: Causes, Consequences, Correctives* National Academy of Science, Washington, DC.
Roelofs, T. D. and Ogelsby, R. T. (1970) Ecological observations on the planktonic cyanophyte *Gleotrichia echinulata. Limnol. Oceanogr.*, **15**, 224–9.
Rorslett, B., Berge, D. and Johansen, S. W. (1986) Lake enrichment by submersed macrophytes: a Norwegian whole-lake experience with *Elodea canadensis. Aquat. Bot.*, **26**, 325–40.
Rother, J. A. and Fay, P. (1977) Sporulation and the development of planktonic blue-green algae in two Salopian meres. *Proc. R. Soc. Lond. B*, **196**, 317–32.
Russell-Hunter, W. D. (1970) *Aquatic Productivity* Macmillan, New York.
Ryding, E. (1996) Experimental studies simulating potential phosphorus release from municipal sewage sludge deposits. *Water Res.*, **30**, 1695–1701.
Ryding, S.-O. (1985) Chemical and microbiological processes as regulators of the exchange of substances between sediments and water in shallow, eutrophic lakes. *Int. Rev. Ges. Hydrobiol.*, **70**, 657–702.
Ryding, S.-O. and Forsberg, C. (1976) Six polluted lakes: a preliminary evaluation of the treatment and recovery processes. *Ambio*, **5**, 151–6.
Ryther, J. H. (1960) Organic production by planktonic algae and its environmental control, in *The Ecology of Algae* Spec. Pub. No. 2, Pymatuning Lab. of Field Biol., University of Pittsburgh, pp. 72–83.
Ryther, J. H. and Dunstan, W. M. (1971) Nitrogen, phosphorus, and eutrophication in the coastal marine environment. *Science*, **171**, 1008–13.
Saether, O. A. (1979) Chironomid communities as water quality indicators. *Holarctic Ecol.*, **2**, 65–74.
Sakamoto, M. (1971) Chemical factors involved in the control of phytoplankton production in the Experimental Lakes Area, northwestern Ontario. *J. Fish Res. Board Canada*, **28**, 203–13.
Salonen, K., Jones, R. I. and Arvola, L. (1984) Hypolimnetic phosphorus retrieval by diel vertical migrations of lake phytoplankton. *Freshwater Biol.*, **14**, 431–8.
Sargent, J. R., Bell, J. G., Bell, M. V., Henderson, R. J. and Tocher, D. R. (1995) Requirement criteria for essential fatty acids. *J. Appl. Ichthyol.*, **11**, 183–98.
Sarnelle, O. (1986) Field assessment of the quality of phytoplanktonic food available to *Daphnia* and *Bosmina. Hydrobiologia*, **131**, 47–56.
Sarnelle, O. (1996) Predicting the outcome of trophic manipulation in lakes – a comment on Harris (1994). *Freshwater Biol.*, **35**, 339–42.
Sas, H. (1989) *Lake Restoration by Reduction of Nutrient Loading: Expectations, Experiences, Extrapolations.* Academia-Verlag Richarz, St Augustin.
Savage, N. L. and Rabe, F. W. (1973) The effects of mine and domestic wastes on macroinvertebrate community structure in the Coeur d' Alene River. *Northwest Sci.*, **47**, 159–68.

Scheffer, M. (1998) *Ecology of Shallow Lakes*, Chapman & Hall, New York, NY, 357pp.
Scheffer, M., Hosper, S. H., Meijer, M. L., Moss, B. and Jeppesen, E. (1993) Alternative equilibria in shallow lakes. *Trends Ecol. Evol.*, **8**, 275–9.
Schell, W. R. (1976) *Biogeochemistry of Radionuclides in Aquatic Environments*. Annual progress report to Energy Research and Development Administration RLO-2225-T18-18.
Schelske, C. L. and Stoermer, E. F. (1971) Eutrophication, silica depletion, and predicted changes in algal quality in Lake Michigan. *Science*, **173**, 423–4.
Schelske, C. L., Stoermer, E. F., Fahnenstiel, G. L. and Haibach, M. (1986) Phosphorus enrichment, silica utilization and biogeochemical silica depletion in the Great Lakes. *Can J. Fish. Aquat. Sci.*, **43**, 407–15.
Schiff, J. A. (1964) Protists, pigments, and photosynthesis, in *Principles and Applications in Aquatic Microbiology* (eds Heukelekian, H. and Doudero, N. C.), John Wiley & Sons, New York, pp. 298–313.
Schindler, D. W. (1971a) A hypothesis to explain the differences and similarities among lakes in the Experimental Lakes Area, northwestern Ontario. *J. Fish Res. Board Canada*, **28**, 295–301.
Schindler, D. W. (1971b) Carbon, nitrogen, and phosphorus and the eutrophication of freshwater lakes. *J. Phycol.*, **7**, 321–9.
Schindler, D. W. (1974) Eutrophication and recovery in experimental lakes: implications for lake management. *Science*, **184**, 897–99.
Schindler, D. W. (1986) The significance of in-lake production of alkalinity. *Water, Air Soil Pollut.*, **18**, 259–71.
Schindler, D. W. (1997) Widespread effects of climatic warming on freshwater ecosystems in North America. *Hydrol. Proc.*, **11**, 1043–67.
Schindler, D. W. and Fee, E. J. (1974) Experimental lakes area, whole lake experiments in eutrophication. *J. Fish Res. Board Canada*, **31**, 937–53.
Schmitt, M. R. and Adams, M. S. (1981) Dependence of rates of apparent photosynthesis on tissue phosphorus concentrations in *Myriophyllum spicatum* L. *Aquat. Bot.*, **11**, 379–87.
Schofield, C. L. and Trojnar, J. R. (1980) Aluminium toxicity to brook trout (*Salvelinus fontinalis*) in acidified waters, in *Polluted Rain* (eds T. Y. Toribara, M. W. Miller and P. E. Morrow), Plenum Press, New York, pp. 341–63.
Schriver, P., Bøgestrand, J., Jeppesen, E. and Søndergaard, M. (1995) Impact of submerged macrophytes on fish–zooplankton–phytoplankton interactions: Large-scale enclosure experiments in a shallow eutrophic lake. *Freshwater Biol.*, **33**, 255–70.
Schultz, D. W. and Malueg, K. W. (1971) Uptake of radiophosphorus by rooted aquatic plants. *Proc. 3rd Natl Symposium on Radioecology*, pp. 417–24.
Schumacher, G. J. and Whitford, L. A. (1965) Respiration and ^{22}P uptake in various species of freshwater algae as affected by current. *J. Phycol.*, **1**, 78–80.
Seehausen, O., van Alphen, J. J. M. and Witte, F. (1997) Cichlid fish diversity threatened by eutrophication that curbs sexual selection. *Science*, **277**, 1808–11.
Seip, K. L. (1994) Phosphorus and nitrogen limitation of algal biomass across trophic gradients. *Aquat. Sci.*, **56**, 16–28.
Seip, K. L., Sas, H. and Vermij, S. (1992) Nutrient-chlorophyll trajectories across trophic gradients. *Aquat. Sci.*, **54**, 58–76.

Seligman, K., Enos, A. K. and Lai, H. H. (1992) A comparison of 1988–1990 flavor profile analysis results with water conditions in two northern California reservoirs. *Water Sci. Technol.*, **25**, 19–25.

Shannon, C. E. and Weaver, W. (1948) *The Mathematical Theory of Communication* University of Illinois Press, Urbana.

Shapiro, J. (1970) A statement on phosphorus. *J. Water Pollut. Control Fed.*, **42**, 772–5.

Shapiro, J. (1973) Blue-green algae: why they become dominant. *Science*, **179**, 382–4.

Shapiro, J. (1979) The importance of trophic level interactions to the abundance and species composition of algae in lakes. *Dev. Hydrobiol.*, **2**, 161–7.

Shapiro, J. (1984) Blue-green dominance in lakes: the role and management significance of pH and CO_2. *Int. Rev. Ges. Hydrobiol.*, **69**, 765–80.

Shapiro, J. (1990) Current beliefs regarding dominance by blue-greens: the case for the importance of CO_2 and pH. *Verh. Int. Verein Limnol.*, **24**, 38–54.

Shapiro, J. (1997) The role of carbon dioxide in the initiation and maintenance of blue-green dominance in lakes. *Freshwater Biol.*, **37**, 307–23.

Shapiro, J., Forsberg, B., LaMarra, V., Lindmark, G., Lynch, M., Smeltzer, E. and Zoto, G. (1982) *Experiments and Experiences in Bio-Manipulation*. Interim Rept. No. 19, Limnological Res. Center, Univ. of Minnesota, EPA-600/3-82-096.

Shapiro, J., Lamarra, V. and Lynch, M. (1975) Biomanipulation: An ecosystem approach to lake restoration, in *Water Quality Management through Biological Control* (eds Brezonik, P. L. and Fox, J. L.), Department of Environmental Engineering Sciences, University of Florida, Gainesville, pp. 85–96.

Shapiro, J. and Wright, D. I. (1984) Lake restoration by biomanipulation: Round Lake, Minnesota, the first two years. *Freshwater Biol.*, **14**, 371–83.

Sherman, R. E., Gloss, S. P. and Lion, L. W. (1987) A comparison of toxicity tests conducted in the laboratory and in experimental ponds using cadmium and the fathead minnow (*Pimephales promelas*). *Water Res.*, **21**, 317–23.

Shin, E. and Krenkel, P. A. (1976) Mercury uptake by fish and biomethylation mechanisms. *J. Water Pollut. Control Fed.*, **48**, 473–501.

Shumway, D. L., Warren, C. E. and Doudoroff, P. (1964) Influence of oxygen concentration and water movement on the growth of steelhead trout and coho salmon embryos. *Trans. Am. Fish. Soc.*, **93**, 342–56.

Shuster, J. I., Welch, E. B., Horner, R. R. and Spyridakis, D. E. (1986) Response of Lake Sammamish to urban runoff control. *Lake Reservoir Management*, **2**, 229–34.

Siefert, R. E. and Spoor, W. A. (1973) Effects of reduced oxygen on embryos and larvae of the white sucker, coho salmon, brook trout and walleye, in *The Early Life History of Fish* (ed. Baxter, J. H. S.), Proc. of Symposium, Marine Biol. Assn., Avon, Scotland, Springer-Verlag, Heidelberg, Germany, pp. 487–95.

Silvey, J. K., Henley, D. E. and Wyatt, J. T. (1972) Planktonic blue-green algae: growth and odor-production studies. *J. Am. Water Works Assoc.*, **64**, 35–9.

Simons, T. J. (1973) *Development of Three-Dimensional Numerical Models of the Great Lakes* Environ. Canada Water Mgt., Scientific Series No. 12, Environment Canada, Ottawa.

Singh, H. N. and Sunita, K. M. (1974) A biochemical study of spore germination in the blue-green alga *Anabaena doliolum*. *J. Exp. Bot.*, **25**, 837–45.

Sivonen, K. and Jones, G. (1999) Cyanobacterial toxins, in *Toxic Cyanobacteria in Water: A Guide to their Public Health Consequences, Monitoring and Management*

(eds I. Chorus and J. Bartram), Published on behalf of WHO by E & FN Spon, London, pp. 41–111.
Sivonen, K., Niemela, S. I., Niemi, R. M., Lepisto, L., Luoma, T. H. and Rasanen, L. A. (1990) Toxic cyanobacteria (blue-green algae) in Finnish fresh and coastal waters. *Hydrobiologia*, **190**, 267–75.
Skulberg, O. M. (1965) Algal cultures as a means to assess the fertilizing influence of pollution. *Adv. Water Pollut. Res.*, **1**, 113–27.
Sládeček, V. (1973) The reality of three British biotic indices. *Water Res.*, **7**, 995–1002.
Sládečková, A. and Sládeček, V. (1963) Periphyton as indicator of the reservoir water quality. I. True periphyton. Prague, *Technol. Water*, **7**, 507–61.
Smayda, T. J. (1974) Bioassay of the growth potential of the surface water of lower Narragansett Bay over an annual cycle using the diatom *Thalassiosira pseudonana* (oceanic clone, 13–1). *Limnol. Oceanogr.*, **19**, 889–901.
Smith, B. W. and Reeves, W. C. (1986) Stocking warm-water species to restore or enhance fisheries, in *Fish Culture in Fisheries Management* (ed. R. H. Stroud), American Fisheries Society, Bethesda, MD, pp. 17–29.
Smith, C. S. and Adams, M. S. (1986) Phosphorus transfer from sediments by *Myriphyllum spicatum*. *Limnol. Oceanogr.*, **31**, 1312–21.
Smith, L. L., Jr., Kramer, R. H. and MacLeod, J. C. (1965) Effects of pulpwood fibers on fathead minnows and walleye fingerlings. *J. Water Pollut. Conf. Fed.*, **37**, 130–40.
Smith, M. W. (1969) Changes in environment and biota of a natural lake after fertilization. *J. Fish Res. Board Canada*, **26**, 3101–32.
Smith, S. V. (1984) Phosphorus versus nitrogen limitation in the marine environment. *Limnol. Oceanogr.*, **29**, 1149–60.
Smith, V. H. (1982) The nitrogen and phosphorus dependence of algal biomass in lakes: an empirical and theoretical analysis. *Limnol. Oceanogr.*, **27**, 1101–12.
Smith, V. H. (1983) Low nitrogen to phosphorus ratios favor dominance by blue green algae in lake phytoplankton. *Science*, **221**, 669–71.
Smith, V. H. (1986) Light and nutrient effects on the relative biomass of blue-green algae in lake phytoplankton. *Can. J. Fish. Aquat. Sci.*, **43**, 148–53.
Smith, V. H. (1990a) Nitrogen, phosphorus, and nitrogen fixation in lacustrine and estuarine ecosystems. *Limnol. Oceanogr.*, **35**, 1852–9.
Smith, V. H. (1990b) Effects of nutrients and non-algal turbidity on blue-green algal biomass in four North Carolina reservoirs. *Lake Reservoir Management*, **6**, 125–32.
Smith, V. H. and Shapiro, J. (1981) Chlorophyll-phosphorus relations in individual lakes: their importance to lake restoration strategies. *Environ. Sci. Technol.*, **15**, 444–51.
Smith, V. H., Sieber-Denlinger, J., deNoyelles, F., Jr., Campbell, S., Pan, S., Randtke, S. J., Blain, G. T. and Strasser, V. A. (2002) Managing taste and odor problems in a eutrophic drinking water reservoir. *Lake Reserv. Manage.*, **18**, 319–23.
Smith, W. E. (1970) Tolerance of *Mysis relicta* to thermal shock and light. *Trans. Am. Fish. Soc.*, **99**, 418–21.
Soballe, D. M. and Kimmel, B. L. (1987) A large-scale comparison of factors influencing phytoplankton abundance in rivers, lakes and impoundments. *Ecology*, **68**, 1943–54.
Soballe, D. M. and Threlkeld, S. T. (1985) Advection, phytoplankton biomass, and nutrient transformations in a rapidly flushed impoundment. *Arch. Hydrobiol.*, **105**, 187–203.

Sodergren, A., Bengtsson, B. E., Jonsson, P., Lagergren, S., Larson, A., Olsson, M. and Renberg, L. (1988) Summary of results from the Swedish project 'Environment/Cellulose'. *Water Sci. Technol.*, **20**, 49–60.

Solbe, J. F., de, L. G. and Shurben, D. G. (1989) Toxicity of ammonia to early life stages of rainbow trout (*Salmo gairdneri*). *Water Res.*, **23**, 127–9.

Søndergaard, M. (1988) Seasonal variations in the loosely sorbed phosphorus fraction of the sediment of a shallow and hypereutrophic lake. *Environ. Geol. Water Sci.*, **11**, 115–21.

Søndergaard, M. and Moss, B. (1998) Impact of submerged macrophytes on phytoplankton in shallow freshwater lakes, in *The Structuring Role of Submerged Macrophytes in Lakes* (eds E. Jeppesen, M. Søndergaard, M. Søndergaard and K. Christoffersen), Springer-Verlag, New York, NY, pp. 115–32 (Chapter 6).

Søndergaard, M., Jensen, J. P. and Jeppesen, E. (1999) Internal phosphorus loading in shallow Danish lakes. *Hydrobiologia*, **408/409**, 145–52.

Søndergaard, M., Jeppesen, E., Jensen, J. P. and Lauridsen, T. (2000) Lake restoration in Denmark. *Lakes Reserv. Res. Manage. 2000*, **5**, 151–9.

Søndergaard, M., Jeppesen, E., Mortensen, E., Dall, E., Kristensen, P. and Sortkjær, O. (1990) Phytoplankton biomass reduction after planktivorous fish reduction in a shallow eutrophic lake: a combined effect of reduced internal P-loading and increased zooplankton grazing. *Hydrobiologia*, **200/201**, 229–40.

Søndergaard, M., Kristensen, P. and Jeppesen, E. (1992) Phosphorus release from resuspended sediment in the shallow and wind-exposed Lake Arresø, Denmark, *Hydrobiologia*, **228**, 91–9.

Søndergaard, M., Wolter, K.-D. and Ripl, W. (2002) Chemical treatment of water and sediments with special reference to lakes, in *Handbook of Ecological Restoration, Volume 1, Principles of Restoration* (eds M. Perrow and A. J. Davy), Cambridge University Press, Cambridge, pp. 184–205.

Sonnichsen, J. D., Jacoby, J. and Welch, E. B. (1997) Response of cyanobacterial migration to alum treatment in Green Lake. *Arch. Hydrobiol.*, **140**, 373–92.

Sonzogni, W. C. and Lee, G. F. (1974) Diversion of wastewaters from Madison lakes. *J. Environ. Eng. Div.*, **100**, 153–70.

Sorokin, C. (1959) Tabular comparative data for the low- and high-temperature strains of *Chlorella*. *Nature (Lond.)*, **184**, 613–14.

Sorokin, C. and Krauss, R. W. (1962) Effects of temperature and illuminance on chlorella growth uncoupled from cell division. *Plant Physiol.*, **37**, 37–42.

Sosiak, A. J. (2002) Long-term response of periphyton and macrophytes to reduced municipal nutrient loading to the Bow River (Alberta, Canada). *Can. J. Fish. Aquat. Sci.*, **59**, 987–1001.

Spalding, M. G., Bjork, R. D., Powell, G. V. N. and Sundlof, S. F. (1994) Mercury and cause of death in great white herons. *J. Wildlife Manage.*, **58**, 735–9.

Spehar, R. L., Anderson, R. L. and Fiandt, J. T. (1978) Toxicity and bioaccumulation of cadmium and lead in aquatic invertebrates. *Environ. Pollut.*, **15**, 195–208.

Spence, D. H. N. (1967) Factors controlling the distribution of freshwater macrophytes with particular reference to the lochs of Scotland. *J. Ecol.*, **55**, 147–70.

Spencer, C. N. and King, D. L. (1984) Role of fish in regulation of plant and animal communities in eutrophic ponds. *Can. J. Fish. Aquat. Sci.*, **41**, 1851–5.

Spencer, C. N. and King, D. L. (1987) Regulation of blue-green algal buoyancy and bloom formation by light, inorganic nitrogen, CO_2, and trophic interactions. *Hydrobiologia*, **144**, 183–92.

Spencer, C. N., McClelland, B. R. and Stanford, J. A. (1991) Shrimp introduction, salmon collapse, and bald eagle displacement: cascading interactions in the food web of a large aquatic ecosystem. *Bioscience*, **41**, 14–21.

Spencer, D. F. (1990) Influence of organic sediment amendments on growth and tuber production by sago pondweed (*Potamogeton pectinatus* L.). *Freshwater Ecol.*, **5**, 255–63.

Spry, D. J. and Wiener, J. G. (1991) Metal bioavailability and toxicity to fish in low-alkalinity lakes: a critical review. *Environ. Pollut.*, **71**, 243–304.

Stadelman, P. (1980) *Der Zustand des Rotsees bei Luzern, Kantonales amt fur Gewasserschutz*, Luzern.

Stanford, J. A. and Ward, J. V. (1988) The hyporheic habitat of river ecosystems. *Nature (Lond.)*, **335**, 64–6.

Stauffer, R. W. and Lee, G. F. (1973) The role of thermocline migration in regulating algal blooms, in *Modeling the Eutrophication Process* (eds E. J. Middlebrooks, D. H. Falkenborg and T. E. Maloney), Proc. Workshop at Utah State University, Logan, pp. 73–82.

Steele, J. H. (1962) Environmental control of photosynthesis in the sea. *Limnol. Oceanogr.*, **7**, 137–50.

Stefan, H. G. and Hanson, M. J. (1981) Phosphorus recycling in five shallow lakes. *J. Am. Soc. Civil Eng., EE Div.*, **107**, 713–30.

Stensen, J. A. E., Bohlin, T., Henriksen, L., Nilsson, B. I., Oscarson, A. G. and Larsson, P. (1978) Effects of fish removal from a small lake. *Verh. Int. Verein. Limnol.*, **20**, 794–801.

Stepczuk, C., Martin, A. B., Effler, S. W., Bloomfield, J. A. and Auer, M. T. (1998a) Spatial and temporal patterns of THM precursors in a eutrophic reservoir. *Lake Reserv. Manage.*, **14**, 356–66.

Stepczuk, C., Martin, A. B., Longabucco, P., Bloomfield, J. A. and Effler, S. W. (1998b) Allochthonous contributions of THM precursors to a eutrophic reservoir. *Lake Reserv. Manage.*, **14**, 344–55.

Sterner, R. W. and Hessen, D. O. (1994) Algal nutrient limitation and the nutrition of aquatic herbivores. *Annu. Rev. Ecol. Syst.*, **25**, 1–25.

Stevenson, J. (1999) Personal Communication with E. B. Welch, Department of Zoology, Michigan State University, Michigan.

Stewart, K. M. (1976) Oxygen deficits, clarity, and eutrophication in some Madison Lakes. *Int. Rev. Ges. Hydrobiol.*, **61**, 563–79.

Stewart, N. E., Shumway, D. L. and Doudoroff, P. (1967) Influence of oxygen concentration on the growth of juvenile largemouth bass. *J. Fish Res. Board Canada*, **24**, 475–94.

Stockner, J. G. (1972) Paleolimnology as a means of assessing eutrophication. *Verh. Int. Verein. Limnol.*, **18**, 1018–30.

Stockner, J. G. and Shortreed, K. R. S. (1978) Enhancement of autotrophic production by nutrient addition in a coastal rain-forest stream on Vancouver Island. *J. Fish Res. Board Canada*, **35**, 28–34.

Stockner, J. G. and Shortreed, K. S. (1985) Whole-lake fertilization experiments in coastal British Columbia lakes: empirical relationships between nutrient inputs and phytoplankton biomass and production. *Can. J. Fish. Aquat. Sci.*, **42**, 649–58.

Stockner, J. G. and Shortread, K. S. (1988) Response of *Anabaena* and *Synechococcus* to manipulation of nitrogen: phosphorus ratios in a lake fertilization experiment. *Limnol. Oceanogr.*, **33**, 1348–61.

Storr, J. F. and Sweeney, R. A. (1971) Development of a theoretical seasonal growth response curve of *Cladophora glomerata* to temperature and photoperiod. *Proc. 14th Conf. Great Lakes Res.*, **14**, 119–27.

Stumm, W. (1963) *Proc. Int. Conf. Water Poll. Res.* Vol. 2. Pergamon Press, Elmsford, NY.

Stumm, W. and Leckie, J. O. (1971) Phosphate exchange with sediments: its role in the productivity of surface waters. *Proc. 5th Int. Conf. Water Pollut. Res.* III-26/1-16, Pergamon Press, Elmsford, NY.

Sumpter, J. P. (1998) Xenoendocrine disruptors – environmental impacts. *Toxicol. Lett.*, **103**, 337–42.

Sundborg, A. (1956) *The River Klarälven. A Study of Fluvial Processes*. Geografiska Annaler Häfte 2–3/1956, University of Uppsala, Sweden, pp. 127–316.

Sverdrup, H. U. (1953). On conditions for the vernal blooming of phytoplankton. *J. Conseil.*, **18**, 287–95.

Sverdrup, H. U., Johnson, M. W. and Fleming, R. H. (1942) *The Oceans, their Physics, Chemistry, and General Biology*, Prentice Hall, Englewood Cliffs, NJ.

Sweeney, B. W., Jackson, J. K., Newbold, J. D. and Funk, D. H. (1992) Climate change and the life histories and biogeography of aquatic insects in eastern North America, in *Global Climate Change and Freshwater Ecosystems* (eds P. Firth and S. G. Fisher), Springer-Verlag, New York, NY, pp. 143–76.

Szumski, D. S., Barton, D. A., Putnam, H. D. and Polta, R. C. (1982) Evaluation of EPA un-ionized ammonia toxicity criteria. *J. Water Pollut. Control Fed.*, **54**, 281–91.

Talling, J. F. (1957a) Photosynthetic characteristics of some freshwater plankton diatoms in relation to underwater radiation. *New Phytol.*, **56**, 29–50.

Talling, J. F. (1957b) The phytoplankton population as a compound photosynthetic system. *New Phytol.*, **56**, 133–49.

Talling, J. F. (1965) The photosynthetic activity of phytoplankton in East African lakes. *Int. Rev. Ges. Hydrobiol. Hydrograph*, **50**, 1–32.

Talling, J. F. (1971) The underwater light climate as a controlling factor in the production ecology of freshwater phytoplankton. *Mitt. Int. Verein. Limnol.*, **19**, 214–43.

Taylor, E. W. (1966) *Forty-Second Report on the Results of the Bacteriological Examinations of the London Waters for the Years 1965–6*, Metropolitan Water Board, New River Head, London.

Templeton, W. L., Becker, C. D., Berlin, J. D., Coutant, C. C., Dean, J. M., Fujihara, M. P., Nakatani, R. E., Olson, P. A., Prentice, E. F. and Watson, D. G. (1969) *Biological Effects of Thermal Discharges. Progress Report for 1968* AEC Research and Development Report, BNWL-1050 reprint, Battelle Northwest Lab, Richland, WA.

Tett, P., Gallegos, C., Kelly, M. G., Hornberger, G. M. and Cosby, G. J. (1978) Relationships among substrate, flow and benthic microalgal pigment density in the Mechuma River, Virginia. *Limnol. Oceanogr.*, **3**, 785–97.

Thomas, E. A. (1969) The process of eutrophication in Central European lakes, in *Eutrophication: Causes, Consequences and Correctives*. Natl Acad. Sci., Washington, DC.

Thomas, R. H. and Walsby, A. E. (1986) The effect of temperature on recovery of buoyancy by *Microcystis*. *J. Gen. Microbiol.*, **132**, 1665–72.
Thornton, J. A. (1987a) Aspects of eutrophication management in tropical/subtropical regions. *J. Limnol. Soc. S. Africa*, **13**, 25–43.
Thornton, J. A. (1987b) A review of some unique aspects of limnology of shallow southern African man-made lakes. *Geojournal*, **14**, 339–52.
Thut, R. N. (1969) A study of the profundal bottom fauna of Lake Washington. *Ecol. Monogr.*, **39**, 79–100.
Todd, D. K. (ed.) (1970) *The Water Encyclopedia. A Compendium of Useful Information on Water Resources*, Water Information Center, Port Washington, NY.
Townsend, C. R. and Hildrew, A. G. (1994) Species traits in relation to a habitat templet for river systems. *Freshwater Biol.*, **31**, 265–75.
Townsend, C. R., Scarsbrook, M. R. and Dolédec, S. (1997) Quantifying disturbance in streams: alternative measures of disturbance in relation to macroinvertebrate species traits and species richness. *J. N. Am. Benthol. Soc.*, **16**, 531–44.
Traaen, T. S. and Lindström, E. A. (1983) Influence of current velocity on periphyton distribution, in *Periphyton of Freshwater Ecosystems* (ed. R. G. Wetzel), *Developments in Hydrobiology*, Vol. 17, pp. 97–9.
Trembly, F. J. (1960) *Research Project on Effects of Condenser Discharge Water on Aquatic Life*. Progress Report 1956–9. Inst. of Research, Lehigh University, Bethlehem, PA.
Trimbee, A. M. and Harris, G. P. (1984) Phytoplankton population dynamics of a small reservoir: use of sedimentation traps to quantify the loss of diatoms and recruitment of summer bloom-forming blue-green algae. *J. Plankton Res.*, **6**, 897–918.
Trimbee, A. M. and Prepas, E. E. (1988) The effect of oxygen depletion on the timing and magnitude of blue-green algal blooms. *Verh. Int. Verein. Limnol.*, **23**, 220–6.
Trussell, R. R. (2001) Endocrine disrupters and the water industry. *J. Am. Water Works Assoc.*, **93**, 58–65.
Tsai, C. (1973) Water quality and fish life below sewage outfalls. *Trans. Am. Fish. Soc.*, **102**, 281–92.
Turback, S. C., Olson, S. B. and McFeters, G. A. (1986) Comparison of algal assay systems for detecting waterborne herbicides and metals. *Water Res.*, **20**, 91–6.
Tyler, C. R., Jobling, S. and Sumpter, J. P. (1998) Endocrine disruption in wildlife: a critical review of the evidence. *Crit. Rev. Toxicol.*, **28**, 319–61.
Uhlmann, D. (1971) Influence of dilution, sinking and grazing rate on phytoplankton populations of hyperfertilized ponds and microecosystems. *Mitt. Inte. Verein. Limnol.*, **19**, 100–24.
UK Environment Agency (1998) *Aquatic Eutrophication in England and Wales*. Environmental Issues Series Consultative Report.
United Nations (2001) *World Population Prospects: The 2000 Revision*. Population Division of the Department of Economic and Social Affairs of the United Nations Secretariat. http://esa.un.org/unpp (accessed 18 June 2003).
USEPA (1982) Secondary treatment information, in *Code of Federal Regulations* Vol. 40, US Government Printing Office, Washington, DC.
USEPA (1983) *Assessing Water Quality in Streams* Environmental Monitoring Systems Laboratory, Las Vegas, NV.

USEPA (1987) *Clean Lakes Program* 1987 Annual Report, North American Lake Management Society for USEPA.
USEPA (1996) *National Water Quality Inventory: 1996 Report*. US Environmental Protection Agency, Office of Water, EPA-841-R-97-008, Washington, DC.
USEPA (1998a) *Lake and Reservoir Bioassessment and Biocriteria – Technical Guidance Document*. US Environmental Protection Agency, Office of Water, EPA 841-B-98-007, Washington, DC.
USEPA (1998b) *National Strategy for the Development of Regional Nutrient Criteria*. U.S. Environmental Protection Agency, Office of Water, EPA-822-R-98-002, Washington, DC.
USEPA (1999) *Ammonia Fact Sheet: 1999 Update*. US Environmental Protection Agency, Office of Water, EPA-823-F-99-024, Washington, DC.
USEPA (2000) *Nutrient Criteria Technical Guidance Manual – Rivers and Streams*. US Environmental Protection Agency, Office of Water, EPA-822-B-00-002, Washington, DC.
USEPA (2001a) *Mercury Update: Impact on Fish Advisories*. U.S. Environmental Protection Agency, Office of Water, EPA-823-F-01-011, Washington, DC.
USEPA (2001b) *Revision to the Interim Enhanced Surface Water Treatment Rule (IESWTR), the Stage 1 Disinfectants and Disinfectant By-Products Rule (Stage 1 DBPR), and Revision to State Primacy Requirements and Implementation of the Safe Drinking Water Act (SDWA)*. Final Rule. Federal Register, Vol. 66, No. 10, 16 January 2001.
USEPA (2002a) *National Recommended Water Quality Criteria: 2002*. US Environmental Protection Agency, Office of Water, EPA-822-R-02-047, Washington, DC.
USEPA (2002b) *National Water Quality Inventory: 2000 Report*. US Environmental Protection Agency, Office of Water, EPA-841-R-02-001, Washington, DC.
USEPA/ILSI (1993) *A Review of Evidence on Reproductive and Developmental Effects of Disinfection Byproducts in Drinking Water*. US Environmental Protection Agency and International Life Sciences Institute, Washington, DC.
Usinger, R. L. (1956) *Aquatic Insects of California*. University of California Press, Berkeley.
Vallentyne, J. (1972) Nutrients and Eutrophication. Special Symposium, *Limnol. Oceanog.*, **1**, 107.
Vance, B. D. (1965) Composition and succession of cyanophycean water blooms. *J. Phycol.*, **1**, 81–6.
Vander Zanden, M. J., Casselman, J. M. and Rasmussen, J. B. (1999) Stable isotope evidence for the food web consequences of species invasions in lakes. *Nature*, **401**, 464–7.
van Donk, E., Grimm, M. P., Gulati, R. D. and Klein Breteler, J. P. G. (1990) Whole-lake food-web manipulation as a means to study community interactions in a small ecosystem. *Hydrobiologia*, **200/201**, 275–89.
van Donk, E. and Gulati, R. D. (1995) Transition of a lake to turbid state six years after biomanipulation: mechanisms and pathways. *Water Sci. Technol.*, **32**, 197–206.
van Duin, E. H. S., Frinking, L. J., van Schaik, F. H. and Boers, P. C. M. (1998) First results of the restoration of Lake Geerplas. *Water Sci. Technol.*, **37**, 185–92.
Van Nieuwenhuyse, E. E. and Jones, J. R. (1996) Phosphorus–chlorophyll relationship in temperate streams and its variation with stream catchment area. *Can. J. Fish. Aquat. Sci.*, **53**, 99–105.

Vannote, R. L., Minshall, G. W., Cummins, K. W., Sedell, J. R. and Cushing, C. E. (1980) The river continuum concept. *Can. J. Fish. Aquat. Sci.*, **37**, 130–7.
van Steenderen, R. A., Scott, W. E. and Welch, D. I. (1988) *Microcystis aeruginosa* as an organohalogen precursor. *Water South Africa*, **14**, 59–62.
Van't Hoff, J. H. (1884) *Etudes de Dynamique Chimique*. Amsterdam.
Viessman, W. Jr. and Welty, C. (1985) *Water Management: Technology and Institutions* Harper & Row, New York.
Visser, P. M., Ibelings, B. W., Van der Veer, B., Koedood, J. and Mur, L. R. (1996) Artificial mixing prevents nuisance blooms of the cyanobacterium *Microcystis* in Lake Nieuwe Meer, The Netherlands. *Freshwater Biol.*, **36**, 435–50.
Vitousek, P. M., D'Antonio, C. M., Loope, L. L. and Westbrooks, R. (1996) Biological invasions as global environmental change. *Am. Sci.*, **84**, 468–78.
Vitousek, P. M., Mooney, H. A., Lubchenco, J. and Melillo, J. M. (1997) Human domination of Earth's ecosystems. *Science*, **277**, 494–9.
Vollenweider, R. A. (1968) *Scientific Fundamentals of the Eutrophication of Lakes and Flowing Waters, with Particular Reference to Nitrogen and Phosphorus as Factors in Eutrophication*, Paris, Rept. Organization for Economic Cooperation and Development. DASCSI/68.27, 182 pp.
Vollenweider, R. A. (1969a) Possibilities and limits of elementary models concerning the budget of substances in lakes. *Arch. Hydrobiol.*, **66**, 1–36.
Vollenweider, R. A. (1969b) A manual on methods for measuring primary production in aquatic environments. *Int. Biol. Program Handbook 12*, Blackwell Scientific Publications, Oxford, 213 pp.
Vollenweider, R. A. (1975) Input-output models with reference to the phosphorus-loading concept in limnology. *Schweizerische Zeitschrift Hydrol.*, **37**, 53–84.
Vollenweider, R. A. (1976) Advances in defining critical loading levels for phosphorus in lake eutrophication. *Mem. Inst. Ital. Idrobiol.*, **33**, 53–83.
Vollenweider, R. A. and Dillon, P. J. (1974) *The Application of the Phosphorus-Loading Concept to Eutrophication Research*, National Res. Council, Canada. Tech. Rept. 13690, 42 pp.
Vorburger, C. and Ribi, G. (1999) Aggression and competition for shelter between a native and an introduced crayfish in Europe. *Freshwater Biol.*, **42**, 111–19.
Wagner, K. J. (1986) Biological management of a pond ecosystem to meet water use objectives. *Lake Reservoir Management*, **2**, 54–9.
Walker, W. E., Jr., Westerberg, C. E., Schuler, D. J. and Bode, J. A. (1989) Design and evaluation of eutrophication control measures for the St. Paul water supply. *Lake Reserv. Manage.*, **5**, 71–83.
Walker, W. W. Jr. (1977) Some analytical methods applied to lake water quality problems. PhD dissertation, Harvard University.
Walker, W. W. Jr. (1981) *Empirical Methods for Predicting Eutrophication in Impoundments*. Report 1, Phase 1: Data base development. Technical Report E-81-9, EWQOS, US Army Corps of Engineers, Vicksburg, MS.
Walker, W. W. Jr. (1985) Statistical bases for mean chlorophyll *a* criteria. *Lake Reservoir Management*, **1**, 57–67.
Walker, W. W. Jr. (1987a) Empirical methods for predicting eutrophication in impoundments. Report 4, Phase III: Applications Manual. Tech. Rept. E-81-9. US Army Engineer Waterways Exp. Sta., Vicksburg, MS.

Walker, W. W. Jr. (1987b) Phosphorus removal by urban runoff detention basins. *Lake Reservoir Management*, **3**, 314–26.
Waller, K., Swan, S. H., DeLorenze, G. and Hopkins, B. (1998) Trihalomethanes in drinking water and spontaneous abortion. *Epidemiology*, **9**, 134–9.
Walshe, B. M. (1948) The oxygen requirements and thermal resistance of chironomed larvae from flowing and from still waters. *J. Exp. Biol.*, **25**, 35.
Walton, S. P. (1990) Effects of grazing by *Dicosmoecus gilvipes* (caddisfly) larvae and phosphorus enrichment on periphyton. MS thesis, University of Washington.
Walton, S. P., Welch, E. B. and Horner, R. R. (1995) Stream periphyton responses to grazing and changes in phosphorus concentration. *Hydrobiologia*, **302**, 31–46.
Warnick, S. L. and Bell, H. L. (1969) The acute toxicity of some heavy metals to different species of aquatic insects. *J. Water Pollut. Control Fed.*, **41**, 280–4.
Warren, C. E. (1971) *Biology and Water Pollution Control*. W. B. Saunders, Philadelphia.
Warren, C. E., Wales, J. H., Davis, G. E. and Doudoroff, P. (1964) Trout production in an experimental stream enriched with sucrose. *J. Wildlife Management*, **28**, 617–60.
Warwick, W. F. (1980) Paleolimnology of the Bay of Quinte, Lake Ontario: 2800 years of cultural influence. *Can. Bull. Fish. Aquat. Sci.*, **206**, 118 pp.
Washington Department of Wildlife (1990) *Grass Carp Use in Washington* Fish Management Division, FM No. 90-4.
Washington, H. G. (1984) Diversity, biotic and similarity indices, a review. *Water Res.*, **18**, 653–94.
Watanabe, M. F., Park, H.-D. and Watanabe, M. (1994) Compositions of *Microcystis* species and heptapeptide toxins. *Verh. Int. Verein. Limnol.*, **25**, 2226–9.
Watson, V. and Gestring, B. (1996) Monitoring algae levels in the Clark Fork River. *Intermountain J. Sci.*, **2**, 17–26.
Watson, V. J., Perlind, P. and Bahls, L. (1990) Control of algal standing crop by P and N in the Clark Fork River, in *Proceedings of the Clark Fork River Symposium*, Montana Academy of Science, Missoula, Montana.
Weibel, S. R. (1969) Urban drainage as a factor in eutrophication, in *Eutrophication: Causes, Consequences and Correctives*. Natl Acad. Sci., Washington, DC.
Welch, E. B. (1969a) Factors controlling phytoplankton blooms and resulting dissolved oxygen in Duwamish River estuary, Washington, US Geol. Survey Water Supply Paper 1873-A, 62 pp.
Welch, E. B. (1969b) Discussion of ecological changes of applied significance induced by the dischange of heated waters, in *Thermal Pollution, Engineering Aspects* (eds F. L. Parker and P. A. Krenkel), Vanderbilt University Press, Nashville, Tennessee, pp. 58–68.
Welch, E. B. (1988) In-lake eutrophication control: progress and limitations, in *Water Pollution Control in Asia* (eds T. Panswad, C. Polprasert and K. Yamamoto), Pergamon Press, New York, pp. 37–44.
Welch, E. B., Anderson, E. L., Jacoby, J. M., Biggs, B. J. F. and Quinn, J. M. (2000) Invertebrate grazing of filamentous green algae in outdoor channels. *Verh. Int. Verein. Limnol.*, **27**, 2408–14.
Welch, E. B., Barbiero, R. P., Bouchard, D. and Jones, C. A. R. (1992) Lake trophic state change and constant algal composition following dilution and diversion. *Ecol. Eng.*, **1**, 173–97.

Welch, E. B., Brenner, M. V. and Carlson, K. L. (1984) Control of algal biomass by inflow NO_3, in *Lake and Reservoir Management* EPA 440/5/84/001, pp. 493–7.

Welch, E. B., Buckley, J. A. and Bush, R. M. (1972) Dilution as an algal bloom control. *J. Water Pollut. Control Fed.*, **44**, 2245–65.

Welch, E. B. and Burke, T. (2001) *Lake Elevation and Water Quality, Upper Klamath Lake, Oregon.* Interim Summary Report. Prepared by R2 Resources, Inc., Redmond, WA. Prepared for the US Bureau of Indian Affairs.

Welch, E. B., Bush, R. M., Spyridakis, D. E. and Saikewicz, M. B. (1973) *Alternatives for Eutrophication Control in Moses Lake, Washington*, Dept of Civil Engineering, University of Washington, Seattle, 102 pp.

Welch, E. B. and Cooke, G. D. (1995) Internal phosphorus loading in shallow lakes: Importance and control. *Lake Reserv. Manage.*, **11**, 273–81.

Welch, E. B. and Cooke, G. D. (1999) Effectiveness and longevity of phosphorus inactivation with alum. *Lake Reserv. Manage.*, **15**, 5–27.

Welch, E. B., DeGasperi, C. L. and Spyridakis, D. E. (1988a) Sources for internal P loading in a shallow lake. *Verh. Int. Verein. Limnol.*, **23**, 307–14.

Welch, E. B., DeGasperi, C. L., Spyridakis, D. E. and Belnick, T. J. (1988b) Internal phosphorus loading and alum effectiveness in shallow lakes. *Lake Reservoir Management*, **4**, 27–34.

Welch, E. B., Horner, R. R. and Patmont, C. R. (1989b) Prediction of nuisance periphytic biomass: a management approach. *Water Res.*, **23**, 401–5.

Welch, E. B. and Jacoby, J. M. (2001) On determining the principle source of phosphorus causing summer algal blooms in western Washington lakes. *Lake Reserv. Manage.*, **17**, 55–65.

Welch, E. B., Jacoby, J. M., Horner, R. R. and Seeley, M. R. (1988c) Nuisance biomass levels of periphyton algae in streams. *Hydrobiologia*, **157**, 161–8.

Welch, E. B., Jacoby, J. M. and May, C. W. (1998) Stream quality, in *River Ecology and Management: Lessons from the Pacific Coastal Ecoregion* (eds R. J. Naiman and R. E. Bilby), Springer-Verlag, New York, pp. 69–94.

Welch, E. B. and Jones, C. A. (1990) Predicting phosphorus in a dendritic, shallow prairie lake following dilution and diversion. *Verh. Int. Verein. Limnol.*, **24**, 427.

Welch, E. B., Jones, C. A. and Barbiero, R. P. (1989a) *Moses Lake Quality: Results of Dilution, Sewage Diversion and BMPs — 1977 through 1988*, Dept. of Civil Eng., University of Washington, Water Resources Series Tech. Rept. No. 118.

Welch, E. B. and Kelly, T. S. (1990) Internal phosphorus loading and macrophytes: an alternative hypothesis. *Lake Reservoir Management*, **6**, 43–8.

Welch, E. B. and Patmont, C. R. (1980) Lake restoration by dilution: Moses Lake, Washington. *Water Res.*, **14**, 1316–25.

Welch, E. B. and Perkins, M. A. (1979) Oxygen deficit rate as a trophic state index. *J. Water Pollut. Control Fed.*, **51**, 2823–28.

Welch, E. B., Quinn, J. M. and Hickey, C. W. (1992) Periphyton biomass related to point-source nutrient enrichment in seven New Zealand streams. *Water Res.*, **26**, 669–75.

Welch, E. B., Rock, C. A., Howe, R. C. and Perkins, M. A. (1980) Lake Sammamish response to wastewater diversion and increasing urban runoff. *Water Res.*, **14**, 821–8.

Welch, E. B., Spyridakis, D. E., Shuster, J. I. and Horner, R. R. (1986) Declining lake sediment phosphorus release and oxygen deficit following wastewater diversion. *J. Water Pollut. Control Fed.*, **58**, 92–6.

Welch, E. B. and Wojtalik, T. A. (1968) *Some Aspects of Increased Water Temperature on Aquatic Life*, Tennessee Valley Authorily, Chattanooga, TN, 48pp.
Wetzel, R. G. (1964) A comparative study of the primary productivity of higher aquatic plants, periphyton, and phytoplankton in a large shallow lake. *Int. Rev. Geol. Hydrobiol.*, **49**, 1–64.
Wetzel, R. G. (1972) The role of carbon in hard-water marl lakes, in *Nutrients and Eutrophication: The limiting nutrient controversy* (ed. G. E. Likens), Sec. Symp. Amer. Soc. Limnol. Oceanogr., **1**, 84–91.
Wetzel, R. G. (1975, 1983) *Limnology*, W. B. Saunders, Philadelphia.
Wetzel, R. G. and Hough, R. A. (1973) Productivity and role of aquatic macrophytes in lakes: an assessment. *Pol. Arch. Hydrobiol.*, **20**, 9–19.
Wezernak, C. T., Lyzenga, D. R. and Polcyn, F. C. (1974) *Cladophora* distribution in Lake Ontario. *Ecol. Res. Ser.* EPA-660/3-74-022, 84 pp.
White, E. (1983) Lake eutrophication in New Zealand – a comparison with other countries of the Organization for Economic and Co-operative Development. *N. Z. J. Marine Freshwater Res.*, **17**, 437–44.
Whitford, L. A. (1960) The current effect and growth of freshwater algae. *Trans. Am. Microcop. Soc.*, **79**, 302–9.
Whitford, L. A. and Schumacher, G. J. (1961) Effect of current on mineral uptake and respiration by a fresh-water alga. *Limnol. Oceanogr.*, **6**, 423–5.
WHO (1998) *Guidelines for Drinking-Water Quality*, 2nd edn, Addendum to Vol. 2, Health Criteria and other Supporting Information, World Health Organization (WHO), Geneva.
WHO (2000) *Disinfectants and Disinfectant By-Products*. World Health Organization, International Program on Chemical Safety (IPCS), Environmental Health Criteria 216, World Health Organization (WHO), Geneva.
Wiederholm, T. (1974) Bottom fauna and eutrophication in the large lakes of Sweden. PhD dissertation abstract. University of Uppsala, Sweden, 11 pp.
Wiederholm, T. (1980) Use of benthos in lake monitoring. *J. Water Pollut. Control Fed.*, **52**, 537–47.
Wilander, A. and Persson, G. (2001) Recovery from eutrophication: experiences of reduced phosphorus input to the four largest lakes of Sweden. *Ambio*, **30**, 475–85.
Wildman, R. B., Loescher, J. H. and Winger, C. L. (1975) Development and germination of akinetes of *Aphanizomenon flos-aquae*. *J. Phycol.*, **11**, 96–104.
Wile, I. (1975) Lake restoration through mechanical harvesting of aquatic vegetation. *Verh. Int. Verein. Limnol.*, **19**, 660–71.
Wilhm, J. (1972) Graphical and mathematical analyses of biotic communities in polluted streams. *Annu. Rev. Entomol.*, **17**, 223–52.
Wilhm, J. and Dorris, T. (1968) Biological parameters for water quality criteria. *Bioscience*, **18**, 477–81.
Williams, J. D. H., Syers, J. K. and Walker, T. W. (1967) Fractionation of soil inorganic phosphate by a modification of Chang and Jackson's procedure. *Soil Sci. Soc. Am.*, **31**, 736–9.
Williams, L. G. and Mount, D. I. (1965) Influence of zinc on periphyton communities. *Am. J. Bot.*, **52**, 26–34.
Williamson, M. (1997) *Biological Invasions*, Chapman & Hall, London.

Wilson, B. (1994) Personal Communication with Welch, E. B. Minnesota Pollution Control Agency, Minneapolis, MN.

Wilson, B. and Musick, T. A. (1989) *Lake Assessment Program 1988, Shagawa Lake, St. Louis County, Minnesota* unpublished manuscrpt Minnesota Pollution Control Agency, 85pp.

Winner, R. W., Boesel, M. W. and Farrell, M. P. (1980) Insect community structure as an index of heavy-metal pollution in lotic ecosystems. *Can. J. Fish. Aquat. Sci.*, **37**, 647–55.

Winner, R. W., Scott Van Dyke, J., Caris, N. and Farrel, M. P. (1975) Response of the macroinvertebrate fauna to a copper gradient in an experimentally-polluted stream. *Verh. Int. Verein. Limnol.*, **19**, 2121–7.

Winter, D. F., Banse, K. and Anderson, G. C. (1975) The dynamics of phytoplankton blooms in Puget Sound, a fjord in Northwestern United States. *Marine Biol.*, **29**, 139–76.

Winter, T. C. (1981) Uncertainties in estimating the water balance of lakes. *Water Res. Bull.*, **17**, 82–115.

Wissmar, R. C., Richey, J. E. and Spyridakis, D. E. (1977) The importance of allochthonous particulate carbon pathways in a subalpine lake. *J. Fish Res. Board Canada*, **43**, 1410–18.

Witting, R. (1909) Zur Kenntnis des vom Winde erzeugten Oberfläckenstromes. *Ann. Hydrogr. Berlin.*, **37**, 193–203.

Woelke, C. E. (1968) Application of shellfish bioassay results to the Puget Sound Pulp Mill pollution problem. *Northwest Sci.*, **42**, 125–33.

Wojtalik, T. A., Hall, T. F. and Hill, L. O. (1971) Monitoring ecological conditions associated with wide-scale applications of DMA 2, 4-D to aquatic environments. *Pesticides Monitoring J.*, **4**, 184–203.

Wong, S. L. and Clark, B. (1976) Field determination of the critical nutrient concentrations for *Cladophora* in streams. *J. Fish Res. Board Canada*, **33**, 85–92.

Woodwell, G. M. (1970) The energy cycle of the biosphere. *Sci. Am.*, **223** (3), 64–97.

WQC (1968) Water Quality Criteria. Tech. Advisory Comm. to EPA, US Government Printing Office.

WQC (1973) Water Quality Criteria-1972. Tech. Advisory Comm., Nat. Acad. of Sci. and Acad. of Engineers. US Government Printing Office.

WQC (1986) *Quality criteria for Water* – 1986 EPA 440/5-86-001.

Wrenn, W. B. (1980) Effects of elevated temperature on growth and survival of smallmouth bass. *Trans. Am. Fish. Soc.*, **109**, 617–25.

Wright, J. C. (1958) The limnology of Canyon Ferry Reservoir. *Limnol. Oceanogr.*, **3**, 150–9.

Wright, R. F. (1983) *Predicting Acidification of North American Lakes*, Report 4, Norwegian Inst. Water Res., Oslo, 165pp.

Wright, R. F., Dale, T., Gjessing, E. J., Hendrey, G. R., Henriksen, A., Johannessen, M. and Muniz, I. P. (1976) Impact of acid precipitation on freshwater ecosystems in Norway. *Water, Air Soil Poll.*, **6**, 483–99.

Wright, R. F. and Henriksen, A. (1983) Restoration of Norwegian lakes by reduction in sulfur deposition. *Nature*, **303**, 422–4.

Wright, R. F. and Schindler, D. W. (1995) Interaction of acid rain and global changes: effects on terrestrial and aquatic ecosystems, in *Acid Rain Research*, Vol. 39/195, Norwegian Institute of Water Research, pp. 232–43.

Wright, R. F. and Snekvik, E. (1978) Acid precipitation: chemistry and fish populations in 700 lakes in southern-most Norway. *Verh. Int. Verein. Limnol.*, **20**, 765–75.

Young, T. C. and King, D. L. (1980) Interacting limits to algal growth: light, phosphorus, and carbon dioxide availability. *Water Res.*, **14**, 409–12.

Younos, T. M. and Weigmann, D. L. (1988) Pesticides: a continuing delemma. *J. Water Pollut. Control Fed.*, **60**, 1199–205.

Yu, S.-Z. (1989) Drinking water and primary liver cancer, in *Primary Liver Cancer* (eds Z. Y. Tang, M. C. Wu and S. S. Xia), China Academic Publishers/Springer, New York, NY, pp. 30–7.

Zand, S. M. (1976) Indexes associated with information theory in water quality. *J. Water Pollut. Control Fed.*, **48**, 2026–31.

Zevenbroom, W. and Mur, L. R. (1980) N_2-fixing cyanobacteria: why they do not become dominant in Dutch hypereutrophic lakes. *Dev. Hydrobiol.*, **2**, 123–30.

Zisette, R. R., Oppenheimer, J. S. and Donner, R. (1994) Sources of taste and odor problems in Lake Youngs, a municipal drinking water reservoir. *Lake Reserv. Manage.*, **9**, 117–22.

Index

acid mine waste 79, 91, 335
acid precipitation (acidification) 79, 91–2, 394–7
adaptation, see population
aeration, see restoration
agricultural runoff 92–3
algal growth potential (AGP) 18, 106, 139
alkalinity, significance of 80, 84–5, 156–60, 231, 394–7
allelopathic factors 163
allochthonous 4–5, 174, 176, 301, 311, 327
alternate stable states, see macrophytes
alum, effects of 260–5
antagonism, see toxicity
application factor, see toxicity
autochthonous 4–5, 174, 176, 311
autolysis 58, 72
autotrophic 5, 7, 274, 304

bacterial plates 80
 see also sulphur: bacteria
best management practices 93, 258–9
bioaccumulation/biomagnification, see toxicity
bioassay, see toxicity
biochemical oxygen demand (BOD) 87–91, 93, 328–30
biological pollution 11, 93–4
 see also invasive, non-native species
biomanipulation 222–5, 246, 267–8
 see also zooplankton: trophic cascades
biomass, pyramid of 12–13
biosphere 3, 6
bioturbation 68

blue-green bacteria (cyanobacteria), blooms
 prediction 250
 toxicity 162, 166–72, 187
 see also phytoplankton
carbon
 alkalinity, significance of 84, 150–4, 156
 atmospheric input 81–5
 cation/anion balance 394
 CO_2: buffering effect on 83–4; cycling processes 81–2; decomposition of organic 150; dissolved inorganic 81, 151–2; forms 81–3; inorganic, equilibrium 81–3; limitation to algae 81, 150–4; organic nutrients, see periphyton; sewage, see waste; system 81–3; total 81–3
cascading trophic interactions, see biomanipulation, zooplankton: trophic cascades
chemolithotrophs 75–6
chemostat 101
chemosynthesis 6–7
Clean Lakes Program 248
Clean Water Act 86
CO_2
 diffusion 81, 83–4
 effects on photosynthesis 83–5, 156–60, 281–2
coastal jet, see water movement
coliform bacteria 88, 90, 93
compensation depth, see light
continuous culture 101
 see also toxicity: bioassay

Coriolis force, *see* water movement
critical depth, *see* light
current velocity   38, 43–4, 228, 355
   intragravel   360–1
cyanobacteria, *see* blue-green bacteria; phytoplankton
cyanotoxins, *see* blue-green bacteria: toxicity; phytoplankton

decomposers   5–7
deforestation, effects of   336–7, 360–2
denitrification, *see* nitrogen
density currents, *see* water movement
detritus food web   4, 327
diatoms, *see* phytoplankton
diffusion, *see* water movement
dilution, *see* population growth; restoration
dimictic   32
dinoflagellates, *see* phytoplankton
disinfection by-products   173–5
dispersion   21, 49–55
diversion, sewage, *see* restoration
dredging, *see* restoration
drinking water quality
   access to safe water   4
   cyanotoxins   166–72
   disinfection by-products   173–5
   taste and odours   172–3

ebullition gas   68
eco-oestrogens, *see* endocrine-disrupting chemicals
ecoregions   350
ecosystem
   biosphere   3, 6
   boundaries   3
   chemical cycle   3; *see also* nutrients
   complexity   7
   diversity   8–10
   efficiency   8–10
   energy flow   5, 212
   function   5–7
   grazing food web   4, 212
   instability   8–9
   management conflicts   10
   maturity   8–9
   net productivity   8–9, 13
   organization   4–8
   production   8–9, 12
   resistance to change   10
   stability   8–9
   steady state   8
   structure   8
   trophic levels   4, 8, 12–15
   waste assimilation   10
   watershed   3–5
Ekman drift/spiral   38, 40
endocrine-disrupting chemicals   392–4
energy
   ATP   5–6
   content of organisms   6–7
   efficiency   8, 14–15
   entropy   8–9
   solar   3, 5
   sources for ecosystems   4–5
   thermodynamic laws   3, 5
   transfer   3, 7–8, 14–15
   turbulence, exchange of water, *see* water movement
epilimnion   30
epiphytes, effects of, *see* macrophytes
erosion, effects of   24, 336–7, 360–2
estuaries
   salinity   122
   stratification   50
   tidal prism   55, 124
   tides, effects of   39–40, 55, 95, 123–8
eutrophication
   coastal marine waters   210–11
   cultural   176, 248–9, 271
   depth, relation to   182, 186, 200–1, 257–8
   fertilization, artificial, *see* nutrients
   fish, effects on   200–3, 249
   food web, effects on   212, 220
   macroinvertebrates, effects on   332–4
   N/P ratio, effect on   141–7
   natural ageing, *see* lakes, ageing
   nutrients, controlling, *see* nutrients: limitation
   rate of, factors affecting   177
   recovery from, *see* restoration
   zooplankton, role of   219–21, 253, 267–8
experiments
   small container   99
   stream   374; *see also* periphyton
   whole-lake   99, 152–3
extinction coefficient, *see* light: attenuation, water

fertilization, artificial, *see* nutrients
fish
   acclimation   363, 370
   activity and growth   354; *see also* temperature
   concentrating capability, toxicants, *see* bioaccumulation/biomagnification
   deforestation, effect of   360–1
   development, embryos and larvae   355–6, 370
   fecundity, hatching   378
   food consumption and growth   357–60, 365
   intragravel flow, *see* current velocity
   invasive, non-natives   397–9
   macrophytes   230
   on metabolism   354–5, 365
   oxygen, effect of   200–3, 249, 353–62; on metabolism   354
   pH, effect of   382–3
   piscivorous   222–5, 267–8
   planktivorous   212, 222–5, 246, 267–8
   reproduction   308–10, 353, 355–6
   sedimentation, effects of   360–1
   swimming performance, effect of oxygen   357
   temperature, effect of   362–76
   tissue hypoxia   383
   urban runoff, effect of   90, 360–1
   warm versus cold water   353–4, 364–8, 370–4
   winter kill   200
fixation, nitrogen, *see* nitrogen
flow, water   21–3, 381
flushing, effects of   25, 40
   see also phosphorus; phytoplankton
fungi   274, 298–302

gas vesicles   98, 160
genetic capacity   15, 100
geosmin   172–3
   see also phytoplankton: blue-greens: taste and odours
global warming (effect on)   96
   fish   369, 374–6
   macroinvertebrates   325
   phytoplankton   136

gravity wave   32–3
grazing   106–8, 218–21, 296–8, 311, 326
   food web   4, 212, 222, 224
   growth equation   15, 17, 100–5
   growth limitation   100–1
   growth rate   100–8

haemoglobin   318
harmful algal blooms (HABs)   210
heated discharge, *see* temperature
heating lakes, *see* temperature: stratification
henry's law   81
herbicides   268
   see also toxicity
heterotrophic   5, 274, 305
hormonally active agents   392
   see also endocrine-disrupting chemicals
hypolimnion   30

impervious area   90, 361
information theory   344
invasive, non-native species
   fish   397–9
   macroinvertebrates   350–3
   macrophytes   230–1, 246–7
   zooplankton   227
iron, oxidation-reduction effects   58–61, 74

Kelvin waves, *see* water movement

lakes, ageing   176, 230
   origin   28–9
   quality, nation's lakes   10–11, 248; *see also* trophic state
   thermal classification   32–5
laws
   of minimum   17–18
   of thermodynamics   3, 5
   of tolerance   19
lentic   95, 271
light
   absorbance, chlorophyll   108–11
   adaptation, photosynthetic   114, 279–81
   attenuation, water   28, 108, 111
   compensation depth   112, 117–19
   critical depth   117–21
   extinction (attenuation) coefficient   111, 120

inhibition, photosynthesis 110, 112–13
mixing depth 122
nutrient interaction with 116, 160–2, 281–2
penetration, visible 109, 229–30
photic zone 112, 275
photooxidation 110
productivity 113–21, 279–80
quality 108–9
quantity 110–12, 279–80
saturation 109–12, 114–15
Secchi disc 111–12
self-absorption (shading) 111, 120, 279–82
sunlight 108, 111, 280
temperature 116, 128
visible spectrum 109, 111, 113
limnology 3
lotic 271
luxury storage, *see* nutrients

macroinvertebrates
adaptation 318
biotic indices 332, 341–4, 349
composition: insects 312; marine versus freshwater 312, 325
current velocity, effect of 314, 320
development, effects of urban 90, 338, 344
disturbance theory 316
emergence 317, 322, 325, 336
erosion, effect of 336–7, 360–1
experimental controls 317
feeding types 265, 311, 326
flow refugia 316–17
food supply, effect of 326
grazing, *see* periphyton
habitat requirements 313–16
Index of Biotic Integrity (IBI) 342–4
indices, formulas 341–50
invasive, non-natives 350–2
lakes 317, 332–4
life history 312, 338–9
metal tolerance 335, 348–9
midges 299, 308, 314, 318, 324, 326–7, 330, 332–4
organic waste, effects on 319, 327–32, 339, 341, 346–8
oxygen, tolerance of 318–20, 326–8
Rapid Bioassessment Protocols 342–3
recycling, effect on 332
repopulation rate 312, 318, 330–1
samplers 313
sampling: control 312, 317; diversity 350; frequency 312; problems 350; qualitative/quantitative 313–14
sediment, effect of 55, 327, 336–7
species: diversity 311, 323–4, 327, 341–9; dominance (shifts) 330–4; evenness (equability) 344–5, 348–9; richness 344–5, 350
statistical treatment 350
stream purification 311
substrata: depositing 311, 315, 317, 330, 337; eroding 314–15, 337; natural and artificial 313; requirements 314–17; stability 315–17, 337
succession, organic waste 330
taxonomy problem 312, 339, 341, 349
temperature; tolerance of 320–5
macrophytes
algae, effect of 240–4
alternate stable states 242–6
colonization, depth prediction 232–3
control techniques 265, 268–9
fish, significance to 230
invasive, non-native species 230–1, 246–7
lake ageing 230
light: effect on growth 229; limitation, effect of eutrophication 240–1
littoral distribution 229
milfoil problem 231
nuisance growth 228, 230–1, 247
nutrient limitation, CO_2/HCO_3 231; critical tissue levels 238
nutrient recycling 68, 228, 230, 241–2
nutrient sources, water versus sediment 235–9
physical substrata, effect of 229, 239–40
productivity, contribution 229–30, 275
reproduction 228
running water, effects of environmental factors 228–9
sewage effluent, response to 296–6
shading effect, milfoil 231

498 Index

macrophytes (*Continued*)
  slope, effect of   239–40
  standing water   229
  submergent/emergent   228–9
  substratum   228, 235–9
  temperature, effect of   235
  wave action   38
meromictic   35
metalimnion   30
metals, in wastewater   89–90
  see also toxicity
Michaelis–Menten kinetics   100–5
microorganisms
  autotrophic   7
  bacterial plates   80
  chemolithotrophs   75
  denitrifiers   76
  fungi, *see* periphyton
  growth: equation, limitation   100–5
  heterotrophic (decomposers)   5, 274, 305; *see also* periphyton
  mercury toxicity, role in   386
  mineralization (regeneration) of nutrients   62, 72; *see also* nutrients: cycling
  nitrifiers   76
  substrate   100
  sulphur bacteria   79–80
milfoil (Eurasian), *see* macrophytes
mixing zone   49
models, *see* population; phosphorus; periphyton; toxicity; zooplankton
monomictic   33

niche   19
nitrification   6, 76, 88, 389
nitrogen
  budget   78, 250
  concentration range   76
  cycling processes   75–9
  denitrification   75–8, 88, 250
  fixation, rate of   75, 77–8, 142–4, 250
  forms   76, 78–9
  limiting factor   138–49, 212–14, 255, 259
  nitrification   6, 76, 88, 389
  nitrogenous oxygen demand (NOD)   88
  NO_3 reserve   145
  preference of forms by plants   76
  relation with chl *a*   145, 195

  sink   75, 78–9
  solubility, relative   78
  sources   75, 78–9, 87–90, 176–7, 250
  waste treatment, removal by   76–7, 88
nutrients
  alkaline phosphatase   140
  allochthonous   4, 176
  autochthonous   4, 176
  cell content   138
  chl:biomass ratios   140
  concentration, effect of, *see* phytoplankton
  current velocity, interaction with   285–6
  cycling   3, 56, 75, 80, 82, 100
  demand:supply ratio   138
  EDTA, effect of   141
  fertilizer addition   69–70, 73, 152–3, 287–91
  growth rate limiting levels   17–18, 137, 148–9, 288–91
  inhibition   142, 149
  limitation: algae   16–18, 78–9, 84, 100–1, 136–54; interaction of N and P   78–9, 148–9; long- versus short-term   154; macrophytes   235–41; marine versus freshwater   147; related to trophic state   141–7; surface/volume of cell   13
  luxury storage   140
  molybdenum   149
  multiple limitation   148–9
  organic factors, substrate   149, 274, 298–303
  point versus non-point   86
  potassium   235–8
  ratios, significance of   138–47, 163, 165
  recycling   58, 69–74, 78, 84, 108, 147, 176
  Redfield ratio   138, 140
  requirements   155, 163, 300–3
  silica, limitation by   136–7, 141, 163
  sources, *see* wastewater
  supply rate, effect of   107–8, 289–90
  trace elements   138, 149
  uptake rate, transport   6–7; *see also* growth rate
  vitamins   149
  wastewater   89

oligomictic 34
organic matter
  cultural 327–32
  natural effect of 311, 327
  sewage, see wastes; restoration;
      macroinvertebrates
  sources for ecosystems 4
  stream purification 311
  tolerance of groups
      (invertebrates) 330, 332
  wastes, effect of, see
      macroinvertebrates; periphyton
oxygen
  ammonia, effect on 76, 88
  anoxic factor (AF), 192, 199
  BOD, secondary 196, 273, 298–302,
      308
  concentration versus percent
      saturation 354
  consumption rate
      (macroinvertebrates) 318
  criteria standards 353–4, 360
  critical concentration 354
  deficit rate (areal hypolimnetic
      oxygen deficit rate,
      AHOD) 188, 192, 196–200,
      203, 253–4
  hypoxia 210
  minimum versus average daily, effect
      on fish 360
  nitrogen cycle, effect on 75–7, 88
  partial pressure 354
  phosphorus cycle, effect on 57–69
  requirements for fish 200–3; for
      macroinvertebrates 318–20
  sulphur cycle, effect on 79–80
  winter kill 200

periphyton
  accumulation (accrual) rate 276–7,
      285–6, 288–90
  blue-greens (cyanobacteria) 273,
      276, 278, 290, 298
  current velocity, effect of 274, 283–6
  diatoms 273, 277, 283, 286–7, 290,
      294–8, 303–4
  enrichment, benefits versus
      detriments 308–11
  filamentous bacteria 273, 285,
      298–304
  filamentous green algae 273–4, 276,
      283, 285–96, 298, 303, 307

  grazing, effect of 273, 296–8
  habitat selection 286–7
  indicator organisms 303–5
  light, effects of 279–83
  mat thickness 284–6, 290
  measurement 276–7
  nuisance: algae 273, 287–8, 291–6,
      308–10; bacteria 273, 298–303
  nutrients, effect on growth
      rate 286–91;
      concentration 291–6
  organic nutrients, effect of 274,
      298–303
  productivity, relative
      contribution 274–6
  protozoans 273, 299, 303–5
  respiration 277
  scouring, effect of 283–4, 286
  sewage, effect of 295, 298, 301,
      303–5
  slimes, bacterial 286, 298–302
  species composition 286, 298–9,
      303–5
  substrata 275, 277, 298
  succession, species 299, 302–3
  temperature, effects of 277–8
  toxicants, effects of 306–8
  turbulence 285–6
phosphorus
  analysis 56
  autolysis 58, 72
  bioavailable 209
  bioturbation 68
  carbon, with respect to 150
  chl a, relation to 144–6, 188–95
  control, point-source 88; see also
      restoration
  controversy 150
  cycling processes 57–8, 73, 78–9
  dissolved 56–7
  equilibrium concentration 69–70
  excretion 58, 68
  flushing, effect on loading 177–9,
      205–8; effect on
      sedimentation 177–9, 184
  forms 56–7, 61, 69
  interception 258
  internal loading 61–9, 182–4, 251,
      253–4, 256–7, 260–2; aerobic
      versus anaerobic, see sediment
      release
  interstitial 59

phosphorus (*Continued*)
  iron, oxidation-reduction, effect of 58–61, 74, 260
  land use, *see* sources
  littoral, source from 73
  loading 177–8; limit critical 205–7
  mass balance 177, 249–50
  metal sorption 58–62
  models, steady state 177–84, 250
  non-steady state 184–6, 250
  organic 58
  oxidation-reduction 58
  oxygen, relation to 196
  pH 61–4, 74
  polymixis 34, 65–6
  precipitation, alum, *see* restoration
  prediction, *see* models
  reactions, mud-water interface 58–69
  regeneration 72
  residence time, water, effect on loading 74, 205–8, 258
  retention coefficient 74, 177–80
  sediment phosphorus 58, 61, 68, 74
  sediment release 58–66, 182–4
  sedimentation 58, 154, 177–80, 208; rate 177–80
  sinks 74, 78–9, 249
  sources 59, 74, 78–9, 88–9, 176–7, 208–9, 249
  steady-state concentration, *see* equilibrium
  thermocline erosion, effect on 66–7, 70–2
  turnover time (rate) 70
  vertical transport, entrainment 67, 185
  wetlands retention 258
photic zone 112, 275
photosynthesis
  bacterial 6, 80
  CO_2 limitation 84, 281
  dark-light reaction 5–6
  efficiency 8, 113–14
  energy for 5–6, 111–12, 280
  equation for 6, 116–21, 196
  glucose 6
  light saturation 109, 111, 128, 232, 235
  mixing, effects of, *see* light; phytoplankton
  oxygen production 117
  photophosphorylation 6

pigments 109
prediction 114
profile 113, 115
quotient 196
self-limitation CO_2 84, 281
phytoplankton
  algal growth potential (AGP) 18, 106, 139
  allelopathic factors, effects of 163
  bloom prediction 99–100, 117
  blue-greens (cyanobacteria) 77, 97–8, 154–66, 215; blooms 154, 163; buoyancy 69, 160–2, 265–7; control of toxic blooms 171–2; King's hypothesis 156–8; migration 163–5; taste and odours 172–3; toxicity of 162, 166–72, 187, 216; water supplies, effects on 172–5
  chlorophyll *a* 109, 111, 120
  diatoms 97, 99, 117, 127, 131–2, 135–7, 154, 161, 204
  dinoflagellates 97, 117
  flushing (retention), effects of 101–5, 126, 259–60, 274
  grazing 106–8, 162, 219–22, 267–8
  green algae, succession of 99–100, 131–3, 154–9
  growth rate 100–2, 128–30
  indicator species, trophic state 186–7, 203–4
  loss rate 106–8, 267
  mixing turbulence, effect of 98, 116–22
  outburst, spring diatom 99–100, 127, 131–2
  pH, effect of 156–60, 266
  productivity, *see* production
  seasonal succession, species 99–100, 131–3, 154–5
  sinking rate, effects of 106–7, 117, 134, 160–2
  thermal bar, effect of, *see* water movement
  thermal optima 131–2
polymictic 34, 65–6
population growth
  age structure, effect of 16
  continuous culture 289; *see also* chemostat
  dilution rate, biomass 106–7
  equation, first order 15, 100

exponential 15, 100
genetic capacity 15
law of minimum 17
law of tolerance 19
limiting nutrient 18
loss rate, biomass 16; see also
  phytoplankton
maximum rate 100
Michaelis–Menten kinetics 100–5
model, see population growth:
  equation, first order
multicellular 15–16
optimum environment 15
phases 16–17
rate of increase 15
steady-state, application of 100–5
unicellular 16
unlimited versus limited
  resources 15–16
population
  adaptation 19
  biomass 12–13, 99
  cell washout, effect on growth 95, 102–4
  human 4, 16
  niche, species 19
  productivity, see production
  standing crop, biomass 12
  steady state 8
  turnover rate 12
predation, fish 221–4, 230, 267–8
primary consumers 13
primary producers 13
production (productivity)
  areal 116
  fish 354, 360
  gross 13, 114–16
  light inhibited, see light
  measurement of 276–7
  mixing, effect of 119, 122–8; see also
    light
  net 8–9, 13, 117–19
  P/R 13, 274, 326
  primary 13, 176, 195–6, 275–6
  secondary 13
  trophic state 195–6
  zooplankton 218
pulping waste 90–1, 299–300, 378, 380, 392
purple and green sulphur bacteria 80
pyramid of biomass 12–13

radiation 28, 109
Rapid Bioassessment Protocols, see
  macroinvertebrates
recovery, see restoration
reference conditions 349
reproduction
  cycle, zooplankton 212–13
  insect, life stage 312
  macrophytes 228
  oxygen requirements 354–7, 360–2;
    see also fish
  parthenogenic, zooplankton 212–13
  rate of, tubificids 331–2
  sexual 212, 228
  temperature and photoperiod 322, 370, 374
residence (detention) time 25, 29, 36, 74, 78, 95, 205–8, 258
  see also water movement; phosphorus
restoration
  advanced waste treatment 88, 251, 256
  aeration, hypolimnetic 260, 263
  biological predator–prey
    effects 267–8; see also
    biomanipulation
  circulation, artificial 266–7
  cost 265
  dilution 101–5, 259–60
  diversion, sewage 66, 251–8
  dredging 66, 262–5
  lake quality status, national 10–11, 248
  macrophytes 268–9
  phosphorus inactivation 73, 260, 263
  Riplox 260, 263–4
  streams, recovery 338–9
  techniques, general 250–1
  toxicity, recovery from 386–7, 390–2
  watershed improvement 258–9
  wetland treatment 258
  withdrawal, hypolimnetic 260, 262
resuspension 24, 55
retention coefficient, see phosphorus
Richardson number 26–7

salinity 122, 231
sampling, see macroinvertebrates;
  periphyton
Secchi disc 111–12, 188–94

secondary consumer 6
sedimentation, effects of 55, 58, 78, 230, 284–5, 315, 332, 336–7, 360–2, 386
  focusing 55; *see also* trophic state
seiche, *see* water movement
self-absorption (shading), *see* light
sewage 86–8, 326–8, 389–90
  see also restoration; wastes
slime bacteria, *see* periphyton
species
  diversity 8–9, 323–4, 354
  succession 131–3, 154–66
specific
  conductance 51
  heat 28
standing crop (biomass) 12–13
steady state 8, 177–84
stormwater 89–90, 258–9
  see also urban runoff
substrata, *see* macroinvertebrates
sulphur
  acidity 80, 91–2, 394–7
  availability 79
  bacteria 79–80
  cycling processes 79–80
  sulphide, odour of 79
suspended solids 87, 284, 327, 360–1
synergism, *see* toxicity

taxonomy, invertebrates 312, 403
temperature
  acclimation 225, 363, 368–70
  adaptation 19–20, 133–5, 324–5, 369
  assimilative capacity 368
  classification of lakes, *see* lakes, ageing: origin
  criteria, standards 321–2, 370–4
  daily mean, significance of 322–3, 367
  deforestation 361
  density, effects on 122
  entrainment, coolant water 321
  equilibrium loss 364
  feeding rate of fish 365
  fish, tolerance 362–70
  global warming 136, 325, 369, 374–6
  growth rate, effect on 128, 226; *see also* Q10 rule
  heated discharge, effects of 136, 225–6, 278, 322, 325, 366–8, 371
  lethal limits, fish 363–4, 368, 371, 373–4; zooplankton 225–6, 321
  longevity, effects on 322, 325
  long-term, short-term effects 136, 321–2, 364
  macroinvertebrates, tolerance 320–5, 371–2
  macrophytes, effect on 235
  maximum: daily 322, 366–8; significance of 370–1
  nutrient interaction 134, 149, 299
  optimized regime 374
  optimum 364
  phosphorus release 64
  preferred temperature 323, 366–8, 374
  productivity, effect on 128–30
  Q10 rule 128–30, 277–8
  rate of rise 321
  reproduction, effects on 369–71, 373
  scope for activity and growth 364–6, 371
  seasonal cycle and standards 369–71
  stenothermal 225
  stratification 29–32, 35, 50–1, 58, 65, 122–3
  sublethal 364–5
  thermocline, barrier 30, 64–6
  toxicity 385
  zones of tolerance, resistance 362–3
Tens Rule 93
  see also biological pollution
thermal
  bar 42–3, 48
  cycle 29–32
  stratification, *see* temperature
thermodynamics, laws of 3, 5
tides, effects of 39, 55
tolerance, *see* population growth
toxicity
  accumulation, *see* bioaccumulation/ biomagnification
  acid mine waste 79–80, 91, 335, 340
  acid rain 80–1, 91–2, 394–7
  acute 377
  additive model, *see* model
  aluminium 395–7
  ammonia 88, 387–9
  antagonism 380–1
  application factor 378–9
  bioaccumulation/ biomagnification 376, 390–2

Index 503

bioassay: fish 377–8; static/
 continuous 378;
 zooplankton 385
blood 382, 390, 397
blue-greens (cyanobacteria), *see*
 phytoplankton
carcinogenicity 391
chemical form, importance of 336,
 383, 385–6
chlorine 88, 379, 386, 388–90
chloroform 173
chronic 377–80
copper 306–7, 335, 348, 378, 380–4
criterion continuous and maximum
 concentration 382, 384–5; *see
 also* water quality: criteria
cyanide 387–9
DDT 92, 306, 340, 376, 390–1, 393
dioxin 92, 392, 393
DO, effect of 385, 389
dumps 376
endocrine-disrupting chemicals 392–4
fly ash 339–40
gill effect 382–3
growth 378
hardness, effect of 384–5
hazardous substances 376
heavy metals 306–7, 335, 381–7
herbicides 268, 391–2
lead 387, 393
long-term 376, 378, 391–2
macroinvertebrates, general effect
 334–6
median, lethal concentration 378–9
mercury 376, 381–2, 385–7, 393
Microtox 385
mode of action (mechanism), metals
 382–3
model (additive, antagonistic,
 synergistic) 380–1
oil 91, 340, 380
organic compounds 92, 334, 376,
 383, 386, 390–2, 393–4
organochlorine insecticides 92, 376,
 390–1, 392
PCBs 92, 376, 391–2, 393
pH effect 388–9
pulp mill 90–1, 380, 392
reproduction 379–80, 389, 390–4
residues, organic 390–4
safe concentration 377, 381
sedimentation, effect of 336–8

short-term 334, 377
solubility of substance 383, 386
standards, criteria 353, 377, 382–5,
 389–91
subchronic effects 379
sulphide 387, 389
synergism 380–1
threshold 377, 380
trace metals 149
trihalomethanes 173, 249; *see also*
 disinfection by-products
waste, sources 92
water quality, effect of 336, 377, 382–4
zinc 306–7, 335, 378–80, 381, 383–5
trace elements, *see* nutrients
transparency, *see* Secchi disc
trophic state
 criteria 186–95, 202, 205
 effect of depth 186, 205, 253
 hypereutrophic 192, 195
 indices 190–3
 mesotrophic 188, 190, 192, 195
 oligotrophic 186–7, 190, 192, 195
 probabilistic 193
 sediment chronology 204, 250
 ultraoligotrophic 205
tropical lakes 100, 193–4
turbidity 93, 229, 231, 233, 242–6, 274
 current 35, 49–50
 see also suspended solids; Secchi disc
turnover time (rate) 12–13, 71

uptake rate, *see* growth rate nutrients
urban runoff, effects of 89–90, 258,
 315–16, 336–8, 344, 347–8, 360–1

vitamins 149
vulcanism 29

washout 103
waste
 assimilation 10
 treatment: advanced 88; on-site
  88, 209, 259, 348;
  secondary 88, 348, 389; stream
  recovery 271, 311, 338–40
wastewater
 agricultural 92–3
 combined systems 89
 definition 86; point and non-point
  86, 93, 176–7, 209, 211, 249, 258,
  271, 376; *see also* urban runoff

wastewater (*Continued*)
  food   91, 307
  petroleum   91, 349
  pulping   44, 90–1, 299–300, 378, 380, 392
water layers
  epilimnion   30, 36
  hypolimnion   30, 36
  metalimnion   30
  thermocline   30, 36
water movement
  advection   105
  amictic   33
  boundary layers   23–4, 285–6
  circulation periods   29–31, 35
  coastal jet   40
  coastal zones   36, 39
  Coriolis force   25, 40
  current velocity   38–9, 43, 283–6
  density currents   27–8, 30, 49–50
  detention time   36
  dispersion   21, 43–4, 49–51; modeling   43, 50; plume   49–51
  eddy diffusion   44, 47, 185
  Ekman drift, spiral   38, 40
  flow, laminar   21–2, 27; shooting and tranquil   23; steady and unsteady   22; turbulent   21–7; uniform and non-uniform   22
  gravity wave   22–3
  large lakes   40
  mixing   49–51; resistance to   65, 122–3; zone, wastes   51, 55
  molecular diffusion   47
  plume, dye   44
  remote sensing   44
  Reynolds number   22
  Richardson number   26–7
  seiche   36–7
  shear stress   22
  surface waves   37–8
  swash zone   29
  thermal bar   42–3, 48
  tidal effects   39
  turbidity current   35, 49–50
  turbulence, lakes   21, 49, 95, 98
  upwelling   30
  velocity gradient   23
  viscosity   22
  wind, effect on   28
  wind-driven current   38–9

water quality, *see also* trophic state
  assessment   339
  BOD, *see* oxygen; biochemical oxygen demand (BOD)
  criteria, standards   335, 337, 353–4, 370–1, 382–5, 387, 389–91
  diffusers, waste water   21
  diffusion, pollutants   44
  indicator organisms   203–4, 307–8, 330, 332–4, 339, 341
  management   10
  pollution indices   277, 304–5, 332, 341–4, 347, 353
  seiche, effect of   36
  standards, *see* oxygen temperature
  taste and odours   172–3, 186–7, 248, 308
  waste assimilation   10
water supplies, effects of cyanobacteria (blue-greens) on   172–5
watershed management   4, 258–9
weathering   394
wind   28, 51, 65
winter kill, *see* fish

zebra mussels   350–1
zooplankton
  enrichment, effects of   212, 218–21, 253
  feeding rate   218
  filtering rate   218–19
  food web model   225
  foods, quality of   212–13, 215–17
  grazing   215, 218–22, 267–8
  growth   213–15
  invasive, non-natives   227
  life cycle   212–3
  migration   215
  oxygen, effects on   226–7
  phytoplankton, grazing, *see* phytoplankton
  population dynamics   213–18
  predation by fish   212, 214–15, 221–5, 267–8
  rotifers   212
  selectivity by fish   221, 267–8
  size   213
  temperature, tolerance of   225–6
  toxicity   188, 217, 221
  trophic cascades   221–5; *see also* biomanipulation
  trophic-decoupling   224

Printed in the United States
128244LV00002B/62/A